HYDRAULIC AND ENVIRONMENTAL

COASTAL WATERS

Hydraulic and Environmental Modelling: Coastal Waters

Proceedings of the Second International
Conference on Hydraulic and Environmental
Modelling of Coastal, Estuarine and River Waters
Volume 1

Edited by
R. A. Falconer
S. N. Chandler-Wilde
S. Q. Liu

Routledge
Taylor & Francis Group
LONDON AND NEW YORK

First published 1992 by Ashgate Publishing

Reissued 2018 by Routledge
2 Park Square, Milton Park, Abingdon, Oxon, OX14 4RN
711 Third Avenue, New York, NY 10017, USA

Routledge is an imprint of the Taylor & Francis Group, an informa business

Copyright © R A Falconer, S N Chandler-Wilde and S Q Liu 1992

Publisher's Note
The publisher has gone to great lengths to ensure the quality of this reprint but points out that some imperfections in the original copies may be apparent.

Disclaimer
The publisher has made every effort to trace copyright holders and welcomes correspondence from those they have been unable to contact.

A Library of Congress record exists under LC control number: 92031775

ISBN 13: 978-1-138-33037-5 (hbk)
ISBN 13: 978-1-138-33041-2 (pbk)
ISBN 13: 978-0-429-44791-4 (ebk)

Contents

Editors' foreword

In recent years there has continued to be an increasing interest in the development and application of numerical hydraulic models as design and management tools for flow and pollutant and sediment transport simulation studies in hydraulic and environmental engineering. Such models also depend heavily on laboratory and/or field measuring programmes for calibration, verification and interpretation of flow phenomena and water quality evaluation. This increasing use of deterministic models by engineers and scientists, in both private and public organisations, can be partly attributed to:- (i) the growing public perception and concern over the pollution of the coastal, estuarine and river environment, (ii) the adoption of a more rational basis for coastal defence works due to an improved understanding of surf zone hydrodynamics, (iii) the more widespread use of sophisticated field and laboratory data analysis techniques and recording equipment, and (iv) the availability of cheaper and more powerful personal computers and work stations and the increasing availability of computer software relating to coastal, estuarine and river flow and solute transport processes.

In continuing to recognise this increasing interest in the use of hydraulic, water quality and sediment transport models, and following the success of the first international conference on Hydraulic and Environmental Modelling of Coastal, Estuarine and River Waters, held in September 1989 and opened by HRH The Princess Royal, the Computational Hydraulics and Environmental Modelling Research Group at the University of Bradford have organised a similar conference to discuss recent advances in the field. The main aim of the conference was to provide a forum whereby engineers, scientists and planners involved in the application, research and development of multi-disciplinery models could share experiences.

The excellent response to the call for papers resulted in a large number of contributions, with the conference proceedings of the edited papers being included in two volumes namely:- Hydraulic and Environmental Modelling : Coastal Waters and Hydraulic and Environmental Modelling : Estuarine and River Waters. This volume contains the papers relating to coastal waters and has been subdivided into the following parts:- Tidal Current Modelling, Water Quality Modelling, Sediment Transport Modelling, Wave Kinematics and Modelling and Computational Methods.

In organising the conference and in the preparation of this volume the editors would like to thank the following:-

- The keynote speakers, including Dr Stuart Reed, Director of Environmental Protection, Hong Kong Government Environmental Protection Department and Prof Donald Harleman, Ford Professor of Engineering (Emeritus), Massachusetts Institute of Technology, USA

- The authors for their time and effort in preparing the papers for this volume and the delegates and all those who participated in the organisation of the conference.

- The members of the UK and Overseas Technical Advisory committees, the paper referees and Mrs Christine Dove (Conference Co-ordinator)

- The Co-sponsors including: Institution of Civil Engineers, Institution of Water and Environmental Management and International Association for Hydraulic Research

- The staff of the publishers - particularly Mr John Hindley - for their ready help and encouragement

- Staff of the University of Bradford who assisted with the organisation of the conference.

RAF, SNCW, SQL
Bradford 1992

Acknowledgments

The editors' wish to acknowledge the valuable assistance of the other members of the Organising Committees, the Conference Co-ordinator and the Panel of Referees.

U K ORGANISING COMMITTEE

Prof R A Falconer (Chairman)	University of Bradford
Mr P Ackers	Private Consultant
Dr S N Chandler-Wilde	University of Bradford
Mr Y Chen	University of Bradford
Mr N E Denman	ABP Research & Consultancy Ltd
Mr C Evans	Wallace Evans Ltd
Dr G P Evans	WRc
Dr C A Fleming	Sir William Halcrow & Partners
Mr A Hooper	W S Atkins Engineering Sciences
Dr M W Horner	Bullen & Partners
Dr D W Knight	University of Birmingham
Prof D M McDowell	Private Consultant
Mr A McLean	Yorkshire Water plc
Dr R G S Matthew	University of Bradford
Dr K Shiono	University of Bradford
Mr D V Smith	Wessex Water
Mr G Thompson	Binnie & Partners
Mr M F C Thorn	HR Wallingford

List of contributors

Abdel-Aal, F M	Cairo University, Egypt
Aldridge, J N	MAFF, UK
Al-Mashouk, M	Sir William Halcrow & Partners Ltd, UK
Amos, C L	Bedford Institute of Oceanography, Canada
Apparao, D	R & D Department Vizianagaram, India
Ashrafi, R A	King Saud University, Saudi Arabia
Bach, H K	Ecological Modelling Centre, Denmark
Balah, M I	Suez Canal University, Egypt
Bando, K	Kajima Technical Research Institute, Japan
Barber, R W	University of Salford, UK
Benn, J R	Bullen & Partners, UK
Borelli, H Rodriguez	Institute of Fluid Mechanics (IMFIA), Uruguay
Cawley, A M	MCS International, Ireland
Chandler-Wilde, S N	University of Bradford, UK
Chen, Y	University of Bradford, UK
Chesher, T J	H R Wallingford, UK
Christian, H A	Bedford Institute of Oceanography, Canada
Cooker, M J	University of Bristol, UK
Cooper, A J	HR Wallingford, UK
Cueva Peidra, J C I	Institute of Fluid Mechanics (IMFIA), Uruguay
Dou, G	Nanjing Hydraulic Research Institute, China
Elahi, K Z	King Saud University, Saudi Arabia

Falconer, R A   University of Bradford, UK
Fleming, C A   Sir William Halcrow & Partners Ltd, UK
Fukumoto, T   Nishimatsu Construction Co Ltd, Japan

Galland, J C   EDF, France
Grant, J   Bedford Institute of Oceanography, Canada

Hall, M   University of Bradford, UK
Hapoğlu, L   Middle East Technical University, Turkey
Harleman, D R F   Massachusetts Institute of Technology, USA
Hartnett, M   MCS International, Ireland
Hodder, J P   BMT Ceemaid Ltd, UK

Janin, J M   EDF, France
Jin, H S   Hohai University, China

Karinao, S   CTI Engineering Ltd, Japan
Karssen, B   Delft Hydraulics, The Netherlands
Kaya, Y   Babtie Shaw & Morton, UK
Keiller, D C   Binnie & Partners, UK
Koike, T   Kajima Technical Research Institute, Japan
Komatsu, T   Kyushu University, Japan
Kudo, K   Japan Marine and Science Technology Center, Japan
Kurita, H   CTI Engineering Ltd, Japan

Lai, C J   National Chen-Kung University, Taiwan
Li, B   Sir William Halcrow & Partners Ltd, UK
Li, C W   Hong Kong Polytechnic, Hong Kong
Li, G   Posford Duvivier, UK
Lin, B   University of Bradford, UK
Liu, S Q   Tongji University, China
Lloret, A   CEPYC, Spain

MacDonald, N   University of Liverpool, UK
Mackinnon, P A   Binnie & Partners, UK
Mansur, L   HR Wallingford, UK
Mead, C T   HR Wallingford, UK
Miles, G V   HR Wallingford, UK
Montero, M   CEPYC, Spain
Morfett, J C   Brighton Polytechnic, UK
Mostafa, H K   Suez Canal Authority, Egypt
Murakami, K   Port and Harbour Research Institute, Japan
Muraoka, K   Osaka University, Japan
Murphy, D G   HR Wallingford, UK

Nakamura, T   Nagasaki University, Japan
Nakatsuji, K   Osaka University, Japan
Narasimham, M L   Andhra University, India
Nicholson, J   University of Liverpool, UK

O'Connor, B A	University of Liverpool, UK
Odd, N V M	HR Wallingford, UK
O'Shea, K	University of Liverpool, UK
Osment, J	Sir William Halcrow & Partners Ltd, UK
Özhan, E	Middle East Technical University, Turkey
Parkinson, N J	Howard Humphreys and Partners Ltd, UK
Paterson, D M	Bedford Institute of Oceanography, Canada
Price, D M	HR Wallingford, UK
Reed, S B	Hong Kong Government Environmental Protection Department, Hong Kong
Reeve, D E	Sir William Halcrow & Partners Ltd, UK
Roelfzema, A	Delft Hydraulics, The Netherlands
Ruiz-Mateo, A	CEPYC, Spain
Sagara, M	Inc. FUJITA, Japan
Shiono, K	University of Bradford, UK
Tanaka, M	Kajima Technical Research Institute, Japan
Togashi, H	Nagasaki University, Japan
Toole, T H	Yorkshire Water Engineering, UK
Wang, Z B	Delft Hydraulics, The Netherlands
Winterwerp, J C	Delft Hydraulics, The Netherlands
Wright, P	Posford Duvivier, UK
Yano, S	Kyushu University, Japan
Ye, J	Nanjing Hydraulic Research Institute, China
Yen, J W	National Chen-Kung University, Taiwan
Zheng, Y M	Hohai University, China
Zhu, L	Hong Kong Polytechnic, Hong Kong

KEYNOTE ADDRESSES

1   Sewage: A mixture of technology politics and economics

S. B. Reed

ABSTRACT

The technology and the techniques needed to deal with the treatment and disposal of sewage and other wastewaters have been available for several decades. Despite this, there has been chronic under investment in sewage infrastructure and a history of declining water quality around most major population centres in the world. The paper discusses the reasons for this and describes the development and implementation of the strategy for dealing with the wastewaters from Hong Kong's concentrated urban development.

Introduction

One of the important goals of environmental hydraulics is the prediction of conditions for the satisfactory conveyance and dispersion of sewage and other wastewaters. There have been tremendous advances in this field over the past two decades, much of it facilitated by the ready availability of high speed computers. There have been also useful advances in wastewater treatment, although the basic processes and technology have been well established for many years.

Despite this ready availability of reliable technology and powerful techniques for ensuring the satisfactory treatment and disposal of sewage, there has been a progressive deterioration in water quality in and around many, if not most, major centres of population, over the past century.

In the UK, for example, many beaches are contaminated by sewage outfalls that reach no further than the low water mark. Major rivers like the Mersey and Trent are still heavily polluted over a significant part of their length.

In Hong Kong, there is widespread surcharging of foul sewers and over 75% of the

Territory's sewage flows directly into the spectacular Victoria harbour, which forms a focal point for the community in Hong Kong, not to mention the 6 million tourists who visit the Territory each year.

Elsewhere in the world there are cities with many millions of inhabitants, and still growing rapidly, which are almost completely devoid of mains sewerage or of any satisfactory means of disposing of wastewaters.

WHY SHOULD THIS BE SO?

We have the technology. For the most part it has been available for decades, and is robust. We have the techniques and these are being continually refined, as reflected in the many papers presented to this Conference.

There are, of course, many reasons for the widespread under-investment in sewage collection and disposal facilities.

Politics

A major reason for chronic under investment in sewage infrastructure can often be found in defects in the institutions that are responsible, within an Administration, for making appropriate arrangements for the disposal of wastewaters. These defects can arise historically, or simply as a result of small internal politics.

A guide book to Hong Kong [3], published nearly a hundred years ago, gives some clues. This guide book records that, in the years leading-up to 1883, the sanitation of the Colony was in the hands of the Colonial Engineer and the Colonial Surgeon, and goes on to remark "..... owing to the frequent changes which took place in the incumbents of these two offices, there was no continuity of action in the sanitary administration of the Colony. Each new incumbent appears to have at once set to work to remedy what appeared to him to be the defects in the administration of his predecessor. The result was confusion."

This unsatisfactory state of affairs led to a special commissioner - Mr Osbert Chadwick - being sent from England to "enquire and report on the sanitary condition of the Colony". Mr Chadwick's report led to the establishment of a Sanitary Board in 1883, and the enactment of a short Ordinance to enable the Board to carry out the more pressing of the many sanitary improvements which were identified in the report as requiring urgent attention. The Colonial Surgeon was appointed to take charge of the Board, along with his other onerous duties. At this point Mr Chadwick returned to England.

When the Sanitary Board tried to implement Chadwick's recommendations, it met with strong opposition from vested interests. Little progress was made and in 1894 bubonic plague broke out, causing nearly 1,000 deaths in what was at that time a quite small community.

As is the case today, there is nothing like a few environmental martyrs to stir an Administration into action : Chadwick was recalled. Having reassessed the situation, one of his first recommendations related to the Sanitary Board. This was that ".... It is essential that it should have its own head, not someone whose main job in the Government keeps his attention elsewhere."

This is golden advice in the environmental field today, where long established bodies, such as Public Works Departments in developing countries, often still see environmental work as a natural extension of their existing activities. But if responsibility for some aspects of environmental work is vested in Public Works-type departments, then when it comes to the crunch, and a choice has to be made between a new road and a new sewer, or between a new freshwater reservoir and a new wastewater treatment plant, or between a new harbour facility and a long sea outfall to deal with existing wastewaters, the sewage item usually loses out.

So a lack of sufficiently focused institutional arrangements and ambiguity of responsibilities are two of the reasons why available technology in the sewage field is not put to its intended use.

It was a century after Chadwick's insights that Hong Kong accepted his advice. This took the form of assigning to the Environmental Protection Department (EPD), established in 1986, the role of client for the sewage programme [7], and creating in 1989 a Drainage Services Department with sole responsibility for implementing the sewage programme, operating and maintaining the sewerage system and dealing with the Territory's storm water and flood control system.

Another reason for the lack of implementation of well-established technology and ever improving techniques in the sewage field, is the community's priorities as reflected in the politics of the day, and the natural sympathies of those in positions of influence and power. Hong Kong has been fortunate in very recent years in having a Governor and Chief Secretary who appreciated the long term implications of failing to invest in the unattractive business of sewage and its treatment and disposal.

As in many other parts of the world, however, the community in Hong Kong has been slow to appreciate the occult virtues of sewerage. Some progress has been made in this regard in Hong Kong in the past 3 or 4 years, mainly as a result of a TV publicity campaign and a programme of so-called "Pollution Black Spot" visits.

The most effective item in the TV campaign was one which portrayed an elegant Chinese society dinner party. The vivacious and animated guests seat themselves for dinner. Dinner is served and the traditional silver cover is removed to display the main dish - a magnificent garoupa. When the waiter starts to cut along the side of the splendid fish to serve it to the diners, raw sewage bubbles-out and the conviviality is transformed rapidly into horror and revulsion.

The "Pollution Black Spot" visits were more targeted and aimed at a narrow sector of opinion formers and decision makers in the community. On various occasions, the most senior politicians and civil servants, media editors and business chief executives were taken to see examples of pollution which were the worst known to the staff of the Environmental Protection Department. Often, simply as a result of the nature of their daily work schedules and living habits, and the fact that VIPs are normally taken to see "successes" rather than "failures", such people are unaware of the full extent of the threat posed by environmental pollution.

As a result of these campaigns, and no doubt influenced by the growing attention paid world-wide to environmental matters, there has been a noticeable increase in the environmental awareness and concern on the part of the community in Hong Kong in the past 3 or 4 years. Activities to increase community awareness of environmental matters have received added impetus recently as a result of the establishment of an Environmental Campaign Committee [2], with members appointed by the Governor from various sectors of the community and from the Government itself.

Despite this newly emerging environmental awareness on the part of the community in Hong Kong, and a rather more long-standing awareness of such matters in many developed countries around the world, there is still limited enthusiasm for the large capital investment in sewage infrastructure that is needed in order to recover decades of neglect, both in Hong Kong and elsewhere in the world. In these circumstances, it is encouraging that, as may be seen in Figure 1, there has been a more than 700% increase in capital investment in Hong Kong's environmental infrastructure during the past five years.

Economics

Another major factor standing in the way of the timely implementation of measures to control water pollution, and for that matter most other forms of pollution, is that the cost of the adverse effects on the environment are not readily quantifiable in monetary terms.

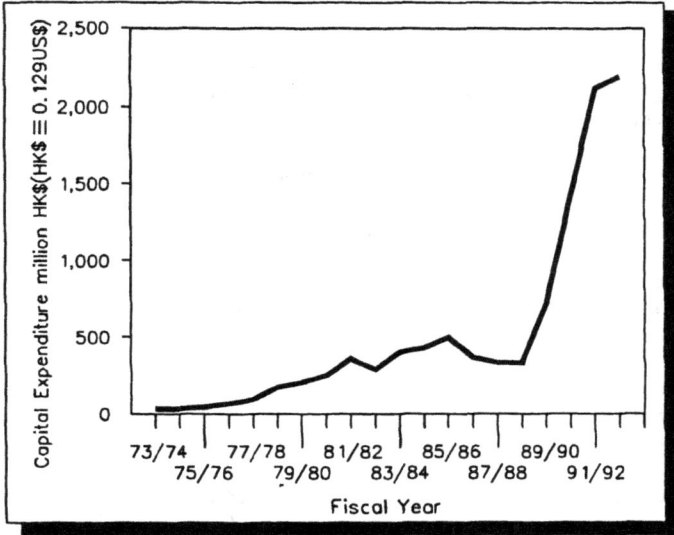

FIGURE 1   *Capital expenditure on environmental infrastructure facilities in Hong Kong*

Conventional economics still regards economic capital, such as fresh air and clean water, as free goods. So the cost to the environment is never fully reflected in the price of products on the market. Rather, it is normally borne by everyone through general taxation, irrespective of the nature or level of their personal consumption. It is an unfortunate fact that, in many cases, this general charge will fall on future generations.

The lack of any real direct coupling between economics and the environment is one of the most significant factors militating against sustainable development.

Whilst it is no more easy in the sewage field than in any other environmental field to quantify the full cost implications for the community of sewage disposal, it is relatively easy to establish the direct costs of providing a service for sewage collection, treatment and disposal. The least that can be done, therefore, is to internalise these costs by charging for the provision of this service. Yet this practice is by no means the general rule. A recent survey by Arthur D. Little (Asia Pacific Inc.) [6] showed that, out of the eight major cities in different parts of the world that were surveyed, only Los Angeles, Singapore, Taipei and Tokyo base sewage charges on a parameter, measured water consumption, that bears some relation to the cost of providing a service for the collection, treatment and disposal of sewage. In London, New York and Mexico City, where a comparatively small proportion of water consumers are metered, sewage charges are based on a variety of criteria - rateable property value in London, property street frontage in New York and socio-economic characteristics of the district in which the user lives, in the case of Mexico City - that bear no direct relationship to the cost of the service provided. In some cases, however, the charging for commercial and industrial effluents in these places is related in one way or another to the cost of the service provided.

In Hong Kong, there are at present no direct charges whatsoever for the provision of sewage services. It is not surprising that, in these circumstances, the cost of such services do not figure at all in the decisions made by industry and commerce in relation to calls on this particular public service. Consideration is being given at the present time in Hong Kong

6

to charging for effluent discharges, with charging based on the volume of water supplied, with a surcharge for C.O.D. concentration exceeding specified levels. Whilst a scheme of this nature does not reflect the full cost to the community and to the environment, of sewage discharges, it is likely to encourage, as a result of some internalisation of the costs, much more rational decisions from an environmental point of view.

There are, of course, other potential advantages that follow from the creation of an hypothecated cash flow arising from a system of charges of this nature. One is that the cash flow provides a viable financial basis for borrowings in the market to fund the implementation of a strategic plan for the provision of sewage services, rather than proceeding by way of a succession of ad hoc projects, related to specific developments, and often subject to the vicissitudes of annual budgeting. Also, the availability of an hypothecated cash flow makes it easier to create an independent statutory entity to be responsible for the provision of sewage services. Among other advantages, this renders the organisation responsible for the provision of sewage services subject to the normal laws relating to pollution control. More directly stated, a private sector sewage service company or corporation can be controlled, through sanctions such as court fines, more readily than can a department, offering the same services, which is a part of the Government. This is equally true in Hong Kong, despite the fact that all the pollution control legislation in Hong Kong applies to the Crown as well as the private sector [9].

A Sewage Strategy for Hong Kong

Following the implementation of Chadwick's recommendations referred to earlier in this paper, further improvements were made gradually, with the provision of more extensive foul sewerage and, in some cases, screening.

Most effluents, however, continued to be discharged through seawall outlets around Victoria Harbour.

With the rapid growth of population in the 1950's and 1960's, it became evident that the sewage programme was not being successful in preventing the deterioration of water quality, and a consultancy firm was commissioned in 1969 to investigate the state of the waters of Hong Kong and to recommend a programme of improvement works.

The recommendations made in the consultants' report included the construction of 15 submarine outfalls to improve the dispersion of effluents and the provision of three plain sedimentation, one high rate biological treatment and one conventional activated sludge treatment plant.

Despite the expenditure of over HK$4 billion*, important elements of the consultants' proposals still remained to be implemented 20 years after their recommendations were made, largely for the reasons outlined in the previous sections of this paper.

The projects that were implemented certainly avoided an otherwise catastrophic deterioration of water quality. Nevertheless, those elements that were not implemented, combined with greater than expected growth in population and industrial activity, led to continued deterioration in water quality [8].

A reorganisation within the Hong Kong Government in 1986 consolidated various environmental activities, being carried out in a number of departments, with those of the existing Environmental Protection Agency, to establish an Environmental Protection Department. Among other things, the EPD became the client department for the sewage programme.

*HK$ ≡ 0.129 US$

Upon assuming the responsibility for the sewage programme, the newly established EPD carried out an urgent review of sewage facilities and by the end of 1986 two lines of action were established. One of these was aimed at improving the sewerage reticulation system, with a series of Sewerage Master Plans (SMPs) being drawn-up as a first step. The second programme was aimed at developing an environmentally acceptable and cost effective means of treating and disposing of the wastewaters collected in the various SMP areas.

Since it was envisaged that these planning activities would soon give rise to a large increase in the amount of sewerage construction that would need to be undertaken, a new Drainage Services Department was established in 1989 to provide a sharper focus for this work.

Sewerage Master Plans

For planning purposes, the Territory was divided into a series of 16 sewage catchment areas covering the entire Territory and a programme established for the preparation of a Sewerage Master Plan for each of these areas [4]. When complete, each SMP provides an inventory of the existing sewerage system and its condition, a number of engineering packages aimed at rectifying existing problems and catering for any increased loading from future urban and industrial development, and an hydraulic model to provide a means of forecasting the flow implications of any future changes to the sewerage in the SMP area.

Progress to date in implementing the SMP programme is shown in Table 1, which provides also the physical and cost details of each SMP completed so far.

A total of HK$1.25 billion in follow-up engineering works on the Sewerage Master Plan areas is in progress at the present time as well as a further HK$ 6.0 billion in other sewerage related projects. These works include sewerage rehabilitation, First Aid measures that can be instituted quickly and at low cost whilst achieving proportionately large reductions in polluting emissions, a novel effluent export scheme and new sewerage and sewage treatment plants,

Sewage Disposal Strategy

As the SMP programme is implemented incrementally, Hong Kong will be provided with the necessary modern and efficient infrastructure to receive all of it's industrial, commercial and domestic wastewaters.

The next step is to determine how the wastewaters collected in this way, and any increases expected in the foreseeable future, can be disposed-of in an environmentally satisfactory and cost effective manner. This issue was addressed in the so called Sewage Strategy Study. The options studied [5] ranged from very high levels of treatment with discharge close inshore at one extreme, to limited treatment and discharge through a very long sea outfall at the other. An additional consideration was the need to provide for an interim discharge to the Harbour that would achieve early improvements despite incomplete implementation of controls on heavy metals under the Water Pollution Control Ordinance.

The study concluded that the most environmentally satisfactory and robust, whilst at the same time the most economic solution, would be to provide a limited degree of chemical treatment to remove E-coli and heavy metals, followed by an approximately 30 km long outfall discharging into 30 m depth of water in the oceanic currents of the Lema Channel. The main elements of this strategy are shown in Fig. 2.

TABLE 1 *Sewerage Master Plans*

SMP No.	Title	Study Cost (HK$x10⁶)	Study Area (sq. km)	Existing Population	Duration (months)	Estimated Implementation Cost (late 91 price) (HK$x10⁶)
Completed						
1.	East Kowloon	14.4	16.0	940,000	29	1,235.0
2.	Hong Kong Island South	7.0	33.0	32,000	14	756.0
3.	Tolo Harbour	12.0	170.0	708,000	17	410.0
4.	Tsuen Wan, Kwai Chung & Tsing Yi	16.0	49.0	690,000	26	778.0
5.	North West Kowloon	15.0	20.0	820,000	27	562.0
6.	Port Shelter	12.0	150.0	44,000	16	367.0
7.	Central, Western & Wanchai West	12.0	8.5	307,000	25	820.0
8.	Yuen Long & Kam Tin	15.0	176.0	223,000	21	874.0
In Progress						
9.	North & South Kowloon	17.2	12.0	580,000	21	575.0
10.	Chai Wan & Shau Kei Wan	12.0	10.0	320,000	18	335.0
11.	Tuen Mun	15.0	81.0	372,000	17	295.0
12.	Wanchai East & North Point	16.0	12.5	390,000	18	309.0
13.	Aberdeen, Ap Lei Chau & Pok Fu Lam	20.0	18.5	220,000	23	258.0
Proposed Between 1992-1995						
14.	North District	12.0	142.0	165,000	18	200.0
15.	Outlying Islands	15.0	160.0	48,000	18	258.0
16.	Tseung Kwan O	10.0	17.5	83,000	18	50.0

It is expected that collector tunnels, linking the existing preliminary treatment works, will be driven in sound rock 150 m below ground level in order to avoid interfering with the existing congested services, including the Mass Transit Railway, and the foundations of the many tall buildings. Wastewater flows from the various Sewerage Master Plan areas will enter the tunnels through vertical shafts located at nodal points in the present system.

The overall system of vertical shafts, long tunnels and the single very long outfall, which will have a diameter of over 5 m, presents a series of formidable problems in environmental hydraulic design [1]. For example :

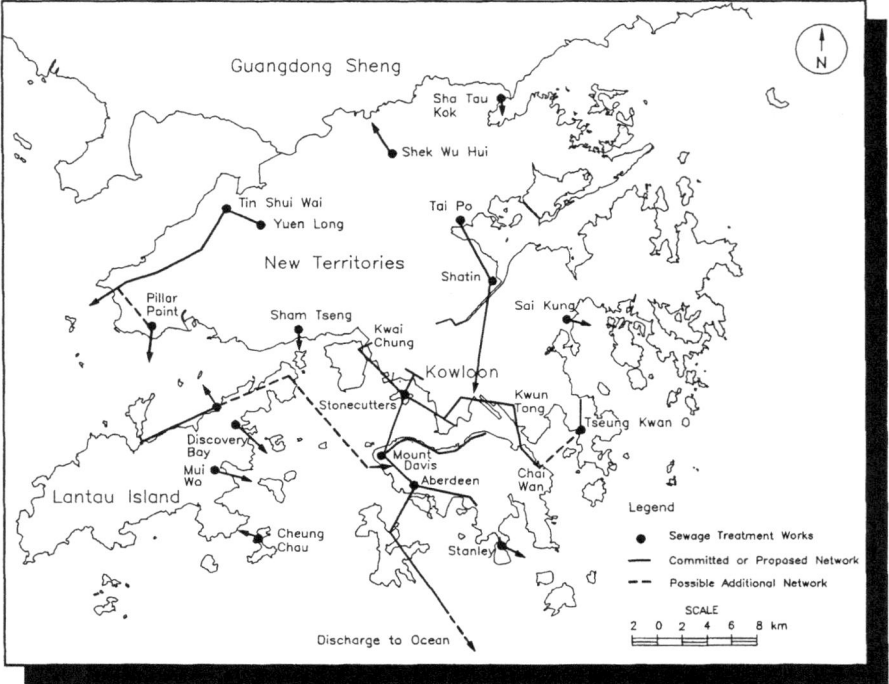

FIGURE 2 *Main elements of Hong Kong's Strategic Sewage Disposal Scheme*

- In order to reduce pumping costs, the tunnels and shafts need to be surcharged and pressurised rather than operate under gravitational force. In these circumstances, access for inspection and maintenance will be extremely difficult. The tunnels must therefore be designed to be maintenance-free. In particular, progressive deposition of sediments must be avoided, even with the relatively low flows obtaining during the first few years of operation.
- As a whole, the system is effectively a series of giant U-tubes. Any abrupt changes in flow of wastewaters into or out of the system will lead to oscillations of the liquid columns. Thus, careful attention needs to be given to pumping arrangements and control systems.
- With the planned depth of tunnels at 150 m, air entrained in the system could be under pressures of up to 15 bar. Any accumulations of air or other gases that find their way into the vertical shafts could expand explosively, as they rise, and propel part of the contents of the shafts out of the system. Selection of tunnel gradients, the use of vortex drop shafts, deaeration chambers and other means are being considered to avoid the problems associated with air entrainment and its eventual release.
- Despite the fact that the outfall will discharge into 30 m depth of oceanic waters, it is estimated that a diffuser about 2,000 m in length will be required to achieve the necessary dispersion. Since the outfall will be 150 m or so below the sea bed, it will be necessary to pay particular attention to the prevention of seawater intrusion and to the arrangements for purging the system should seawater intrusion occur.

10

- Since very large flows of wastewaters are to be discharged from a single outfall, it will be essential that the location of the outfall ensures that the wastewaters are rapidly dispersed and adequately treated through natural processes well offshore.

The hydraulic design and dispersion aspects of the Strategic Sewage Disposal Scheme are being tackled with the help of consultants commissioned by the Drainage Services Department and the Environmental Protection Department of the Hong Kong Government, using a combination of physical and mathematical models. Notable amongst these are a physical model to determine the likely extent of seawater intrusion and a very extensive and flexible water quality and hydraulic mathematical model (WAHMO) to explore dispersion aspects of the outfall.

The physical model comprises a controlled salinity tank 15 m long by 1.5 m wide and 2 m deep, standing up to 10 m above a distribution manifold. The rig is provided with dye injection as well as pressure and flow metering at salient points. An indication of the size of the model can be gained from Figure 3.

FIGURE 3 *Hydraulic model for seawater intrusion studies in relation to Hong Kong's Strategic Sewage Disposal Scheme*

The suite of mathematical models, known as WAHMO, that is being used to determine the best location of the outfall, was developed originally for analysing the water quality implications of land reclamations for container terminals and commercial and residential development, Fig. 4. The flow modelling components of WAHMO produce information on water surface elevation and current velocity, while the water quality models simulate the interaction of 13 parameters including dissolved oxygen, E-coli, BOD, various forms of nitrogen and suspended solids. In order to use WAHMO to optimise the outfall location, it has been necessary to extend the boundaries of the present model into the Pearl River estuary and other parts of the waters of the People's Republic of China. The hydrological and water

FIGURE 4 *Mathematical modelling for the 30 km outfall which is part of Hong Kong's Strategic Sewage Disposal Scheme*

quality data needed to validate the extended model is being obtained in a collaborative exercise with China's State Oceanic Adminstration.

The Future

It is envisaged that, with full implementation of the Sewerage Master Plans, the collector tunnels and the long sea outfall, the water quality in Victoria Harbour can be brought within the statutory Water Quality Objectives and maintained at acceptable levels despite the construction of a major port peninsular, the introduction of a massive breakwater and some further land reclamation for urban development purposes.

There is, however, one uncertainty that could adversely affect the otherwise satisfactory outcome of the comprehensive long term strategy for the disposal of Hong Kong's wastewaters. This is the influence of the Pearl River.

As may be seen from Fig. 5, Hong Kong lies within the Pearl River estuary. The influence of the Pearl River on Hong Kong waters is readily observed from the brown silt laden waters to the West of the Territory as compared with the translucent blue waters on the Eastern side. Less evident is the considerable nitrogen load carried by the Pearl River flows.

Economic growth in the Pearl River delta has averaged over 15 % per annum in recent years, much of it in manufacturing industry. During a recent visit to the region, the Chinese elder statesman Deng Xiao Ping forecast that Guangdong would join the "little dragons of Asia" by the year 2000, which assumes an annual growth rate of nearly 13 % continuing to the end of the millennium. It would be reasonable to expect that this economic growth will be accompanied by increased generation of industrial and other wastewaters in the Pearl River catchment area. The amount of this increase, and the degree to which such wastewaters are

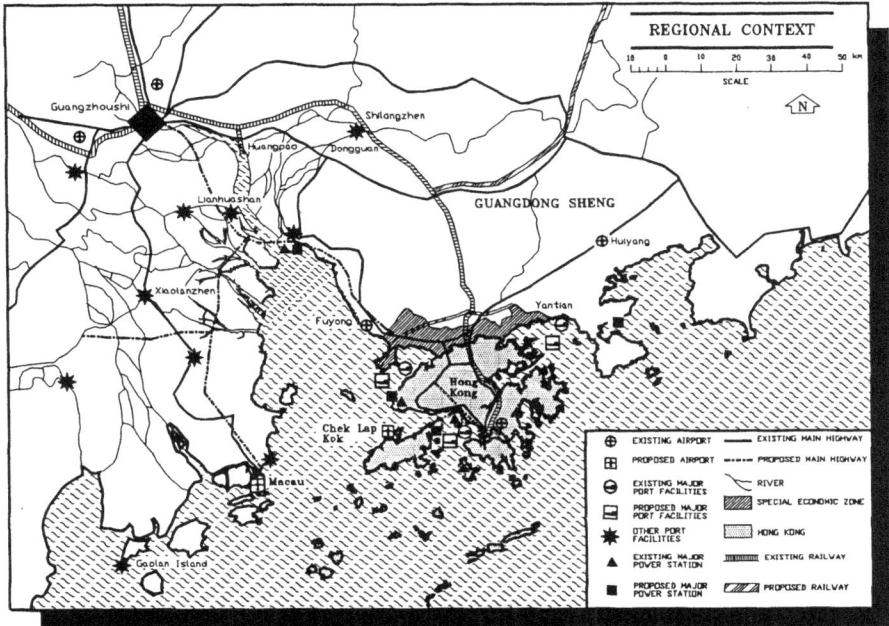

FIGURE 5 *Hong Kong's position within the Pearl River Estuary*

treated prior to discharge, will determine the extent of any possible adverse effects on water quality in Hong Kong.

It was with issues such as the above in mind, as well as the need to provide a channel for agreeing effluent limits for the oceanic outfall that will discharge into the territorial waters of the People's Republic of China, that a protocol was signed in 1990, between the authorities in Guangdong and the Hong Kong Government, to establish the Hong Kong-Guangdong Environmental Protection Liaison Group.

Another initiative that could have a bearing on any future influence of Pearl River flows on Hong Kong waters, is a project sponsored by the British Council. This project involves the preparation of a draft brief for a wastewater strategy for the Pearl River delta. The draft brief is intended to provide the main agenda item for consideration by a workshop with participants drawn from relevant bodies and organisations in Guangdong, Macau and Hong Kong. The intention is that, if the brief can be endorsed, international funding will be sought for the development of the strategy.

Acknowledgements

The preparation of a paper of this nature inevitably draws on the work of many others. I would like to acknowledge here the help and the constructive comments of my colleagues in the Environmental Protection Department and the contribution of the many consultants involved in the programmes I have mentioned.

References

[1]   Clark P.B., Gray S.N. & Baker R.E., (1991), Hydraulic design aspects of the Strategic Sewage Disposal Scheme for Hong Kong, International Symposium on Environmental Hydraulics, Hong Kong University.

[2]   Environment Hong Kong 1991, Hong Kong Government, pp 129-135.

[3]   Lethbridge H.J., (1982), The Hong Kong Guide 1883, Oxford University Press, Hong Kong, pp 121-122.

[4]   McNally V. & McKenzie H.S., (1988), Urban Sewerage : The Neglected Sector in Development Planning, Proceedings of Polmet '88, Vol 1, Hong Kong Institution of Engineers, Hong Kong, pp 538-543.

[5]   Oswell M.A. & Stokoe M.J., (1989), Sewage Strategy for Hong Kong, Proceedings of WPCF Asia/Pacific Rim Conference on Water Pollution Control, Hawaii, USA.

[6]   Private communication, (1991), Arthur D. Little (Asia Pacific Inc.).

[7]   Reed S.B., (1987), Ènvironmental Protection - Hong Kong 1987, Journal of the Hong Kong Institution of Engineers, May, pp 9-17

[8]   Reed S.B., (1988), Sewage and a City by the Sea - Hong Kong : A case in point, Journal of the Institute of Water & Environmental Management, Vol 2, No. 1, pp 13-21.

[9]   Reed S.B., (1991), The Government's Role in Managing Hong Kong's High Pressure Environment, Proceedings of Polmet '91, Vol 1, Hong Kong Institution of Engineers, Hong Kong, pp 579-598.

2 Recent developments in water quality modelling and management

D. R. F. Harleman

ABSTRACT

Advances in hydrodynamics, mass transport and water quality modeling have reached the point where it is feasible to use models to make comparative risk assessments of alternative environmental management scenarios. Progress has been rapid in computational efficiency -- thereby making three-dimensional modeling practical in situations where previously overly simplified one- or two-dimensional models were the norm.

Interaction is needed among scientists, engineers, regulators and managers in order to establish local priorities and to achieve cost effective solutions to environmental problems.

1. Introduction

1.1 Water quality modeling

A survey of two recent conferences on estuarine and coastal modeling [1,2] indicates a definite shift from two-dimensional to three-dimensional models, especially hydrodynamic models. Cheng and Smith [3] point out that for a fixed cost, computing power increases 5 to 10 times every 5 years, thereby making 3-D models computationally practical alternatives to 2-D models. In river and estuarine applications, 2-D laterally averaged models were traditionally chosen to simplify modeling. However, the laterally averaged model makes it impossible to resolve impacts of point and non-point sources entering at lateral banks. In many coastal models 2-D depth averaged models are used, although this requires neglecting all temperature and salinity stratification effects and restricts the modeling of sediment-water column interaction.

Cheng and Smith [3], after justifying the practicability of 3-D computation, observe that the data requirements for 3-D modeling are increased greatly over its 2-D counterpart. The writer does not agree with this commonly held position. Because of the dynamic nature of the estuarine and coastal environment, simultaneous observations at a sufficient number of discrete points to permit determination of either lateral or depth averaged conditions for comparison with the 2-D model are very rare. A significant advantage of the 3-D model is its ability to compare model output with field data at whatever spatial location and time the data was obtained. The same is true in regard to field data requirements for boundary conditions (such as open-ocean boundaries).

Another frequently voiced objection to the 3-D model is that the model results are unnecessarily complex for use by decision makers. Here again, the writer does not agree because model output can be averaged spatially, temporally or both, to suit the needs of management users. In fact, there is much to be said for averaging the output of a model rather than averaging the input and the model itself. Spatial or temporal averaged models (whether 1-D or 2-D) introduce dispersive model parameters that increase the data required for calibration.

While significant advances have been made in 3-D hydrodynamic modeling, parallel advances in 3-D mass transport models and in interfacing 3-D hydrodynamic and mass transport models have not kept pace. Coupled hydrodynamics (HY) and mass transport water quality (WQ) models are being used to study complex eutrophication problems in estuarine and coastal regions. The occurrence of nutrient fed algal blooms are important environmental issues and the understanding and control of these events increases the need for coupled 3-D, HY, and WQ models. For example, depth-wise variations in light, temperature, dissolved oxygen and nutrient concentrations are crucial to the simulation of algal uptake and growth. The short (10 minutes) time step required for HY model stability together with the need to simulate eutrophication over one or many years means that it is impractical to couple HY and WQ models at the same time step. Another important WQ modeling problem is the simulation of point sources and their associated near-field entrainment induced flows into a numerical grid whose horizontal dimensions are larger than that of the source. Some recent developments in modeling and interfacing are discussed below.

1.2 *Water quality management*

The past 25 years has seen significant shifts and changes in legislation aimed at water quality control and management. For example, in the USA in the early 1970's there was widespread dissatisfaction with local efforts to improve water quality on the basis of local receiving water standards. This was largely a consequence of a lack of confidence in the ability of scientists and engineers to forecast the degree of wastewater treatment or the type of pollution control facilities needed to achieve local goals. In 1972 the U.S. Congress passed the first Clean Water Act (CWA) requiring uniform treatment plant effluent standards as well as uniform treatment levels (e.g., 85% removal of suspended solids and BOD) for the entire country regardless of receiving water type or size. The uniform 1972 effluent standards (e.g., 30 mg/L for suspended solids and BOD) and treatment levels were set at what was known to be achievable by conventional primary settling followed by activated sludge secondary treatment. This was landmark legislation and resulted in significant improvements in the nation's water quality. An unfortunate consequence of the technology-based standards in the CWA has been the hindering of research and innovation on wastewater treatment technology. The 1972 act is now due for reauthorization and one of the issues being debated is whether the rigid and uniform, technology-based effluent standards should be retained. Arguments for change suggest that water quality modeling science has progressed to the point where local needs and priorities can and should be addressed and that the technique of comparative risk assessment should be used in local decision making. Case studies illustrating important water quality management issues and controversies are discussed below.

2 Eutrophication modeling

In the late 1970's it became evident that low dissolved oxygen problems in lakes, reservoirs, estuaries and coastal embayments would not be solved by BOD type models. Nutrients, such as nitrogen and phosphorus, remaining after conventional wastewater treatment, as well as nutrient inputs from agricultural non-point sources promote the growth of algae which increase the complexity of the dissolved oxygen problem. Initial efforts at modeling algae uptake and growth and the eutrophication process focussed on the biogeochemical side of the mass transport equation (i.e., sources and sinks). Nutrient models were coupled to existing tidally-averaged [4] and real-time 1-D models [5] as well as to two-dimensional "link-node" depth averaged models [6].

Another group of models, developed for the Great Lakes [7], dealt with three-dimensional water quality problems using sophisticated algae models and very crude hydrodynamics in what became known as "large box" models. Empirical transports between the "large boxes" were often calibrated on a quasi-steady state basis by attempting to reproduce seasonal average temperature distributions.

2.1 Chesapeake Bay eutrophication model

In 1985 the U.S. Environmental Protection Agency (EPA) decided on a comprehensive eutrophication study of Chesapeake Bay and its tributaries. Chesapeake Bay, one of the largest and most productive estuaries in the world, extends 300 km north from the ocean entrance to the Susquehanna River. The average depth in the main bay is 8m and a natural channel with depths greater than 15m covers about two-thirds of its length. Its width is as great as 50 km in the middle bay, and it has five major tributaries, the largest being the Potomac. Over the past 50 years the bay has experienced a dramatic decline in living resources concurrent with a decline in water quality. Probable causes are increased algae growth due to increases in nutrient inputs coupled with sediment nutrient interactions caused by the absence of dissolved oxygen in the deep channel during the summer months. Strong salinity gradients exist in the longitudinal, vertical and lateral directions. With the cost of potential pollution control measures in the billions of dollars, it was decided to undertake a major eutrophication modeling effort to evaluate the effectiveness of proposed pollution control strategies. The major management decisions would be between point source nutrient removal at municipal waste treatment plants and, or, non-point sources nutrient removal associated with agricultural inputs from farming along the tributaries and main bay. It is known that algal production (depending on the season and location) can be either nitrogen or phosphorus limited. Therefore a major concern of the modeling effort would be to resolve individual nutrient control impacts.

EPA in conjunction with the Waterways Experiment Station (WES) of the U.S. Army Corps of Engineers began the 3-D model development in 1985. The writer is chairman of an independent review board that meets 3 to 4 times a year to monitor the modeling effort.

Hydrodynamic model The hydrodynamic model, called CH3D (Curvilinear Hydrodynamics in Three Dimensions) is a modification [8] of an earlier model [9] in which the vertical dimension employed a sigma stretched grid. With this type of model the bottom layer in one column communicates with the bottom layer in a laterally adjacent column. Thus, if the depth changes are rapid, as in the main bay, channel stratification cannot be maintained. As a result, the model was changed to a fixed vertically-layered grid. The numerical grid used for Chesapeake Bay, shown in Figure 1, consists of more than 700 surface elements and a total of 4000 computational cells. The hydrodynamic model is baroclinic in that it includes the 3-D mass and heat transport equation for salinity

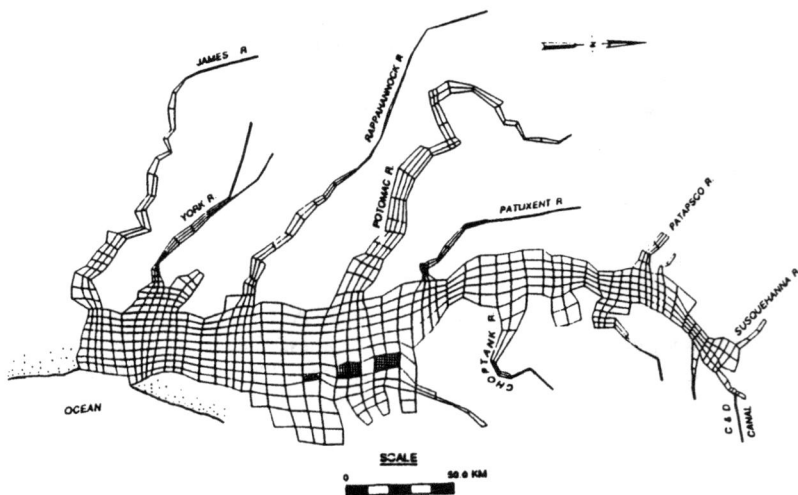

FIGURE 1   *Planform grid for 3-D model of Chesapeake Bay*

18

and temperature as well as an equation of state relating salinity and temperature to density. The turbulence closure is an algebraic model accounting for vertical stratification in a relatively simple manner. Nevertheless, the model shows reasonably good ability to reproduce wind-mixing de-stratification and re-stratification events in the main bay [10,11]. A one-year simulation of the full momentum, salinity and temperature model operating at a $\Delta t = 5$ min. takes about 6 hours on a CRAY-YMP computer. The hydrologic variability of Chesapeake Bay was approximated by choosing three years representing normal, wet and dry, average annual freshwater inflows to the Bay.

Interfacing the hydrodynamic and water quality models   The Chesapeake Bay eutrophication modeling effort was one of the first to explore the 3-D interfacing problem in a systematic manner. The water quality model has 22 state variables, including three types of algae. The mass transport algorithm uses a third-order scheme called QUICKEST that is capable of resolving advection dominated transport [12,13]. However, even with the use of a super-computer, it was considered impractical to simultaneously solve the momentum equations and the 22 mass transport equations at the required time step of 5 minutes given that continuous water quality simulations for periods of 10 years or more were envisioned. The water quality model uses the same grid as the hydrodynamic model; therefore interfacing involves only temporal considerations.

The following options for interfacing the baroclinic hydrodynamic and the water quality mass transport models were investigated:
1. Intratidal mode: the 5-minute output of advective and diffusive transports from the hydrodynamic model would be processed into water quality model input at 2-hour intervals that preserves the highly advective intratidal velocities. This requires a mass transport model capable of resolving the high Peclet numbers such as QUICKEST.
2. Intertidal mode: the hydrodynamic model would be processed at the 12-hour tidal period interval. However, conventional temporal averaging would result in Eulerian residual velocities and loss of important intratidal mass transport processes. This would require a calibration of dispersion coefficients to account for transport information lost in the tidal averaging process. The alternative is to use some form of Lagrangian residual transport processing. The form chosen for the Chesapeake Bay model calculates a first-order estimate of the Lagrangian residual velocity as the sum of the Eulerian residual and Stokes' drift [14]. Salinity is calculated both by the baroclinic hydrodynamic model at 5-minute intervals and by the water quality model operating at the intertidal model. Thus a direct indication of the validity of the Lagrangian averaging is possible by comparison of the two salinity calculations.

This comparison was found to be satisfactory for the weakly non-linear bay system, in which the tidal amplitude to depth ratio is about 0.05. The intertidal mode was adopted because it was found to be computationally more efficient than the first option. A difficulty with the QUICKEST numerical scheme is that even though it is capable of resolving advection-dominated transport, it must still satisfy the Courant number stability criterion, which requires relatively small time steps. In the Chesapeake Bay application

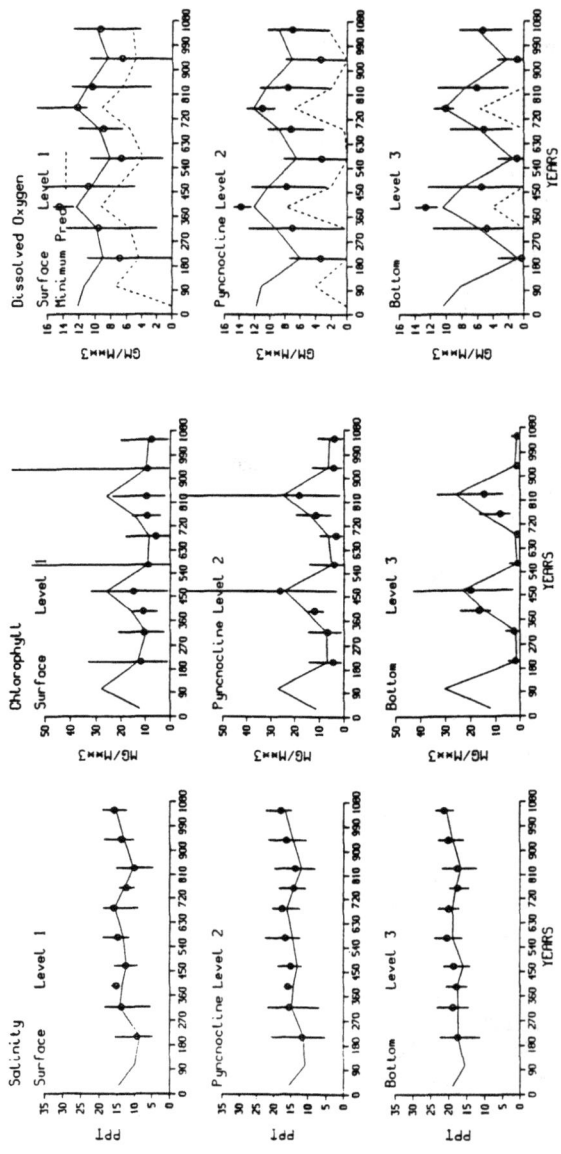

FIGURE 2   *Calibration of Chesapeake Bay eutrophication model, zone 3. 1984-1986.*
(From C. Cerco, WES)

this time step was 2 hours. Although the hydrodynamic model was averaged over 12 hours, this information had to be repeated at 2-hour intervals for WQ model stability.

Water quality model The 22 state variables of the water quality model include the physical group (salinity, temperature and particulates), the biological cycles, carbon-oxygen, nitrogen and phosphorus, the silica cycle, deposition of particulate organic matter and a sediment model. The sediment model computes the diagenesis of sediment organic matter (both aerobically and anaerobically), sediment oxygen demand and fluxes of nitrogen, phosphorus and silica to and from the water column [15]. The model is sensitive to long term changes due to seasonal inputs of dead algae.

The WQ model was calibrated to a three-year time period, 1984-86, involving wet, dry, and normal years. A typical result for a section near the middle of the main bay is shown in Figure 2. The combined HY and WQ models has been run for continuous periods as long as 30 years [16] in order to look at long term trends in summer anoxia conditions in the main bay. The three hydrodynamic model years (wet, dry, normal) were repeated in varying sequences to simulate the historical hydrologic conditions. Two interesting phenomena were evident. In wet years more nutrients enter the bay to provide nourishment to algal which grow, die, and sink to the bottom where decay produces anoxia. Wet years are more highly stratified than dry years and this tends to isolate bottom waters from reaeration and produces more anoxia than in dry years. The bottom anoxia reinforces the algal cycle because it triggers additional sediment nutrient releases to the water column. The three-dimensional nature of the predicted surface salinity distribution is shown in Figure 3.

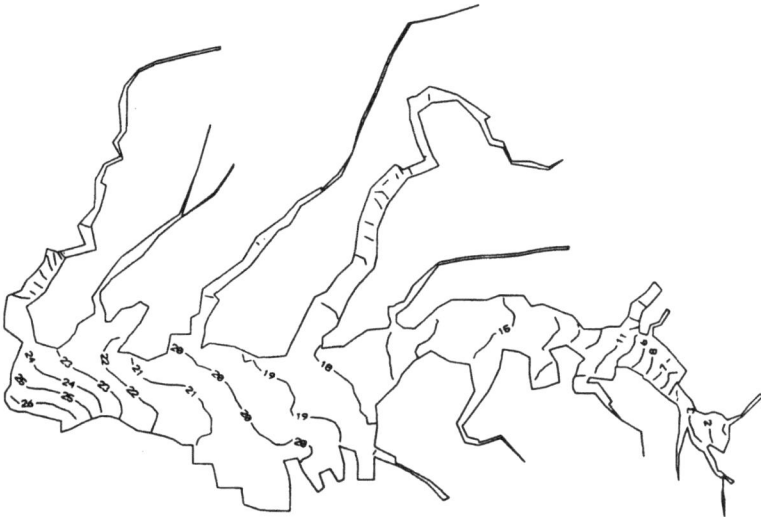

FIGURE 3   *Predicted surface salinity in Chesapeake Bay. Season 4 1986.*
                (From C. Cerco, WES)

The model is currently being used to assess the impact of a number of point and non-point source nutrient control scenarios for the bay and tributaries. The enormous 3-D, time-varying model output is made tractable for management purposes by spatially averaging the model results over major bay and tributary segments and by temporally averaging over one or more months.

2.2 Additional 3-D mass transport model developments

A recent study at MIT [17] addresses a number of problems associated with the Chesapeake Bay transport modeling. This approach uses Eulerian-Lagrangian methods based on a decomposition of the transport equation into an advective and diffusive part. The advective part is solved with the backwards method of characteristics using the Lagrangian approach whereas the diffusive part is solved using an Eulerian approach. Thus the Lagrangian residual circulation is modeled directly without resort to the Stokes drift approximation. There is no Courant number restriction requiring subdividing the water quality model time step. Additionally the method is applicable in cases where the tidal non-linearity is large such as the North Sea.

Another interesting interfacing problem arises in linking near-field point-source dilution models and far-field eutrophication models using concentration as a state variable. The representation of point sources in concentration models is difficult because they cannot resolve large gradient concentration fields whose spatial extent is small compared to the grid size. Studies [18,19,20] have demonstrated two useful approaches. One method is to treat the continuous plume near the source as the superposition of discrete puffs that are advected and diffused forward in time until they are large enough to be resolved into the numerical grid. This method works well in relatively uniform 2-D flow fields, but encounters difficulties in representing 3-D entrainment near an outfall. The other method uses particle tracking, where mass loading is represented by the introduction of discrete particles. At each timestep the displacement of each particle consists of an advective, deterministic component and an independent, random walk Markovian component within the framework of an Eulerian-Lagrangian transport model. When the particles have spread over a sufficiently large spatial region, they are mapped onto the numerical grid by calculating the concentration (particle density) associated with each node [17].

3.   Estuary and coastal water quality management issues

Through advances in modeling we are now in a position to make reasonably reliable estimates of estuary and coastal water quality and to assess the environmental impact of various levels of wastewater treatment and effluent delivery. By means of coordinated studies, marine outfall length, location, multi-port diffuser length and orientation can be optimized. Much more interaction is needed among scientists, engineers, environmental regulators and managers in order to achieve cost effective solutions to local problems. Those of us who are primarily concerned with environmental modeling must be aware of the range of wastewater treatment technologies available -- not only those in common use, but innovative processes as well. Regulators and managers must be more willing to adopt innovative technology, especially where satisfactory results have been conclusively demonstrated.

Assessments of the state-of-the-art of wastewater treatment processes at MIT [22,23,24] indicate that it is appropriate to revisit physical-chemical treatment process. It is well known that the addition of positively charged metallic salts to the influent to primary sedimentation basins causes precipitation, coagulation, and flocculation of colloids and larger particles. The increased particle size results in increased settling velocity and increased removal efficiency. This type of physical-chemical wastewater treatment was widely practiced in the early 1900's, but the large quantities of lime or alum that were added as primary coagulants caused difficulties in the processing and disposal of the sludge residue. Physical-chemical wastewater treatment was largely displaced by 1950 due to the growing reliance on conventional primary settling followed by the biological activated sludge process. In the 1980's concern with eutrophication, particularly in the US-Canada Great Lakes and in Scandinavia, led to efforts to reduce nutrients. Chemical additives began to be used in tertiary treatment processes to remove phosphorus. At the same time, very small quantities, less than 1 mg/l, of polymers were found to be effective in coagulation and flocculation. This reduced the concentrations of primary coagulants such as alum or ferric chloride to less than 50 mg/l and greatly alleviated sludge quantity and processing problems [21]. This type of treatment, known as chemically enhanced primary treatment (CEPT), has largely been practiced in the USA by retrofitting existing primary plants.

The following case studies illustrate some environmental conflicts involving this innovative technology.

3.1 *San Diego, CA*

A number of large treatment plants in Southern California which have not as yet come into compliance with full secondary discharge requirements of the Clean Water Act, began to use chemical additives in their conventional primary sedimentation tanks in the mid-1980's to meet the 75% removal of suspended solids (SS) required by the State of California Ocean Plan (COP). It is interesting to note that the COP said nothing about BOD removal for ocean discharges, instead it addressed dissolved oxygen directly by allowing a decrease of no more than 10% of the ambient value. Figure 4 shows annual average SS and BOD5 removals with chemically enhanced primary treatment at San Diego (Pt. Loma) and Los Angeles (Hyperion). The removal rates are plotted as a function of the overflow rate (the flow through the plant divided by the surface area of the primary settling basins) the usual measure of treatment plant efficiency. Also shown is the expected treatment efficiency for conventional primary treatment [25]. In addition to the increased treatment efficiency it is evident that chemical additives allow overflow rates 2 to 3 times greater than generally used in conventional primary plants. This has important consequences in expanding existing plant capacity or in reducing the capital cost of new plants.

Despite San Diego's seven year successful experience with CEPT, during which time the Pt. Loma plant has achieved average annual suspended solids removal of 80% and close to 60% for BOD5, the city has been under a Federal Court order to achieve full

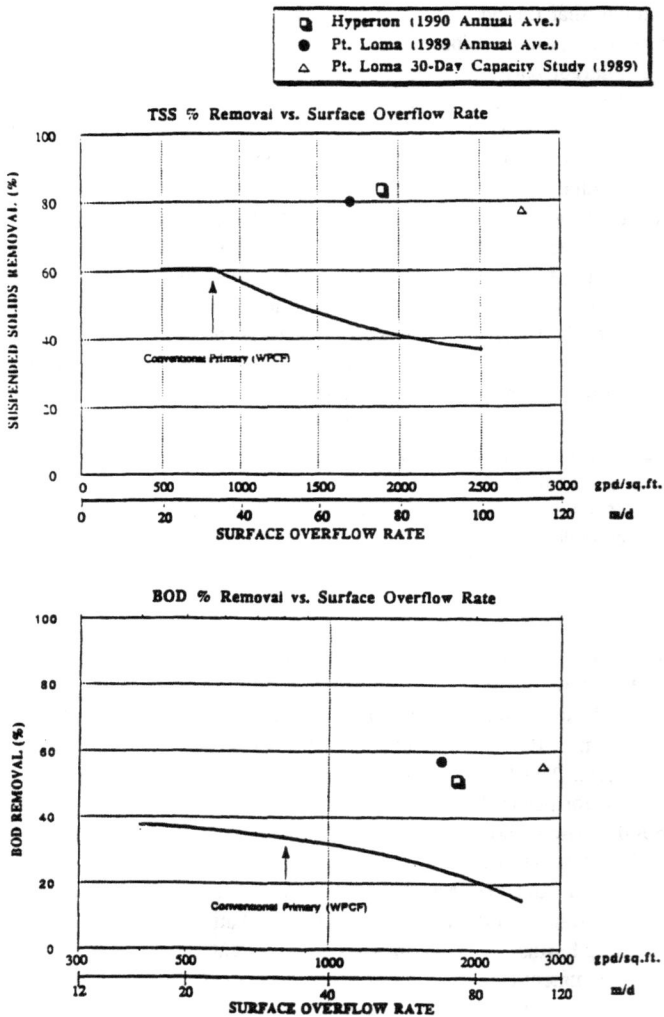

FIGURE 4   Removal efficiency at Pt. Loma (San Diego) and Hyperion (Los
Angeles) chemically enhanced primary treatment plants

secondary treatment by building (at a cost of several billion dollars) an activated sludge plant to provide the additional treatment needed to comply with the 85% removal required by the CWA. Scientists affiliated with local universities, after years of monitoring ocean conditions in the vicinity of the 4 km ocean outfall extending to a depth of 60m, have concluded that there are no significant coastal environmental impacts of the CEPT waste effluent or its delivery system. Waste resource experts contend that San Diego's potable water supply cannot be expanded, and probably will be reduced in the future. They maintain that the highest priority issue for San Diego is water conservation and that the region should construct a number of tertiary wastewater treatment plants for a portion of the total waste flow to allow water reclamation and reuse. Civic leaders contend that there is not enough time or money to do both. Environmental organizations state that the environmental impacts of processing and disposing of the additional biological sludge generated by upgrading Pt. Loma to full secondary treatment would far outweigh the oceanic benefits of additional treatment.

In 1991 the San Diego federal judge was persuaded by the several arguments and postponed for one to two years the implementation of the secondary treatment requirement in order to allow the city to test methods of improving on the chemically enhanced primary treatment currently in use.

3.2 Sydney, NSW

About ten years ago Sydney began an ambitious program to protect their bathing beaches. Three existing high-rate (detention time less than 1 hour) primary plants discharging onto the beaches were not adequate to protect the public health or esthetics. Three ocean outfalls 2 to 4 km in length shown in Figure 5 were designed and constructed; however, there is concern that outfalls will not be sufficient and that additional treatment will be required for these coastal treatment plants. At issue are competing demands for funding combined sewer overflow controls in Sydney Harbor and nutrient removal upgrading of 23 inland plants to control eutrophication of fresh water. A 1989 study in which the writer participated [26] recommended improving the treatment efficiency of the coastal plants by retrofitting them for chemically enhanced primary treatment. Because of high cost of real estate in the Sydney coastal zone, expansion of the treatment plants to secondary treatment is impractical and chemical enhancement represents the cost effective and fastest way of obtaining additional beach protection. The CEPT process is effective in removing floatables such as fats, oil and grease. In addition, heavy metals tend to adsorb to particulates and the high suspended solids removal capability of CEPT gives effective protection against these hazardous substances [27].

3.3 Boston, MA

Boston began discharging untreated waste into its harbor more than a hundred years ago. It was not until 1968 that all dry weather sewage began receiving conventional primary treatment. The findings of successive water quality studies have shown that the major pollution problem for the harbor is the combined sewer overflows (CSO). These overflows derive from sources on the perimeter of the harbor and result from the

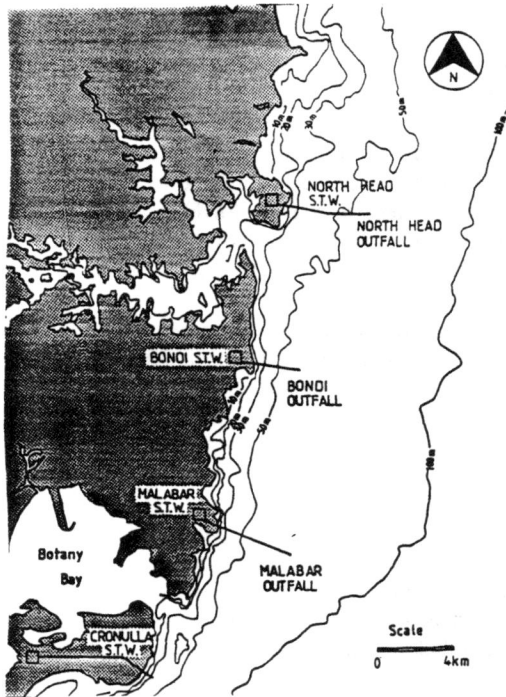

FIGURE 5   *Sydney, NSW ocean outfalls*

collection of storm water and sewage in the same pipes throughout much of the older inner city area. Raw sewage is discharged from these sources during wet weather when the treatment plant capacity is exceeded. These discharges of raw sewage occur about 60 times a year. It is generally agreed that the combined sewer overflows are responsible for the frequent closing of the shellfishing and bathing areas within the harbor.

Local interests argued for a 15 km ocean outfall and diffuser with less than secondary treatment combined with stringent controls on CSO. Federal regulators insisted on the highest priority for full secondary treatment and in 1986 this was mandated by a Federal court order together with the 15 km outfall. This 6 billion dollar plan will result in one of the largest (50 m^3/s) conventional primary and secondary activated sludge treatment plants in the country. It is scheduled for completion in 1999. The case has been made

[28] that chemically enhanced primary treatment should be implemented during the first phase of the harbor cleanup. This would permit the use of less costly secondary treatment options, such as aerated biofilters, and provide financial resources needed to control the combined sewer overflows that are the major source of shell fishing and beach closures.

REFERENCES

[1] Spaulding, M.L. (Ed.) (1989), "Estuarine and coastal modeling," *Proceedings*, Conference on Estuarine and Coastal Modeling, ASCE, Newport, RI.
[2] Spaulding, M.L. (Ed.) (1991), "Second international conference on estuarine and coastal modeling," *Conference Proceedings*, ASCE, Tampa, FL, November.
[3] Cheng, R.T., and Smith, P.E. (1989), "A survey of three-dimensional numerical estuarine models," *Proceedings*, Conference on Estuarine and Coastal Modeling, ASCE, Newport, RI.
[4] Thomann, R.V. and Fitzpatrick, J.J. (1982), "Calibration and verification of a model of the Potomac Estuary," HydroQual, Inc., Final report to D.C. Department of Environmental Services, Washington, DC.
[5] Najarian, T.O. and Harleman, D.R.F. (1977), "Real time simulation of nitrogen cycles in an estuary," *Proceedings, ASCE*, 103(EE4).
[6] DiToro, D.M., O'Connor, d.J. and Thomann, R.V. (1971), "A dynamic model of the phytoplankton population in the Sacramento-San-Joaquin Delta," *Advances in Chemistry*, No. 106, American Chemical Society.
[7] DiToro, D.M. and Connolly, J.P. (1980), *Mathematic models of water quality in large lakes, Part 2: Lake Erie*, EPA-600/3/80-065, EPA Laboratory, Duluth, MN.
[8] Johnson, B.H., Heath, R.E., Hsieh, B.B., Kim, K.W. and Butler, H.L. (1991), *User's guide for a three-dimensional numerical hydrodynamic, salinity, and temperature model of Chesapeake Bay*, W.E.S., U.S. Army Corps of Engineers, T.R. HL-91-20, Vicksburg, MI.
[9] Johnson, B.H., et al. (1989), "Development of three-dimensional model of Chesapeake Bay," *Proceedings*, Conference on Estuarine and Coastal Modeling, ASCE, Newport, RI.
[10] Kim, K.W., Johnson, B.H. and Sheng, Y.P. (1989), "Modelling a wind-mixing and fall turnover event on Chesapeake Bay," *Proceedings*, Conference on Estuarine and Coastal Modeling, ASCE, Newport, RI.
[11] Johnson, B.H., Kim, K.W., Heath, R.E., and Butler, H.L. (1991), "Verification of a three-dimensional hydrodynamic, salinity, and temperature model of Chesapeake Bay," *Environmental Hydraulics*, J.H.W. Lee and Y.K. Cheung (editors), Balkema, Rotterdam.
[12] Dortch, M.S., et al. (1989), "Interfacing three-dimensional hydrodynamic and water quality models of Chesapeake Bay," *Proceedings*, Conference on Estuarine and Coastal Modeling, ASCE, Newport, RI.
[13] Dortch, M.S. (1990), "Three-dimensional Lagrangian residual transport computed from an intertidal hydrodynamic model," TR No. EL-90-11, Waterways Experiment Station, Corps of Engineers, Vicksburg, MI.

[14]   Dortch, M.S. (1991), "Long-term water quality transport simulations for Chesapeake Bay," *Environmental Hydraulics*, J.H.W. Lee and Y.K. Cheung (editors), Balkema, Rotterdam.

[15]   DiToro, D.M. (1990), "Sediment oxygen demand model: methane and ammonia oxidation," *J.Envr. Eng. Div.*, ASCE, Vol. 116, No. 5.

[16]   Cerco, G. and Cole T. (1991), "Thirty-year simulation of Chesapeake Bay dissolved oxygen," *Environmental Hydraulics*, J.H.W. Lee and Y.K. Cheung (editors), Balkema, Rotterdam.

[17]   Dimou, K.N. (1992), "3-D hybrid Eulerian-Lagrangian particle tracking model for simulating mass transport in coastal water bodies," Ph.D. thesis, Department of Civil Engineering, Massachusetts Institute of Technology, Cambridge, MA.

[18]   Adams, E.E., Kossik, R., and Baptista, A.M. (1986), "Source representation in a numerical transport model," *Proceedings*, VI International Conference on Fin. Elements in Water. Res., Lisbon.

[19]   Dimou, K.N. and Adams, E.E. (1989), "2-D particle tracking model for estuary mixing, *Proceedings*, Conference on Estuarine and Coastal Modeling, ASCE, Newport, RI.

[20]   Dimou, K.N. and Adams, E.E. (1991), "Representation of sources in a 3-D Eulerian-Lagrangian mass transport model," *Proceedings*, Water Pollution 91, Wessex Institute of Technology, Southampton, U.K.

[21]   Heinke, G., Tay, J.A. and Qazi, M. (1980), "Effects of chemical addition on the performance of settling tanks," *Journal Water Pollution Control Federation*, 52(12).

[22]   Harleman, D.R.F. and Morrissey, S. (1990), "Chemically enhanced treatment: an alternative to biological secondary treatment for ocean outfalls," *Proceedings* 1990 ASCE National Conference, San Diego, CA.

[23]   Morrissey, S. (1990), "Chemically-enhanced wastewater treatment," M.S. thesis, Department of Civil Engineering, Massachusetts Institute of Technology, Cambridge, MA.

[24]   Murcott, S. (1992), "Performance and innovation in wastewater treatment," M.S. thesis, Department of Civil Engineering, Massachusetts Institute of Technology, Cambridge, MA.

[25]   Water Pollution Control Federation (1985), Clarifer design manual.

[26]   Camp, Dresser and McKee (CDM) (1989), "Review of Sydney's beach protection programme," Report to Minister of Environment, New South Wales.

[27]   Young, D.R., Baumgartner, D.J., Snedaker, S.C., Udey, L., Brown, M.S. and Corcoran, E.F. (1990), "Effects of wastewater treatment and seawater dilution in reducing lethal toxicity of municipal wastewater to sheepshead minnow (Cyprinodon variegatus) and pink shrimp (Penaeus duorarum), *Research Journal Water Pollution Control Federation*, Vol. 62, No. 6., pp. 763-770.

[28]   Harleman, D.R.F., Morrissey, S., and Murcott, S. (1991), "The case for using chemically enhanced primary treatment in a new cleanup plan for Boston Harbor," *Civil Engineering Practice*, Journal of Boston Society of Civil Engineers, Vol. 6, No. 1.

Part 1
TIDAL CURRENT MODELLING

3 Three-dimensional numerical modelling of coastal hydrodynamics and pollutant dispersal

C. T. Mead and A. J. Cooper

ABSTRACT

Three-dimensional models of coastal hydrodynamics and bacterial dispersion have been developed to assist in the design of submarine outfalls to disperse treated sewage effluent with the aim of meeting EC directives. The models provide a representation of the effects of vertical density variations on pollutant mixing, making them applicable in areas where plumes from submarine outfalls are trapped below a thermocline. Other important physical processes simulated include the three-dimensional structure of wind- and density-driven currents and diurnal and vertical variations of bacterial mortality rates. In combination with initial dilution models, the models provide a more physically realistic representation of bacterial dispersion in three dimensions than has generally been possible before the advent of the present generation of powerful computers.

1 Introduction

In recent years, random walk pollutant dispersion models have been successfully calibrated against field data in a number of coastal applications, and used to predict the effects of changes in sea outfall location and effluent treatment on pollution levels. Such models usually use hydrodynamic fields generated by depth-averaged flow models as input data. These hydrodynamic fields can then be modified to simulate three-dimensional effects by applying analytical profiles representative of tidal/seasonal and wind-driven current structure [6].

Plumes in regions of strong currents, where pollutant rapidly becomes mixed through the water column, can be simulated well by random walk models based on depth-averaged flows. In regions of relatively weak vertical mixing, however, pollutant concentrations can vary strongly with depth over wide areas. Outfall plumes may, for example, be prevented from reaching the sea surface by strong saline or thermal stratification. In order to simulate pollutant plumes in areas of strong vertical variations, a random walk pollutant dispersion model based on hydrodynamic fields computed by a three-dimensional flow model has been developed. The advantages of this model over depth-averaged modelling approaches are that vertical mixing is represented in a physically realistic way by deriving vertical diffusivities from densities computed at many depths in the water column, and that the three-dimensional structure of wind-driven currents is included in the simulated hydrodynamic fields by forcing the flow model with a wind stress term, in addition to water surface level variations. Geostrophic currents associated with the interaction of gravitational and Coriolis forces can also be simulated.

Section 2 of this paper describes the three-dimensional numerical hydrodynamic and pollutant dispersion models. In Section 3, the application of the three-dimensional modelling approach is illustrated with reference to a modelling study of submarine outfall options for Barcelona in southern Spain. The summary and discussion given in Section 4 include consideration of the coastal environments in which three-dimensional models are required, and those which can be adequately simulated using depth-averaged techniques.

2 The mathematical models

2.1 The hydrodynamic model

The hydrodynamic model is based on the three-dimensional Navier Stokes equations, including turbulent viscosity, simplified using the hydrostatic pressure approximation. This approximation is normally valid for marine and estuarine flows, and simplifies the solution because it eliminates the need to solve an elliptic equation to obtain the pressure [4]. The model includes the transport of salinity and temperature by the flow field, and the simulation of the flows driven by the resulting density gradients.

The model grid consists of squares in the horizontal plane, and in the vertical direction the cells are separated by horizontal planes. The numerical method used to solve the finite difference equations is explicit in the horizontal plane and vertically implicit. The equations are solved on an AMT DAP parallel processing computer, with 4096 processors which carry out calculations simultaneously. For applications involving small numbers of model cells, a whole layer of cells is updated at the same time.

The model includes the contribution to the pressure from the varying density of water, so that density-driven flows can be simulated. A mixing-length formulation of vertical turbulent mixing is incorporated, which includes empirical suppression of vertical mixing by vertical density gradients, such as arise near the thermocline in the ocean or near the interface in a buoyant plume [7]. Wind effects can also be simulated by incorporating a surface wind stress, which can vary in space and time.

When used in combination with the bacterial dispersion model (Section 2.2), the hydrodynamic model enables the effects of wind-driven flows on pollutant dispersion to be simulated. This can be important in producing upwelling and downwelling, as well as complicated patterns of flow in the horizontal plane. The effects on dispersion of density-driven flows, such as cooling water plume spreading and regional oceanographical flows, are also included in the models. The hydrodynamic model generates three-dimensional flow and density fields which drive the pollution model.

The hydrodynamic model has been applied in many studies, including simulations of the dispersion of cooling water from power stations, secondary and gravitational flows in estuaries and flows in the sea where density and wind forcing dominate.

2.2 The bacterial dispersion model

The three-dimensional bacterial dispersion model is based on the well-known random walk representation of turbulent dispersion [5,6]. Continuous pollutant discharge is represented as a regular release of discrete particles, with each particle representing a given number of bacteria. The particles move in three dimensions in response to the layer-averaged horizontal and vertical velocities computed by the hydrodynamic model. In addition to this ordered motion, particles experience displacements in the horizontal and vertical planes which represent the effects of turbulent dispersion. Whilst the directions of the turbulent displacements are random, the displacement lengths are calculated from eddy diffusivities. The horizontal eddy diffusivity is specified as a spatially-invariant value, as these diffusivities are generally approximately constant over large areas of coastal waters. Vertical diffusivities in coastal waters, however, vary significantly horizontally and over the depth of the water column and, for this reason, the vertical diffusivities used in the dispersion model are derived from the vertical current shear and densities computed by the hydrodynamic model. In this way, the effects of saline and thermal stratification on the vertical mixing of pollutant are represented.

During the course of a dispersion model simulation, the number of bacteria represented by particles released into the model changes, in order to represent diurnal variations of pollutant discharge. The number of bacteria represented by each particle can also decrease after release to represent bacterial mortality (Section 3.2.1). At regular intervals during the simulation, the total number of bacteria represented by the model particles in each layer of each cell of the output grid are divided by the appropriate layer volumes in the cells to derive layer-averaged bacterial concentrations. The model results are, therefore, time series of three-dimensional bacterial concentration fields. Simulations are usually performed over several diurnal cycles, in order for an equilibrium diurnal cycle of concentrations to be computed.

3 The Barcelona outfall study

The three-dimensional hydrodynamic and bacterial dispersion models have been used to assist in the assessment of outfall options to disperse treated effluent from Barcelona in southern Spain discharged into the sea near the mouth of the Rio Besos (Figure 1). The model study was part of an engineering investigation carried out by Watson Hawksley Ltd for EMSSA Barcelona. At present, the bathing beaches at Barcelona are sometimes contaminated by effluent discharged from an outfall pipe near the Rio Besos extending 600m from the shore, but broken along its length. The models were used to assist in the evaluation of various design options for a new submarine outfall to be constructed to ensure compliance with bathing water directives on the nearby beaches. Due to the strong, seasonally-variant thermal stratification of the coastal waters near Barcelona, it is expected that the plume from the submarine outfall will be trapped below the sea surface in the vicinity of the diffuser, and the resulting strong vertical variations in bacterial concentrations necessitate the three-dimensional modelling approach. The study variables were the distance of the diffuser from the shore and the diffuser length. Each of the outfall options was examined under various seasonal conditions, in order to determine the likely worst-case pollution levels resulting from the outfall construction.

3.1 Hydrodynamic modelling

The dominant features of the oceanography of the Barcelona region are the lack of significant tidal motion (range less than 0.3m) and the presence of the Catalan Current; a regional, southwestward, offshore flow along the southeastern coast of Spain from the Gulf of Lions. This current is in approximate geostrophic balance, with the landward Coriolis force associated with the flow being balanced by the offshore force resulting from the horizontal density gradient (lighter, less saline water is located inshore). The peak current speeds are typically of the order of 0.2 to 0.3ms^{-1}, and the current is about 30km wide. The currents in the area are affected by winds.

In order to model the flows near Barcelona, both a regional model (to simulate the main features of the offshore Catalan Current) and a local model (to simulate conditions in more detail near the planned outfall and along the shoreline at Barcelona) were set up. Both flow

33

models were three-dimensional (Section 2.1); the outer model in order to simulate the important density gradients offshore, and the inner model in order to simulate the three-dimensional flows generated by the wind in the presence of stratification. The boundary conditions for the local model were taken from the regional model.

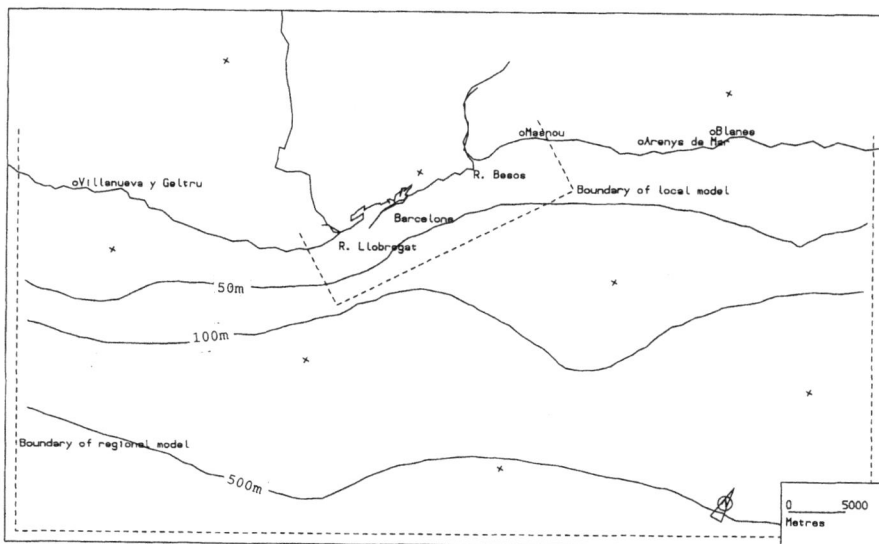

Figure 1   The Barcelona coastal area showing flow model boundaries

3.1.1 The regional model The regional model had a grid size of 1km, and extended along 80km of coast and 30km offshore (Figure 1). The grid size was chosen to enable resolution of important features in an evolving density field, as well as to allow a reasonable area to be covered. There were 10 layers in the model, and the vertical grid size was 30m. The model was started from a density field based on the results of a survey of salinity and temperature along a cross-section of the Catalan Current near Barcelona [8], and was run without any tide. The model was run for a density field for the month of May, for which some observations of current speed were available [3]. The resulting model surface currents are shown in Figure 2. The results showed variations in time, and a radiation boundary condition was imposed to allow surface waves to escape. Subsequently, the model was run with density fields corresponding to the months of July and January, for which corresponding current fields were not available, but for which it was desired to predict the bacterial dispersion. For these cases, velocity boundary conditions were based on the density cross-sections assuming geostrophic balance.

3.1.2 The local model The boundary conditions for the local model were extracted from the regional model. The local model had a grid size of 150m and a vertical grid dimension of 5m. It extended along about 23km of coast and a distance of 6km offshore (Figure 1). This model was used to simulate the local stratification and water movement off Barcelona due to combined seasonal flows and local wind conditions. The flow fields generated by the local model formed the basis for the plume dispersion modelling. The initial temperature distribution in the local model was taken from observations made within the area of the model

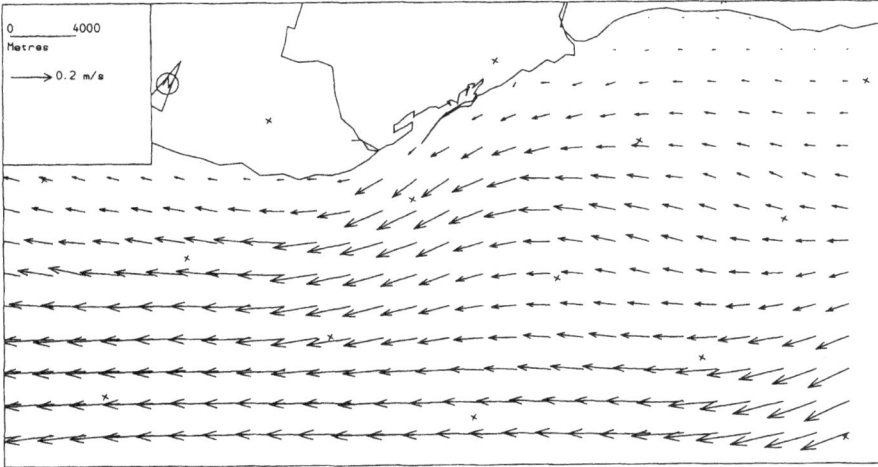

Figure 2   Regional flow model velocity vectors at the sea surface simulated for May

during Proyecto SPIO [1]. Thermal stratification is present in summer, with a thermocline at 5m to 10m depth, but in winter the water column is well mixed. Flows simulated by this model for a summer condition (with a daily onshore-offshore wind variation) are shown in Figure 3 for the onshore wind period, and in Figure 4 for the offshore wind period. The flows are shown at the sea surface and at 22m depth.

3.2 Bacterial dispersion modelling

The bacterial dispersion model was based on the results of the local hydrodynamic model simulations.

3.2.1 Representation of bacterial mortality The rate of mortality of E. coli bacteria in sea water is a function of incident sunlight levels. Mortality varies, therefore, with a diurnal period, with minimum T_{90} values occurring near midday, and maximum values occurring during darkness [2]. In order to represent this variation in the bacterial dispersion model, the number of bacteria represented by each model particle was reduced according to T_{90} values which varied with a diurnal period, as shown in Figure 5. Seasonal variations of incident sunlight were also included in the simulations by using lower day-time minimum T_{90} values in the summer simulations than in the winter simulations.

In addition to the temporal T_{90} variations described above, variations occur with depth below the sea surface, due to light attenuation by the water column. The depth below the sea surface, d (m), at which light intensity is reduced from the surface value, I_o, to a value, I_d, is given by:

$$I_d = I_o e^{-kd} \qquad (1)$$

where k is a light extinction coefficient that depends on turbidity and chlorophyll-A concentrations. Due to this dependence of k on local factors, the functional form of the vertical T_{90} variation is site-specific, and must be derived uniquely from observations for all model applications. k is also related to Secchi disc measurements through the equation:

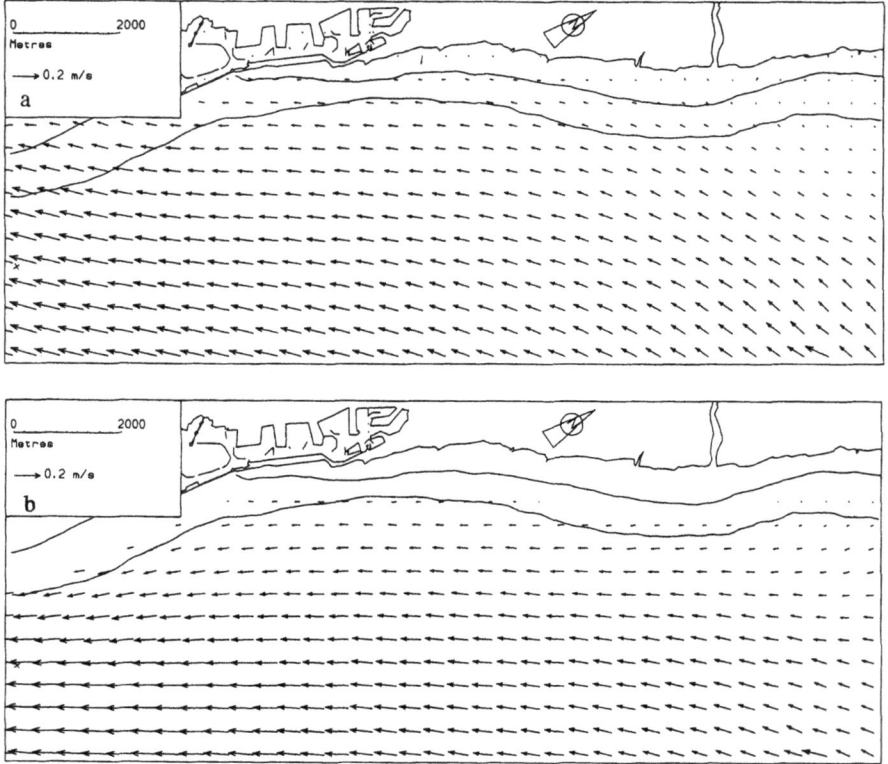

Figure 3   Local flow model velocity vectors at (a) the sea surface and (b) 22m depth during an onshore wind in summer.

$$k = 1.7/s \qquad (2)$$

where s is Secchi disc depth (m). Combining equations (1) and (2) gives:

$$1.7d = s.\log_e(I_d/I_o) \qquad (3)$$

which was used with Secchi disc measurements made in the coastal waters near Barcelona to determine depths at which I_d/I_o decreases to 0.25, 0.10 and 0.01. T_{90} data indicate that bacterial mortality is approximately constant for light intensities between 25% and 100% of the incoming radiation, whilst for intensities less than 1% die-off rates are approximately equal to total darkness values [2]. At intermediate intensities, mortality rates are between the high intensity and zero intensity values. This indicates that a typical vertical T_{90} profile would show little depth-variation in the near-surface waters, with values increasing at greater depths to the point at which mortality no longer decreases with increasing depth. The light intensity calculations based on equation (3) enabled such a profile to be derived for the coastal waters near Barcelona, with diurnally-variant T_{90} values (Figure 5) near the sea surface, night-time values at depth, and a weighted average of the day- and night-time values at intermediate depths. This profile was used in the bacterial dispersion model simulations.

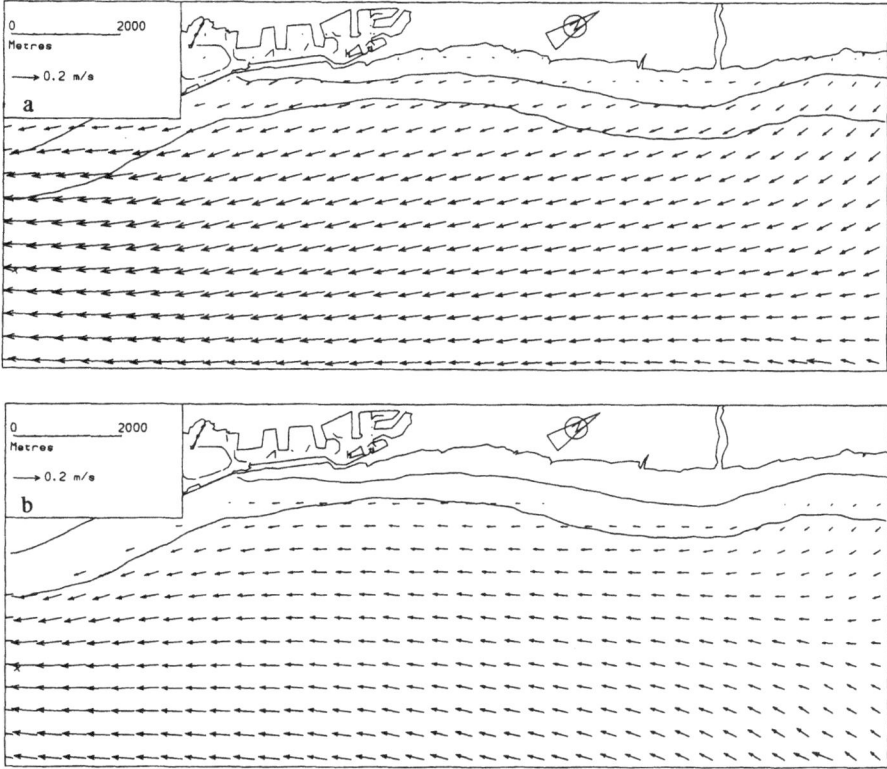

Figure 4   *Local flow model velocity vectors at (a) the sea surface and (b) 22m depth during an offshore wind in summer.*

3.2.2 Initial dilution   The modelling approach adopted enabled three-dimensional plume structure to be simulated. It was, therefore, necessary to estimate the vertical distribution of effluent in the water column near the planned diffuser locations for use as initial conditions for the bacterial dispersion model. In winter, vertical density gradients in the coastal waters near Barcelona are small, and it is expected that effluent discharged from a submarine outfall will be approximately evenly distributed over the full depth of the water column in the vicinity of the diffuser. However, in summer, strong thermal stratification exists, and it is believed that the outfall plume will be trapped below the thermocline near the outfall. In order to derive initial conditions for the bacterial dispersion model simulations in such a way that the likely seasonal variation of those conditions could be represented, the EPA initial dilution models [9] were run for each planned outfall design option. These model runs were based on observations of the vertical density gradient in the coastal waters near Barcelona, and provided estimates of the plume trapping depth for various ambient current speeds, which could be used in combination with outfall discharge volumes to calculate the depth range occupied by effluent at the end of the initial dilution phase. Particles in the bacterial dispersion model were released over the depth ranges so computed, and over the lengths of the planned diffusers, in order to take a realistic plume position at the end of the initial dilution phase as initial conditions.

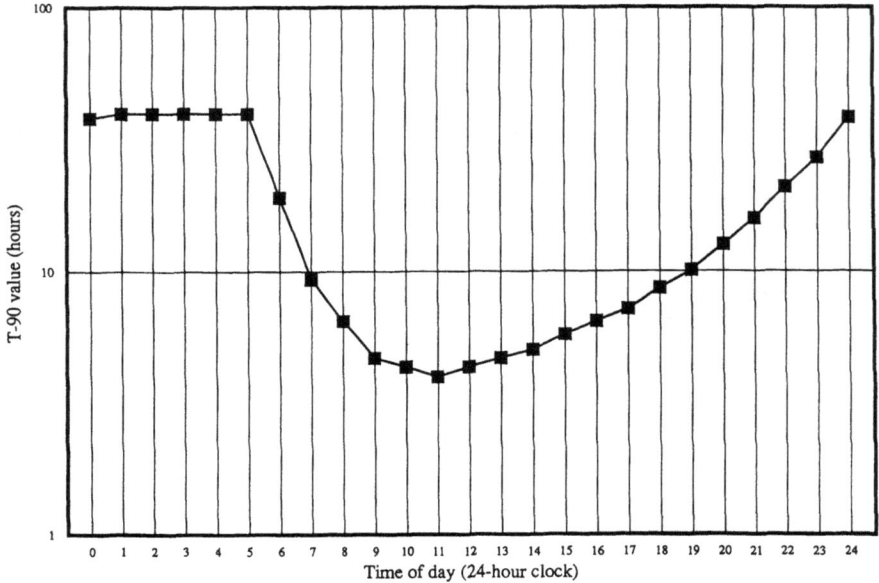

Figure 5   Summer diurnal T_{90} variation at the sea surface in the bacterial dispersion model.

3.2.3 Model calibration Due to the sparsity of E. coli concentration observations in the Barcelona area, the dispersion model could not be calibrated in detail. A limited data set was, however, available, and the model was run to simulate effluent dispersal from the existing short, broken outfall. Small adjustments were made to the model T_{90} values and horizontal diffusivity, until the trends in bacterial concentrations indicated by the data were simulated approximately. Having derived the model parameters in this way, the model could be run to simulate bacterial dispersion associated with the planned outfall design options.

3.2.4 Assessment of outfall options Each of the outfall design options was tested under various seasonal conditions, in order to ensure that probable worst-case pollution levels on the area's bathing beaches were simulated. For each simulation, the initial dilution model was run with the appropriate seasonal density profile to derive initial conditions for the bacterial dispersion model. The dispersion model was then run for several days of simulated time, until the diurnal variation of the simulated bacterial concentration fields repeated approximately. By comparison of the results for the various design options, the shortest outfall needed to achieve compliance with the bathing water directives was selected. Figures 6 and 7 show simulated diurnal geometric mean E. coli concentrations at three levels in the water column for two outfall options under the same summer wind, bacterial mortality and density stratification conditions. Detailed analyses of the model results indicated that the longer outfall option (Figure 7) would achieve the required water quality conditions, but that the shorter option (Figure 6) would not.

Figures 6 and 7 demonstrate the effect of the initial conditions on the bacterial dispersion model solution. The plume discharged from the shorter outfall reaches the sea surface in the immediate vicinity of the diffuser (Figure 6(a)). However, as the longer outfall lies in deeper water, the initial dilution model predicted that the effluent plume would not surface over the outfall, due to the strong summer stratification, and the dispersion model plume only surfaces

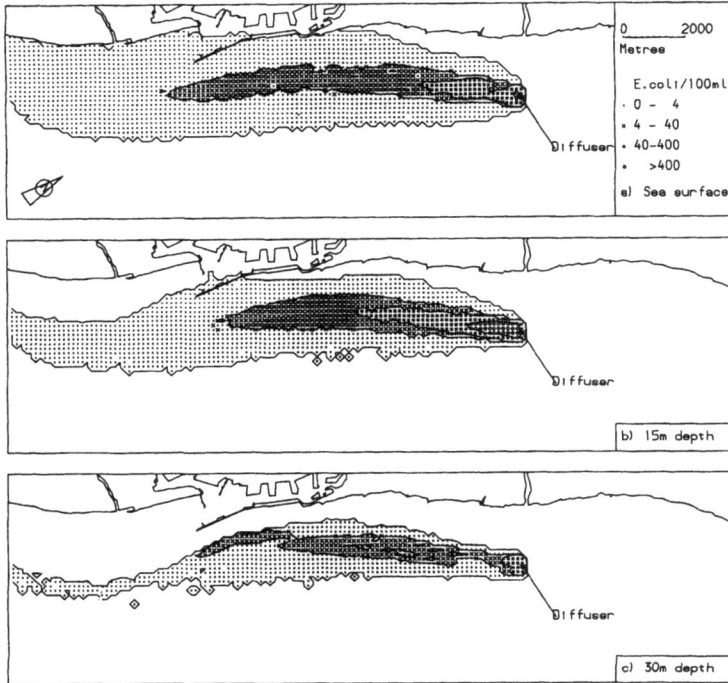

Figure 6   Summer diurnal geometric mean E. coli concentrations simulated by the bacterial dispersion model at (a) the sea surface, (b) 15m depth and (c) 30m depth for a relatively short outfall design option.

over 1000m downstream of the diffuser (Figure 7(a)). Figures 6 and 7 demonstrate the strong three-dimensional variation of the simulated plume, and the effects on predicted water quality of relatively small increases in the lengths of the outfall and diffuser.

4 Summary and discussion

Three-dimensional models of water movements and bacterial dispersion in coastal waters have been described, and their application has been demonstrated by reference to a study of effluent dispersion from various planned outfall design options for Barcelona in southern Spain. The models were validated approximately by comparison with limited bacterial concentration data, and have been used to assist in the selection of an outfall design which will meet the required standards set by bathing water directives.

One of the usual reasons for applying depth-averaged flow and bacterial dispersion models when planning outfall locations or levels of effluent treatment is the relatively limited use of computing resources. In such two-dimensional modelling approaches, the effects of winds on pollutant dispersion can be parameterised by superimposing wind-driven current vectors onto tidal or seasonal flow vectors in the bacterial dispersion model, with the vertical structure of wind-driven currents being approximated as a parabolic profile [6]. This parameterisation allows pollutant dispersion for a given outfall configuration to be assessed under various com-

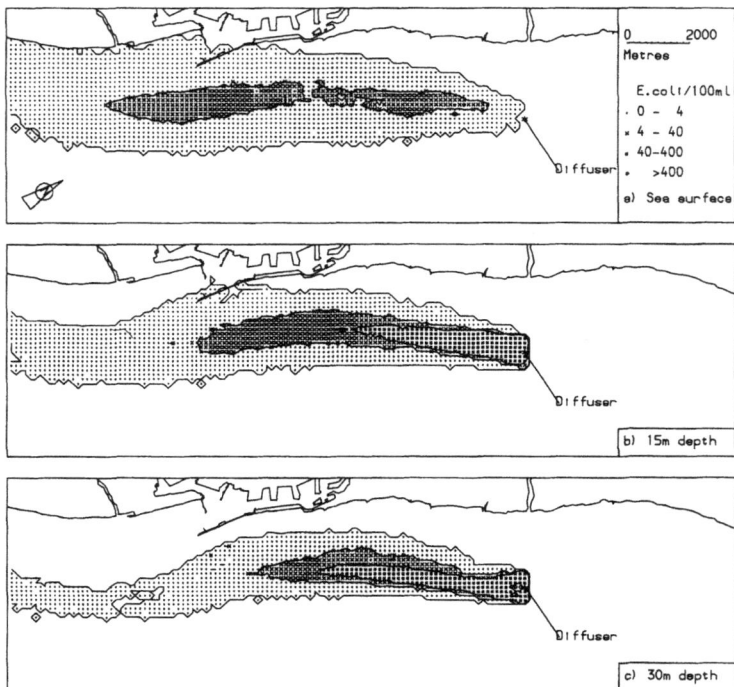

Figure 7   Summer diurnal geometric mean E. coli concentrations simulated by the bacterial dispersion model at (a) the sea surface, (b) 15m depth and (c) 30m depth for a relatively long outfall design option.

binations of tidal/seasonal and wind conditions with a relatively small number of flow model runs.   In contrast, the three-dimensional techniques require both the flow and dispersion models to be run for each combination of conditions studied, as the wind forcing is included as a surface stress in the flow model.

In many applications, the use of depth-averaged models provides a highly satisfactory means of utilising computing resources, as they provide good approximations of the important coastal processes.   However, the Barcelona study has demonstrated the importance of simulating the three-dimensional structure of plumes from submarine outfalls in stratified coastal waters. Structure of this kind, which is dominated by the trapping of the plume below the thermocline at the end of the initial dilution phase, requires a three-dimensional modelling approach, which can resolve the thermocline and represent the effects of vertical density variations on vertical mixing.   Furthermore, in open coastal waters, the parabolic wind-driven current profile approximation often does not provide an adequate representation of three-dimensional flow structure, and a more complete description of the hydrodynamic processes is required.   There are, therefore, numerous coastal regions where a three-dimensional modelling approach is recommended during the assessment of sewage disposal options, in order to fully consider the dominant physical processes.

With the advent of increasingly powerful computing systems, three-dimensional modelling studies can be completed on time scales which were, until recently, only achievable with two-dimensional techniques.   The techniques described in this paper, therefore, provide a

40

modelling approach which allows the disposal of sewage effluent in complex coastal environments to be planned in detail, whilst remaining within time limits usually imposed by commercial budget constraints.

Acknowledgements

The authors are grateful to EMSSA Barcelona and Watson Hawksley Ltd for permission to show the results of the Barcelona outfall study, and to Mr. G. Toms for his work on the initial dilution modelling and on the formulation of T_{90} variations. Dr. J. Font and Mr. D. Osorio also provided valuable advice during the Barcelona outfall study.

References

[1]   Amengual, P., Borras, G., Julia, A. and Navarro, R (1987), *Proyecto Integrado Para el Estudio del Efecto del Deposito Submarino de Lodos de La Zona del Rio Besos Sobre La Zona Costera de Barcelona*. Instituto de Ciencias del Mar, 142 pp.

[2]   Caldwell Connell Engineers (1979), *Environmental Impact Statement. Malabar. Water Pollution Control Plant*. Report to Metropolitan Water Sewerage and Drainage Board, Sydney, NSW, Australia.

[3]   Castellon, A., Font, J. and Garcia-Ladona, E. (1990), "The Liguro-Provencal-Catalan Current (NW Mediterranean) observed by doppler profiling in the Balearic Sea", *Scientia Marina*, **LIV**.

[4]   HR Wallingford (1983), *Formulation and Development of a Three-Dimensional Numerical Model of Estuaries*. Report No. IT 254.

[5]   Hunter, J.R. (1987), "The application of Lagrangian particle-tracking techniques to modelling of dispersion in the sea", in *Numerical Modelling: Applications to Marine Systems*, J. Noye (Ed), Elsevier, North Holland, pp 257-270.

[6]   Mead, C.T. (1991), "Random walk simulations of the dispersal of sewage effluent and dredged spoil", in *Proceedings of the 10th Australasian Conference on Coastal and Ocean Engineering*, Auckland, New Zealand, pp 477-480.

[7]   Odd, N.V.M. and Rodger, J.G. (1978), "Vertical mixing in stratified tidal flows', *Journal of the Hydraulics Division, ASCE*, **CIV**, 3, pp. 337-351.

[8]   Salat, J., Font, J. and Cruzado, A. (1978), *Datos Oceanographicos Frente a Barcelona (1975-1976)*. Instituto de Investignaciones Resqueras, 72 pp.

[9]   U S Environmental Protection Agency (1985), *Initial Mixing Characteristics of Municipal Ocean Discharges: Volume 1, Procedures and Applications*. Newport, Oregon.

4 Impact of large-scale reclamation on tidal current system and water quality structure in Osaka Bay

K. Nakatsuji, S. Karino, H. Kurita and K. Muraoka

ABSTRACT

A finite element procedure with SGS eddy viscosities is used to simulate depth-integrated tidal flow and dissolved pollutant transport in Osaka Bay, Japan. The procedure enables intricate geographical features to be represented smoothly. The SGS eddy viscosities based on Smagorinsky's assumption are applied for simulating momentum transfer due to turbulent motion. Favorable modelling results are obtained by comparing the computed time-varying tidal flow and tide-induced residual current to the observed ones. The model is used to make an environmental impact assessment of two hypothetical 22,600 ha reclamation projects and of a barrier island plan.

1. Introduction

In urban coastal areas in Japan large-scale land-fill operations have recently increased, causing drastic changes in the geographical features of the coastline. In Osaka Bay, for example, coastal industrial zones of 5,200 ha have been reclaimed in the past 40 years and many development projects are under construction including a 511 ha island for the new airport. As a natural consequence the sea water is polluted due to the poor exchange of water, and algal blooms and anoxia have had adverse effects on fishery and recreation in many enclosed coastal seas. Strategies for environmental management must be examined thoroughly to ensure the utilization of enclosed coastal seas in a sustainable manner.

In the present study the impacts of large-scale reclamation on the tidal current system and water quality transfer are examined in Osaka Bay, in which 10-m depth shallow reclamation has been completed and more large-scale reclamation in deeper areas is planned. First of all, finite element formulation with SGS eddy viscosities is employed to simulate depth-integrated tidal flow and dissolved pollutant transport. The finite element procedure can represent irregular boundaries smoothly without computational difficulties and the SGS eddy viscosities can simulate momentum transfer due to turbulent motion heavily dependent upon tidal flow intensity. After model calibration with the observed data obtained in Osaka Bay, the model is used to make environmental impact assessments of two hypothetical 22,600 ha reclamation projects and of a barrier island plan.

2. Model description

A finite element technique is applied to calculate time-varying tidal flow and pollutant transport to simulate changes in the tidal current system due to reclamation, since the rapidly-varying flow around the strait and the slowly-varying flow in the inner part of the basin must be calculated simultaneously with the same precision.

2.1 Basic equations

By use of the depth-integrated Reynolds equation for incompressible viscous fluid the following two-dimensional equations of continuity, motion and convective diffusion are obtained:

$$\frac{\partial \zeta}{\partial t} + \frac{\partial ((h+\zeta) U_i)}{\partial x_i} = 0 \quad \dots\dots\dots\dots\dots\dots\dots\dots\dots\dots\dots\dots\dots\dots\dots \quad (1)$$

$$\frac{\partial U_i}{\partial t} + U_j \frac{\partial U_i}{\partial x_j} = -g \frac{\partial \zeta}{\partial x_i} - \frac{\gamma_b^2 U_i \sqrt{U_1^2 + U_2^2}}{h+\zeta} + \frac{\partial}{\partial x_j}\left(\nu_T \frac{\partial U_i}{\partial x_j} \right) \dots\dots\dots\dots\dots \quad (2)$$

$$\frac{\partial C}{\partial t} + U_j \frac{\partial C}{\partial x_j} = \frac{\partial}{\partial x_j}\left(K_j \frac{\partial C}{\partial x_j} \right) \dots\dots\dots\dots\dots\dots\dots\dots\dots\dots\dots\dots\dots\dots \quad (3)$$

where x_i: space coordinates ($x_1=x$, $x_2=y$ represent east-west and north-south directions respectively), t: time, U_i: x_i component of velocity, ζ: water level, h: water depth, g: gravitational acceleration, γ_b: coefficient of bottom shear stress, ν_T: eddy viscosity coefficient, C: concentration of material and K_j: eddy diffusivity coefficient.

2.2 SGS eddy viscosities

The idea of applying an averaging operator to the Navier-Stokes equation, with averaging typically being over the grid volume of the calculations to filter out the subgrid scale (SGS) motions, has been known since the early work of Reynolds. Explicit calculations can then be made for the filtered variables after assumptions are made for the SGS Reynolds stresses which arise from the averaging process. Although various turbulence models have been proposed in engineering calculations, many calculation methods for tidal flows use constant eddy viscosity/diffusivity coefficients for the whole flow field, the values of which are determined empirically from field surveys and hydraulic experiments. Some calculations have taken it as an invariant, or as a single-variable function of the ⅓th power of the grid spacing. This approach, however, cannot represent the motions of eddies ranging in size down to the subgrid scale accurately because the SGS Reynolds stresses may change not only in time but also in space depending on tidal motions.

In particular, it causes several contradictions in the computation of a finite element procedure in which the grid spacing is determined arbitrarily. A more rigorous expression is obtained by explicitly incorporating the rate of turbulent kinematic energy dissipation, as Smagorinsky (1963) and other meteorologists have employed for the general circulation of the atmosphere with considerable success. This leads to an expression represented in terms of the velocity deformation as well as the grid spacing Δ, i.e.,

$$v_T = (c\Delta)^2 \left[\frac{\partial U_i}{\partial x_j} \left(\frac{\partial U_i}{\partial x_j} + \frac{\partial U_j}{\partial x_i} \right) \right]^{\frac{1}{2}} \quad \dots\dots\dots\dots\dots\dots\dots\dots\dots\dots\dots\dots \quad (4)$$

in which c is a constant value. Deadroff (1970) et al. recommended that the constant c is in the range of 0.1 to 0.2. In the present study, $c=0.12$ is adopted based on the field results obtained in Tokyo Bay by Tsuji (1991).

2.3 Finite element formulation

A formulation based on the Weighted Residual Method is applied because it permits free selection of size and shape of the computational grid system (Sawaragi, et al., 1976, Murota, et al., 1985). Curve elements are introduced to smoothly represent complicated land shapes. A quadratic polynomial isoparametric element is used to space-discrete variables. The resultant equation system consists of the following discrete simultaneous linear equations regarding $3 \times M$ unknown values of $\zeta_j(t)$, $U_j(t)$, and $V_j(t)$ at all nodes, $j=1$ to M.

$$\left\{ \theta\,[A] + \frac{1}{\Delta t}\,[C] \right\} \Phi_j^{n+1} = \left\{ (\theta - 1)\,[A] + \frac{1}{\Delta t}\,[C] \right\} \Phi_j^n - \left\{ (\theta - 1)\,b^n + \theta \cdot b^{n+1} \right\} \quad \dots\dots \quad (5)$$

where Φ_j^n: $\{\zeta_j, U_j, V_j\}$ at time step n, [A], [C]: coefficient matrix, b: coefficient vector, Δt: time increment, and θ: parameter of time discretization. $\theta=0.5$ is used for the Crank-Nicolson method, $\theta=\frac{2}{3}$ is used for the Galerkin method. The above equations can successfully be solved by the aid of a wave front scheme even for a large time increment.

2.4 Model setup

Ocean model As shown in Figure 1, the computational domain of Osaka Bay is in the north of the Kii Channel at lat. 33° 50'N. and in the east of the Sea of Harima at long. 134° 19'E. This domain is determined according to the Report by the Ministry of Transportation, MOT (1980). The water depth data is read from a published marine chart. Osaka Bay is a typical semi-enclosed bay and has two openings at the Straits of Akashi and Kitan. Finite element discretization must be determined carefully, taking the geographical features and the tidal flow pattern into account. The size of the elements of the present situation of Osaka Bay ranges from 800 m to 5000 m. The computationed domain consists of 583 elements with 728 nodes. As a result, the simultaneous linear equations with 2184 unknown values must be solved repeatedly in step by step over time. Time increment Δt is set to 600 seconds.

Calculation parameters Cosine waves having the amplitude and phase lag of a semidiurnal tide, M_2 shown in Table 1 are given for the tide elevation at both ends of the open boundaries: the west end of the Sea of Harima and the south end of the Kii Channel. The tidal cycle is set to 12 hours for convenience. The variation of water level with time at any location on the boundaries is yielded by interpolated values. The total flow rate from rivers located in the interior of Osaka Bay is assumed to be 377 m³/s. The boundary conditions for the velocity are set to $\partial Un/\partial n = 0.0$ for the open boundary and $Un = 0.0$ for the coastline.

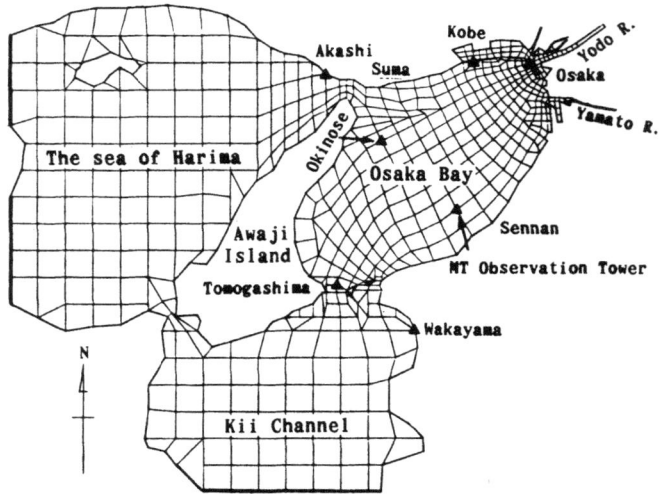

FIGURE 1   *Computational domain and finite element discretization*

TABLE 1   *Boundary conditions of amplitude and phase of M_2 tide*

Open boundary		H: Amplitude (cm)	*: Phase (degrees)
The Sea of Harima North Side	North end	38.0	319.0
	South end	39.1	332.1
The Sea of Harima South Side	North end	38.5	324.4
	South end	35.7	333.7
Kii Channel	West end	48.0	172.0
	East end	45.9	174.1

Three numerical experiments are carried out: (1) the present situation of Osaka Bay for model calibration, (2) simulation model I based on the large-scale reclamation project, and (3) simulation model II based on the barrier island project. The present coastal geography is based on a survey for 1980 from MOT. Model I assumes a large-scale reclamation project which plans to connect Kansai International Airport Island and Kobe Port Island and reclaim an area of about 22,600 ha to the east of the connection line. As a result, a waterway some 40 km long and 3 km wide will be sandwiched in between the coastline and the reclaimed land. River flow rate is not taken into account in this calculation but it is in the numerical experiment for the present situation and in the barrier island project. Model II assumes a barrier island project which plans to reclaim land 5 to 6 km off the coastline, forming a 22 km arc-shaped barrier with islands about 1.3 km wide. The assumed coastline in both cases is based on the reclamation project "Basic Concept of Osaka Bay Planning" proposed by the 3rd Construction Bureau of MOT in 1985.

3.   Model calibration

First, in order to validate the model's potential for reproducing and predicting real flow, the tidal current in the present situation is simulated for a period of 8 cycles. Figure 2 shows the velocity

46

vector fields at the time of the maximum eastward/westward flow through the Akashi Strait. Although inflow from the Akashi Strait to Osaka Bay at the maximum eastward flow disperses all over the bay, the major flow travels through a deep water area off the coast of Awaji Island and exits via the Kitan Strait. A flow running eastward off Kobe makes a large clockwise turn and flows southwest parallel to the Sennan coast. Two hours before the maximum westward flow, the inflow from the Kitan Strait to Osaka Bay becomes a maximum. Inflow current tends to disperse all over the bay. Precisely speaking however, it branches into two currents: one flowing directly to the Akashi Strait along the east coast of Awaji Island, and the other running off Sennan toward the inner bay. At the maximum westward flow, both currents merge and flow out through the Akashi Strait. By drawing tidal current ellipses it has been verified that the tidal calibration is excellent as a whole although the absolute velocities within the inner bay are smaller than the measured values.

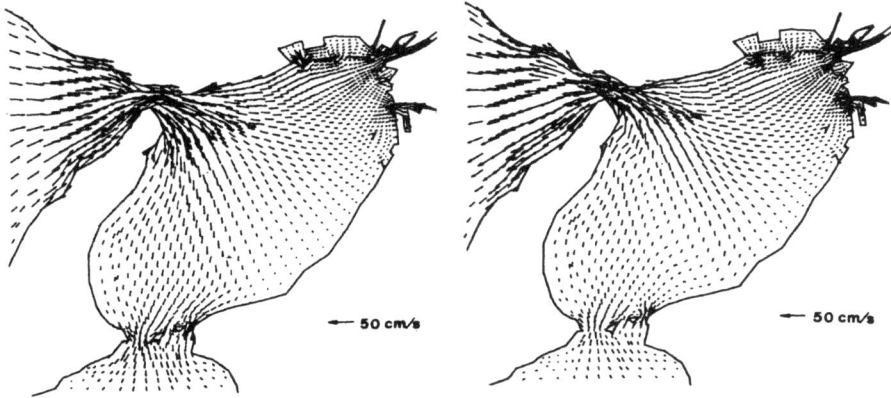

FIGURE 2   *Velocity vector fields at the maximum eastward/westward flow at the Akashi Strait*

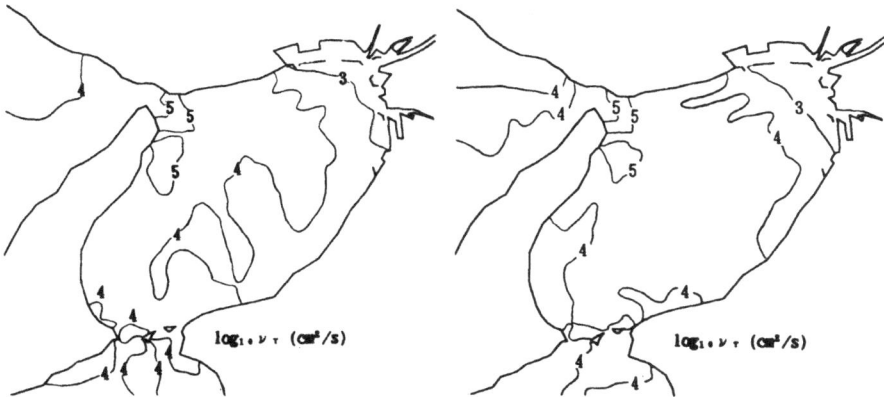

FIGURE 3   *Distribution of SGS eddy viscosity coefficients corresponding to Figure 2*

47

Figure 3 shows the distributions of the order of SGS eddy viscosity coefficients which correspond to the velocity fields as shown in Figure 2. It indicates that the value in the stagnant water region located in the interior of Osaka Bay is of the order of 10^3 cm^2/s, while in the central area of the bay it is of the order of 10^4 cm^2/s. On the other hand, in the high-velocity narrow streams such as the Akashi and Kitan Straits, the value reaches the order of 10^5 cm^2/s or more. It is evident from the figure that the eddy viscosity coefficients depend heavily upon the tidal flow intensity and the simulated values agree approximately with those used empirically.

Figure 4 compares simulated values with observed values of the tidal amplitude and phase at the tidal gauge observatories shown in Figure 1. The capital letters on the horizontal axis correspond with the first letters of the observatories. The simulated values agree with the measured values well, except that the amplitude in the inner bay is on average only 85% of the measured values.

Figure 5 shows the tidal residual velocities averaged over the last tidal cycle of an 8-cycle run. Two large-scale residual circulations can be clearly seen in Osaka Bay, which are induced by the strong tidal flow through the Akashi Strait or the Kitan Strait. Of the two, the large-scale clockwise Okinose circulation which appears near the Akashi Strait is mainly responsible for governing the tidal current system and the transport of water quality in Osaka Bay. It is worth noting that the residual flow velocity of the Okinose circulation is a little faster than 0.5 m/s, which is equivalent to the maximum tidal velocity. The agreement between the simulated and observed residual circulations would appear to be entirely satisfactory.

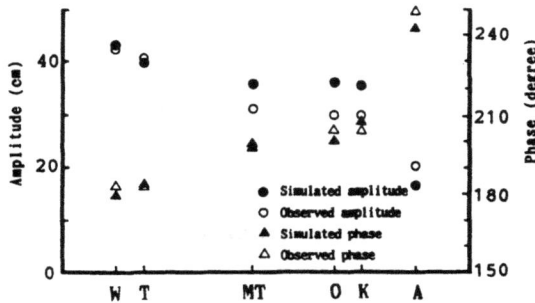

FIGURE 4   Comparison of simulated amplitude and phase to observed ones

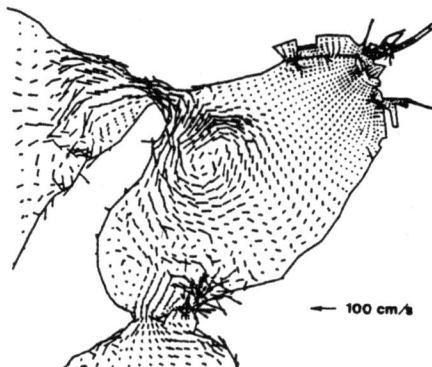

FIGURE 5   Tidal residual current vectors in the present situatin of Osaka Bay

4. Impact assessment

4.1 Simulation model I: Large-scale reclamation project

Figure 6 shows the flow velocity vector fields calculated for simulation model I. Focusing on the variation of the tidal current system caused by the large-scale reclamation, it can be seen that the current running eastward off the Kobe coast is accelerated at the north end of the reclamation zone at the maximum eastward flow in the Akashi Strait, and then flows eastward. This phenomenon is caused by a reduction in cross-section due to reclamation. A current from the mouths of the Yodo and Yamato Rivers in the inner bay can be seen to flow southward along the waterway between the reclaimed land and the coastline. Even 4 hours later (two hours before the maximum westward flow in the Akashi Strait), a current running from the north end of the reclamation zone to the inner bay can still be observed. The current along the Kitan Strait to the Sennan coast becomes more powerful in the direction from the south end of the reclamation zone to the inner bay. A minor current, not shown in the figure, is also found which runs along the waterway in the direction of the inner bay.

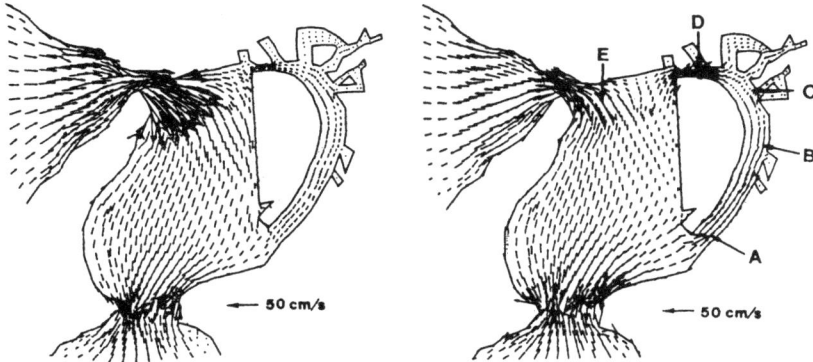

FIGURE 6   *Velocity vector fields at the maximum eastward flow and at two hours before the maximum westward flow in the Akashi Strait on the large-scale reclamation model*

Figure 7 shows the time-varying tide elevation and tidal velocity at locations within the waterway between the coastline and reclaimed land. The component of velocity V in the direction into the inner bay along the waterway is represented by a plus sign. The variation of tidal elevation is slight; only a 0.6-hour phase lag is observed at E on the Osaka side of the Akashi Strait. At four locations in the waterway the tide elevation varies with the same phase and amplitude. The tidal velocity variation is ¼ cycle behind the water elevation variation since the tidal current driving force is generated from the water surface slope. This trend is remarkable at A and D, the ends of the reclamation zone, and the amplitudes at these two locations are almost the same. Considering that the eastward flow at the Akashi Strait becomes a maximum at 9:20, an inflow current from both ends of the reclamation zone to the inner bay is generated in the waterway during the first half cycle from 1.5 hours before the maximum eastward flow. An inverse flow from the inner bay to both ends appears in the other half cycle.

49

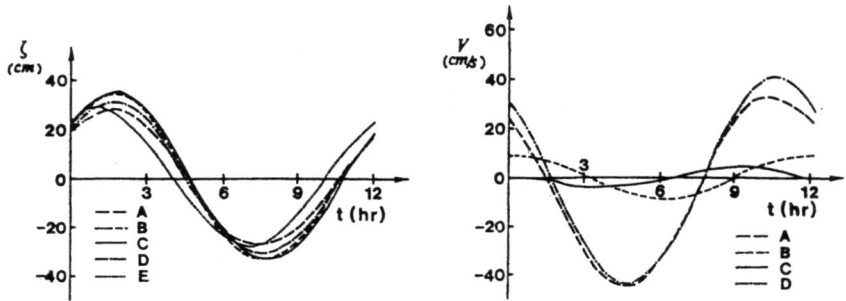

FIGURE 7   *Time-varying tidal current system on the large-scale reclamation model*

On the other hand, at the mouths of the Yamato River (B) and the Yodo River (C) the phase lag of tidal velocity does not correspond to the tide elevation phase and its absolute values are very small. This may be because two water masses flow concurrently from both ends of the reclamation zone into the enclosed waterway and hit each other. The travel distance obtained by integrating the velocity at location E at the north end of the reclamation zone is about 6 km for both inflow and outflow. This denies the possibility of exchange between polluted water masses from rivers in the inner bay and the water mass in the central bay.

To compare the effects of reclamation on the tidal residual current, the calculation result is shown in Figure 8 in the same manner as in Figure 5. On the large-scale reclamation model the Okinose circulating residual current develops into a 20 km diameter circulation, whose east end is the reclaimed land and west end is Awaji Island. Reclamation will therefore expand and stabilize the Okinose circulation which means that pollutants once entrapped in the Okinose circulation will reside in Osaka Bay for a considerable period. No tidal residual current is found within the waterway between the coast and reclaimed land. The flow in the waterway is generated only by the inflow rate of the rivers, which is not taken into account in this simulation. Assuming that the average depth is 6 m and river water flows toward both ends of the reclamation zone, the transport velocity is calculated to be approximately 0.01 m/s. This means that the transfer of pollutants from the river mouths to the ends of the reclaimed land takes more than 20 days. Of course, river water

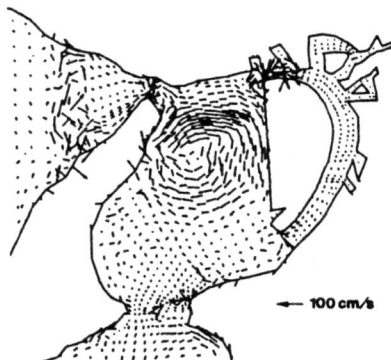

FIGURE 8   *Field of tidal residual current vectors on the large-scale reclamation model*

50

spreads over the surface of the sea due to density current effect, which must reduce the required number of days for water exchange. Another point to be noted is a change in the tidal residual current in the east of the Sea of Harima. This phenomenon suggests that impact from large-scale reclamation like this project is not limited to within Osaka Bay.

4.2 Simulation model II: Barrier island project

The barrier island model (see the solid portions in Figure 9) is devised to improve water movement in the stagnant enclosed waterway between the coast and reclaimed land and thus to enhance the possibility of flow control of the tidal current system. Figure 9 illustrates the velocity vector fields at the maximum eastward/westward flow in the Akashi Strait. At the maximum eastward flow, a flow running eastward off the Kobe coast is faster than under the present situation, promoting flow from the north end of the barrier islands and among the islands to the inner bay. At the maximum westward flow, flow from the mouths of the Yodo River and the Yamato River in the inner bay to the Akashi Strait seems to be accelerated to some extent. A characteristic common to both tidal flows is a flushing flow among the barrier islands. This means that the barrier islands cause a water level difference between the central part of the bay and the inner bay. This flow can be expected to promote mixing and exchange of water at a local level.

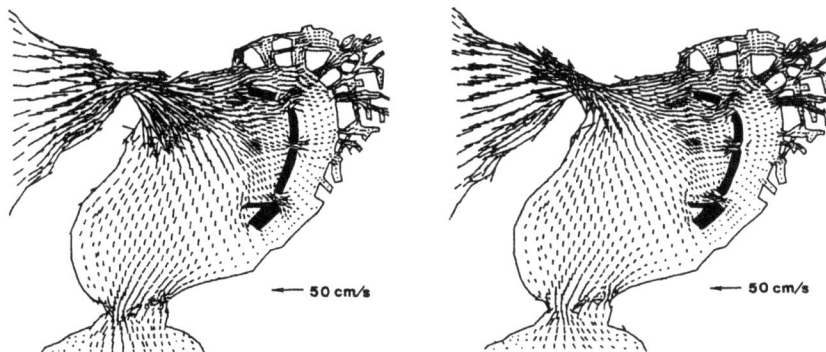

FIGURE 9   *Velocity vector fields at the maximum eastward/westward flow in the Akashi Strait in the barrier island model*

The tidal residual current system as shown in Figure 10 is more complicated compared to Figures 5 and 8. A strong residual current is generated circulating clockwise on the periphery of the Okinose circulation current from the Akashi Strait to the south. As a result, the center of the Okinose circulation is a little closer to Awaji Island, and the velocity of residual current along the Awaji Island coast is reduced. Also, a weak counterclockwise circulation arises in the waterway between the Osaka coast and the barrier islands. This circulation may cause retention in the exchange of pollutants transported from rivers.

Figure 11, therefore, compares the distributions of COD concentration based on the present situation to that based on the barrier island project by considering the tidal residual current as advection current. The inflow load of COD transported from 5 rivers and the concentration at the open boundary are set according to the MOT Report (1980). The eddy diffusivity coefficient is set to the same value as the eddy viscosity coefficient, the time increment is set to 12 hours and diffusion is computed for 50 tide cycles. It can be seen that the iso-concentration lines above 2 mg/l on the

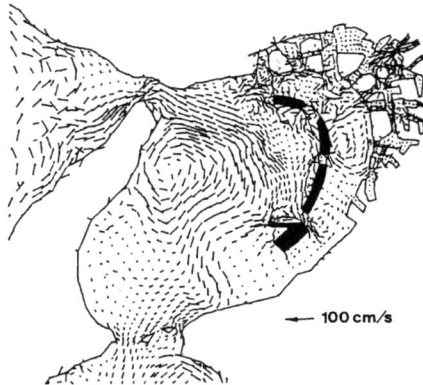

FIGURE 10   *Field of tidal residual current vectors in the barrier island model*

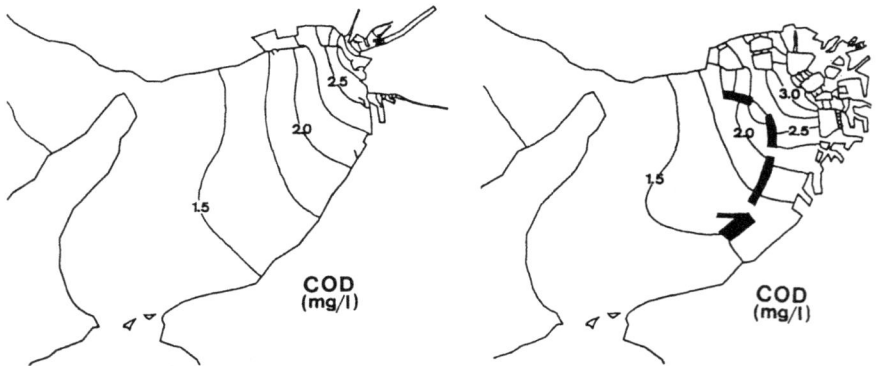

FIGURE 11   *Distributions of COD concentration after 50 tidal cycles in the present situation of Osaka Bay and in the barrier island model*

model of the barrier island project expand from the inner bay when compared to the present situation. However, the 1.5 mg/l line almost coincides between the two cases except that, in the barrier island model, the iso-concentration lines tend to extend slightly toward the center of the bay along the current circulating on the periphery of the Okinose circulation from the Akashi Strait.

5.   Conclusions

Two-dimensional numerical experiments have been conducted using a finite element procedure with SGS eddy viscosities to examine the impact of reclamation on the tidal current system and water quality structure. The model calibration has been further checked by simulating tidal flow and residual current systems. The computed SGS eddy viscosity coefficient is of the order of $10^3 \sim 10^5$ cm^2/s, which depends heavily on the tidal flow intensity.

It has been clarified that large-scale reclamation as assumed in this study affects the Okinose circulation that governs the water quality transfer in Osaka Bay. Since the pollutant source is the inflow load from the rivers located in the inner bay, problems of water exchange and water quality within the waterway between the coast and reclaimed land will arise. Based on this idea, the case of the barrier island project is examined. Compared to the present situation, a slight variation is seen in the tidal current system in the simulation model. However, the enhancement effects of the water mass flowing from the Akashi Strait into Osaka Bay and those of flushing among barrier islands upon mixing in the bay are smaller than expected. In order to promote further optimum control of the tidal current system, study of a different reclamation model should be encouraged.

REFERENCES

[1] Deadroff, J.W., (1970), A Numerical Study of Three-dimensional Turbulent Channel Flow at Large Reynolds Numbers, *Jour. Fluid Mech.*, Vol. **41**, part 2, pp. 453-480.

[2] Murota, A. and Nakatsuji, K., (1985), Environmental Assessment of Waste Water Disposal from Sewage Treatment Plant, *Proc. Ocean Space Utilization '85*, Springer-Verlag, pp. 213-220.

[3] Sawaragi, T., Nakatsuji, K. and Wate, N., (1976), Application of Weighted Residual Method to Tidal Flow Analysis, *Proc. 23rd Japanese Conf. on Coastal Engineering*, pp. 488-492.

[4] Smagorinsky, J., (1963), General Circulation Experiments with Primitive Equations, I. The Basic Experiment, *Monthly Weather Review*, Vol. **91**, No. 3, pp. 99-164.

[5] The 3rd Construction Bureau of the Ministry of Transportation (1980), Environment Assessment Report of the Kansai International Airport island Project.

[6] Tsuji, M., (1991), private communication.

5 Numerical simulation of nearshore residual currents

D. E. Reeve

ABSTRACT

The depth integrated equations of motion are solved over the southern North Sea using an implicit finite difference scheme on body-fitted coordinates. Computed residual currents and eddy statistics are discussed.

1   Introduction

This paper gives details of a depth-integrated hydrodynamic model used for predicting the tidal flow over large areas of shallow coastal waters. The primary aim of designing the model was to obtain a reliable picture of the tidally forced current residuals in the East Anglian nearshore region.

The importance of residual currents in the evolution of this coastline and their role in nearshore sediment circulation patterns have been highlighted by previous studies, eg. [2,15]. Knowledge of the residual currents aids the prediction of the likely impact of proposed coastal works on nearshore sediment movement.

2   Model Description

2.1   Governing equations

Attention is restricted to domains for which the f-plane approximation is valid, [6]. Vertical variations in fluid density are neglected. Using the hydrostatic approximation, conservation of momentum implies:

$$(H\rho <u>)_t + \nabla.(H\rho <u><u>) = -H\rho g\nabla\eta - \tfrac{1}{2}gH^2\nabla\rho - 2H\rho\omega\wedge<u>$$

$$+\nabla.[H(<T> - \rho <u^\bullet u^\bullet>)] + \tau_s - \tau_b \tag{1}$$

with the depth integrated continuity equation

$$(H\rho)_t + \nabla.(H\rho <u>) = 0 \tag{2}$$

Here, H is total depth, ρ is density, u is fluid velocity, t is time, T is the viscous stress tensor, τ_s and τ_b are boundary stresses, $u^\bullet = u - <u>$, g is the acceleration due to the Earth's gravity, η is fluid elevation of the free surface above zero datum, ω is the Earth's angular velocity and angle brackets denote a depth mean quantity. Atmospheric pressure is assumed uniform over the model domain.

Equations (1) and (2) may be cast into a form analogous with equations for compressible flow [14]. This allows the use of general purpose finite difference Navier-Stokes solvers. The code used, [1], has the following features: the temporal discretisation uses unconditionally stable backward differences; the equations are solved on a boundary-fitted, non-orthogonal, collocated grid using a coordinate transformation approach; and within each time step the coupled equations are solved iteratively using the SIMPLEC algorithm.

2.2 Wetting/drying algorithm

The process of wetting and drying in nearshore intertidal regions has a profound effect on tidal residual currents, and thus a suitable method of representing it must be incorporated into the model. An algorithm for use with the Navier-Stokes solver described in section 2.1 has been developed [14]. It has the following two key features. Firstly, mass flow across cell faces occurs only with the elevation gradient. Secondly, mass outflow from any individual cell does not exceed the mass available. Grid cells are adjudged dry if the water level falls below 0.02m and wet for water levels greater than 0.08m.

2.3 Boundary stress

The following bottom stress law is used:

$$\tau_b = K\rho |u_b| u_b \tag{3}$$

where K is the bottom friction coefficient and u_b is fluid velocity at the bed. The latter is evaluated using a spectral technique, [3], modified to allow efficient implementation within the implicit scheme outlined in section 2.1. The velocity field u is expanded in terms of vertical eigenfunctions to derive equations for the temporal evolution of the series coefficients. These equations are solved and u_b determined from a summation over the eigenfunctions, [14].

2.4 Boundary conditions

At open sea boundaries tidal elevation and depth mean normal velocity are prescribed. The latter is used to apply a radiation condition with free tangential slip [5]. The tidal input is expressed as:

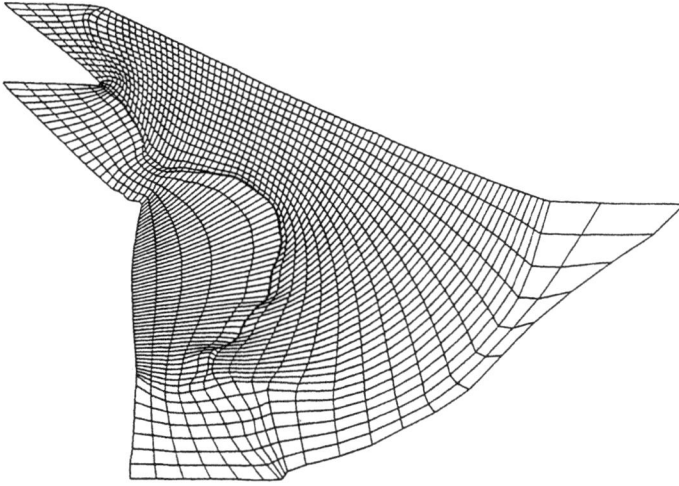

Figure 1: *Southern North Sea model coarse grid.*

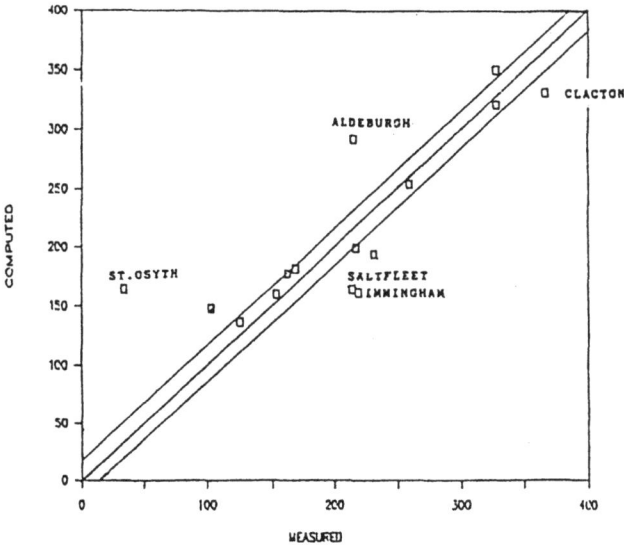

Figure 2: *Comparison of elevation data from the model and observations for the phase of tidal constituent M_2 at a number of coastal measuring stations.*

$$Z_0(\mathbf{x}) + \Sigma_i A_i(\mathbf{x}) f_i \cos(e_i t + V_i + u_i - g_i(\mathbf{x})) \qquad (4)$$

where A_i and g_i are the Greenwich amplitude and phase of constituent i, e_i is the angular speed of constituent i, f_i and u_i are the amplitude and phase of the nodal modulation, V_i the phase of the equilibrium constituent at Greenwich at $t=0$, Z_0 denotes the mean value and $\mathbf{x}$ is the horizontal position vector. The summation in (4) was taken over the five predominant constituents: M_2, S_2, N_2, O_1, K_1.

2.5   Body-fitted grid

Two grids were prepared: a fine grid (394 x 104 cells), and a coarse grid (78 x 20 cells) which is shown in Figure 1. Both grids cover the region enclosed by Dover, Cap Gris Nez, Texel and Flamborough Head. Bathymetric data was obtained from Admiralty charts and nearshore bathymetric surveys. The coarse grid model was run on a desktop workstation, and results from this model are presented here.

3   Calibration and Validation

Model elevations and velocities were calibrated against detailed spatial descriptions of each constituent as published in [13]. Good agreement with observations was obtained [14]. The time step was 15 minutes, the horizontal viscosity was zero, the bottom friction coefficient was assigned the value of 0.0025 [5], and the wetting/drying algorithm outlined in section 2.2 proved reliable and exhibited negligible oscillatory behaviour.

The calibrated model was validated against elevation and current data gathered in an independent survey. Figure 2 shows a comparison of model and observed elevation data for the phase of tidal constituent M_2. The ± 15 degrees lines are drawn on the plot. If agreement were perfect all the points (representing different locations) would lie on a line inclined at 45 degrees to the vertical. Overall agreement was good for results predicted by a depth-integrated model. Largest discrepancies occured in estuaries, where stratification and river flows can be significant.

4   Application

A simulation of a 32 day period with the calibrated and validated model was used to generate current velocity time series at each grid cell. These time series were analysed to determine the tidally induced time average, or residual, moments of the velocity:

$$\{|\mathbf{v}|^{n-1}\mathbf{v}\} \quad n = 1,2,3 \qquad (5)$$

Where $\mathbf{v} = (u,v) = <\mathbf{u}>$, braces denote a time average and a prime denotes a deviation from this average. The first moment is the tidal residual current; the second and third moments may be related to the bed shear stress and sediment mass transport respectively, [10]. Figure 3 shows the computed tidal residual currents. Good large scale agreement is found both with earlier diagnostic model results [11] and dynamical modelling studies [12]. However, the present model provides finer resolution of the coastal region and includes the important wetting and drying

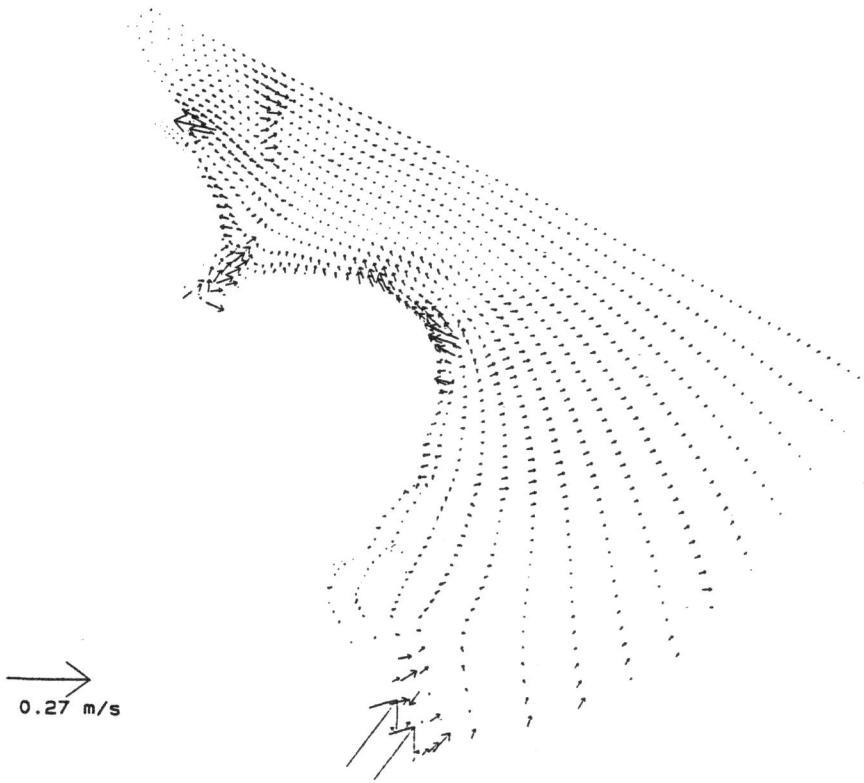

0.27 m/s

Figure 3:   *Tidally induced residual currents, computed over 32 days. Vectors are plotted at grid cell centres.*

59

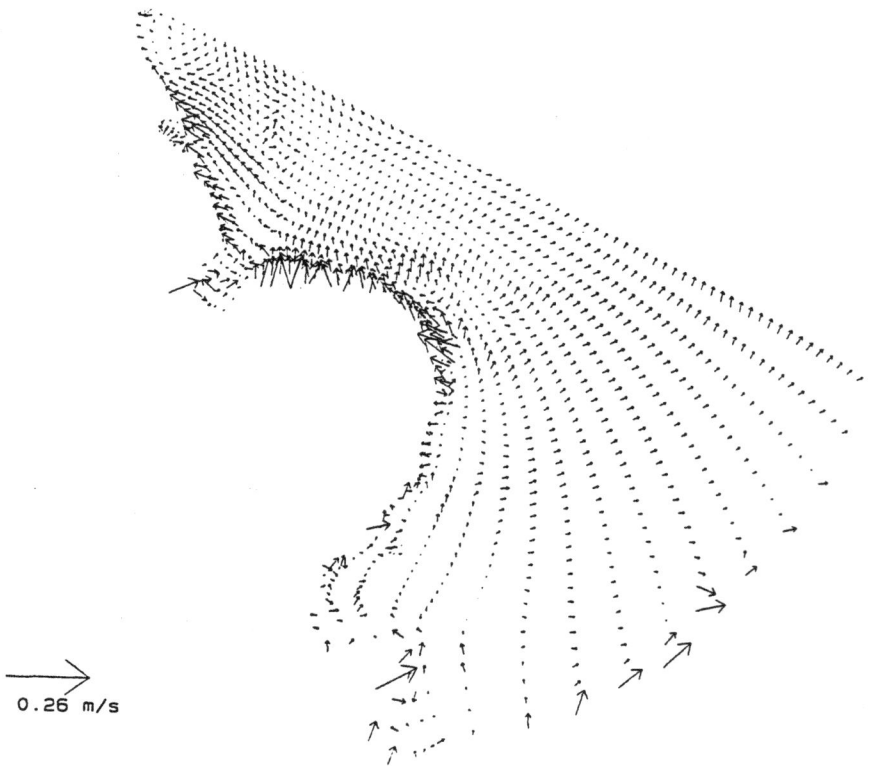

Figure 4: *As in Figure 3, except vectors show the third moment of the computed residual currents.*

processes.

Some of the main features of the tidal residual flow captured by the model are: the generally south-westerly drift over a large part of the central Southern Bight, complicated circulation patterns off the Norfolk and Suffolk coasts, pronounced seaward flow out of The Wash, confused circulations over Silver Pit and near Spurn Head. Recent numerical studies [4] have provided a detailed picture of residual currents within the Humber estuary.

Figure 4 shows the third moment of the tidal residual flow. It shows a southwesterly component over most of the central Southern Bight with the notable exception of the region between Orfordness and Rotterdam. Also the potential transport in the vicinity of Silver Pit is predominantly to the southwest. Further, there is well organised potential transport out of The Wash with a region of strong convergence off Gibraltar Point. Finally, very complex implied transport patterns in the region of the sand bank system off Gt.Yarmouth and to the north over Hewett Ridges are apparent.

5   Discussion

The momentum equation for the residual velocity, derived from equation (1), includes the effects of advection, Coriolis acceleration and the momentum flux due to tidal eddies. The latter can provide a significant contribution as the currents associated with tidal eddies may have magnitudes several orders of magnitude greater than the residual currents.

The symmetric eddy correlation tensor may be written as the sum of isotropic and anisotropic parts, and whose elements are: $K = \{u'^2 + v'^2\}/2$, $M = \{u'^2 - v'^2\})/2$ and $N = \{u'v'\}$. If the eddies are non-divergent their forcing of the time mean vorticity field may be expressed in terms of the quantities M and N, [7]. The link between the asymmetry of the tidal oscillations and their contribution towards the vorticity and momentum balance of the residual flow is thus explicit in the dynamical equations.

Tidal current asymmetry is generally accepted as being important to net sediment transport. At a local scale dynamical considerations have shown that the eddy vorticity flux plays a crucial role in the growth and maintenance of sand banks (eg. [8,16]).

Figure 5 is a map of the quantity $\alpha = (M^2 + N^2)^{1/2}/K$, which lies between 0 and 1 and provides a dimensionless measure of eddy anisotropy, derived from averages computed over a simulation of a 32-day period. It shows a large area off the Norfolk, Suffolk and Essex coast, stretching across to the Dutch coast, where tidal current ellipses are highly asymmetrical (major to minor axis ratios of 10:1 are typical). The same is found in a smaller region running from the mouth of The Wash northwards along the Lincolnshire coast to the north of the Humber estuary. Regions of more symmetrical tidal oscillations include the Thames estuary, the nearshore area of N. France, Belgium and southern Holland, and a large area stretching from Flamborough Head in the west to the northern boundary of the sand wave field near Texel in the east, and southwards over Burnham Flats to the N. Norfolk coast.

This map of eddy anisotropy matches the sand bank morphology, eg. [10], extremely well. In particular, the observed sand wave 'tongue' off the coast of Texel coincides with an area showing values of α close to 1. Furthermore, the northern boundary of the sand wave field has been observed as being fairly abrupt with no gradual decrease in sand wave height. This behaviour is mirrored closely in the rapid decrease in the value of α. Given that the implied mass transport in this region is from the southwest where a plentiful supply of sediment exists it may be inferred that the

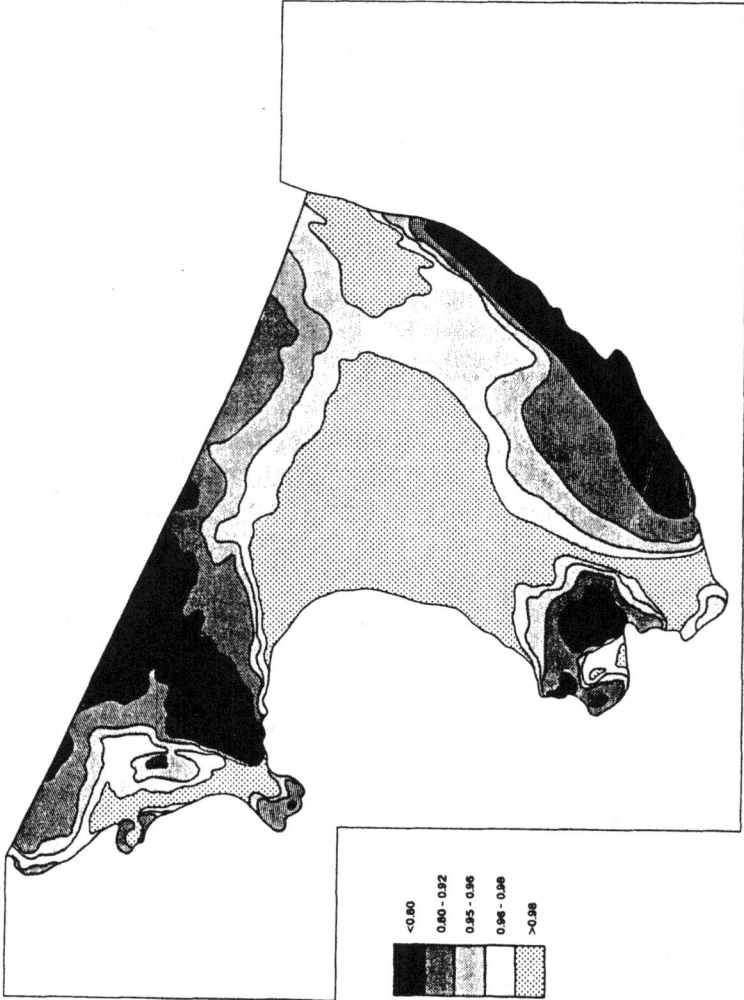

Figure 5: *Map of the eddy anisotropy factor.*

extent of the sandwave field is primarily controlled by the characteristics of the tidal eddies. That is, because the tidal eddies become almost isotropic over a large region to the north, their associated vorticity flux is reduced and thus the mechanism for maintaining sand banks is weakened. In the southern part of the sand wave field symmetrical waves are found. This coincides with an area of negligible net sediment movement and fairly isotropic eddies.

In drawing conclusions from plots such as Figures 3,4 and 5 it should be remembered that the third moment indicates potential rather than actual transport, and its definition includes no acknowledgement of a threshold current magnitude for transport. Further, eddy anisotropy by itself does not necessarily imply the genesis of sand banks, this depending on the sediment supply and the balance of terms in the equations governing the residual flow.

In the preceding discusion it should be borne in mind that the modelling study has dealt only with tidally induced residuals. It is known that significant local changes can be caused by surface wind stress associated with storm conditions. A more complete description of sediment transport trends in the southern North Sea requires further study of the influence of wave action and storm induced residuals, as well as a better understanding of the relationship between grain size and transport rate.

Acknowledgements

The author thanks Rob Hiley and Dr. Chris Whitlow for their inputs to coding and grid generation, and colleagues in the Coastal Engineering Department for many useful discussions. This paper is based on work undertaken as part of the Anglian Sea Defence Management Study on behalf of the National Rivers Authority Anglian Region.

References

1   Burns, A.D. and Wilkes, N.S.,(1987),A finite difference method for the computation of fluid flows in complex three-dimensional geometries, UKAEA Harwell Laboratory AERE R 12342, HMSO.

2   Carr, A.P.,(1981),"Evidence for sediment circulation along the coast of East Anglia", Marine Geology, XL, ppM9-M22.

3   Davies, A.M.,(1988),"On formulating two-dimensional vertically integrated hydrodynamic numerical models with an enhanced representation of bed stress", J.Geophys.Res. XCIII, C2, pp1241-1263.

4   Falconer, R.A. and Owens, P.H.,(1990),"Numerical modelling of suspended sediment fluxes in estuarine waters", Estuarine, Coastal and Shelf Science, XXXI, pp745-762.

5   Flather, R.A.,(1984),"A numerical model investigation of the storm surge of 31 January and 1 February 1953 in the North Sea", Q.J.R.Met.Soc., CX, pp591-612.

6   Gill, A.E.,(1982), Atmosphere-Ocean Dynamics, Academic Press, London.

7   Holopainen, E.O.,(1978),"On the dynamics of the long-term mean flow by the large-scale Reynolds' stresses in the atmosphere", J.Atmos.Sci., XXXV, pp1596-1604.

8   Huthnance, J.M.,(1982),"On the formation of sand banks of finite extent", Estuarine, Coastal and Shelf Science, XV, pp277-299.

9   Huthnance, J.M.,(1991),"Physical oceanography of the North Sea", Ocean & Shoreline Management, XVI, pp199-231.

10   McCave, I.N.,(1971),"Sand waves in the North Sea off the coast of Holland",
      Marine Geology, **X**, pp177-198.

11   Nihoul, J.C.J and Ronday, F.C.,(1975),"The influence of the tidal stress on the
      residual circulation. Application to the Southern Bight of the North Sea",
      Tellus, **XXVII**, 5, pp484-489.

12   Prandle D.,(1978),"Residual flows and elevations in the southern North Sea",
      Proc.R.Soc.Lond., A359, pp189-228.

13   Proudman Oceanographic Laboratory, (1990), Atlas of tidal elevations and
      currents around the British Isles. HMSO, OTH 89293.

14   Reeve, D.E. and Hiley, R.A.,(1992),"Numerical prediction of tidal flow in
      shallow water", to appear in *Proceedings of Seas and Coastal Regions*,
      Computational Mechanics Press.

15   Robinson, A.H.W.,(1980),"Erosion and accretion along part of the Suffolk
      coast of East Anglia", *Marine Geology*, **XXXVII**, pp133-146.

16   Zimmerman, J.T.F.,(1981),"Dynamics,diffusion and geomorphological
      significance of tidal residual eddies", *Nature*, **CCXC**, pp549-555.

6 Numerical modelling of tidal currents along the West European continental shelf in order to predict the movements of polluted matters

J.M. Janin and J.C. Galland

ABSTRACT

A tidal simulation, with a 2D code (TELEMAC) has been performed by the L.N.H. (Laboratoire National d'Hydraulique of Electricité de France) during a complete moon cycle in order to calculate current fields in the English Channel and the close Atlantique. These tidal currents combined with wind-induced-currents are aimed at predicting the movements of polluted matters. They will be used also as boundary conditions for modelling local areas.

The study area as it ends farther off the continental shelf encloses an exceptionnal number of elements (20,000). However, the recent improvements in numerical methods and computers allow this calculation with a reasonable cost. The finite element technic is also used very favourably as the surface of an element decreases from 500 square miles along the west boundary to 1 square mile along the coast. Consequently, a new performance has been reached: modelling tidal currents inside a very large oceanic area with an extreme precision along the coast.

1   Introduction

The most commonly encountered type of tide is that in which two high waters and two low waters occur each day, the times of high water and low water becoming later by approximately 50 minutes from one day to the next. This type of tide is found around nearly all shores bordering on the Atlantic Ocean. It is dominated by the lunar semi-diurnal constituent, denoted by M2, having a period of 12h25, which is half of a lunar day. It is also well known that the range of the tide, defined as the difference in height between a high water and the following low water, varies periodically with the phases of the moon. Tides of maximum range, known as spring tides, occur within a day or two of a new or a full moon. This sequence arises from the solar semi-diurnal constituent denoted by S2 and having a period of half a solar day, alternately reinforcing and opposing the lunar semi-diurnal constituent. Both constituents described above, and also a third one called N2 due to the elliptic trajectory of the moon and which effect is similar to S2 but weaker, are taken into account in the L.N.H model.

Horizontal movements of water, the tidal currents, are totally associated with the vertical rise and fall of the sea surface and are very sensitive to variations of the range of the tide. They are represented by the same harmonic constituents as elevation but not necessarily with the same phases. The phase relationship between elevation and current varies from place to place and is dependant on whether the tidal wave behaves rather like a progressive wave or rather like a stationnary wave. Thanks to the numerical code TELEMAC, the L.N.H model. estimates all these parameters all over the domain regarding an average current : the depth-mean current. However, variation of tidal currents with depth are evaluated afterwards.

2   Description of the numerical code TELEMAC

The code TELEMAC [2] solves the bidimensional shallow-water equations with a finite element method. We assume that the pressure is hydrostatic and the velocities uniform throughout the vertical. We estimate afterwards the variation of tidal current with depth. A celerity-velocity formulation of the continuity and momentum equations is employed, written in the Mercator plan [3], including a diffusion term and two source terms, the bottom friction and the Coriolis force :

$$
\begin{cases}
2c\, \dfrac{\partial c}{\partial t} + 2c\, \vec{u} \cdot \dfrac{1}{\cos \lambda}\, \overrightarrow{grad}\, c + \dfrac{c^2}{\cos \lambda}\, \text{div}\, \vec{u} - \dfrac{c^2}{R}\, tg\, \lambda \cdot \vec{v} = 0 & (1) \\[3mm]
\dfrac{\partial \vec{u}}{\partial t} + \left(\dfrac{\vec{u} \cdot \overrightarrow{grad}}{\cos \lambda} \right) \vec{u} + 2c\, \dfrac{1}{\cos \lambda}\, \overrightarrow{grad}\, c = \dfrac{1}{\cos^2 \lambda}\, \text{div}\left(v\, \overrightarrow{grad}\, \vec{u} \right) - g\, \dfrac{1}{\cos \lambda}\, \overrightarrow{grad}\, Z_f + \vec{S} & (2)
\end{cases}
$$

$\overline{u}$, $\overline{v}$: components of the depth-mean current $\vec{u}$. $\vec{S}$: source terms.
c : velocity of propagation (celerity).   g : gravity.
v : coefficient of horizontal eddy viscosity. λ : latitude.
Z_f : bottom level

2.1   Source terms

The bottom stress :   The Coriolis force :

$$
\vec{S_f} = -\, \dfrac{g}{h\, C_z^2}\, \left|\left| \vec{u} \right|\right| \vec{u} \qquad (3)
$$

$$
\vec{S_c} = -\, 2\omega_T \sin \lambda\, \vec{k} \wedge \vec{u} \qquad (4)
$$

C_z : Chézy coefficient. ω_T : angular rate ot rotation of the earth.

h : depth of water. $\vec{k}$: vertical unitary vector.

2.2   The algorithm

The numerical method solves the equations by means of a decomposition in fractionals steps. Each numerical operator can this way be treated by an adequat method. The resolution is achieved in two steps : an advection step and a diffusion-free surface-continuity-pressure step.

The advection is computed through a characteristic curve method which is unconditionally stable. Both, the celerity and the velocity can be advected. However, in this tidal model, the current number is always lower than 0.2 and the advection of the velocities is quite negligible. The advection of the variable "c" induces in our model a lack of conservativity especially along the shelf edge where the gradient of "c" is important. This effect is due to the interpolation of this variable at the foot of the characteristic. It has been greatly reduced by refining the mesh around the edge.

Then, diffusion terms, source terms and free surface-continuity terms are taken into account. The adopted space discretization is a finite element discretization in triangles. A temporal discretization is performed before so that non linearities are removed. The obtained linear system is finally solved by iterative operations of the conjugate gradient method.

3   Study domain and boundary conditions

3.1   The choice of the domain

An international working group [5] was formed in 1984 to assemble and disseminate a data set for tidal model testing. This group selected an area that encloses the English Channel and the meridional part of the North Sea. They supplied along the open boundaries a complete data field regarding amplitude and phase of elevation of the main tidal waves. Consequently, we have fixed our northern boundary in order to coincide with the one they had chosen.

However, as we wanted also to model at least the Atlantic coast of France, we needed a complete data field in the Bay of Biscay too. Unfortunately the amount of data is very small there and obliged us to move this boundary far away from the coast to damp the effect of the poor quality of the boundary conditions we could fix there. Besides, it should be noted that, in the Bay of Biscay, the shelf width ranges between more than 500 km in its northern part to almost 0 km close to the Spain border. It means that, if the west boundary is not sent far enough it will necesseraly cross the shelf edge, a weak spot of the model. In order to avoid that, we finally decided to fix this boundary along the 12.5 West meridian where the height of water is about 4,500 m. We cross-connected it to the coast, in the south along the parallel of Cape Finistere (Spain), in the north along the parallel of Mizen Head (Ireland). The depth of water is about 4,500 m throughout the first link as there is no shelf there. The second link is crossed by the edge in its middle but it induces no effect as the main tidal waves, which propagate from South to North, are coming out the model there.

To enclose the shelf edge inside the model is favourable concerning the problem of the quality of the boundary conditions. As a matter of fact, the edge acts like a filter for the incoming waves : they are both reflected and diffracted by the steep bottom gradients, and the results of calculation near the coast are less dependent on the boundary conditions.

Two other open boundaries appear in the model. One is crossing the Irish Sea from Carnsore Point (Ireland) to Cape St David (Wales), the other one the Severn estuary. Both are fairly well documented [9].

The whole domain is shown on figure 3.

3.2   The tested boundary conditions

Three kinds of boundary conditions have been successively tested; we called them :
- incident wave
- total wave
- specified elevation with correction

3.2.1 Incident wave
The so called boundary condition has been tested first. It prescribes some relationship between elevation and current at the boundary such as a radiation condition. It takes the following form for a tidal wave with an angular frequency ω in a point M at the time t :

$$c\,\vec{u}.\vec{n} = g\,S - g\,(\,1 - \vec{k}.\vec{n}\,)\,\hat{S} \tag{5}$$

$\vec{n}$: unitary vector outgoing normally to the boundary in M.

S : free surface elevation at the previous time step in M.

$\vec{k}$: unitary vector in the direction of propagation of incident wave in M.

$\widehat{S}$ = A(M) cos (ωt - φ(M)) : sinusoïdal signal of incident wave elevation.

A(M) and φ(M), respectively the amplitude and the phase of the incident wave in M, have to be estimated, which is the deficiency of this approach. As a matter of fact, the measures give the amplitude and the phase of the total wave and it leads to unacceptable phase shifts to consider that incident and total waves are equal even in deep water.

Therefore, we developped an iterative calculation with feedback in order to evaluate A(M) and φ(M) at every open boundary nodes. However, we got undesirable oscillations along the boundaries that we could not remove. This calculation reduced partially the shifts but we still had an important phase shift of -23° in M2 elevation in the middle of the English Channel.

Note that this approach developped by the L.N.H. gave excellent results for several schematic test cases without any reflection along the open boundaries. Finally, this approach seems to be very interesting as long as we can get data about the incident wave.

3.2.2 Total wave

Another radiation condition formulated by Flather [7] has been used and quite successfully tested in this model. It is given for a wave with an angular frequency ω by :

$$c\,\vec{u}.\vec{n} = c\,\widehat{u} + g\left(S - \widehat{S}\right) \tag{6}$$

$\widehat{u}$ = U(M) cos (ωt + φ(M)) : sinusoïdal signal homogeneous to a velocity.

$\widehat{S}$ = A(M) cos (ωt - φ(M)) : sinusoïdal signal of total wave elevation.

On the contrary to the previous approach, this one supposes that $\widehat{S}$ is the signal of the total wave elevation which is given by the measurements. However, it is necessary to define a new signal $\widehat{u}$ obtained by Fourier analysis of $\vec{u}.\vec{n}$ (t). We achieved this calculation thanks to iterations with a duration multiple of the period of the considered wave.

We tested first this boundary condition in the main tidal wave M2. The iterative process presented before converged after 8 iterations of 4 periods of M2 wave each. Then we obtain amplitudes and phases regarding elevation of M2 in very good agreement with the measurements. The generated non linear wave M4 was also in reasonable agreement except in Saint-Malo gulf where it was slightly increased.

This formulation used for a single wave can be applied also for n waves of different angular frequencies ω_i as it follows :

$$c\,\vec{u}.\vec{n} = c\,\widehat{u}_{tot} + g\left(S - \widehat{S}_{tot}\right) \quad \text{with} \quad \widehat{u}_{tot} = \sum_{i=1}^{n}\widehat{u}_i \quad \text{and} \quad \widehat{S}_{tot} = \sum_{i=1}^{n}\widehat{S}_i \tag{7}$$

However, in order to calculate $\widehat{u}_{tot}$ at the end of each iteration with a Fourier analysis of $\vec{u}.\vec{n}$ (t), it is necessary for the iteration duration to be long enough to seperate the waves.

The minimal duration is about : $\dfrac{2\pi}{\min\mid \omega_i - \omega_j\mid_{i\neq j}}$

It means that for the two main waves on the west European shelf, M2 and S2, 15 days of simulations are necessary for each iteration. If we suppose that we still need 8 iterations for the convergence, 4 months of simulation are expected to complete the calculation, that is to say several hours of CPU time of a Cray computer. Consequently, we decided to look for a third formulation of the boudary condition especially for the model calibration with several tidal waves.

3.2.3 Specified elevation with correction

To specify elevation on open boundaries for a scheme without correction yields an unsteady calculation around the shelf edge after a couple of tidal cycles. As we said before this is a critical point of the model especially because of the advection of the velocity of propagation.

A recommended solution (see [1]) is to build a scheme that achieves each time step twice. As a matter of fact, at the end of the first lap we modify the advector current field used for the advection of the velocity of propagation in the second lap in order to centre it on time, whereas for the first lap we utilize the current field obtained at the previous time step as we used to do for a scheme without correction.

The effect of this correction is extremely positive : we get for the wave M2, after 4 tidal cycles, a result as good as the one we got with the radiation condition formulated by Flather. We decided finally to consider this approach for the whole study.

Another point is worth to be emphasised. As we split the resolution into two steps, with only one lap there is no influence of the second one to the first one. With the correction this influence exists. Usually such interdependence is positive to the stability and consequently the quality of a numerical model.

4   Simulated components

We decided to simulate and analyse with an adapted harmonic method a few waves all together. We wanted also to limit the duration of simulation to about one month. The harmonic method that we chose, adapted for short simulations, consists in minimizing the root-mean-square deviation. We applied it to the following 4 waves :

- M2 : principal lunar, semi-diurnal, period 44 714 s (12h 25' 14")
- S2 : principal solar, semi-diurnal, period 43 200 s (12h)
- N2 : larger lunar elliptic, semi-diurnal, period 45 570 s (12h 39' 30")
- M4 : first harmonic of M2, quarter-diurnal, period 22 357 s (6h 12' 37")

The last wave is in fact specified only on shallow water open boundaries like the one in the middle of the North Sea. However, M4 is mostly generated inside the domain thanks to non linear effects applied to M2, such as the bottom stress.

The choice of these 4 waves can be explained as follows :

The first 3 ones are the strongest ones all over the English Channel and the close Atlantic Ocean. The average semi-amplitude of these waves obtained from ten offshore stations along the west European continental shelf are respectively 220, 75 and 40 cm.

The fourth wave in descending order is K2 with an average semi-amplitude of 22 cm. Unfortunately, this wave which effect is to amplify the tidal range during equinoctial periods cannot be separated from S2 unless a simulation of 6 month is achieved. As a matter of fact, S2 and K2 periods (43 082 s) are very close to one another. It must be added that the behaviour of this two waves is quite similar and K2 components may be roughly deduced from S2.

The fifth wave is M4 with an average semi-amplitude of 18 cm. Three reasons justify the choice we made in considering it in the harmonic analysis :

First, the average amplitude of this wave equals almost K2 but, as their spatial distributions are totally different, there exist many areas where M4 is much more important than K2 (like Seine and Somme Bays or around Wight Island).

Also, M4 plays a prominent part in the asymmetry of the tide. One of its most remarkable effects is to create a deviation between the duration of the rising level and the duration of the falling level. Therefore, the amplitudes of the associated flood and ebb currents can slightly differ too. It seems interesting to reproduce this asymmetry whenever we are engaged in following polluted matters.

Last but not least, M4 is a test wave for the code because as we said before this wave is mainly produced inside the domain by the non linear terms of the shallow water equations. Now, usually the treatment of these terms is critical for numerical codes. TELEMAC, which is a fully non linear code produces a wave M4 really in agreement with the mesures.

5 Variation of tidal current with depth

From the depth mean-current computed by TELEMAC code, an evaluation of the current in the three-dimensional domain has been achieved subject to a few hypothesis (see [12]). Such a kind of calculation is judicious since the fluctuation of the tidal current along the vertical is low as compared with the depth-mean current. If we neglect density variations and suppose that the wind is weak, the fluctuation of current with depth is due entirely to the frictionnal stresses arising from bottom friction.

With this three-dimensional current field, it will be possible to appreciate more precisely the movements of polluted matters since their immersion depth is known. The vertical profiles of the current will be used also favourably as boundary conditions for local three-dimensional models and lastly, the estimation of the current at the surface will allow more realistic comparisons with the measures given by the Hydrographic Service (SHOM) [14].

5.1 Hypothesis and equations

Subsequently, we take into account the current generated by a single tidal wave with an angular frequency ω. In comparison with what is commonly made in studying turbulent flow, we divide the current into a depth-mean current and a fluctuation as following :

$$\vec{u} = \vec{\bar{u}} + \vec{u}' \qquad \text{with} \qquad \int_{-H}^{0} \vec{u}' \, dz = 0 \tag{8}$$

The momentum equations in the x and y directions at any point x, y, z in the water are written to $\vec{u}$ with the following hypothesis :

- advection terms are neglected for $\vec{u}'$ like for the depth-mean current $\vec{\bar{u}}$.
- horizontal viscosity of the fluctuation is neglected.
- we suppose that the coefficient of vertical eddy viscosity is independant of both z and t. This hypothesis is not valid inside the bottom boundary layer that accounts for 10 to 20 % of the depth of water.
- in order to be consistent with the harmonic analysis achieved on the depth-mean current at the considered angular frequency, we assume that the associated $\vec{u}'$ is a sinusoïdal signal with the same frequency. Therefore, just for the needs of this calculation, we use in the equation written to $\vec{\bar{u}}$ a partial bottom stress as follows :

$$\vec{S}_c = - \frac{g \, U}{H C_z^2} \vec{\bar{u}} \tag{9}$$

U the average amplitude during a tidal period (12h25) of the entire depth-mean current.
H mean water depth (= $- Z_f$)

Then, by substracting to the equations written to $\vec{u}$, the corresponding ones written to $\vec{\bar{u}}$ and born of shallow water equations, we obtain :

$$\begin{cases} \dfrac{\partial u'}{\partial t} - f\,v' - \nu_z \dfrac{\partial^2 u'}{\partial z^2} = \dfrac{gU}{HC_z^2}\,\bar{u} & (10) \\[4mm] \dfrac{\partial v'}{\partial t} + f\,u' - \nu_z \dfrac{\partial^2 v'}{\partial z^2} = \dfrac{gU}{HC_z^2}\,\bar{v} & (11) \end{cases}$$

ν_z : coefficient of vertical eddy viscosity.   f : Coriolis parameter.

These are Ekman equations for a non permanent flow.
The previous system can be reduced into one equation by introducing the complex variables :

$$Z = u' + i\,v' \qquad \text{and} \qquad \overline{Z} = \bar{u} + i\,\bar{v} \qquad (12)$$

By defining the depth-mean components as :

$$\bar{u} = \bar{u}_x \cos \omega t + \bar{u}_y \sin \omega t \qquad \text{and} \qquad \bar{v} = \bar{v}_x \cos \omega t + \bar{v}_y \sin \omega t \qquad (13)$$

we get :

$$\overline{Z} = \overline{Z}_1 \exp(i\,\omega t) + \overline{Z}_2 \exp(-i\,\omega t)$$

with :

$$\overline{Z}_1 = \frac{1}{2}(\bar{u}_x - i\,\bar{u}_y + i\,\bar{v}_x + \bar{v}_y) \qquad \text{and} \qquad \overline{Z}_2 = \frac{1}{2}(\bar{u}_x + i\,\bar{u}_y + i\,\bar{v}_x - \bar{v}_y) \qquad (14)$$

Finally, the Ekman equation becomes :

$$\frac{\partial Z}{\partial t} + i\,f\,Z - \nu_z \frac{\partial^2 Z}{\partial z^2} = \frac{gU}{HC_z^2}\Big[\overline{Z}_1 \exp(i\,\omega t) + \overline{Z}_2 \exp(-i\,\omega t)\Big] \qquad (15)$$

5.2   Resolution

In order to solve this differential equation we take into account the following remarks.
- we neglect the effect of the wind on the surface and assume that stress there is null. It means that :

$$\forall t ,\ \frac{\partial Z}{\partial z}(0) = 0 \qquad (16)$$

- we recall the definition of $\vec{u'}$ (8), the integration of which from the bottom to the surface equals zero :

$$\forall t ,\ \int_{-H}^{0} Z\,dz = 0 \qquad (17)$$

Let us now introduce a characteristic coefficient of vertical eddy viscosity ν_0, and a joint dimensionless variable K :

$$\nu_0 = \frac{g}{C_z^2}\,H\,U \qquad \text{and} \qquad K = \frac{\nu_0}{\nu_z} \qquad (18)$$

71

Let us also introduce the dimensionless variables :

$$\alpha_H = H (1 + i) \sqrt{\frac{\omega + f}{2\nu_z}} \quad ; \quad \beta_H = H (1 - i) \sqrt{\frac{\omega - f}{2\nu_z}} \quad ; \quad z^* = \frac{-z}{H} \tag{19}$$

(note that ω - f is positive since we are dealing with waves, the periods of which are lower or equal than the period of a semi-diurnal wave).

In that case, we obtain :

$$Z = \frac{K}{\alpha_H^2} \left\{ 1 - \frac{\alpha_H}{Sh\ \alpha_H} Ch\ (\alpha_H z^*) \right\} \overline{Z}_1 \exp\ (\ i\ \omega t\ )$$
$$+ \frac{K}{\beta_H^2} \left\{ 1 - \frac{\beta_H}{Sh\ \beta_H} Ch\ (\beta_H z^*) \right\} \overline{Z}_2 \exp\ (\ -i\ \omega t\ ) \tag{20}$$

The variables α_H and β_H are directly connected to both Ekman number ($E = fH^2/\nu_z$) and Rossby number ($R = \omega/f$) :

$$\alpha_H^2 = i\ (\ R + 1\ )\ E \qquad \text{and} \qquad \beta_H^2 = -\ i\ (\ R - 1\ )\ E \tag{21}$$

We finally define two modified Ekman numbers, E_1 and E_2, as follows :

$$\alpha_H = (\ 1 + i\ )\ E_1 \qquad \qquad 2\ E_1^2 = (\ R + 1\ )\ E$$
$$\qquad \qquad \qquad \text{and} \tag{22}$$
$$\beta_H = (\ 1 - i\ )\ E_2 \qquad \qquad 2\ E_2^2 = (\ R - 1\ )\ E$$

Therefore :

$$Z = f\ (E_1, 1)\ \overline{Z}_1 \exp\ (\ i\ \omega t\ ) + f\ (E_2, -1)\ \overline{Z}_2 \exp\ (\ -i\ \omega t\ ) \tag{23}$$

with :

$$f\ (E_k, \varepsilon) = \frac{K}{2\varepsilon\ i\ E_k^2} \left\{ 1 - \frac{(1+\varepsilon i)\ E_k}{Sh\ [(1+\varepsilon i)\ E_k]} Ch\ [(1+\varepsilon i)\ E_k z^*] \right\} \tag{24}$$

5.3 Solution analysis

In this paragraph, let us have a look at the solution in two special cases; a complete analysis would be too long and is not the purpose of this paper. The first case will deal with very shallow water, the second one with the current at the surface.

5.3.1 If $E_1 < 0.7$

In that case, as the modulus of α_H and β_H are lower than 1, we can substitute the hyperbolic terms in Z for their Taylor expansions at the third order. The expression of Z is mainly simplify to :

$$Z = \frac{K}{6} \left(1 - 3\ z^{*2}\right) \overline{Z} \tag{25}$$

Bowden, from mesurements, established the following formula valid outside the boundary layer that, according to him, accounts for 14.5 % of the depth of water :

$$\vec{u} = \vec{u}(0) \left(1 - 0.37 \frac{z^2}{H^2}\right) = 1.14\ \vec{\bar{u}} \left(1 - 0.37 \frac{z^2}{H^2}\right) = \vec{\bar{u}} + 0.14\ \vec{\bar{u}} \left(1 - 3 \frac{z^2}{H^2}\right) \tag{26}$$

Davies [10] suggested, outside the boundary layer, 2 formulations of the eddy viscosity :

$$v_z = 0.0025 \; H \; U \qquad so : \qquad v_z \approx v_0 \qquad (27)$$

and

$$v_z = 2.0 \; 10^{-5} \frac{U^2}{f} \approx 0.2 \; U^2 \qquad (28)$$

The first formulation which is rather valid for shallow water fits perfectly our expression to the mesures as it gives to K the value 1. The interest of the calculation is that it fixes the limits of validity of the law. If we introduce the parameter $\tau = H/U$, and choose for C_z^2 the value 4 000 m/s^2, we obtain that the law is valid until : $\tau < 10$ s.

(note that this law which is valid near the coast, set $\vec{u}$ to be colinear along the profile)

5.3.2 If $z^* = 0$

As the current at the free surface is the one usually measured, we have a special interest in evaluating it in order to achieve reliable comparisons. Furthermore, this current induces the movements of a large family of polluted matters : the ones lighter than sea surface water (oil products in particular).

For low values of Ekman number, as it has been established previously, the current at the surface generated by a tidal wave is nearly 15 % greater than the mean-depth current and colinear to it. Further, a global study of f (E_k,ε) is impossible. We resolve it into two functions, a (E_k,ε) and b (E_k,ε), respectively the real part and the imaginery part of f (E_k,ε). Consequently, we get for $z^* = 0$:

$$a\,(E_k,\varepsilon) = K \; \frac{Ch \; E_k \; \sin E_k - Sh \; E_k \; \cos E_k}{E_k \; [Ch\,(2E_k) - \cos\,(2E_k)]} \qquad (29)$$

$$b\,(E_k,\varepsilon) = \varepsilon \; K \; \left\{ \frac{Ch \; E_k \; \sin E_k + Sh \; E_k \; \cos E_k}{E_k \; [Ch\,(2E_k) - \cos\,(2E_k)]} - \frac{1}{2E_k^2} \right\} \qquad (30)$$

As well as we did for $\vec{u}$ (13), we can express $\vec{u'}$ as follows :

$$u' = u'_x \cos \omega t + u'_y \sin \omega t \qquad and \qquad v' = v'_x \cos \omega t + v'_y \sin \omega t \qquad (31)$$

We finally obtain the following matrix that transforms the harmonic components of $\vec{u}$ into the ones of $\vec{u'}$:

$$\begin{pmatrix} u'_x \\ u'_y \\ v'_x \\ v'_y \end{pmatrix} = \begin{pmatrix} a^+ & -b^- & -b^+ & -a^- \\ b^- & a^+ & a^- & -b^+ \\ b^+ & a^- & a^+ & -b^- \\ -a^- & b^+ & b^- & a^+ \end{pmatrix} \begin{pmatrix} \bar{u}_x \\ \bar{u}_y \\ \bar{v}_x \\ \bar{v}_y \end{pmatrix} \qquad (32)$$

with :

$$a^+ = \frac{a\,(E_2,-1) + a\,(E_1,+1)}{2} \qquad\qquad b^+ = \frac{b\,(E_2,-1) + b\,(E_1,+1)}{2}$$

$$\qquad\qquad\qquad\qquad and \qquad\qquad\qquad\qquad (33)$$

$$a^- = \frac{a\,(E_2,-1) + a\,(E_1,+1)}{2} \qquad\qquad b^- = \frac{b\,(E_2,-1) - b\,(E_1,+1)}{2}$$

In order to calculate a^+, a^-, b^+ and b^-, we had also to estimate either v_z or K.

Both formula suggested by Davies [10] have their own validity range. The first one adjusts very well to shallow water as it equilibrates eddy viscosity and bottom stress. However, when

the water depth is getting greater, this formula increases exaggeratedly the value of v_z and the second one fits better. For τ equal to 80 seconds that we call latter τ_0, the two formulae give the same result. Therefore, whenever τ is lower than τ_0, we will try to be close to the first formulation, whenever τ is greater than τ_0, we will try to be close to the second one.
If we consider K we notice that :
- according to formula (27) : $K = 1$
- according to formula (28) : $K = \dfrac{\tau}{\tau_0}$

It exists a simple expression for K that tends quickly to the previous ones when τ deviates from τ_0 and also avoids any kind of discontinuity in τ_0. This expression is :

$$K = \text{Coth} \frac{\tau_0}{\tau} \tag{34}$$

Then, we obtain the values of the two modified Ekman numbers, E_1 and E_2, and consequently the ones of the matrix components (see figure 1) because E_1 and E_2 are related to the Ekman number (22), which is given by :

$$E = K \frac{f H^2}{v_0} \approx K \frac{\tau}{25} \tag{35}$$

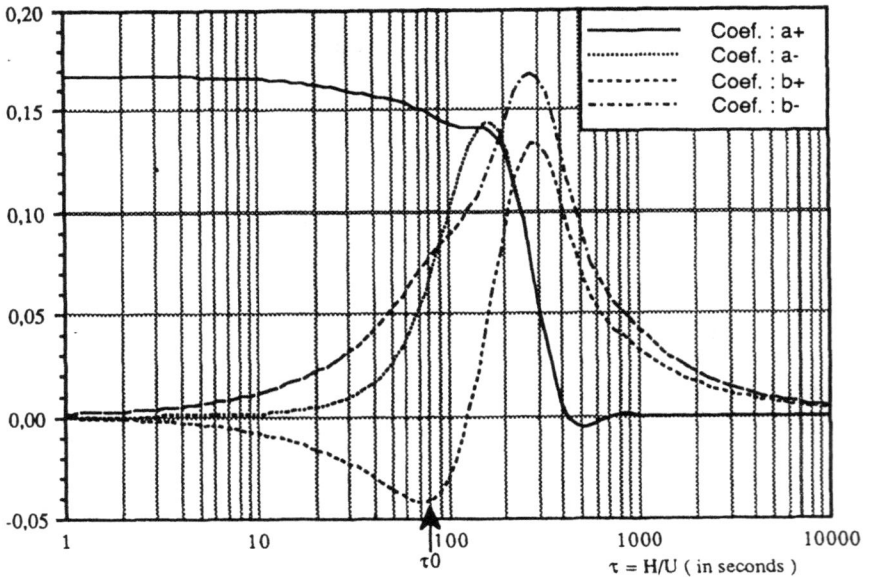

FIGURE 1 *Coefficients of the matrix that transforms depth-mean current into surface current*

On figure 2 one can see two comparisons; on the one hand between measured surface current and calculated depth-mean current; on the other hand between measured and estimated surface current with the exhibit method. The chosen point is situated in Mer d'Iroise in front of Brittany peninsula. The depth of water is about 150 m there and the average current 1 m/s, so the parameter τ is about 150 s. You can note that the current at the surface is stronger than the depth-mean current and also rotates with about an hour's delay; such a remark has already been exposed by Bowden [11].

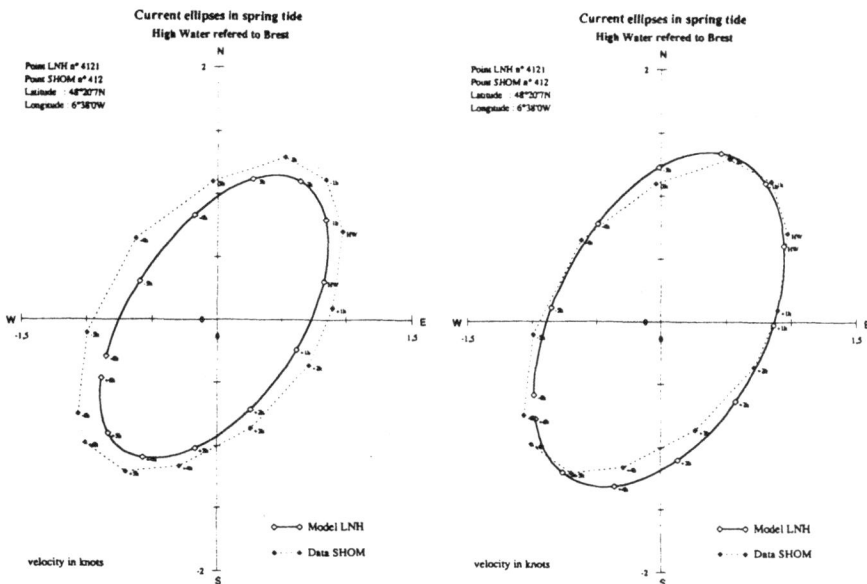

FIGURE 2   *Effect of the transformation : on the left depth-mean current given by Telemac; on the right estimated surface current*

6   Results and comparisons

The results of simulations for each component regarding elevation have been successfully compared with maps made by Chabert d'Hières and Le Provost [4] and measures given by [5] and [6].

The amplitudes and phases of semi-diurnal waves of the model are in good agreement. Regarding the main component M2, the variations are lower than 20 cm for the amplitudes and 10 degrees for the phases all over the area.

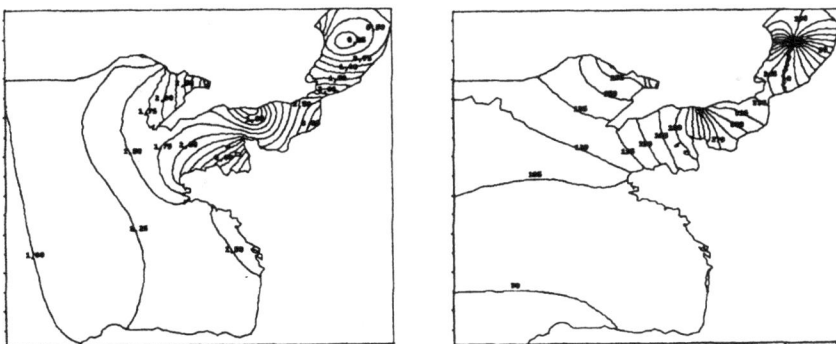

FIGURE 3   *Semi-amplitude and phase of M2 wave elevation all over the domain*

The amplitude and phase of its harmonic M4 are in good agreement also, especially in the North Sea and in the eastern part of the English Channel where the two amphidromic points are correctly situated. More consistent bias appear in the Bay of Biscay and in the western part of the English Channel where the influence of the shelf edge is prevalent. Nevertheless, no amplitude bias is greater than 5 cm there.

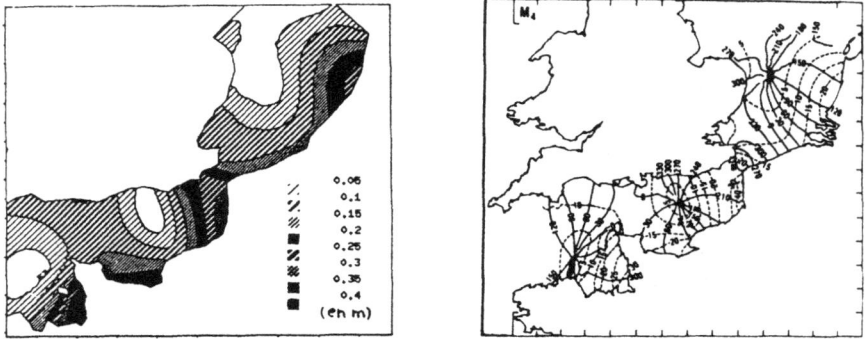

FIGURE 4   *Semi-amplitude of M4 wave elevation in the English Chanel and the North Sea; on the left LNH model; on the right Tidal Flow Forum*

Concerning the current components two kinds of comparaisons has been achieved.
The first one in the English Channel was made with the results obtained by Fornérino [1]. We compared directly the depth-mean currents components by means of a decomposition in 4 characteristic parameters :
   - amplitude of the maximum current
   - phase of the maximum current
   - ratio of the minimum current to the maximum current
   - direction of the maximum current
Qualitatively, no manifeste discordance came into sight. The place and the magnitude of the strongest currents fairly coincide and so do the positions of the amphidromic points (in that case, an amphidromic point is a point where the current vector describes a circle). However, it seems to be hard to go deeply into details as none of this models may be considered as a reference.

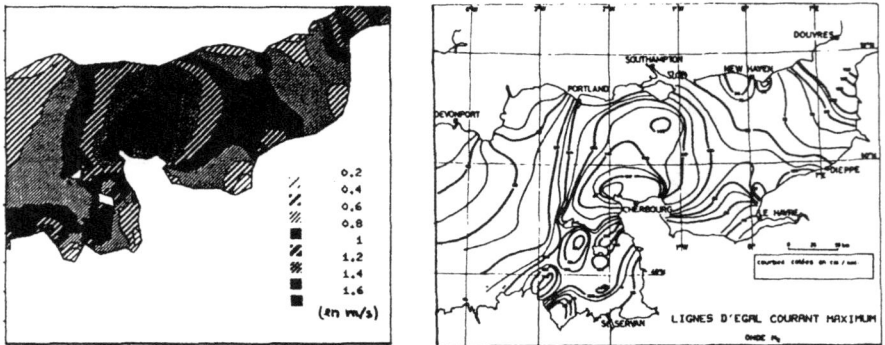

FIGURE 5   *Amplitude of M2 maximum current in the English Chanel; on the left LNH model; on the right Fornérino's model*

Another comparison has been realised with the measures made at the sea surface by the Hydrographic Service (SHOM) [14]. For each analysed point, we superposed the ellipse of the measured tidal mean spring current and the one obtained by the L.N.H. model and modified as described in the previous chapter. We consider that a mean spring current is obtained when M2 and S2 are perfectly in phase. Then we sum M2, S2 and M4 currents. We neglect N2 current as a mean spring tide does not inform about the phase relationship between this wave and the previous ones.

We got very good results in the Bay of Biscay and in the English Channel. At one and the same time, the sharpness of the ellipses, their axes and the phases coincided in these areas for more than 70% of points (see figure 2). The 30% of defective points are scattered all over the domain so we believe that it is due to local problems such as a bad interpolation of the bathymetry.

7   Conclusion

The improvement in computer performances but also in numerical technics by which TELEMAC profits (non assembly of matrices, vectorizable matrix vector product...) are of great benefit for this tidal model, with an important reduction of computation efforts. Consequently, we could refine the mesh size wherever it was important to do it and the results that we obtained for elevation as well as for current components are in very good agreement with the measurements.

We now dispose of a complete data field on a monthly tidal cycle, with estimate vertical velocity profiles that we have already begun to use as boundary condition for local 3D models.

ACKNOWLEDGEMENTS

This study has been partially financed by the Service Technique de la Navigation Maritime et des Transmissions de l'Equipement within the framework of the E.D.F.-Ministries convention.

REFERENCES

[1]   Fornerino, M., (1982), *Modélisation des courants de marée dans la Manche*, Thèse INPG, Grenoble, France.

[2]   Galland, J.C., and Hervouet, J.M., (1990), *Projet Saint-Venant - Code TELEMAC - version 1.0 - Note de principe*, Rapport EDF HE-43/90.04, L.N.H., Chatou, France.

[3]   Lepeintre, F., (1989), *Comment résoudre les équations de Saint-Venant sphériques avecun code résolvant les équations de Saint-Venant cartésiennes - Application au code TELEMAC*, Rapport EDF HE-41/89-28, L.N.H., Chatou, France.

[4]   Chabert d'Hières, G., and Le Provost, C., (1978), *Atlas des composantes harmoniques de la marée dans la Manche*, Institut de Mécanique de Grenoble, France.

[5]   Tidal Flow Forum, (1987), "6th International Conference on Computational Methods in Water Resources", *Adv. in Water Resources*, Vol. 10.

[6]   Cartwright, D.E., Edden, A.C., Spencer, R., and Vassie, J.M., (1980), *The tides of the Northeast Atlantic ocean*, Institute of Oceanographic Sciences, Bidston Observatory, Vol. 298, A. 1436, Birkenhead, Merseyside L43 7RA, U.K.

[7]   Flather, R., (1976), "A tidal model of the north-west european continental shelf", *Mémoires Société Royale des Sciences de Liège*, 6e série, pp. 141-164.

[8]   Galland, J.C., Janin, J.M., (1991), *Simulation des courants de marée en Manche - Première partie*, Rapport EDF HE-42/91.13, L.N.H., Chatou, France.

[9] Robinson, I.S., (1979), "The tidal dynamics of the Irish and Celtic Seas", *Geophys. J. R. astr. Soc*, Vol. 56, pp. 159-167.

[10] Davies, A.M., (1991), "On the accuracy of finite difference and modal methods for computing tidal and wind wave current profiles", *international Journal for Numerical Methods in Fluids*, Vol. 12, pp. 101-124.

[11] Bowden, K.F., (1983), *Physical Oceanography of Coastal Waters*, Ellis Horwood Ltd., Market Cross House, Chichester, West Sussex, PO19 1EB, U.K.

[12] Marchuk, G.I., Kagan, B.A., (1989), *Dynamics of Ocean Tides*, Kluwer Academic Publishers, P.O. Box 17, 3300 AA Dordrecht, The Netherlands.

[13] Schureman, P., (1971), *Manual of Harmonic Analysis and Prediction of Tides*, United States Government Printing Office, Washington D.C., U.S.A.

[14] SH N°550, (1976), *Courants de Marée dans la Manche et sur les Côtes Françaises de l'Atlatique*, Service Hydrographique et Océanographique de la Marine, Paris, France.

7 Tidal residual current in Omura Bay

T. Fukumoto, T. Nakamura and H. Togashi

ABSTRACT

Omura Bay is located in the western extremity of Japan. The current in the bay has fairly complicated characteristic because of the enclosed shape of the bay with a very narrow bay mouth.

The field observation and numerical simulation analysis were curried out in order to comprehend behavior of tidal currents relating to the long-term prediction on bay water quality.

As the result of observation, it is clarified that M_2 tide constituent is dominant and a long-period fluctuation is contained in tidal currents, and futher a tidal residual circulation current exists in the northern part of the bay.

The numerical computation results explain well the observed results at least qualitatively.

1. Introduction

Omura Bay is located in the near center of Nagasaki Prefecture being the western end of Japan, as shown in Fig.1. The surface area is about 330km^2 and the mean water depth is 16m. In particular, the bay mouth called HARIO SETO is very narrow and about 0.2km wide.

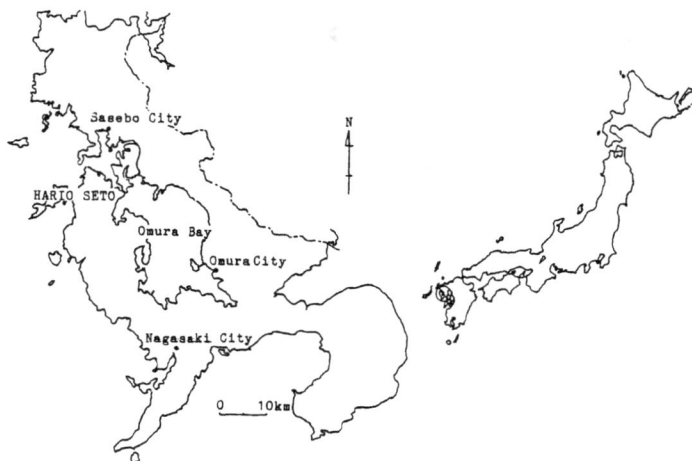

FIGURE 1 *Map showing the location of Omura Bay*

There are two special qualities on tidal currents due to its configuration. One of them is caused by the effect of tide of the enclosed bay and another one is caused by the effect of wind. So the current in the bay is very complicated.

Lately big scale development projects are planed and executed in the bay area, so that the water quality is becoming worse. Therefore, it is very important to understand the tidal currents in the bay as exactly as possible.

Many field observations in the bay had been carried out[1], while the definite result was not established because most of observation periods were short of about one day, etc..

The field observation and numerical simulation analyses have been renewed by the authors since Jan. 1989[2]. This paper deals with the result of observation at 7 points and two dimensional numerical simulations in Omura Bay[3].

2. Observation

The field observation all over the bay was carried out for about 15 days on end. Table 1 and Fig.2 show the observed stations and its periods. The items of measurements are current direction, current velocity, water temperature and water pressure(tide level). They were measured by using the electro-

TABLE 1   *Locations of observed stations and periods*

Station	Latitude(N)	Longitude(E)	Period of observation
P1	32°56'03"	129°52'00"	9 Jan. –26 Jan. '89
P2	32°58'17"	129°52'00"	26 Jan. –14 Feb. '89
P3	33°01'00"	129°52'00"	14 Feb. – 2 Mar. '89
P4	33°00'06"	129°50'24"	2 Mar. –18 Mar. '89
P5	32°58'07"	129°50'24"	18 Mar. – 6 Apr. '89
P6	33°01'52"	129°53'36"	6 Apr. –20 Apr. '89
P7	32°58'37"	129°55'13"	20 Apr. – 9 May '89

FIGURE 2 *Locations of observation station*

magnetic current meter (ALEC ACM-16M3). These data were stored in a memory as machina language every 10 minutes. The current meter was attached at 5m above the sea bed, as shown in Fig.3.

Fig.4 shows the variations in time series of current velocities, water temperature and tide level at the station P2. The semidiurnal tide has been dominant in the fluctuations of both current velocity and tide level. And further, the current velocities contained a fluctuating constituent with a long-period of about one week. On the other hand, water temperature was nearly constant and unaffected by semidiurnal tide because of the winter observation.

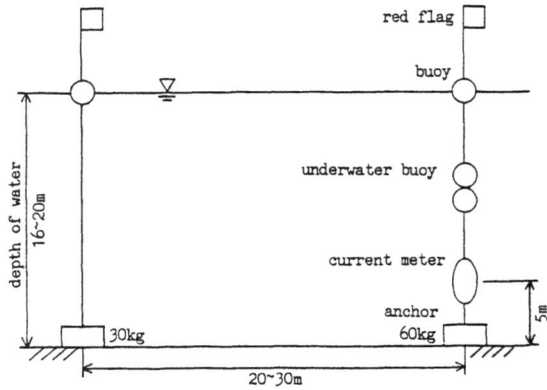

FIGURE 3 *Measurement way of tidal currents*

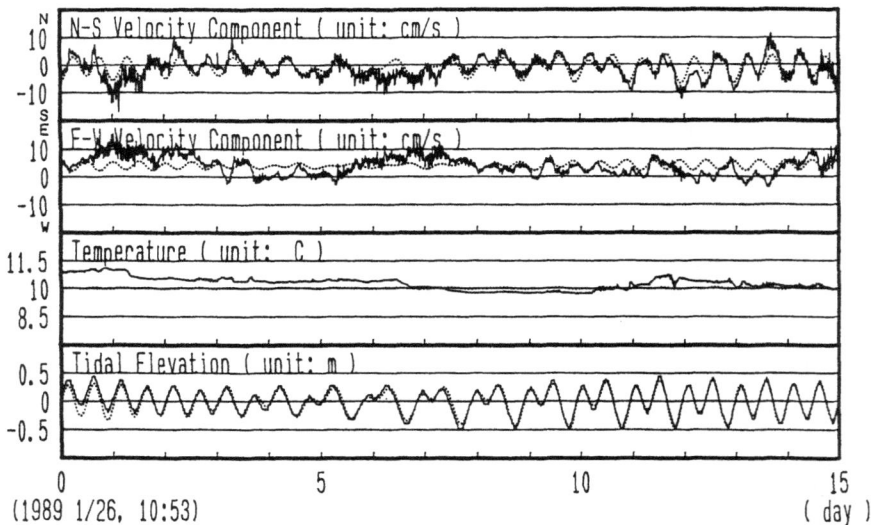

(1989 1/26, 10:53)   (day)

FIGURE 4 *Records of N-S current, E-W current, water
temperature, tide level and resultants of
hormonic analysis(dotted line)
at station P2*

The hormonic analyses for them were carried out by
means of the method of least squares. Dotted lines
shown in Fig.4 indicate analized results. The value
on tide level agrees fairly well with observed

records, while there is quantitative difference
between analytical results and measured records in
the current velocities. Therefore, it is realized
that current velocities are significantly influenced
by wind and atmospheric pressure.

Fig.5(a),(b) show not only the averaged current
velocities(a tidal residual current vector : M_0) but
also tidal ellipses on the four large component
tides(M_2, S_2, K_1, O_1) at the station P2 and P4
respectively. It is obvious from these figures that
the M_2 constituent is dominant as compared with
others in Omura Bay.

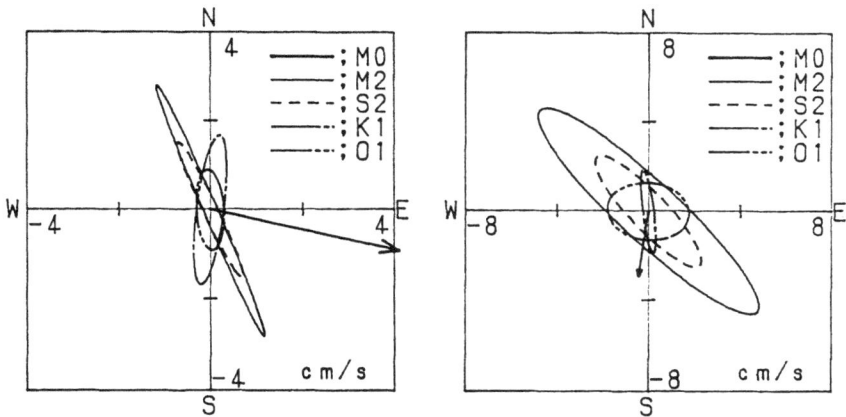

(a) Station P2   (b) Station P4

FIGURE 5 *Tidal ellipses on the four large component
tides(M_2, S_2, K_1, O_1) and the mean of current
velocity(M_0)*

Fig.2 shows vectors of M_0 constituent at 7 stations respectively. These results suggest that there
are an anticlockwise tidal residual circulation
current in the northern part of Omura Bay.

3. Numerical simulation

3.1 Basic equations and boundary conditions

In the numerical simulation, two dimensinal(2-D)
models for shallow water were adopted to investigate
behaviors of tidal currents, as follows.

Continuity equation:

$$\frac{\partial \zeta}{\partial t} + \frac{\partial M}{\partial x} + \frac{\partial N}{\partial y} = 0 \tag{1}$$

Momentum equation along the x-direction:

$$\frac{\partial M}{\partial t} + \frac{\partial (uM)}{\partial x} + \frac{\partial (vM)}{\partial y} - fN = -g(h+\zeta)\frac{\partial \zeta}{\partial x} + \frac{\tau_{sx}}{\rho_w} - \frac{\tau_{bx}}{\rho_w} \tag{2}$$

Momentum equation along the y-direction:

$$\frac{\partial N}{\partial t} + \frac{\partial (uN)}{\partial x} + \frac{\partial (vN)}{\partial y} + fM = -g(h+\zeta)\frac{\partial \zeta}{\partial y} + \frac{\tau_{sy}}{\rho_w} - \frac{\tau_{by}}{\rho_w} \tag{3}$$

in which x, y and t are the independent variables of space and time, h is flow depth, u and v are velocity components in x- and y- directions respectively, $M(=uh)$ and $N(=vh)$ are flow flux in x- and y- directions respectively, ζ is free surface elevation above still water level, f is the Coriolis parameter, g is gravity acceleration, ρ_w is sea water density. Moreover, τ_{sx} and τ_{sy} are wind shear stress at the free water surface, τ_{bx} and τ_{by} are the bottom shear stress at the water bottom in x- and y- directions respectively, as follows.

$$(\tau_{sx}, \tau_{sy}) = \rho_a \gamma_a^2 (W_x, W_y)\sqrt{W_x^2 + W_y^2} \tag{4}$$

$$(\tau_{bx}, \tau_{by}) = gn^2 (u,v)\sqrt{u^2 + v^2} / (h+\zeta)^{1/3} \tag{5}$$

in which W_x and W_y are the wind velocity component in x- and y- directions respectively, ρ_a is air density, γ_a^2 is the drag coefficient at the free water surface, n is Manning's roughness coefficient.
   These parameter are taken here: $g=9.8\mathrm{m/s^2}$, $\rho_w=1.02$ g/cm^3, $\rho_a=1.293\times10^{-3}$g/cm^3, $f=7.943\times10^{-5}$s^{-1}, $n=0.023$, $\gamma_a^2=1.3\times10^{-3}$.
   The Leap-frog method transforms these basic equation into finite difference representations[4].
   The various conditions and parameters in computations are as follows: The grid interval(Δx and Δy) is 1 km. The time increase(Δt) is every 20 second. And one tidal period is 12 hours and 24 minutes. As boundary condition, the tide level is given on the grid of the bay mouth(HARIO SETO).

84

3.2 Numerical simulation results

Fig.6 shows the averaged current velocities of M_2 constituent for 30 tidal periods. It is known that the magnitude of calculated velocities are very smaller than observed values, because it seems that the observed results have all factors(ex. tide, wind, river etc.). On the other hand, the numerical results have only one factor of M_2 constituet. The anticlockwise tidal circulation is also confirmed in the northern part of the bay.

Fig.7 is the tidal ellipse at the station P4. The solid line indicates a calculated result, and the dotted line shows a observed result. the both are forward to the NW-SE direction on the line of apsides and in fairly good agreement. In other words, the numerical results are in accord with the observed one from a qualitative viewpoint.

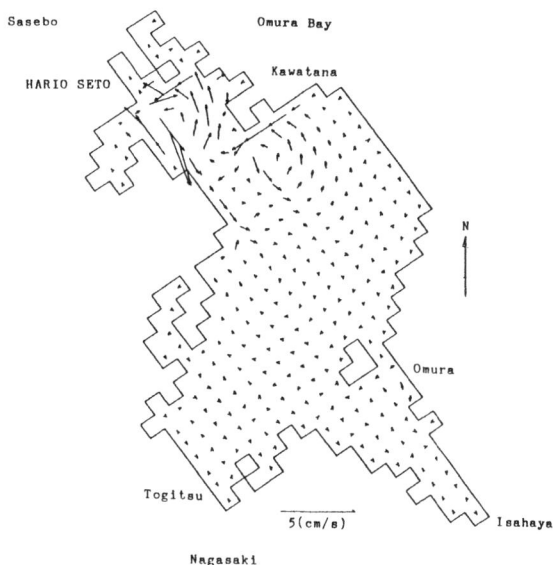

FIGURE 6 *Averaged current velocities on M_2 constituent(Tidal residual currents)*

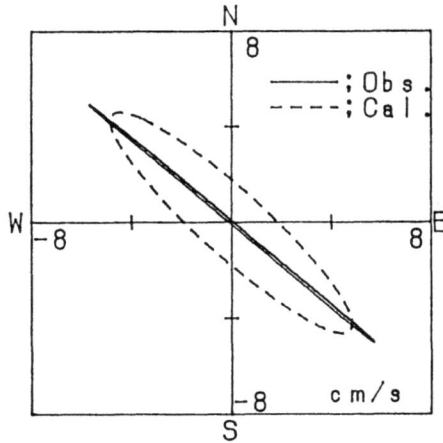

FIGURE 7 *Comparison of observed and calculated tidal ellipses on M_2 constituent at station P4*

4. Conclusions

The field observation and the numerical simulation analyses were carried out in order to grasp characteristic of tidal currents and tidal residual current in Omura Bay. The obtained results are summarized as follows:

(1) Observation

1) The principal lunar semidiurnal constituent(M_2) is dominant as compared with others in Omura Bay.
2) A long-period fluctuating of about one week is contained on the current velocities.
3) There is the anticlockwise tidal residual current in the northern part of the bay.

(2) Numerical simulation

1) A tidal circulation is found in the northern part of Omura Bay and it is similar to the observed result in terms of qualitative aspect.
2) Calculated values with only M_2 tide constituent are very smaller than the observed ones containing all factors.

In the near future, it is intended for the

authors to carry out the 3-D numerical simulation analyses with all factors.

REFERENCES

[1] Nakamura,T. et all(1991): Currents in Omura Bay (3), The report of the Faculty of Engineering, Nagasaki University, Vol.21, No.37, pp 179∽187 (in Japanese).
[2] Nakamura,T. et all(1989): Currents in Omura Bay (2), The report of the Faculty of Engineering, Nagasaki University, Vol.21, No.37, pp 167∽177 (in Japanese).
[3] Nakamura,T. et all(1991): Currents in Omura Bay (1), The report of the Faculty of Engineering, Nagasaki University, Vol.19, No.33, pp 69∽75 (in Japanese).
[4] Fukumoto,T.(1990): The analysis of tidal currents in Omura Bay, The thesis for M. Eng. at Nagasaki University, 65p(in Japanese).

8 Tidal modelling of the Arabian Gulf

K. Z. Elahi and R. A. Ashrafi

ABSTRACT

Being the center of the world oil, there has been tremendous trade and industrial activities on the coastal areas of the Arabian Gulf. A two dimensional hydrodynamical numerical model for the tides in the Arabian Gulf (from $25°N$ to $30°N$ and $48°E$ to $56°E$) is presented. An explicit finite difference scheme is developed to solve the partial differential equations of the model representing the tidal dynamics of the region. Tidal elevations for the four major tidal constituents M_2, S_2, K_1 and O_1 are reproduced by the model. Tidal currents for the M_2-tides are presented. Amphidromic points for all the major partial tides are also reproduced by the model.

Introduction

Tidal model is a numerical model which reproduces the tidal dynamics, the tidal elevation and tidal currents. A knowledge of the tidal dynamics is crucial for the decision making of the offshore drilling activities, monitoring pollution and for sea-traffic routes. A two-dimensional shallow-water equations, are derived from the three-dimensional equations by averaging over the depth. An explicit finite-difference scheme is used for the solution of the two-dimensional shallow-water equations. The surface elevation and depth- averaged velocity components in the Arabian Gulf are reproduced.

Bathymetry of the Region and Literature Review
In this study, we consider the Arabian Gulf from $25°N$ to $30°N$ and from $48°E$ to $56°E$. It covers nearly all the major coastal areas of the region. Bathymetry of the region can be found in Lehr [10], Koske [8], Hughes and Hunter [7] and defence mapping agency [3].

Major studies of the area have been reported by Von Trepka [13], Evans-Roberts [6], Lardner et al [9], Danish Hydraulic Institute [2] Le Provost [11] and Elahi [4]. Other major work has been done by Tetra Tech [12], and Beltagy [1].

Numerical Model of the Arabian Gulf
The equations of unsteady flow on which the numerical simulation for the tidal dynamics is based, consists of incompressive three-dimensional Navier-Stokes equations and continuity equation. Approximating the vertical momentum equation by the hydrostatic pressure equation in shallow waters and integrating the equations over the depth of the sea, these are reduced to the following two-dimensional equations.

$$\frac{\partial \zeta}{\partial t} + \frac{1}{R\,cos\phi}[\frac{\partial (HU)}{\partial \lambda} + \frac{\partial}{\partial \phi}(HV cos\phi)] = 0 \tag{1}$$

$$\frac{\partial U}{\partial t} - \Omega V + \frac{1}{H}\tau_b^\lambda - A_h \nabla^2 U + \frac{g}{R\,cos\phi}\frac{\partial \zeta}{\partial \lambda} = 0 \tag{2}$$

$$\frac{\partial V}{\partial t} + \Omega U + \frac{1}{H}\tau_b^\phi - A_h\,\nabla^2 V + \frac{1}{R}\frac{\partial \zeta}{\partial \phi} = 0. \tag{3}$$

where

λ, ϕ : geographical longitude and latitude

U, V : components of vertically averaged velocity in the λ

        : and ϕ directions, resp. $[ms^{-1}]$

t : time $[s]$

ζ : water elevation [m]

h : mean water depth [m]

H : $h + \zeta$(actual depth) [m]

A_h : coefficient of horizontal eddy viscosity $[m^2 s^{-1}]$

g : acceleration due to gravity $[ms^{-2}]$

R : radius of the earth [m]

ω : angular speed of the earth rotation $[s^{-1}]$

Ω : coriolis parameter $[= 2\omega sin\phi]$

∇^2 : horizontal Laplacian operator $[m^{-2}]$

If $u(z)$ and $v(z)$ denote the horizontal velocity components at depth z below the undisturbed sea surface, then

$$U = \frac{1}{h + \zeta} \int_{-h}^{\zeta} u(z)dz \tag{4}$$

$$V = \frac{1}{h + \zeta} \int_{-h}^{\zeta} v(z)dz \tag{5}$$

The component of the bottom frictional stresses $(\tau_b^\lambda, \tau_b^\phi)$, are parametrized by a quadratic law relating bottom stress to the depth mean velocity

$$\tau_b^\lambda = rU(U^2 + V^2)^{\frac{1}{2}} \tag{6}$$

$$\tau_b^\phi = rV(U^2 + V^2)^{\frac{1}{2}} \tag{7}$$

where $r = .003$ is a non-dimensional friction coefficient.

The coefficient of horizontal eddy viscosity A_h is related to grid size $\Delta\ell$ and time step Δt of the numerical model. Considering the Arabian Gulf as shallow water body the empirical value of the horizontal eddy viscosity is $.1 \times 10^3 m^2 s^{-1}$.

The tides in the Arabian Gulf are predominantly semi-diurnal and diurnal. Thus, the tides in the model are generated by prescribing amplitudes and phases of tidal constituents of M_2, S_2, K_1 and 0_1 at the open boundaries. Waterlevels as a function of time for each partial tide are computed by

$$\zeta(t) = Acos(\sigma t - K). \tag{8}$$

Where A is amplitude, K is the phase of incoming tide and σ is the frequency. The effect of the coriolis force in the horizntal plane is given by

$$\Omega = 2\omega \sin \phi \qquad (9)$$

where ω is the angular speed.

There are two types of boundaries, a solid boundary and an open boundary. The open boundary is the supposed joining line of the model area with the main sea. On the solid boundary no-slip conditions are considered. On the open boundary, waterlevels are prescribed at everytime step using Eq.(8). Moreover, the velocity gradients in the normal direction are taken equal to zero. For the initial conditions a state of zero displacement and motion is considered.

The system of hyperbolic partial differential equations (1-3) along with initial and boundary conditions are transformed into a system of explicit finite difference equations by replacing the space derivatives by the central difference and time derivative by forward difference.

Results and Analysis

The set of explicit finite difference equations (Elahi [5]) has been solved numerically. Programs were written in FORTRAN IV. Computations were performed on a personal computer PC-AT with RM/FORTRAN compiler. The PC includes the Math Coprocesser 80387 and a hard disk of 40 mb. Graphic work was done by using GEOGRAF VER 4.0.

The tidal elevations for the partial tides M_2, S_2, K_1 *and* O_1 are reproduced, and presented in the form of tidal charts in Fig.1, 2, 3 and 4 respectively. Two amphidromic points appear for the semi- diurnal tides and one amphidromic point appears for the diurnal tides (Table 1). These charts permit the viewer to follow the tidal waves, that is the high water fronts, in forward (or backward) direction.

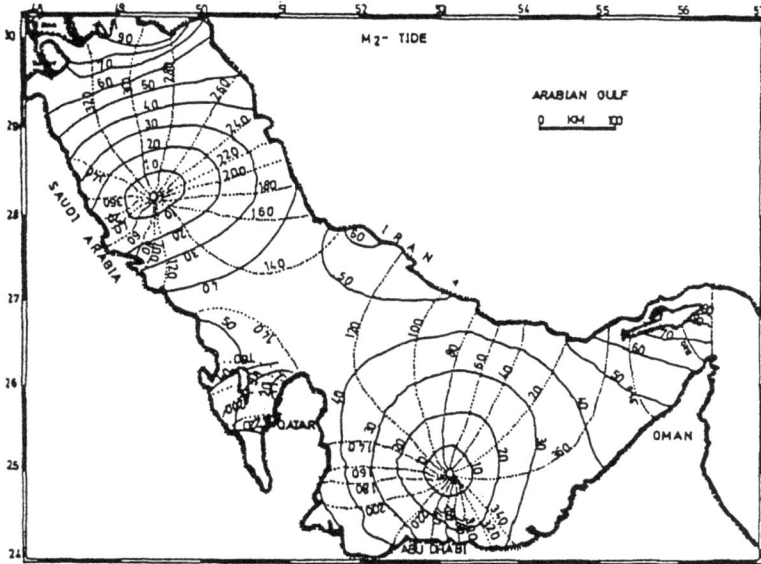

Fig.1: M$_2$ - tide in the Arabian Gulf. Co-range lines
(————) in cm and co-tidal lines (.....) in
degree.

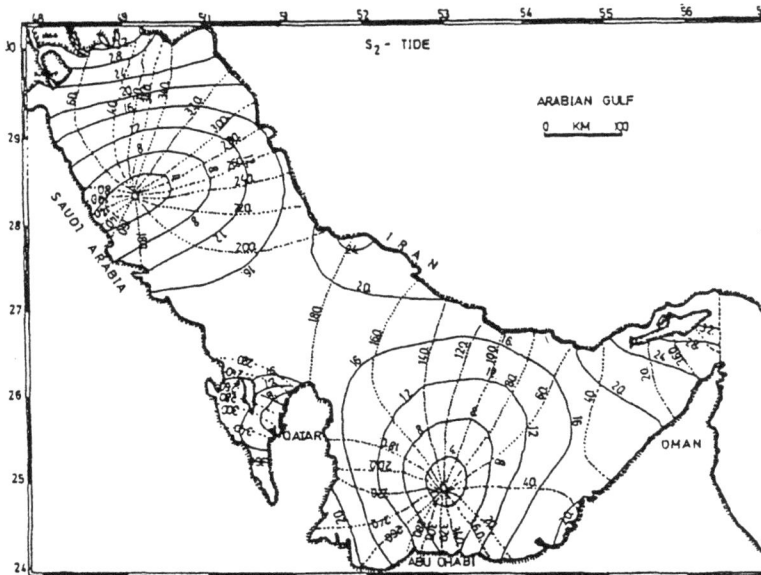

Fig.2: S$_2$ - tide in the Arabian Gulf.

93

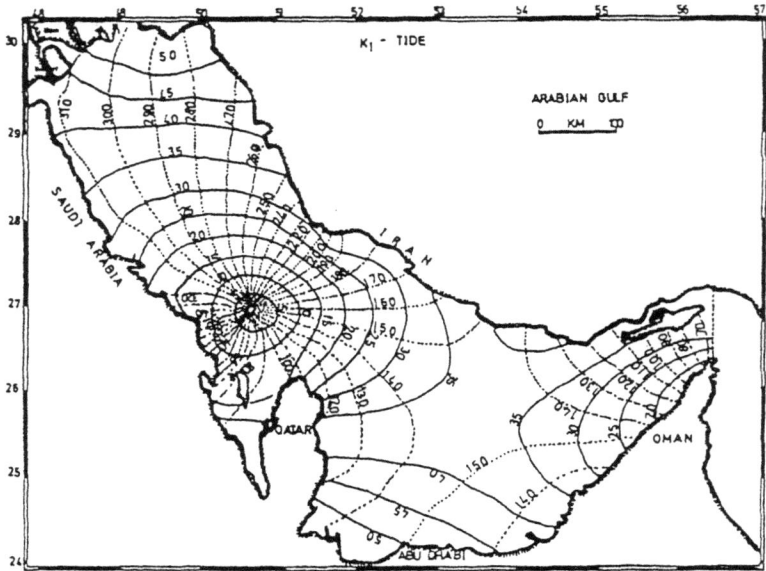

Fig.3: K_1 - tide in the Arabian Gulf.

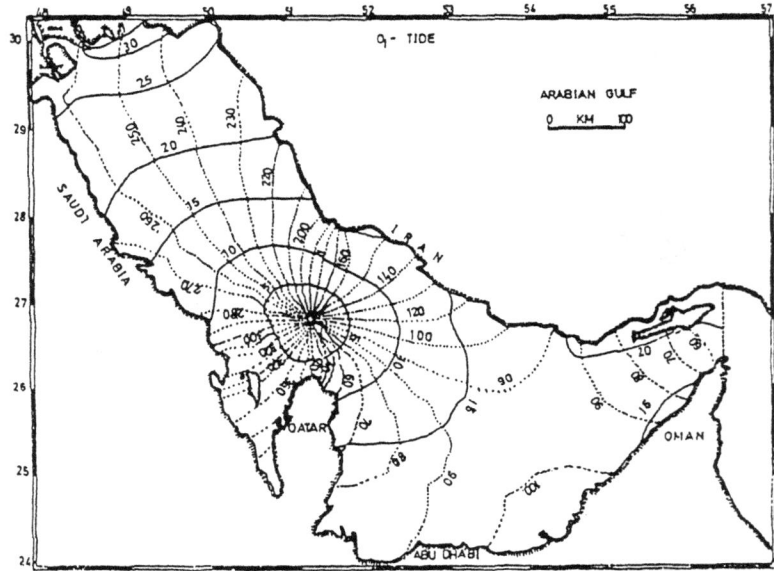

Fig.4: O_1 - tide in the Arabian Gulf.

Table 1. Location of Amphidromic Points

Tide	Location of Amphidromic Points
M_2	$28°.15'N$, $49°.39'E$
	$25°.00'N$, $53°.10'E$
S_2	$28°.25'N$, $49°.10'E$
	$25°.00'N$, $53°.02'E$
K_1	$26°.57'N$, $50°.40'E$
O_1	$26°.50'N$, $50°.10'E$

The locations of the amphidromic points for M_2, S_2 and K_1 tides are compared with the charts constructed from experimental observations of tidal heights and these are in reasonable agreement with each other. Patterns of the co-tidal and co-phase lines are also quite similar. Results are also compared with the observational data of IHO Tidal Constituents Bank at 18 different tidal gauges around the coast of the Arabian Gulf. Around the coast of Bahrain the degree of accuracy of the results is not very encouraging, whereas in the rest of the Arabian Gulf, amplitude and phases are reproduced with maximum errors of $\pm 10cm$, and $\pm 5deg.$ respectively.

Velocity fields for the all four partial tides are reproduced. Velocity fields are mean depth currents. It is not possible to analyse the tidal currents as there is no observational data available in the Gulf. Von Trepka [13] based on numerical model, has given the magnitude of the maximum velocity field of the M_2 - tide. These results are also not so well presented that these can be used for the comparison purposes.

Velocity fields for the M_2 - tide at the end of the first quarter $(T/4)$ and second quarter $(T/2)$, where T is the period of the M_2 - tide, are presented in Figs. 5, 6. Velocities are of the order of 15 cm/sec and do not exceed 40 cm/sec. Velocities are strong in the strait of Hormuz and at the end of the Gulf.

In the middle of the Gulf, velocities are very weak. The Middle part of the Gulf (Fig. 5) has appeared as a source. The currents are moving away from middle Northwestwards towards the Iraqi coast and southeastwards towards the strait of Hormuz.

At the end of the period $\frac{1}{2}T$, the middle part of the Gulf (Fig. 6) has appeared as a sink. The currents are moving from the Iraqi coast and

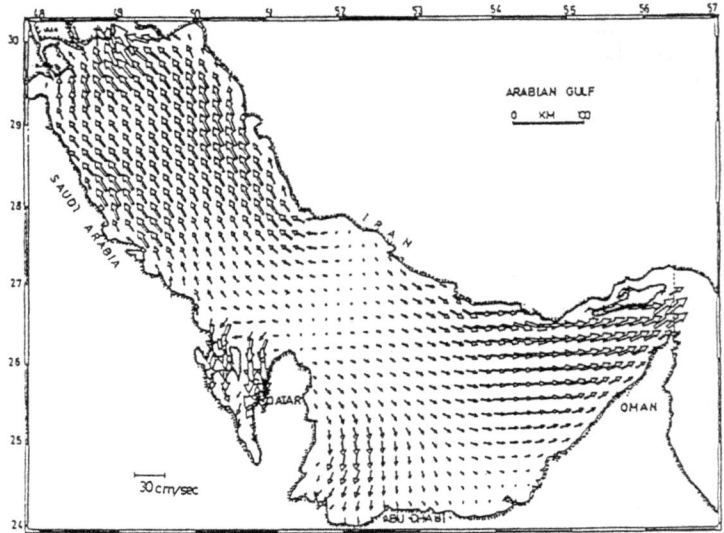

Fig.5. Velocity field of M_2-Tide at the end of the
first quarter.

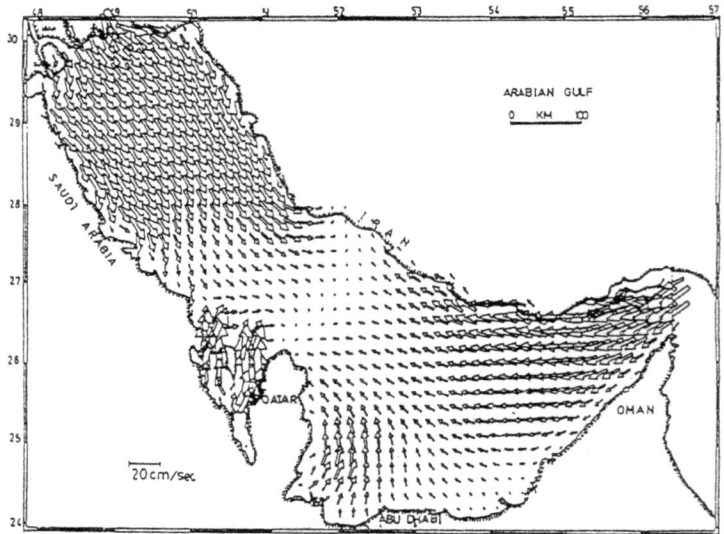

Fig.6. Velocity field of M_2-Tide at the end of the
second quarter.

from the strait of Hormuz to the middle.

References

[1] Beltagy, I. (1980), IMCO/UNEP Workshop on Combating Marine Pollution from oil exploration and transport in the Kuwait Action Plan region, 6 to 10 December 1980, Bahrain.

[2] Danish Hydraulic Institute (1980) Applications of System 21 in the Gulf. Horsholm, Denmark.

[3] Defence Mapping Agency, U.S.A. (1975) Sailing Directions for the Persian Gulf. Hydrographic Center, Washington, D.C., U.S.A.

[4] Elahi, K.Z., (1985) Tidal Charts of the Arabian Sea North of $20°N$. Proc. of the Symp. on Oceanographic modelling of the KAP region, UNEP Report No.70, 1985, PP. 53-62.

[5] Elahi, K.Z., (1991) A numerical model for the tides in the Gulf of Oman, AJSE, Vol. 16, No.24. PP. 173-188, 1991.

[6] Evans-Roberts, D.J.(1979) Tides in the Persian Gulf, Consulting Engineer, Vol. 43, No.6, June 1979, PP. 46-48.

[7] Hughes, P. and Hunter J. (1980) A proposal for a physical oceanography program and numerical modelling of the KAP region. Project for KAP 2/2, Unesco, Paris.

[8] Koske, P. (1972) Hydrographische Verhaltnisse im Persischen Golf Grand von beobachtungen von F.S. Meteor in Fruhjahr 1965. Meteor Forsch Ergnbn, Gebruder Borntraeger, Berlin, PP. 58-73.

[9] Lardner, R., Belen M., and Cekirge, H. (1982), Finite Difference Model for tidal flows in the Arabian Gulf., Comp. and Maths. with applications 8, PP. 425-444.

[10] Lehr, W.J. (1983). A brief survey of oceanographic modelling and oil spill studies in the region. Proc. of the Symp. on Oceanographic modelling of the KAP region, UNEP Report No.70, 1985, PP.175-192.

[11] Le Provost (1985) Models for tides in the Kuwait Action Plan (KAP) region. Proc. of Symp. on oceanographic modelling of the Kuwait Action Plan (KAP) region, UNEP Report No.70, 1985, PP. 205-230.

[12] Tetra Tech. (1977) Marine and Atmospheric surveys, Report for Royal Commission for Jubail and Yanbu, Saudi Arabia.

[13] Von Trepka, L. (1968) Investigation of the tides in the Persian Gulf by means of a Hydrodynamica-numerical model. Proceedings of Symposium on Mathematical-Hydrodynamical Investigations of the Physical Processes in the sea. Institute fur Meereskunde der Universitat Hamburg, PP. 59-63.

9 Hydrodynamic modelling of the River Esk, harbour and coast at Whitby, North Yorkshire

T. H. Toole and J. R. Benn

ABSTRACT

Waste water from the Yorkshire fishing town of Whitby is currently discharged through a short sea outfall into the North Sea. As part of Yorkshire Water's ongoing investment programme, the existing outfall is to be replaced by a new treatment works and long sea outfall, designed to meet the requirements of the EC Bathing Water Directive and the recently introduced Urban Waste Water Treatment Directive.

To enable the design of the new outfall, and also to demonstrate compliance with water quality standards, a mathematical model of the river, harbour and coastline has been developed using the DIVAST and ONDA hydrodynamic modelling packages. The completed model has been used to simulate the tidal characteristics within the bay and harbour and the dilution and dispersion of effluent at various discharge sites. The model includes discharges direct from the outfall and the intermittent flows from the various storm sewage overflows on the sewerage system. Solutes modelled include faecal and total coliforms, ammoniacal nitrogen, biological oxygen demand and dissolved oxygen. The completed model has been verified using survey data and has proved to be a useful integrated design tool incorporating most aspects of the waste water system.

1 Introduction

Whitby is a small fishing port located on the estuary of the River Esk on the North Yorkshire Coast of England. The River Esk is Yorkshire's only salmon river and as such is subject to very demanding water quality limits. The river and adjacent coastline are also popular tourist locations and are used extensively for bathing, water sports and general recreation. Any consent for a new discharge is therefore subject to stringent water quality limits.

The existing sewerage system dates from the nineteenth century and is divided into two sub-systems: a sealed 'low level' system which serves all the eastern side of the town and low-lying parts of the western side, and a 'high level' system serving the areas of the west side above the 7m (25 feet) contour. There are numerous inter-connections between these sub-systems and 26 storm overflows which discharge either directly into the River Esk or other minor watercourses. At present all wastewater flows, except for stormwater overflows, are discharged to a short sea outfall located immediately east of the harbour breakwater with pumping being required during high tide to maintain flow.

2   Scheme Objectives

2.1   EC Bathing Water Directive

The new outfall design is to be consistent with the standards defined in the EC Bathing Water Directive[1]. In addition to the consented continuous discharges, coastal and estuarine water quality is influenced by storm sewerage overflows (SSOs). The regulatory authorities therefore require that appropriate account is taken of storm overflows in the design of schemes to achieve the requirements of the Bathing Water Directive and that the location, available dilution in the receiving waters and/or frequency and duration of operation of the overflows are consistent with achieving the requirements of the Directive. The Directive allows for 5% of samples failing the coliform standards and a proportion of these failures can be allocated to the operation of SSOs. Abnormal weather conditions, unless exceeding a 1 in 5 year return period, would not normally lead to exclusion of failed samples from the compliance total.

2.2   EC Waste Water Treatment Directive

In the Spring of 1991, a new Directive, called the Urban Waste Water Treatment Directive[2] was agreed by the EC member states. The Directive is aimed at bringing about an improvement of the environment through the reduction of pollution of rivers, estuaries and seas caused by the inadequate collection and treatment of urban waste water and some industrial discharges. In the past, along the tidal estuaries and coasts where there was considered to be massive dilution of any waste and treatment works were not considered necessary. Now, under the Urban Waste Water Treatment Directive, this issue is being tackled and waste water treatment works are being built for the major population centres along the Yorkshire Coast. The Directive also lays down requirements for the various types of works, according to several criteria based primarily on the size of population involved and the ecological sensitivity of the river, estuary or sea.

2.3   Proposals for Whitby

Based on the current understanding of the requirements of the new Directives a new compact primary treatment works is proposed for Whitby on a derelict riverside site some 1½ km from the harbour area. The works will be designed to deal with future peak summer loadings. The works will be totally enclosed in a building which will be landscaped to minimise the visual intrusion in such a sensitive area.

Recognising that further treatment may be required in the future sufficient land is being acquired to enable Yorkshire Water to respond to any such change. The sewerage system is being thoroughly investigated and a computer model of the system developed to enable an optimum collection and transfer system to be determined. The new sewerage arrangements will be designed to eliminate the unsatisfactory storm water overflows by the provision of storm detention tanks and pumping systems. Following treatment of the wastewater at the new works the flow will be pumped back to a new long sea outfall. It is envisaged that the existing outfall will be retained for storm water discharges and emergency overflows. The proposals for Whitby which are due for completion by 1998 are estimated to cost in excess of £30 million.

In order to achieve the most effective design for the new outfall, and to demonstrate compliance with the relevant consent limits, Yorkshire Water commissioned Bullen and Partners in May 1991 to develop a mathematical model of the coastline, harbour and river at Whitby. The first phase of this work is now complete but work on more detailed areas is currently continuing.

3 Model Details

3.1 Computational scheme

The area of interest was modelled using the two-dimensional, depth-integrated tidal model DIVAST originally developed by Bradford University[3]. This model has been further developed by Bullen and Partners to include the equations of constituent transport for biological oxygen demand (BOD), ammoniacal nitrogen, dissolved oxygen (DO), and faecal and total coliforms. DIVAST uses a regular mesh of square grids over the area of interest, with the governing hydrodynamic differential equations being re-formatted using Taylor's series and solved for each potentially wet grid cell. For reasons of computational efficiency and for minimum artificial diffusion, a finite difference scheme is adopted. Hence, for each wet grid square, there are four or more governing equations - depending on the number of water quality parameters being modelled - including the equations of continuity, momentum in two mutually perpendicular (x and y) directions and the advective-diffusion equation for each solute. The finite difference equations are then solved implicitly for each grid square at each timestep with the timestep being constrained by the Courant condition[4].

3.2 Computational grids

The coastline between Kettle Ness and Maw Wyke Hole was specified in the numerical model using a regular mesh of 137 x 109 grid squares with a grid spacing of 75m (Figure 1). Nested within this grid was a further 114 x 24 grid square model using a 15m grid of the River Esk from the harbour mouth upstream as far as Whitby bridge (Figure 2). At the corner of each grid square a representative depth below chart datum was included using digitised data from the Admiralty Charts and bathymetric data specially surveyed for the study, with the use of a digital ground modelling package to transform the data onto a regular grid. Bed roughness was specified at the centre of each square. As no data was available concerning bed roughness characteristics, initial model runs were undertaken for assumed Manning roughness coefficients in the range 0.040-0.055. At the calibration stage, bed levels and roughness values were modified slightly, as required to correct interpolation errors and to take into account the local knowledge of fishermen and other boat operators.

Within the computational domain of the model, an outfall location can be specified at any location, and a time-varying flow and loading can be specified to match the diurnal changes which occur in the sewerage system. Also, time-varying coliform die-off rates, expressed as a T_{90} value (i.e. the time taken for 90% of the coliforms to die), can be specified in order to simulate the effect of sunlight on coliform mortality.

3.3 Boundary conditions

The hydrodynamic boundary conditions used in the 75m grid model were derived from a 333m coarse grid coastal model of the Yorkshire Coast developed for Yorkshire Water by the Water Research Centre. Of the three open seaward boundaries, the northern and eastern were read in as flow, with the western boundary specified as a water level. Data on velocities and water levels from the 75m grid model were, in turn, fed into the 15m grid model of the harbour at the common boundaries between the two models. These boundaries consisted of the open harbour entrance and the three overflow weirs in the harbour breakwaters which allow flow to enter and leave the harbour at high tide. The remaining open boundary, within the 15m grid model, was at Whitby bridge. Here, data on flow and level were fed in for specified fluvial and tidal flow conditions from a one-dimensional river model of the River Esk for the reach between the bridge and the tidal limit at Sleights Weir, some 2.3 km upstream. For this river model, the ONDA[5] package was used and the completed model was verified using existing river flow and level data.

3.4 Hardware Platforms

The models were run on both a Sun Sparcstation workstation running under UNIX and a 486 personal computer running under OS/2. Graphical plots of model results were produced using a package developed by Bullen and Partners based on the AutoCAD drawing software.

4 Model verification

4.1 Hydrodynamic data

The 75m and 15m grid models were calibrated by varying bed friction and momentum diffusion coefficient so as to produce the best agreement in range and phase between modelled and observed water levels along the coastline and river. Detailed verification of the model hydrodynamics was subsequently carried out using data from a specially commissioned marine survey over both a spring and neap tide in September and October 1991. During this survey, water level was recorded using a transducer monitor installed in Whitby harbour and the existing Proudman Oceanographic Laboratory gauge at the Harbour Master's Office. Current direction and velocity were measured at several locations, including installation of multiple station direct reading current meters at four points; three within the 75m computational domain and one at the harbour entrance on the boundary between the 75m and 15m grid models (Figures 1 and 2). Readings were taken at 30 minute intervals over a 13 hour period of both spring and neap tides. Tidal current velocity and direction were also recorded for a period of 6 weeks using self-recording current meters, but this data was unfortunately lost during adverse weather conditions before the information could be recovered. At the same time as the current measurements, wind speed and direction were recorded at two locations - at the coastguard station (to measure onshore wind) and in the harbour (to measure wind in the Esk valley). Data on flow and velocity were also collected on the River Esk at Sleights Weir and at Whitby Bridge. To establish the general direction and strength of surface currents, drogue tracking was carried out by noting the track followed by 19 cruciform vanes over 13 hour periods of both spring and neap tidal cycles. The cruciform vanes were 1m square with the centre located 1.5m below the sea surface. They were released at 2 hourly intervals from the outfall site, in groups of three.

Table 1 shows the standard error between the observed water levels and velocities and the model predictions. The fit between water level and phase is generally very good and work is currently being undertaken on improving the boundary conditions to improve the correlation. Comparison of velocity is a more severe test of model performance and the results in Table 1 and Figure 3 show good correspondence at the five measurement sites with an average standard error of less than 11.2%. The fit between observed and model data is poorest at site M4 at the harbour entrance which is largely due to measurement error rather than particular model deficiencies. Gathering field data at this location was difficult due to the presence of the harbour walls and the effect of boat traffic.

4.2   Tracer surveys

Coincident with the collection of the hydrodynamic data, several tracer surveys were carried out to establish plume dispersion from a possible site of the new outfall. The first involved the use of Rhodamine WT dye. Two releases of dye were carried out on a Spring and a Neap tide. The raw data were corrected by reference to a drogue moving with the dye patch. The tracer used was 10 litres of *Bacillus globigii* spores mixed with Rhodamine B dye to aid visual tracking. It was injected at the site of the proposed new outfall at low tide. When the measured plume was compared with the model results for the same outfall location correlation between the two was found to be good.

TABLE 1   *Comparison of hydrodynamic survey results with model predictions*

| Location | | Standard error between observed and model results (%) | | |
		Water level	Current velocity	Current direction
Spring Tide				
M1	NW coast	-	18.9	10.3
M2	N coast	-	3.2	14.5
M3	Eastern boundary	-	1.6	16.0
M4	Harbour Entrance	4.8	33.0	9.8
M5	Whitby Bridge	6.3	4.3	19.4
Neap Tide				
M1	NW coast	-	12.2	12.3
M2	N coast	-	6.4	12.3
M3	Eastern boundary	-	6.3	16.0
M4	Harbour Entrance	3.3	21.0	26.9
M5	Whitby Bridge	7.8	5.3	14.8
Average		5.6	11.2	15.2

4.3   Coliform sampling

Numerous water samples have been taken and tested for faecal and total coliforms in order to establish the effect of the current outfall discharges on water quality. Flows at the pumping station were recorded, coliform loadings were measured, and water quality was continuously monitored. This data has been used to verify the coliform diffusion and decay parameters in the model. Peak coliform counts occur on the beaches west of the harbour at low tide and are highest when low tide is more or less coincident with the peak flowrate through the outfall between 11-12 am. The results from this exercise were then used to specify suitable die-off rate (T_{90} values) and diffusion coefficients for the runs with the new outfall. The best-fit values were subsequently adopted in the model runs for the new outfall locations.

5   Model Results

At the time of writing this paper, the model is being used to help define the optimum length and position of the new outfall. The position of the outfall is constrained by the sea-bed topography which drops away sharply offshore and consists in places of loose boulders and rock ridges. Preliminary runs over spring and neap tidal cycles indicate that the new outfall will need to be greater than 900m in order to reach the strong north-west/south-east offshore flow (Figures 4 and 5). An outfall shorter than this will result in poor dispersion of the sewage plume due to the eddies that form in-shore. It is worth noting that such important effects could not have been picked up if a coarse model grid had been used, or if the effects of wetting and drying of the coastline were not modelled effectively.

The model is currently being run for a variety of possible outfall flows, wind conditions, river flow rates and tidal cycles, in order to establish the worst cases for plume dispersion. The worst case condition has been found to be when the peak diurnal flow coincides with mid-tide on a spring cycle or low tide on a neap cycle. The model is also being used to determine the effects of various pumping regimes on dispersion, and the effect of different treatment levels on water quality. The proposed new outfall, when complete, will lead to a significant improvement in water quality as a result of the continuous discharges from the treatment works. However, the study of overall water quality would not be complete without consideration of the 26 overflows within the Whitby sewerage system. The models constructed for Whitby are now being developed further to include the intermittent discharges from these overflows and are being integrated with the separate work being undertaken on developing a WALLRUS computer model of the sewerage system. The water quality parameters currently being considered are BOD, DO, ammoniacal nitrogen, and faecal and total coliforms.

6   Conclusions

The design process has benefited immensely from the use of mathematical models both as a design tool and to provide data which can be assessed by the regulatory authorities. The study has also demonstrated the need for careful choice of the size of the computational grid where inshore hydrodynamic and wetting and drying effects are important. The 75 m grid adopted, together with careful attention to the interpolation of bed levels and the wetting and drying process, has enabled the accurate modelling of the significant in-shore eddy effects in the region of the proposed new outfall. Without such detailed modelling, choice of a new outfall location with the required degree of confidence would be extremely difficult.

ACKNOWLEDGEMENTS

The Authors would like to thank Yorkshire Water for permission to use data from the Whitby study in this paper and to Bullen and Partners for assistance in its preparation. The opinions, views and reasoning are those of the authors and do not necessarily represent those of Yorkshire Water or Bullen and Partners.

REFERENCES

[1]   Council of the European Communities, (1975), *Bathing Water Directive*, 76/160/EEC, Official Journal of the European Communities, 8 December 1975.

[2]   Council of the European Communities, (1991), *Urban Waste Water Treatment Directive*, 91/271/EEC, Official Journal of the European Communities, 21 May 1991.

[3]   Falconer, R.A., (1986), *"A two-dimensional mathematical model study of the nitrate levels in an inland natural basin"*, International Conference on Water Quality Modelling in the Natural Environment, Bournemouth, England, 10-13 June 1986.

[4]   Falconer, R.A., (1991), *"Review of modelling flow and pollutant transport in hydraulic basins"*, Proceedings of First International Conference on Water Pollution: Modelling, Measuring and Prediction, Southampton, UK, Computational Mechanics Publications, September 1991, pp.519-531.

[5]   *Program ONDA: Open Channel Network Modelling User Manual* (1990), Version 3.2, Sir W Halcrow & Partners Limited, Swindon, UK.

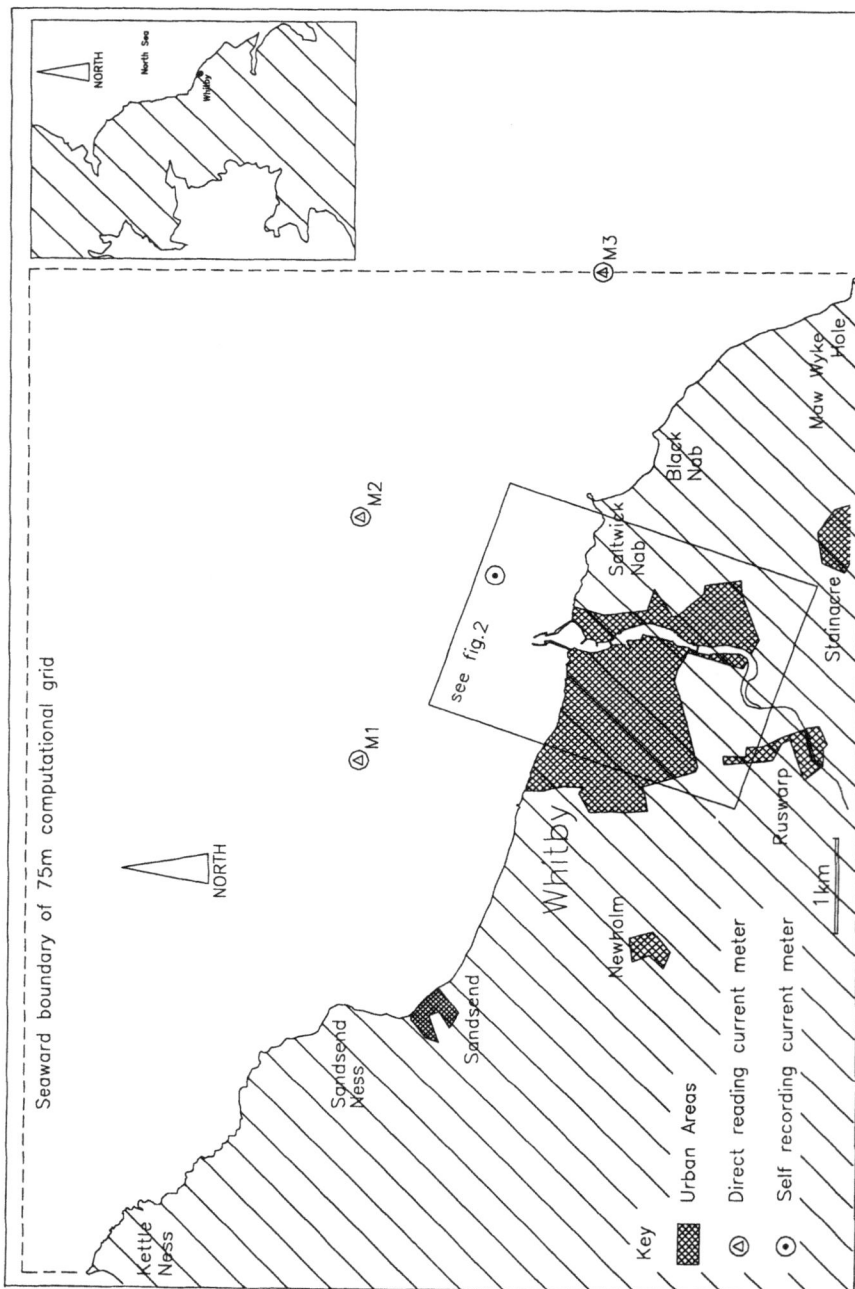

FIGURE 1   *Location Plan of Whitby*

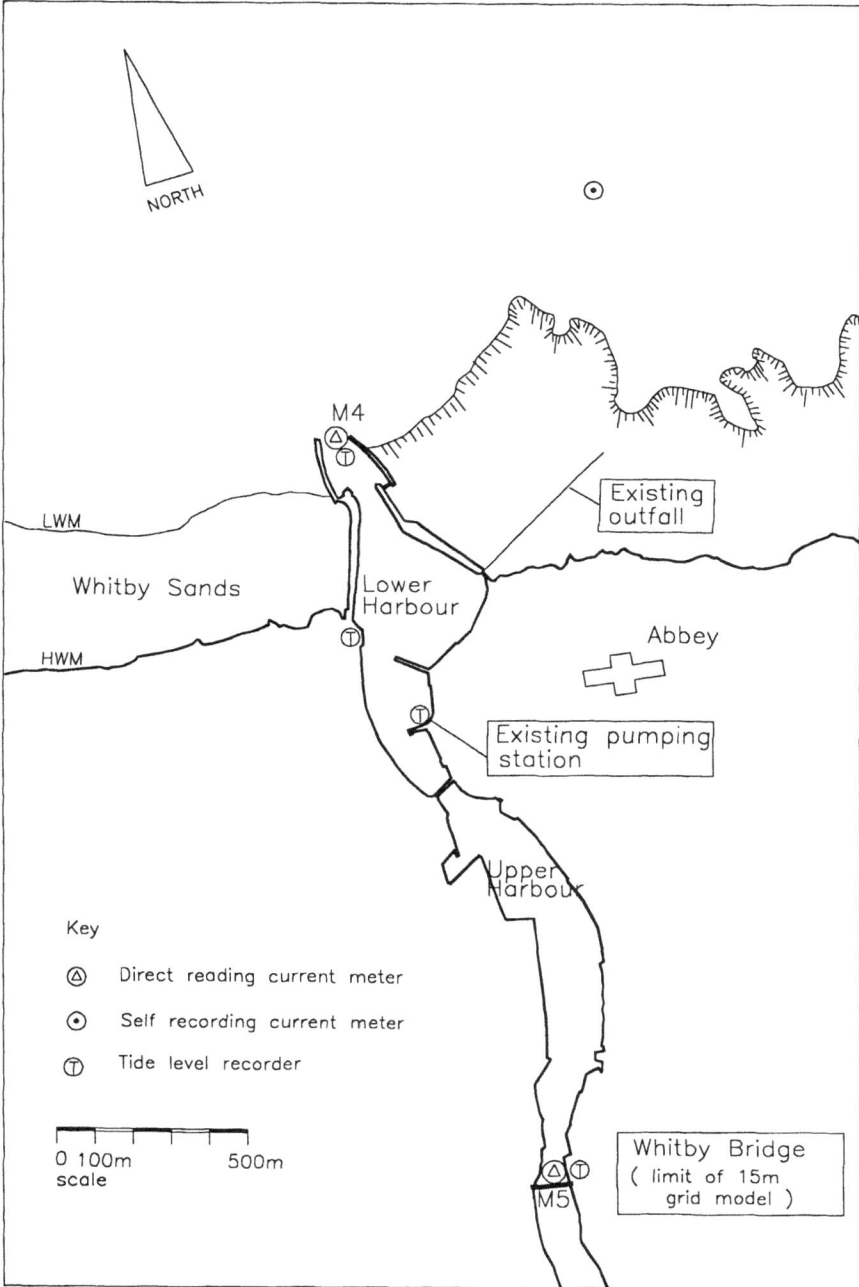

FIGURE 2 *Detailed plan of Whitby Harbour and the River Esk*

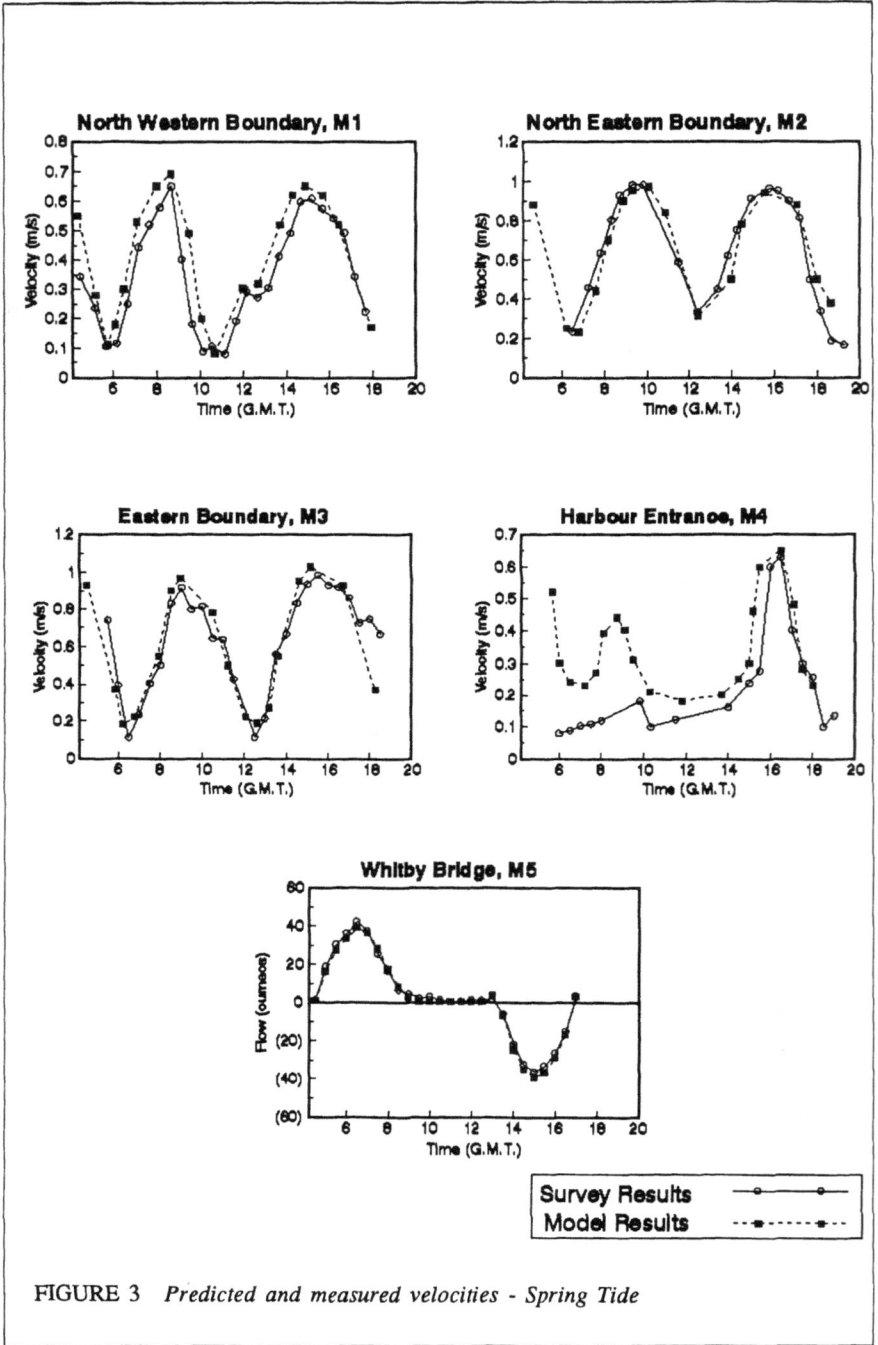

FIGURE 3 *Predicted and measured velocities - Spring Tide*

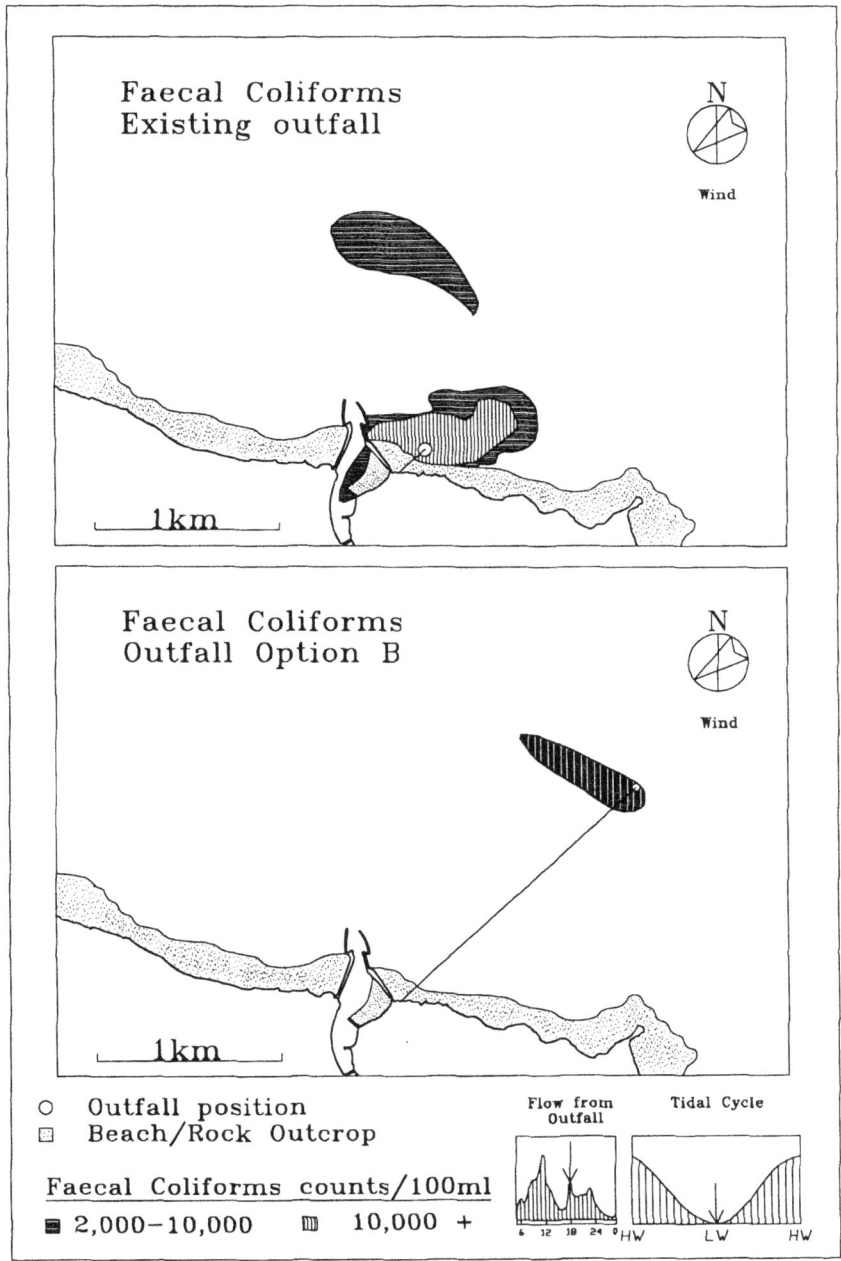

Faecal Coliforms
Existing outfall

N

Wind

1km

Faecal Coliforms
Outfall Option B

N

Wind

1km

○ Outfall position
▣ Beach/Rock Outcrop

Faecal Coliforms counts/100ml

▰ 2,000–10,000   ▥ 10,000 +

Flow from
Outfall

Tidal Cycle

6 12 18 24 0
HW   LW   HW

FIGURE 4   *Comparison of discharge plumes for existing short sea outfall and a*
new long sea outfall: spring tide, low water

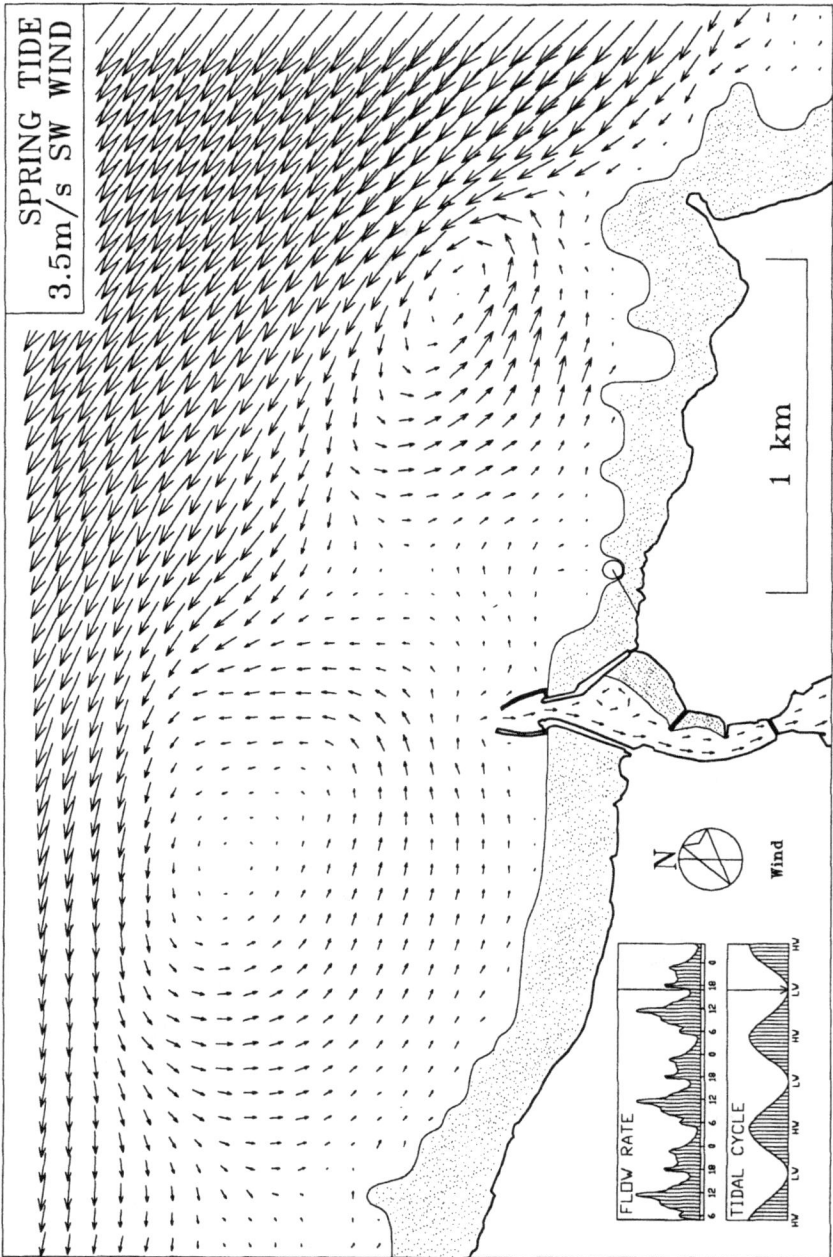

FIGURE 5   *Current structure in the vicinity of the existing outfall - Spring Tide*

10 Tidal exchange mechanism in enclosed regions

K. Murakami

ABSTRACT

This paper describes the investigations of tidal exchange mechanisms in enclosed coastal regions. Enclosed coastal regions are contaminated because of small tidal exchange between inner bay and outer ocean waters. When we construct some marine structures such as tsunami breakwaters, we must investigate the influence on water quality environment due to the structures. From the results of physical and numerical experiments, it is pointed out that the influence of the structure located at the entrance of the bay on tidal exchange has two aspects. Tidal exchange is promoted by horizontal circulation generated by the structure, but is decreased by vertical circulation suppressed by the structure.

1 Introduction

Enclosed coastal regions are utilized for many purposes such as ports, fishery, marine leisure, etc., because the region is tranquil. Many big cities and metropolitans are located in hinterland of the enclosed coastal regions. Since the great amount of sewage is discharged into the region, and tidal exchange between the region and the ocean waters is small, water quality in enclosed coastal region is generally contaminated.

Japanese coastal area was often damaged by tsunami waves and typhoon storm surges. In order to protect the coastal area from these kind of natural disasters, various break- waters were constructed at the entrance of the region. Since the area of entrance section of the bay is decreased by the breakwater, many persons are afraid to degrade the water quality in the region. Prior to the construction of the breakwater, we must investigate the influence of the breakwater on water exchange as well as on tsunami wave height.

Tidal exchange is caused by turbulent mixing, horizontal circulation, and vertical circulation. Effect of these mechanism on tidal exchange is depend upon the flow field of interested area. In the paper, the author described the tidal exchange mechanism in enclosed coastal regions and how to consider the influence of marine structures on tidal exchange between enclosed bay and ocean waters.

2. Tidal exchange mechanism

Tidal exchange is very important phenomenon to dilute the contaminated bay water with clean ocean water. Tidal exchange mechanisms are schematically explained as shown in Fig.1(Murakami,1988). In the figure, the mechanism (a) is caused by turbulent mixing, (b) is caused by horizontal circulation, and (c) is caused by vertical circulation. Horizontal circulation flow has an effect on tidal exchange between two water bodies. The horizontal circulation is caused by tidal residual flow, which is generated by the non-linearity property of tidal current. When the breakwaters are constructed at the entrance of the region, tidal velocity is increased because of the reduction of cross section area. Therefore, horizontal circulation is generated behind the breakwater by the relatively strong tidal velocity. The tidal exchange mechanism caused by horizontal circulation is explained as tidal pumping by Fischer et al.(1979).

Vertical circulation flow also has an effect on tidal exchange between two water bodies. The vertical circulation is generated by wind driven current or density current. When we consider the influence of marine structures such as breakwater on tidal exchange, we must investigate the flow field of interested area such as turbulent flow, horizontal circulation and vertical circulation.

Hydraulic model experiment and numerical simulation techniques are commonly used as the prediction method of influence due to future project such as breakwater construction on tidal exchange behavior in enclosed coastal regions.

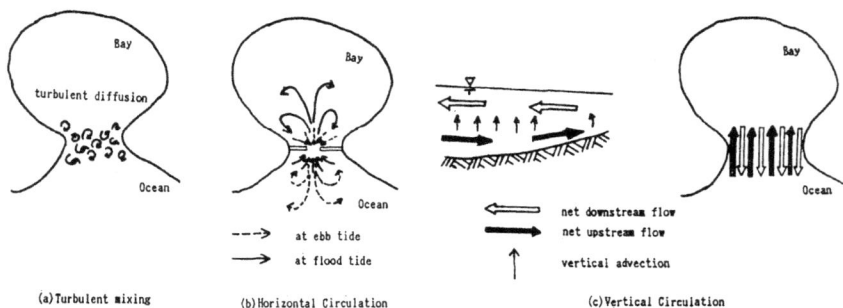

Fig.1 Tidal exchange mechanisms, (a):turbulent mixing,
      (b):horizontal circulation, (c):vertical circulation

Table 1 Hydraulic model experiment conditions

Case	Experimental Conditions
Run-A	Without breakwater and without density current
Run-B	With breakwater and without density current
Run-C	Without breakwater and with density current
Run-D	With breakwater and with density current

3. Hydraulic model experiment

The investigation of tidal exchange mechanism by horizontal circulation and vertical circulation was performed by hydraulic model experiment (Murakami & Shirai,1988). Four experimental runs as shown in Table 1 were carried out. For the experiment without density difference, freshwater is used as both inner bay water and outer ocean water. For the experiment with density difference, seawater drawn from the Tokyo Bay is used as outer ocean water, and composite mixture of seawater and freshwater is used as inner bay water.

Experimental procedure of tidal exchange by hydraulic model is shown in Fig.2. At the first stage, water is filled in the model up to mean water level. Then the water is divided into inner bay water and outer ocean water by the barrier board installed at the cross section of the bay entrance. Fluorescein sodium dye is used as the tracer of contaminated substance. The dye is put into the inner bay water, and mixed up suffi-ciently until the water becomes homogeneous concentration. In case of experiment with

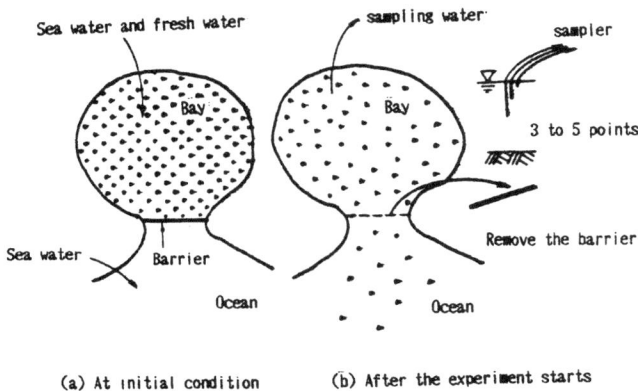

(a) At initial condition   (b) After the experiment starts

Fig.2 Tidal exchange experimental procedure by physical model

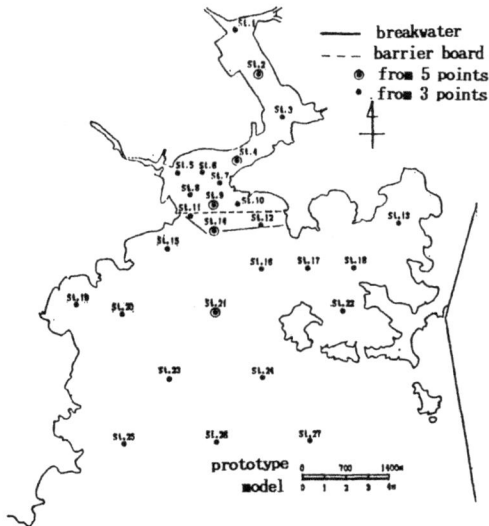

Fig.3 Locations of sampling points, breakwaters and barrier board

density difference, the inner bay water is mixed well with freshwater up to homogeneous salinity, which is 0.05% less than the one of outer ocean water.

At the second stage, the barrier board is removed from the cross section which divides into two regions. At the same time, the operation of pneumatic tide generator starts. After that the mixing between two water bodies begins due to tidal current and turbulent diffusion.

After the experiment starts, water samples are picked up from the model basin to measure the distributions of dye concentration in every several tidal cycles.

The measurements of tidal current for whole area are conducted by float trajectories which are taken by camera.   At the cross section of the entrance, where the tidal velocity is relatively high, the velocity is measured by ultra-sonic current meter. Locations of water sampling and the barrier board are shown in Fig.3. In the figure, double circle means the sampling stations from five various depth, black circle means the sampling stations from three various depth, the real lines show the location of tsunami breakwater, and the dotted line shows the location of the barrier board.

Tidal ranges of each experimental runs are obtained of almost the same values. This fact means that the net volume of water entering during flood tide is almost the same notwithstanding the experimental conditions of breakwater existence or the density difference.

Figures 4 and 5 show the trajectories of floats for two tidal cycles in case of without density current. From the figures, it is shown that the horizontal circulation is generated behind the breakwaters. In case of with density current, the flow pattern behind the breakwater is similar. Table 2 shows the maximum current velocity measured by the current meter installed at the center between two breakwaters. The maximum velocity in case of with breakwater is much higher than the one of without breakwater. This high velocity current generates the circulation flow behind the breakwaters.

114

Fig.4 Float trajectories
(without breakwater Run-A)

Fig.5 Float trajectories
(with breakwater Run-B)

Table 2 Maximum tidal velocity at the entrance measured by ultra-sonic current meter
for each experimental runs

Exp. Case	Density Current	Break water	Flood (cm/s)	Ebb (cm/s)
Run-A	without	without	1.6	0.8
Run-B	without	with	10.5	8.9
Run-C	with	without	2.9	2.8
Run-D	without	with	6.0	13.8

Under the experimental conditions, tidal exchange hydraulic experiments were carried out. Figure 6 shows the decay curve of the averaged dye concentration whole over the inner region in case of without density difference drawn by real lines and with density difference drawn by dotted lines.   According to the figure, it is evident that the tidal exchange with density current is much larger than the case of without one. As to the case of without density current, tidal exchange is promoted due to the existence of the breakwaters at the bay entrance. This is depend upon the horizontal circulation generated by high speed current behind the breakwaters. As to with density current, tidal exchange is decreased due to the existence of the breakwaters. This is mainly due to the reduction of cross sectional area of the bay entrance.

Thus, the existence of breakwaters at the entrance of the bay promotes the tidal exchange by horizontal tidal residual circulation, but decreases the exchange owing to suppressed vertical circulation.

4. Numerical simulations

Similar considerations were carried out by numerical simulations. For hydraulic model experiments, initial condition is not realistic situation, because the tidal velocity is zero for the whole computational area and the salinity distribution is changed suddenly at the cross section.   And the control of boundary condition for salinity is very difficult. From the above mentioned reasons, the author carried out similar

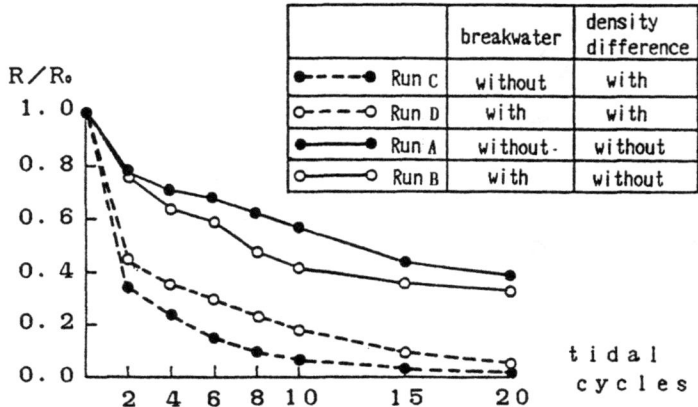

		breakwater	density difference
●----●	Run C	without	with
○----○	Run D	with	with
●----●	Run A	without·	without
●———○	Run B	with	without

Fig.6 Comparisons of the decay curves of averaged dye concentrations obtained by hydraulic model experiments

considerations by numerical simulations. Mathematical model used here is ADI method developed by J.J. Leendertse. In order to consider the tidal exchange mechanism of horizontal circulation, the depth averaged one layer model is used here. Computational conditions are shown in Table 3.

Figure 7 shows the computational result of tidal current in case of without breakwaters, and Fig. 8 shows the one in case of with breakwaters. According to these figures, horizontal circulation is also generated behind the breakwaters. Figure 9 shows the comparison of the decay curves of averaged concentrations calculated by one layer model between the case of large horizontal mixing coefficient and the small one. The magnitude of horizontal circulation radius is dependent upon the horizontal eddy viscosity. If the horizontal mixing coefficient becomes large, horizontal circulation radius becomes small. Then the decay rate of the averaged concentration in a bay slows down owing to small

Table 3 Computational conditions of tidal current and substance diffusion by one layer model

Items	Computational Conditions
Area	as shown in Figs. 7 and 8
Grid interval	$\Delta S = 150m$
Time interval	$\Delta t = 90s$ for current
	$\Delta t = 360s$ for diffusion
Manning Roughness	$n = 0.026$
Horizontal mixing coefficient	$E_h = 1.8 \times 10^3 \, cm^2/s$
	$E_h = 1.8 \times 10^4 \, cm^2/s$
Diffusion coef.	$K_x = 1.0 \times 10^4 \, cm^2/s$
Initial condition	$C_a = 10$ ppm for inner bay
	$C_a = 0$ ppm for ocean
Boundary cond.	$C_b = 0$ ppm for open boundary

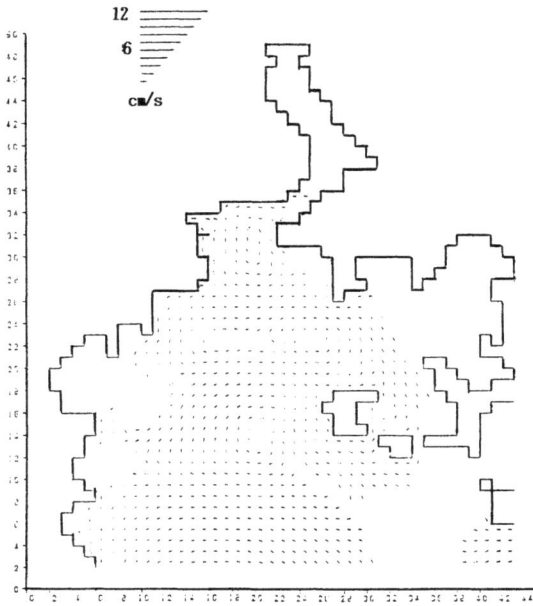

Fig.7 Tidal residual currents calculated by one layer model
(without breakwater)

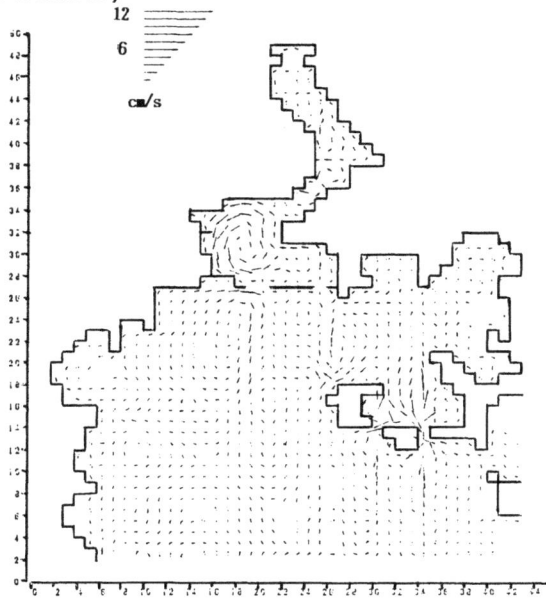

Fig.8 Tidal residual currents calculated by one layer model
(with breakwaters)

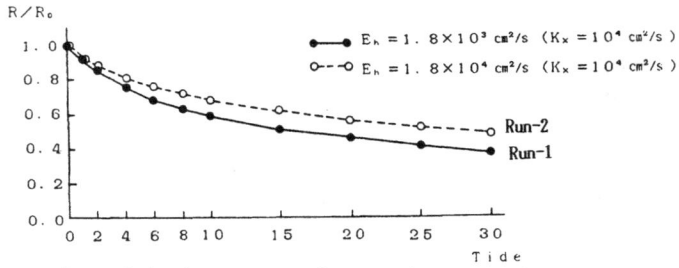

Fig.9 The comparison of the decay curves of averaged concentrations
(Run-1:E_h=1.8x10^{3}cm^2/s, Run-2:E_h=1.8x10^{4}cm^2/s)

horizontal circulation.

From the numerical model experiments, the following results are obtained. When the horizontal mixing coefficient is large, the tidal exchange ratio in case of with breakwater is smaller than the one of without breakwater. When the horizontal mixing coefficient is small, on the other hand, the ratio in case of with breakwater is larger than the one of without breakwater.

5.Discussion

By using the hydraulic model experiment and numerical simulation, tidal exchange mechanisms in enclosed coastal region were investigated. The influence of breakwater on tidal exchange has two aspects in relation with density current condition. The author tried to explain this contradict result of the influence of breakwater on tidal exchange by using the concepts as shown in Fig.1. Tidal exchange mechanisms by hydraulic model experiment are assumed as follows.

In case of Run-A, tidal exchange is caused by turbulent diffusion (a) only. In case of Run-B, tidal exchange is caused by turbulent diffusion (a) and horizontal circulation (b), Run-C is caused by turbulent diffusion (a) and vertical circulation (c), and Run-D is caused by turbulent diffusion (a), horizontal circulation (b) and vertical circulation (c). Furthermore, the author assumes that the tidal exchange volumes by turbulent diffusion and vertical circulation are proportional to the cross sectional area of the entrance. According to these assumptions, the amount of tidal exchange volume of each experimental runs is expressed as following equations.

Run-A : $q_1 = \alpha \times Q$ .... (1)

Run-B : $q_2 = (\alpha \times s'/S + \beta) \times Q$ .... (2)

Run-C : $q_3 = (\alpha + \gamma) \times Q$ .... (3)

Run-D : $q_4 = (\alpha \times s'/S + \beta + \gamma \times s'/S) \times Q$ .... (4)

where, α is the tidal exchange ratio by turbulent diffusion, β is the ratio by horizontal circulation, γ is the ratio by vertical circulation, Q is the volume of water entered into a bay during flood tide or vice versa, s' is the cross sectional area of the

118

Table 4 Averaged residence time τ_d and empirical constants A,B in Eq.(5)

Case	A	B	τ_d (days)
Run–A	0.143	0.63	15.3
Run–B	0.201	0.57	13.2
Run–C	0.688	0.58	1.3
Run–D	0.432	0.62	2.7

entrance in case of with breakwater, and S is the area of without breakwater. The decay curves are expressed approximately by Eq.(5).

$$r(t) = \exp\{-A \times t_d{}^B\} \qquad\qquad \cdots\cdots (5)$$

where, r(t) is the remnant function, t_d is the number of tidal cycles, A, B are empirical constants. Averaged residence time τ_d (Takeoka,1984) expressed by Eq.(6) is obtained in Table 4 of each runs.

$$\tau_d = \int_0^\infty r(t)dt \qquad\qquad \cdots\cdots (6)$$

From the table, tidal exchange ratio in each runs can be estimated. Substituting these exchange ratios into Eqs.(1) to (3), exchange ratios of each mechanisms are obtained as α =0.199 from Run-A, β =0.198 from Run-B, and γ =0.443 from Run-C. These values are substituted into Eq.(4), the tidal exchange ratio in case of Run-D is estimated as 39.1%. This value is relatively close to the value of 43.6% obtained by the result of Run-D.

From these considerations and very rough assumptions, different tidal exchange behaviors between with density current and without one are explained successfully.

6.Concluding remarks

In the paper, the author investigated the tidal exchange mechanisms of horizontal circu-lation and vertical circulation. Especially, the influence of breakwaters located at the bay entrance on tidal exchange is examined. Conclusions obtained from the considera-tions are summarized as follows:
1. The influence of the breakwaters located at the bay entrance on tidal exchange has two aspects. As to the case of without density current, the existence of breakwaters promotes the tidal exchange, because the horizontal circulation is generated behind the breakwaters. On the other hand in case of with density current, tidal exchange is decreased due to the existence of the breakwaters, because the vertical circulation is suppressed by the breakwaters.
2. According to very rough assumptions, this contradict results of influence of break water on tidal exchange are explained successfully.
3. When we predict the influence of marine structures on tidal exchange, we must consider both configuration conditions and detailed flow field conditions.

7.Acknowledgements

The author would like to express his appreciation to Mr. Masao Shirai, formerly the member of the Environmental Hydraulics Laboratory of the Port and Harbour Research Institute, who carried out much of the experimental work. The author also thanks the Third District Port Construction Bureau, Ministry of Transport for their support of the project.

References
[1] Fisher,H.B.,List,E.J.,Koh,R.C.Y.,Imberger,J., and Brooks,N.H.,(1979),Mixing in Inland and Coastal Waters, Academic Press, pp.229-278.
[2] Murakami,K.,(1988), Hydraulic Model Experiments on Water Exchange foe Enclosed Inner Bay, Int'l Symp. on Scale Modeling, the Japan Society of Mechanical Engineers, Tokyo, pp.301-307.
[3] Murakami,K. and Shirai M.,(1988), Investigations on water exchange with density current by hydraulic model experiment, Note of the Port and Harbour Research Institute, No.625, pp.1-29.(in Japanese)
[4] Takeoka,T.,(1984), Fundamental Concepts of Exchange and Transport Time Scale in Coastal Sea, Continental Shelf Research, Vol.3, pp.311-326.

11 Three-dimensional layer averaged model for tidal flows

M. Hall, K. Shiono and R. A. Falconer

ABSTRACT

Details are given of the development and preliminary application of a three-dimensional numerical hydraulic model for predicting tidal flow fields in nearshore coastal and estuarine waters. The numerical model is implicit in form and has proved to be computationally more efficient than a number of other three-dimensional models documented in the literature. The model has been applied to a laboratory rectangular harbour and preliminary results have shown good agreement between the laboratory measured and numerically predicted velocities.

1 Introduction

As two-dimensional numerical hydraulic models are increasingly used for tidal current predictions, and relatively cheap computers become more powerful, a widespread need has arisen for efficient three-dimensional hydrodynamic tidal models. Existing depth integrated numerical models are not appropriate for predicting three-dimensional currents, such as those induced by wind shear or where stratification effects are important, and many existing three-dimensional models are not computationally efficient for relatively shallow coastal and estuarine flows.

Details are given herein of a refined implicit three-dimensional hydrodynamic model which has been applied to a laboratory model rectangular harbour (see Figure 1), with an asymetric entrance and a horizontal bed. Simulation results suggest that the new three-dimensional model is more efficient and has greater stability properties than a range of other three-dimensional models documented in the literature and tested to date. Particular emphasis has been focused on the numerical treatment of the advection terms and the vertical diffusion of momentum. Details are also given of the experimental programme undertaken to determine the velocity profiles at certain locations within the model harbour, using a two component laser doppler anemometer. A comparison of numerically predicted and experimentally measured velocity profiles has been undertaken, with comparisons being made for various eddy viscosity variations.

2   Mathematical Model

The mathematical flow model consists of two modules, including the two-dimensional depth integrated and three-dimensional layer integrated equations respectively. A Liebnitz integration over the vertical of the Navier-Stokes and continuity equations gives the depth integrated hydrodynamic equations in the x and y directions respectively:-

$$
\frac{\partial UH}{\partial t} + \int_{-h}^{\eta} \frac{\partial \bar{u}^2}{\partial x} dz + \int_{-h}^{\eta} \frac{\partial \bar{u}\bar{v}}{\partial y} dz + \frac{gH \partial \eta}{\partial x}
$$

$$
-2.0 \left(\frac{\partial \epsilon \frac{\partial UH}{\partial x}}{\partial x} \right) - \frac{\partial \left(\epsilon \frac{\partial UH}{\partial y} + \epsilon \frac{\partial VH}{\partial x} \right)}{\partial y} + \frac{\tau_{bx}}{\rho} = 0 \tag{1}
$$

$$
\frac{\partial VH}{\partial t} + \int_{-h}^{\eta} \frac{\partial \bar{v}\bar{u}}{\partial x} dz + \int_{-h}^{\eta} \frac{\partial \bar{v}^2}{\partial y} dz + \frac{gH \partial \eta}{\partial y}
$$

$$
-2.0 \left(\frac{\partial \epsilon \frac{\partial VH}{\partial y}}{\partial y} \right) - \frac{\partial \left(\epsilon \frac{\partial VH}{\partial x} + \epsilon \frac{\partial UH}{\partial y} \right)}{\partial x} + \frac{\tau_{by}}{\rho} = 0 \tag{2}
$$

$$\frac{\partial \eta}{\partial t} + \frac{\partial UH}{\partial x} + \frac{\partial VH}{\partial y} = 0 \tag{3}$$

where U and V = depth average velocity components in x and y directions respectively, H = total depth of flow, t = time, $\bar{u}$ and $\bar{\upsilon}$ = layer average velocity components in x and y directions respectively, g = gravitational acceleration, η = water surface elevation above datum, h = bed elevation below datum, ϵ = depth average horizontal eddy viscosity, calculated using the method outlined in Falconer(1), τ_{bx}, τ_{by} = bed shear stress components in x and y directions respectively, and ρ = water density. In a similar manner the Navier-Stokes equations are integrated over each individual layer to give the layer integrated hydrodynamic equations accordingly:-

$$\left(\frac{\partial \bar{u} Z}{\partial t}\right)_k + \left(\beta\left(\frac{\partial \bar{u} Z^2}{\partial x} + \frac{\partial \bar{u}\bar{\upsilon} Z}{\partial y}\right)\right)_k + \frac{gZ \partial \eta}{\partial x} - \left(2.0\left(\frac{\partial \bar{\epsilon} \frac{\partial \bar{u} Z}{\partial x}}{\partial x}\right) + \frac{\partial\left(\bar{\epsilon}\frac{\partial \bar{u} Z}{\partial y} + \bar{\epsilon}\frac{\partial \bar{\upsilon} Z}{\partial x}\right)}{\partial y}\right)_k$$
$$-\tau_{xz_k} + (w\bar{u})_{k-\frac{1}{2}} - (w\bar{u})_{k+\frac{1}{2}} = 0 \tag{4}$$

$$\left(\frac{\partial \bar{\upsilon} Z}{\partial t}\right)_k + \left(\beta\left(\frac{\partial \bar{\upsilon} Z^2}{\partial y} + \frac{\partial \bar{\upsilon}\bar{u} Z}{\partial x}\right)\right)_k + \frac{gZ \partial \eta}{\partial y} - \left(2.0\left(\frac{\partial \bar{\epsilon} \frac{\partial \bar{\upsilon} Z}{\partial y}}{\partial y}\right) + \frac{\partial\left(\bar{\epsilon}\frac{\partial \bar{\upsilon} Z}{\partial x} + \bar{\epsilon}\frac{\partial \bar{u} Z}{\partial y}\right)}{\partial x}\right)_k$$
$$-\tau_{yz_k} + (w\bar{\upsilon})_{k-\frac{1}{2}} - (w\bar{\upsilon})_{k+\frac{1}{2}} = 0 \tag{5}$$

where $\bar{u}$ and $\bar{\upsilon}$ = layer average velocity components in x and y directions respectively, Z = layer thickness, $\bar{\epsilon}$ = layer average eddy viscosity, β = momentum correction factor for each layer, w = vertical velocity component, which is calculated explicitly using the layer averaged continuity equation (see Hall(5)), τ_{xz_k}, τ_{yz_k} = interlayer shear stress components in x and y directions – evaluated using a Bousinesq eddy viscosity representation and a simple zero-equation turbulence model and k = layer number. It should be noted that the hydrostatic approximation has been used in deriving these equations, i.e.

$$P + P_0 = \rho g(\eta - z) \tag{6}$$

where P_0 = pressure at the surface, P = pressure at a

123

given depth and z = elevation relative to datum (positive upwards).

3   Numerical Implementation

The methodology used to generate a velocity profile is similar to that outlined in Hall(5). Firstly, the depth integrated equations are solved and the predicted water elevations are then used to drive the layer integrated equations. To ensure consistency between the two modes the non-linear advection terms are replaced by the depth averaged version of their layer averaged counterparts. The depth and layer integrated equations are then solved using a finite difference implementation, with the x direction depth integrated equations being of the following form:-

$$
\frac{\left(UH^{n+\frac{1}{2}} - UH^{n-\frac{1}{2}}\right)_{i+\frac{1}{2},j}}{\Delta t} + \beta\left(\sum_{k-}^{k-K} \frac{\left(\bar{u}^2 Z_{i+1,j,k} - \bar{u}^2 Z_{i,j,k}\right)}{\Delta x}\right)^n
$$

$$
+ \beta\left(\sum_{k-1}^{k-K} \frac{\left(\bar{v} Z_{i+\frac{1}{2},j+\frac{1}{2},k}\,\bar{u}_{i+\frac{1}{2},j+\frac{1}{2}+p,k} - \bar{v} Z_{i+\frac{1}{2},j-\frac{1}{2},k}\,\bar{u}_{i+\frac{1}{2},j-\frac{1}{2}+q,k}\right)}{\Delta y}\right)^n
$$

$$
+ \frac{gH^n_{i+\frac{1}{2},j}}{2\Delta x}\left(\eta^{n+\frac{1}{2}}_{i+1,j} - \eta^{n+\frac{1}{2}}_{i,j} + \eta^{n-\frac{1}{2}}_{i+1,j} - \eta^{n-\frac{1}{2}}_{i,j}\right)
$$

$$
- \frac{2}{\Delta x^2}\left(\epsilon^n_{i+1,j}\left(UH_{i+\frac{3}{2},j} - UH_{i+\frac{1}{2},j}\right)^n - \epsilon^n_{i,j}\left(UH_{i+\frac{1}{2},j} - UH_{i-\frac{1}{2},j}\right)^n\right)
$$

$$
- \frac{1}{\Delta x^2}\left(\epsilon^n_{i+\frac{1}{2},j+\frac{1}{2}}\left(UH_{i+\frac{1}{2},j+1} - UH_{i+\frac{1}{2},j}\right)^n - \epsilon^n_{i+\frac{1}{2},j-\frac{1}{2}}\left(UH_{i+\frac{1}{2},j} - UH_{i,j-1}\right)^n\right)
$$

$$
- \frac{1}{\Delta x^2}\left(\epsilon^n_{i+\frac{1}{2},j+\frac{1}{2}}\left(VH_{i+1,j+\frac{1}{2}} - VH_{i,j+\frac{1}{2}}\right)^n - \epsilon^n_{i+\frac{1}{2},j-\frac{1}{2}}\left(VH_{i+1,j-\frac{1}{2}} - VH_{i,j-\frac{1}{2}}\right)^n\right)
$$

$$
+ \frac{g}{2C^2 H^n_{i+\frac{1}{2},j}}\left(HU^{n+\frac{1}{2}} + HU^{n-\frac{1}{2}}\right)_{i+\frac{1}{2},j} = 0 \tag{7}
$$

$$
\left(\frac{\eta^{n+\frac{1}{2}} - \eta^{n-\frac{1}{2}}}{\Delta t}\right)_{i,j} + \frac{1}{2\Delta x}\left(\left(UH_{i+\frac{1}{2},j} - UH_{i-\frac{1}{2},j}\right)^{n+\frac{1}{2}} + \left(UH_{i+\frac{1}{2},j} - UH_{i-\frac{1}{2},j}\right)^{n-\frac{1}{2}}\right)
$$

$$
+ \frac{1}{\Delta x}\left(VH_{i,j+\frac{1}{2}} - VH_{i,j-\frac{1}{2}}\right)^n = 0 \tag{8}
$$

where $U^n_{i+\frac{1}{2},j} = \dfrac{\left(U^{n-\frac{1}{2}}_{i+\frac{1}{2},j} + U^{n-\frac{1}{2}}_{i+\frac{1}{2},j}\right)}{2}$, $\bar{u}^n_{i+\frac{1}{2},j} = \dfrac{\left(\bar{u}^{n+\frac{1}{2}}_{i+\frac{1}{2},j} + \bar{u}^{n-\frac{1}{2}}_{i+\frac{1}{2},j}\right)}{2}$, C = Chezy

coefficient, calculated using the Colebrook-White formula, K = maximum number of vertical layers, $p = -\dfrac{|\bar{v}|_{i+\frac{1}{2},j+\frac{1}{2},k}}{2\bar{v}_{i+\frac{1}{2},j+\frac{1}{2},k}}$ and $q = -\dfrac{|\bar{v}|_{i+\frac{1}{2},j-\frac{1}{2},k}}{2\bar{v}_{i+\frac{1}{2},j-\frac{1}{2},k}}$. The corresponding layer

integrated x direction difference momentum equation is given as:-

$$\dfrac{\left(\bar{u}Z^{n+\frac{1}{2}} - \bar{u}Z^{n-\frac{1}{2}}\right)_{i+\frac{1}{2},j,k}}{\Delta t} + \beta\left(\dfrac{(\bar{u}^2 Z_{i+1,j,k} - \bar{u}^2 Z_{i,j,k})}{\Delta x}\right)^n$$

$$+ \beta\left(\dfrac{\left(\bar{v}Z_{i+\frac{1}{2},j+\frac{1}{2},k}\,\bar{u}_{i+\frac{1}{2},j+\frac{1}{2}+p,k} - \bar{v}Z_{i+\frac{1}{2},j-\frac{1}{2},k}\,\bar{u}_{i+\frac{1}{2},j-\frac{1}{2}+q,k}\right)}{\Delta y}\right)^n$$

$$+ \dfrac{gZ^n_{i+\frac{1}{2},j}}{2\Delta x}\left(\eta^{n+\frac{1}{2}}_{i+1,j} - \eta^{n+\frac{1}{2}}_{i,j} + \eta^{n-\frac{1}{2}}_{i+1,j} - \eta^{n-\frac{1}{2}}_{i,j}\right)$$

$$- \dfrac{2}{\Delta x^2}\left(\epsilon^n_{i+1,j,k}\left(\bar{u}Z_{i+\frac{3}{2},j,k} - \bar{u}Z_{i+\frac{1}{2},j,k}\right)^n - \epsilon^n_{i,j,k}\left(\bar{u}Z_{i+\frac{1}{2},j,k} - \bar{u}Z_{i-\frac{1}{2},j,k}\right)^n\right)$$

$$- \dfrac{1}{\Delta x^2}\left(\epsilon^n_{i+\frac{1}{2},j+\frac{1}{2},k}\left(\bar{u}Z_{i+\frac{1}{2},j+1,k} - \bar{u}Z_{i+\frac{1}{2},j,k}\right)^n - \epsilon^n_{i+\frac{1}{2},j-\frac{1}{2},k}\left(\bar{u}Z_{i+\frac{1}{2},j,k} - \bar{u}Z_{i,j-1,k}\right)^n\right)$$

$$- \dfrac{1}{\Delta x^2}\left(\epsilon^n_{i+\frac{1}{2},j+\frac{1}{2},k}\left(\bar{v}Z_{i+1,j+\frac{1}{2},k} - \bar{v}Z_{i,j+\frac{1}{2},k}\right)^n - \epsilon^n_{i+\frac{1}{2},j-\frac{1}{2},k}\left(\bar{v}Z_{i+1,j-\frac{1}{2},k} - \bar{v}Z_{i,j-\frac{1}{2},k}\right)^n\right)$$

$$- \tau_{xz_k} + (w\bar{u})_{i+\frac{1}{2},j,k-\frac{1}{2}} - (w\bar{u})_{i+\frac{1}{2},j,k+\frac{1}{2}} = 0 \qquad (9)$$

with similar equations being obtained for the depth and layer integrated equations in the y-direction.

It can readily be shown that these finite difference equations are of second order accuracy in both space and time. With regard to the depth integrated equations the elevation terms are treated implicitly and are centred in time via the Crank-Nicolson method. Such an implicit treatment of the elevation terms allows the use of a larger Courant number than would be the case for an explicit formulation. A maximum Courant number of the order of eight can generally be used, based upon maximum flow depth considerations.

The finite difference equations are defined on a regular staggered grid, with the grid structure in three dimensions and the locations of all variables being illustrated in Figure 2. The use of such a grid structure avoids pressure coupling problems and facilitates the efficient evaluation of space centred derivatives. The depth integrated equations are solved using the ADI (alternating direction implicit) technique, which consists of treating only the non-advective derivatives implicitly in the x and y directions respectively on alternate half time steps. This allows the generation of a TDM (tri-diagonal matrix) which is solved for velocity and water elevation fields using Gauss elimination and back substitution. With regard to the layer integrated equations the spatial derivatives in the increasing k direction are treated implicitly, thereby allowing the definition of comparitively narrow computational layers. The subsequent equations again form a TDM and are solved using the method outlined previously.

3.1 Treatment of advection terms

In developing this three-dimensional model special emphasis has been focused on modelling the advection terms, with these terms having been treated explicitly in the model and requiring the advective Courant number to be less than unity. Such an explicit treatment of these terms is desirable, not simply because of the ease of computational manipulation, but also because the explicit implementation satisfies the important 'transportative property' (see Roache (6)) unlike its implicit counterpart.

A time and space centred leapfrog representation of the advection terms is not suitable for this application. Such a method would allow the onset of numerical instabilities because of the existence of strong lateral velocity gradients within the harbour being modelled. This is due to the widely publicised inability of the leapfrog scheme to dampen unwanted grid scale oscillations. One possible solution to this problem would be to introduce some type of upwinding method centred explicitly in time. However, such a solution can give problems due to the unconditional instability of this method, in the long wave limit. Another possibility would be to use a forward time upwinding method. Although linearly stable this method is only first order accurate in time, hence the time step size would need to be reduced to achieve stable and periodic results. If however, time centring is achieved by using a two step iteration, then stable results are generated.

4   Boundary conditions

In applying the model to the laboratory model rectangular harbour shown in Figure 1, the closed boundaries in the form of harbour walls were represented using the no-slip boundary condition, i.e. all velocities both normal and lateral to the walls were equated to zero. The outer open boundary, located parallel to the entrance plane, was used to drive the flow field within the domain by including a sinusoidally varying tidal elevation along this boundary. For the other open boundaries normal to the plane of the harbour a free slip or streamline velocity boundary condition was imposed, with the normal boundary velocity component being equated to zero. The model simulations were always commenced from low (or high) tide, thereby minimising the initial errors imposed by the initial conditions of zero velocity and a constant water elevation everywhere.

5   Experimental program

All experimental work was undertaken in the Hydraulics Laboratory at Bradford University, using the tidal basin illustrated in Figure 1. Full details of the tidal basin are given in Falconer and Yu(7), with the tidal period and range being set to 1416s and 0.1m respectively. Velocity measurements were carried out near the mouth of the harbour using a portable probe connected to a two component LDA (laser doppler anemometer) device via a fibre optic link. The probe was situated on a movable mount with its position being governed by a software driven automatic control system. A grid of 0.12m was marked on the bed of the harbour, to coincide with the numerical model grid. Aluminium powder in solution was discharged into the harbour at two points, strategically located so as to achieve maximum intrusion at both flood and ebb tides. Measurements of the local velocity were carried out at five different elevations over the total depth, with the tip of the probe being either submerged or located above the water depending upon the layer elevation being studied and the current water depth.

6   Results

A typical comparison of the computed and measured tide varying velocity magnitude is shown in Figures 3 and 4, for one tidal cycle and two layers centred at 1 and 5cm above the bed respectively, and at a location just inside the harbour entrance. The comparisons show an encouraging degree of agreement between the predicted and measured results, with there also being little variation in velocities in the lower and upper layers. Typical velocity field distributions across the domain are shown for the bottom layer in Figures 5 and 6, at mean water level flood tide and high tide respectively. These results were produced for a constant vertical eddy viscosity of $10^{-3}m^2s^{-1}$.

A series of test simulations were undertaken for a range of constant eddy viscosities, varying from $10^{-3}m^2s^{-1}$ to $10^{-5}m^2s^{-1}$. The resulting simulations showed that for a constant vertical eddy viscosity little variation occurred in the horizontal velocity field distribution for the various layers. Following these tests, simulations were undertaken for a parabolic variation in the vertical eddy viscosity as given by:-

$$\overline{\epsilon}_{parb} = u_* \kappa z \left(1 - \frac{z}{H} \right) \qquad (10)$$

where $\overline{\epsilon}_{parb}$ = parabolic eddy viscosity, u_* = shear velocity, κ = von Karman's constant, z = elevation above bed and H = total depth of flow. This eddy viscosity distribution is derived from the assumption of the existence of a linear shear stress distribution and a logarithmic velocity variation over the depth. The predicted vertical velocity distribution is shown for two different times within the harbour in Figures 7 and 8, as compared with the corresponding layer averaged velocity given for the same depth mean velocity. The layer averaged velocity was obtained from the following equation:-

$$\overline{u} = \frac{u^*}{(z_2 - z_1)\kappa} \left(z_2 \ln \frac{z_2}{z_0} - z_1 \ln \frac{z_1}{z_0} - (z_1 - z_2) \right) \qquad (11)$$

where z_0 = thickness of viscous sub-layer, z_1 = elevation of bottom of layer above the bed, z_2 = elevation of top of layer above the bed and $\overline{u}$ = layer averaged velocity. The corresponding comparisons in Figures 7 and 8 show that the numerically predicted variation in the vertical velocity is less than for an assumed logarithmic velocity profile, thereby indicating that further emphasis needs to be focused on the vertical shear stress distribution.

These simulations were undertaken for five layers and for a total simulation period of two tidal cycles, i.e. 2832s. The model runs were carried out on a SUN SPARC workstation and required 3.5 hours of cpu time.

7   Conclusions

The model outlined herein has proved to be both stable and efficient in predicting the three-dimensional velocity fields in a model square harbour. Comparisons with other well documented three-dimensional models has indicated that the current model is computationally highly efficient and produces comparable accuracy with models developed by Leendertse(2), Blumberg(3) and Noye(4), for a time step 10 times greater. Although the model produces velocity field predictions which agree reasonably well with measured data for a model square harbour, current efforts are being focused on improving the vertical sheer stress distribution in the numerical model.

Acknowledgements

This project was undertaken as part of an EC funded contract and using a laser doppler anemometer provided by SERC. The authors would like to thank both the EC and the SERC for their support.

REFERENCES

1. Falconer, R A (1980), "Numerical Modelling of Tidal Circulation in Harbours", Journal of the Waterway, Port and Coastal Division, Vol.106, pp 32-33.
2. Leendertse, J J and Liu, S K (1973), "A Three-dimensional Hydrodynamic Model for Estuaries and Coastal Seas, Vol. 1", Report No.R-1417-OWRR, Rand corporation, pp 1-22.
3. Blumberg, A and Mellor, G (1987), "A Description of a Three-dimensional Coastal Ocean Circulation Model", Coastal and Estuarine Sciences 4, pp 1-16.
4. Noye, J and Stevens, M (1987), "A Three-dimensional Model of Tidal Propagation Using Transformations and

Variable Grids", Coastal and Estuarine Sciences 4, pp 41-69.

5. Hall, P (1983), "Numerical Modelling of Wind Induced Lake Circulation", thesis submitted to the University of Birmingham for the degree of Doctor of Philosophy, pp 1-252.
6. Roache, P J (1972), "Computational Fluid Dynamics", Published by Hermosa, Albuquerque, New Mexico, pp 67-73.
7. Falconer, R A and Yu, G P, (1991), "Effects of Depth, Bed slope and Scaling on Tidal Currents and Exchange in a Laboratory Model Harbour", Proceedings of the Institution of Civil Engineers, Part 2, Research and Theory, Vol.91, pp 561-576.

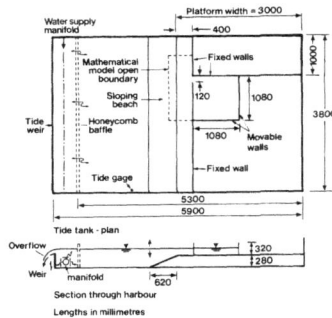

Figure 1. Schematic illustration of laboratory tidal basin and harbour configuration

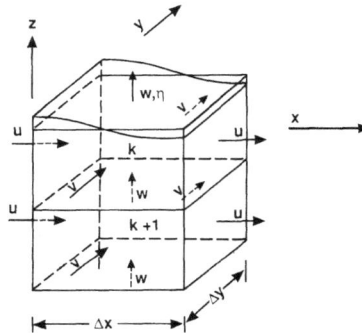

Figure 2. Relative position of the variables in the model

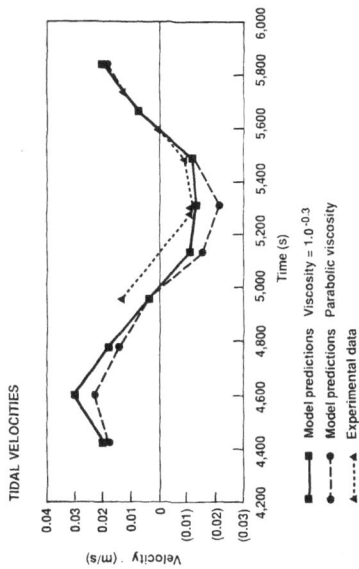

TIDAL VELOCITIES

● Model predictions   Viscosity = 1.0 -0.3
■ Model predictions   Parabolic viscosity
▲······▲ Experimental data

Figure 4.   Comparison of predicted and measured velocity magnitudes for the layer centred 5cm above the bed

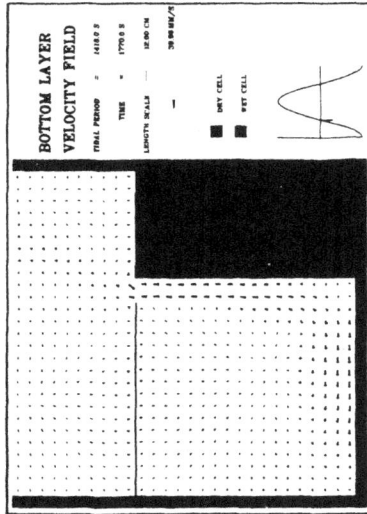

TIDAL VELOCITIES

■ Model predictions   Viscosity = 1.0 -0.3
● Model predictions   Parabolic viscosity
▲······▲ Experimental data

Figure 3.   Comparison of predicted and measured velocity magnitudes for the layer centred 1cm above the bed

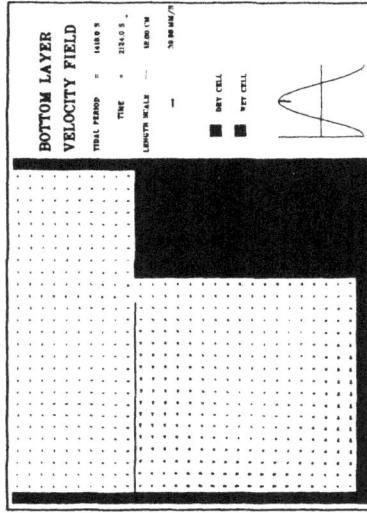

Figure 6.   Bottom Layer Velocity Field Distribution at High Tide

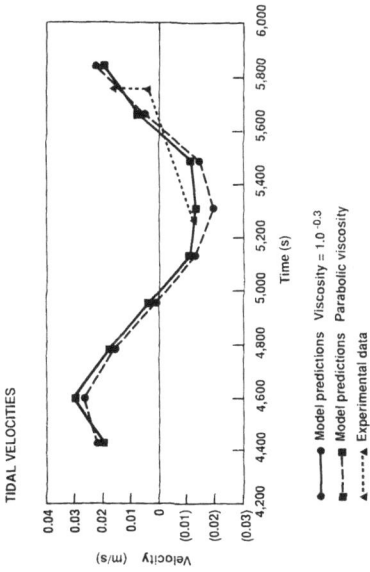

Figure 5.   Bottom Layer Velocity Field Distribution at Mean Water Level Flood Tide

131

vertical velocity profiles

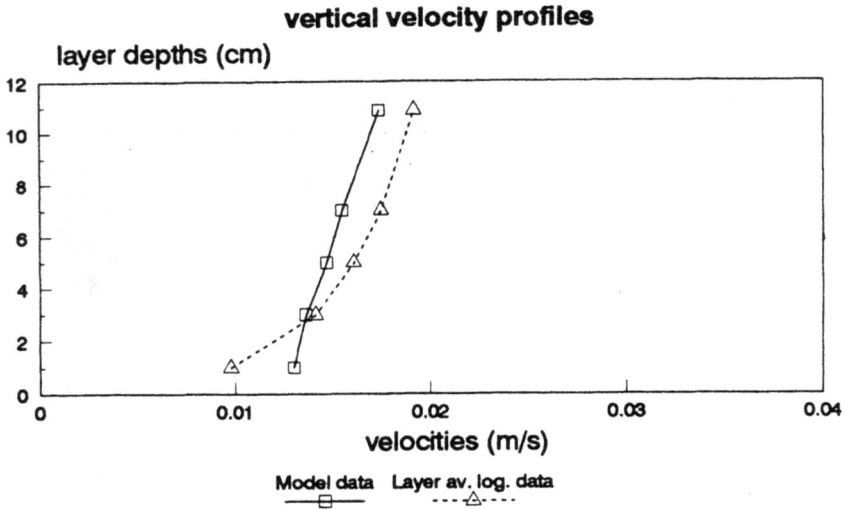

Figure 7. Comparison of Predicted and Logarithmic Velocity
Magnitudes at 1593 seconds

vertical velocity profiles

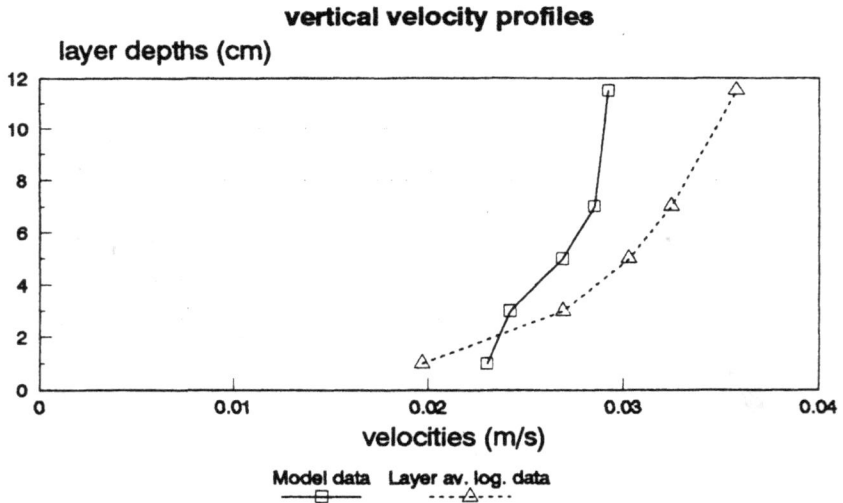

Figure 8. Comparison of Predicted and Logarithmic Velocity
Magnitudes at 1947 seconds.

12 Numerical modelling of the Severn estuary

J. Osment

ABSTRACT

A nested three-grid numerical model of the Severn estuary from Beachley to Avonmouth is constructed and calibrated using field data from a specially commissioned hydrographic survey. The model is applied for environmental impact assessment, hydrodynamic design of permanent and temporary works, ship protection studies, and for programming of on-board computers for use by works vessels.

1   Introduction

A second road crossing of the Severn estuary is to be constructed approximately 6km downstream of the existing bridge at Beachley.

Aside from design of the associated infrastructure, the bridge itself requires a wealth of data, not only for design of the bridge structure and foundations, but also for design of temporary works, environmental impact, and ship protection measures.

Preliminary modelling work by HR (1986) had presented some data to predict the likely hydrodynamic effects due to construction of the proposed bridge, but had also identified a need for a more detailed model to be based upon recent data. As a result, the present model study was commissioned. The model was to be based on detailed topographic/bathymetric survey data of the bridge site obtained by Laing GTM in 1990, and upon a hydrographic survey carried out as part of the model study. Although intended primarily for the assessment of environmental impact, the model also served to provide data for several other aspects of design such as temporary works and ship protection measures.

The bridge design for the new crossing comprises a cable stayed centre span of 456 metres, with approach viaducts supported on oval piers at 98 metre centres. These piers have a width of 10 metres, and therefore represent a potentially significant source of disruption to the flow regime, even when aligned to present a minimum width to the flow. Temporary works such as access bunds and cofferdams represent further sources of flow modification, and the numerical model study was therefore designed to evaluate the likely modification to the hydrodynamic regime arising at two stages of construction, namely the finished condition, and a condition during construction considered to represent a "maximum blockage"situation.

2.   Numerical Model

The two dimensional model is based on the Halcrow in house program DAWN, which uses an ADI solution of the depth integrated equations of continuity and momentum, in conjunction with an explicit Eulerian-Lagrangian treatment of the depth integrated advection diffusion equation.

The hydrodynamic equations may be expressed as:

$$\nabla.\underline{V} = 0 \qquad\qquad (1)$$

$$\underline{V}_t + \underline{V}.\nabla\underline{V} + 2\underline{\Omega}_\Lambda\underline{V} = g + \underline{F} \qquad (2)$$

where $\underline{V}$ is the velocity field, g the Earth's gravitational force, and $\underline{\Omega}$ is the vector angular velocity of the Earth. The term $\underline{F}$ includes pressure gradient, wind stress, bed shear resistance, and turbulence induced shear stress.

The equations are integrated through depth to allow a two dimensional solution, assuming velocity to be a seventh power function of depth. The Coriolis effect (due to the earth's rotation) is included using the f-plane approximation (see eg Gill 1982). Over the geographical area normally considered in coastal studies the atmospheric pressure gradient is generally negligible in comparison to the water surface slope, and is therefore omitted from the equations. Water density may either be assumed constant, or may be modelled as a function of temperature and salinity.

Equations (1) and (2) are solved using an upwinded central implicit finite difference formulation, in conjunction with the "alternating direction" algorithm (see eg Ames 1977).

The advection diffusion equation may be written as:

$$C_t + \underline{V}.\nabla C = \nabla.(D \nabla C) + G \qquad (3)$$

where C is determinand value, D is a dispersion tensor, and G is a source/sink term. The dispersion tensor D is represented in the manner suggested by Preston (1985). Values of longitudinal and transverse dispersion coefficients were taken as 5.93 m^2/s and 0.15 m^2/s as obtained by Elder (1959) and Fischer (1973) respectively. The term G includes both inputs (and abstractions where appropriate) of pollutants, heat etc, and changes in determinand budget due to chemical and physical changes.

A depth integrated form of equation (3) is solved using an explicit method based on the Eulerian-Lagrangian approach described by Cheng et al (1984).

The model was set up as a triple nested grid system, using mesh sizes of 90 metres, 30 metres, and 10 metres for primary, secondary and tertiary grids respectively. The grids were aligned at 33° from National Grid north, and covered the area from Beachley to Avonmouth, with an arrangement as shown in Figure 1. Bathymetric data were obtained from survey data supplied by the Admiralty, by the Port of Bristol Authority, and by Laing GTM from their 1990 survey of the bridge site.

Hydrodynamic effects of bathymetric features of dimensions significantly less than the grid spacing (such as bridge piers and ship protection islands) were approximated by introduction of "blocking" factors. These factors were individually defined for each grid cell in each direction, and reduced the effective widths of each cell according to the values attributed.

In addition to the "existing" situation, two additional bathymetric configurations were constructed: one to represent the situation after completion of the new bridge, and a second to model a condition obtaining during construction of the crossing, selected as that causing the maximum obstruction to flow. This latter configuration therefore included all bridge caissons and ship protection islands, the temporary cofferdam on the Gwent shore, the temporary access bund from the Avon shore, and materials handling jetties, also on the Avon shore.

A common boundary configuration was applied to all three grids, namely an "upstream" current boundary and a downstream "elevation" boundary. Primary grid boundary data were obtained from recording tide gauge and current meter data acquired during a specially commissioned hydrographic survey carried out in December 1990. A recording tide gauge was deployed at Avonmouth for definition of the downstream boundary condition, and currents were measured at Beachley over tidal cycles representing typical spring and neap tides. Boundary conditions were then formulated to represent spring and neap tidal scenarios with 12.6 metre and 7.4 metre range respectively.

The model was calibrated against further current meter data, and verified against float track data also obtained during the hydrographic survey. Calibration was effected against currents measured at two stations sited upstream and downstream of the bridge site. Typical currents predicted by the model are shown plotted with the corresponding field data in Figure 2.

3   Environmental Assessment

Assessment of the environmental effects likely to result from construction of the proposed new crossing was the primary reason for construction of the numerical model, and included the study of such aspects as:

- modification of patterns of dispersion from sewage treatment works and industrial sources
- ecological effects due to modification of tidal phasing
- effects on patterns of fish behaviour and fishing operations
- modification of patterns of sedimentation.

All aspects were studied both for the long and short term conditions using the finished and construction bathymetries respectively. Results were presented directly as predicted differences by use of baseline results obtained from corresponding runs with the existing bathymetry.

Primary (90 metres) grid results were used for assessment of aspects such as dispersion and tidal phasing whose effects are manifest over an area extending beyond the secondary (30 metre) grid. Modifications to patterns of sedimentation and to fishing operations are the result of changes in the current regime, which generally are limited to an area 750 metres either side of the bridge. Results from the 30 metre grid were therefore appropriate for assessment of these aspects.

The energetic nature of the estuary ensures that discharged effluents are dispensed rapidly and effectively. Small modifications caused by the bridge and temporary works therefore have only second order effects on the resulting pollutant concentrations. For instance, Figure 3 shows differences in pollutant concentration in the effluent plume from the Sudbrook paper mill at 3½ hours past high water on a spring tide, for the "finished" condition. In this condition the accelerated currents adjacent to the Gwent shore (see below) have resulted in reduced pollutant concentrations close to the shoreline, whilst concentrations further offshore show corresponding increases in level. This discharge had a nominal concentration within the effluent of 1000 mg/l in a flowrate of 0.126 m^3/sec, and the maximum concentration increase of 0.01 mg/l therefore represents a percentage increase of .001% of the original concentration.

Modelling of the local sewage treatment works at Redwick, Caldicot and Magor assumed slow BOD concentrations of 15mg/l, 30mg/l and 21mg/l in discharges of 0.016m^3/sec, 0.145m^3/sec, and 0.11m^3/sec respectively. Typical maximum differences of 0.01mg/l are predicted at Caldicot, representing a percentage increase of 0.03% over the original concentration.

Sea birds such as dunlin, plover and knot depend for their existence on food available on the tidal flats, and any modification to the periods for which these flats are exposed will affect the success of these species in the estuary. Investigation of the tidal phasing was therefore an important aspect of the environmental study. Time series of water surface elevation at selected points for conditions before, during and after construction show a maximum phase shift of approximately 8 mins, a reduction of 30mm in high water level, and a reduction in the period of exposure of 10 mins.

These effects were not considered to have any environmental significance, and were not exceeded at any of the other nine selected locations.

The Severn estuary has a history of salmon fishing, with methods unique to the estuary. There are existing fisheries in the immediate vicinity of the proposed crossing on both banks of the estuary. Two types of fishing operation are licensed: putcher ranks and lave nets. The last recorded use of putcher ranks in the area likely to be affected by the new bridge was in 1986 (Solomon 1988), but lave net licences were issued in 1990 for fisheries at Gruggy Rocks and English Stones. The annual catch estimated for the former fishery (on the Gwent side) was 20 fish in 1989, and although the English Stones fishery is known to be the more active, no separate figures are available for annual catch from this licence.

Two locations are fished on English Stones: English Lake and Salmon Pool. Both are fished on the ebb tide, and rely upon local bathymetry (in conjunction with leaders in English Lake) to guide the fish into passively operated lave nets. They are, however, only fishable for approximately 40 minutes on each tide, when both tidal elevations and current speeds reduce to manageable levels. Both of these fisheries are situated close to the proposed crossing, and are likely to be affected not only by the completed bridge, but also to an even greater extent by the temporary works on the Avon side of the estuary.

In the finished condition the bridge piers will inevitably cause some disturbance of flow in the immediate vicinity of the bridge, but during the construction phase there will also be an access causeway and materials handling jetties erected as temporary works. Only the Salmon Pool fishery will be affected by the jetties, but both will be affected by the access causeway which runs from the Avon shore to the western side of English Stones, with a top level 1 metre below Ordnance Datum. The causeway will have several culverts but will nonetheless cause a major disruption to flow, especially at low tide.

Time series of current velocity (speed and direction) at Salmon Pool presented as Figure 4 show significant changes in current patterns, particularly during construction. They also show the reversal of current direction that occurs approximately 1½ hours before low water as a result of a local bathymetric feature, which forms a pool with its only outlet at the northern end. It is at this outlet where the lave nets are cast. These and similar results for English Lake suggest that modifications to the hydrodynamic regime could reduce fishing time available at these locations by 10 minutes or more during the construction period. Since this represents approximately 25% of the available period, and no practical measures are available to mitigate these effects, it was concluded that compensation may be payable to the fishing licensees, at least while construction is in progress.

Changes in patterns of fish behaviour due to changes in flow regime and bathymetry are not known, and may not therefore be predicted by interpretation of model results.

The modifications of sedimentation patterns likely to arise as a result of the bridge and its construction were inferred from examination of predicted current vector plots by Dr R Kirby, a specialist in estuarine sedimentation processes, with particular knowledge of the Medway and Severn estuaries. The model results were presented to Dr Kirby as vector plots of velocity differences.

As might be expected, increased resistance to flow generated by the bridge structure and its temporary works causes accelerations through The Shoots channel bridged by the cable stayed centre span. Accelerated flows in this area are of no environmental significance, but increased current speeds were also evident close to the shore near the Gwent abutment. The latter accelerations are due to the inclusion of ship protection devices on the eight caissons to the immediate west of the centre span. These devices were modelled as separate islands aligned with the direction of flow, and situated approximately 100 metres upstream of each protected bridge pier. With a top width of 3 metres and side slopes of 1:1.5 , the mean width presented to the flow was significantly greater than that of the bridge caissons, and the group of islands therefore caused additional accelerations through The Shoots channel, and around its western end, close to the Gwent shore. In the worst case, the model predicts velocity increases of 60cm/sec in an area where the undisturbed flow has a speed of 250cm/sec. It is therefore possible that these locally increased current speeds could lead to some erosion of the tidal flats, exposing more of the over consolidated muds below. However, these areas have a species diversity that is already low as a result of high turbidity and high current speeds, and EAU (1991) accordingly conclude that these increased current speeds are unlikely to have ecological implications.

4   Ship Protection Measures

Currents in The Shoots channel beneath the proposed bridge regularly reach peak speeds of 8 knots. When combined with a ship speed through the water of 10 knots, there is a potential ship impact speed of approximately 20mph. Although the bridge structure could be designed to withstand such impacts, any ship involved in such a collision would undoubtedly be seriously damaged. It was therefore deemed necessary to provide protection for ships on the upstream side of the bridge for appropriate piers selected in consultation with port authorities and mariners' representatives.

Detailed description of the design of the ship protection measures is outside the scope of this paper, but the 10 metre grid model was run to represent two different design approaches: separate islands constructed approximately 100 metres upstream of each selected pier, and protective nosings built around the upstream end of each appropriate pier.

The 10 metre grid model covered only the area defined by the bridge abutments as diagonal points, but nonetheless required a mesh of dimensions 621 x 162. As such, the model was too unwieldy to run on the company work stations, and was therefore taken to Harwell to run on the Cray 2. In view of the expense of running the 10 metre grid model, it was run to cover only the critical period from high water to mid ebb.

These current vector data were used primarily to optimise the position and orientation of the protection devices, but are also expected to be used for detailed design of the scour protection measures to be included at the bridge piers.

The 30 metre grid model was also used to assess the differences in hydrodynamic effect of the islands and pier nosings scheme. However, since the nosings are of similar cross section to the islands, the effective difference is simply that due to a 100 metre displacement of the major point of resistance, and differences in the flow regime were found to be limited to an area immediately local to the bridge site.

5   Navigation Aid Data

The contractor proposes to equip works vessels engaged on construction of the bridge with computerised navigation aids for prediction of current speeds. The purpose designed software would work in conjunction with specially installed position fixing equipment, and would enable ships' masters to obtain predictions of current velocity at a given location at any state of the tide, for a selection of typical tidal ranges.

The 30 metre grid model was used to output current velocity data for the whole grid at 5 minute intervals, over a whole tidal cycle. Corresponding results for the "spring" and "neap" tidal conditions were then interpolated to produce equivalent data for three intermediate tidal ranges. Since these data were to be produced to represent a single bathymetric configuration, that configuration was necessarily selected intuitively to provide the best overall representation of a situation that will change continuously during construction. The configuration selected was that obtaining at the placing of the first caisson ie all temporary works constructed, but no ship protection devices or bridge caissons in position.

Time series data also output during the course of the above runs were used to evaluate current speeds in the vicinity of the bridge site, as an aid for construction planning. For example, predicted current speeds have been used to define the engine power required in the purpose designed catamaran vessels that will be used to place the bridge caissons. The stability of bridge caissons in the newly placed (empty) condition was also assessed by use of the same data.

6   Conclusions

A two dimensional numerical model system of the Severn estuary, representing the area from Beachley to Avonmouth has been constructed, calibrated and verified with recent hydrographic data.

The model has been used primarily for assessment of the hydrodynamic effects associated with construction of a proposed second road crossing of the estuary.

Secondary uses of the model included design of ship protection measures, production of data for navigation aid program, and design of temporary works.

7   Acknowledgement

The data presented in this paper are reproduced by kind permission of Laing GTM.

8   References

(1)   Ames W F "Numerical Methods for Partial Differential Equations", Nelson 1977.

(2)   Cheng R T, Casulli V and Milford S N "Eulerian-Lagrangian solution of the convection-dispersion equation in natural co-ordinates". Water Resources Research, Vol 20 No 7, pp 944-952, July 1984.

(3)   Environmental Advisory Unit. "The second Severn crossing - effects on estuary ecology and fisheries". June 1991.

(4)   Elder J W, (1959). "The dispersion of marked fluid in turbulent shear flow". J Fluid Mech, 5(4), 544-560.

(5)   Fischer H B, (1973). "Longitudinal dispersion and turbulent mixing in open channel flow". Annu Rev Fluid Mech, 5, 59-78.

(6)   Gill A E. "Atmosphere-Ocean Dynamics". Academic Press, 1982.

(7)   Hydraulics Research. "Second Severn crossing, numerical flow model". Report No EX1383, Dec 1985.

(8)   Preston R W (1985) "The representation of dispersion in two-dimensional water flow". Report No TPRD/L/2783/N84, Central Electricity Research Laboratories, Leatherhead, England, May 1985, pp 13.

(9)   Solomon, D J "Commercial fisheries for migratory fish". Severn Barrage Development Project, document ref SBDP/DJS/s.7(ii) d1(ii), November 1988.

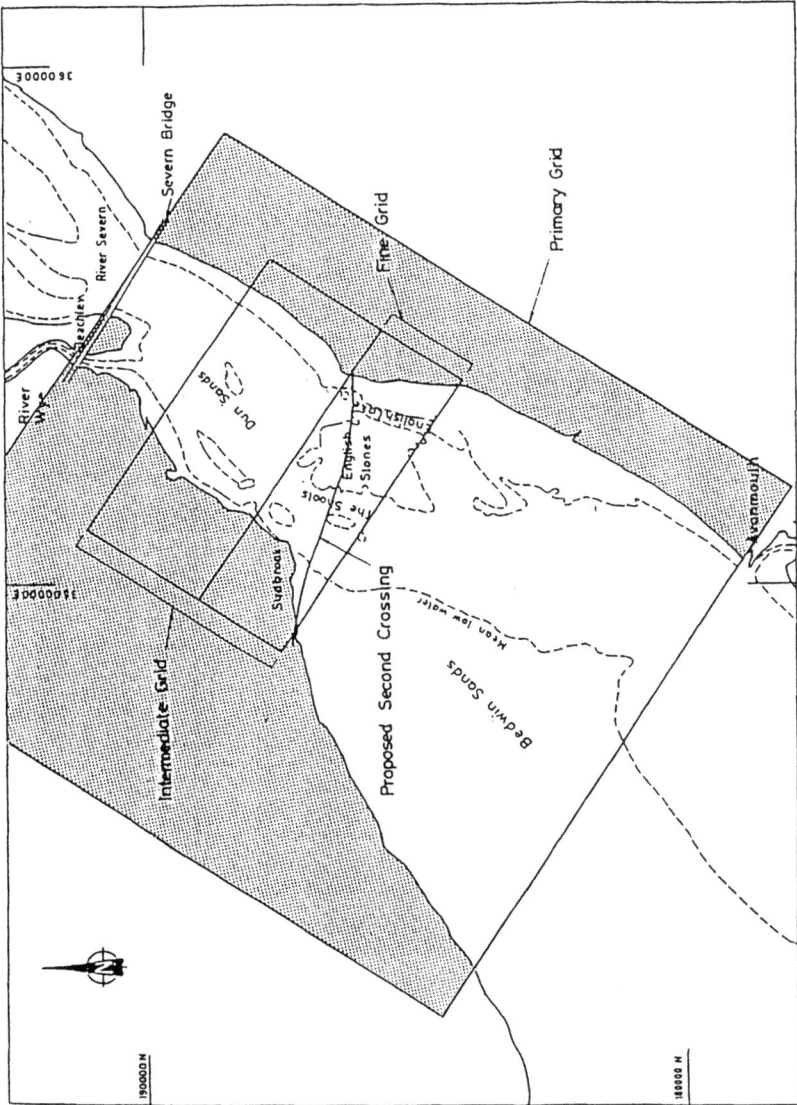

Model Layout

Figure 1

141

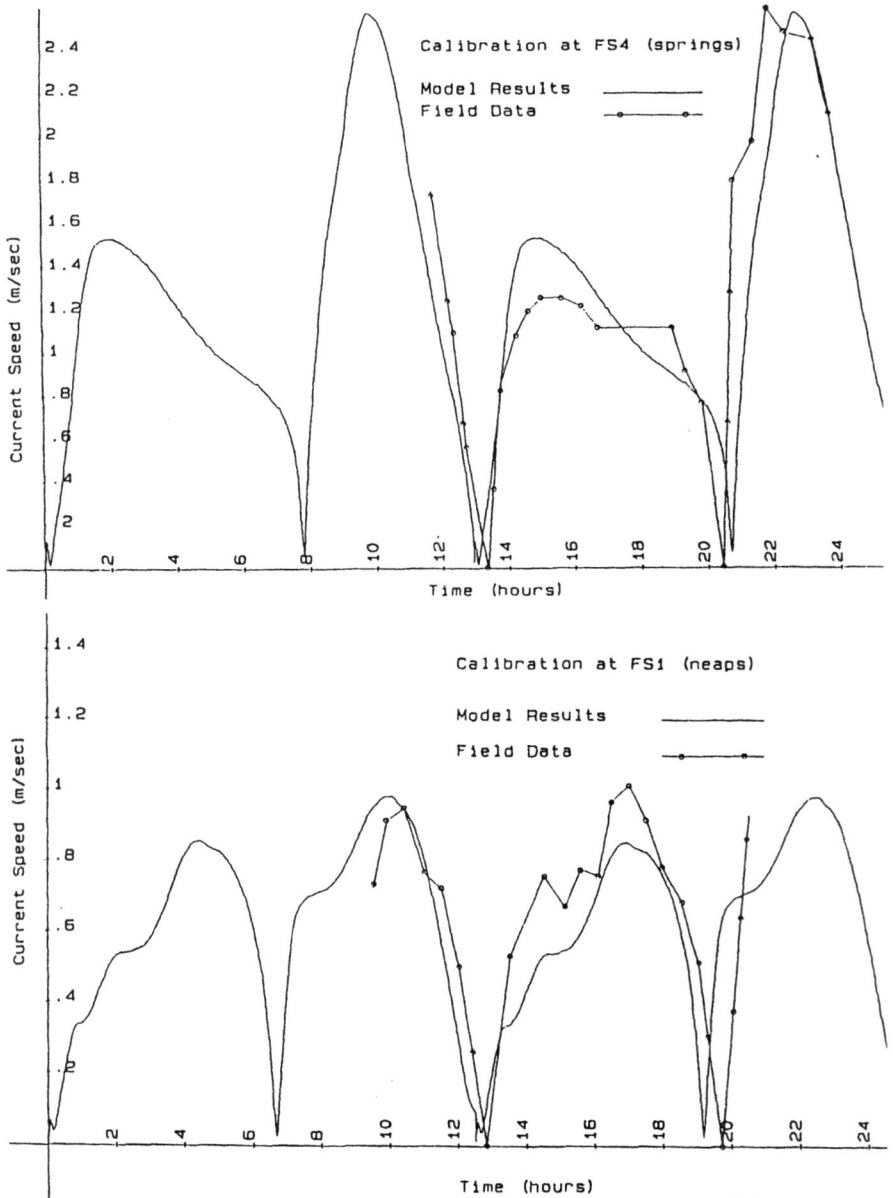

Figure 2

Differences in Sudbrook effluent concentration
@ HW+3.5 Hours (finished condition, spring tide)

Figure 3

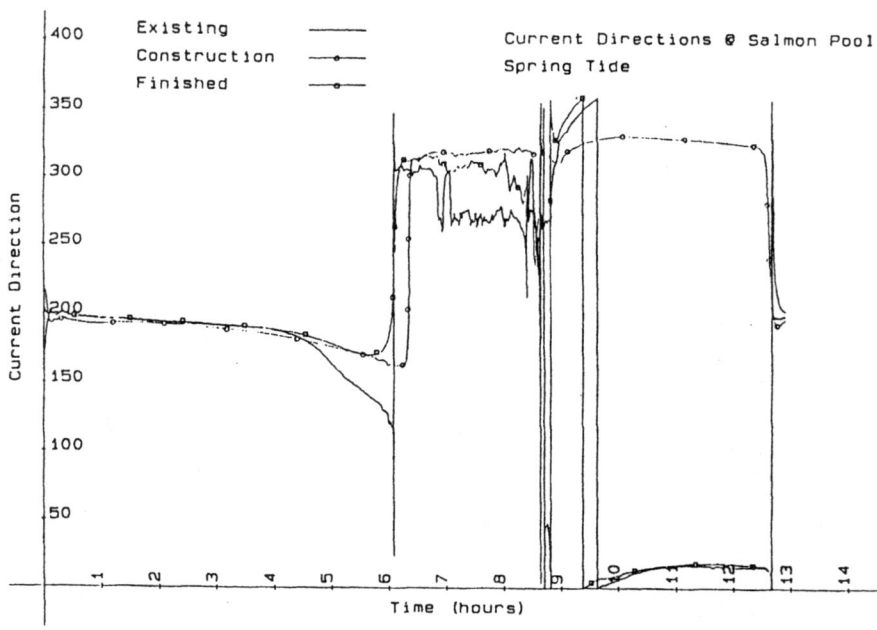

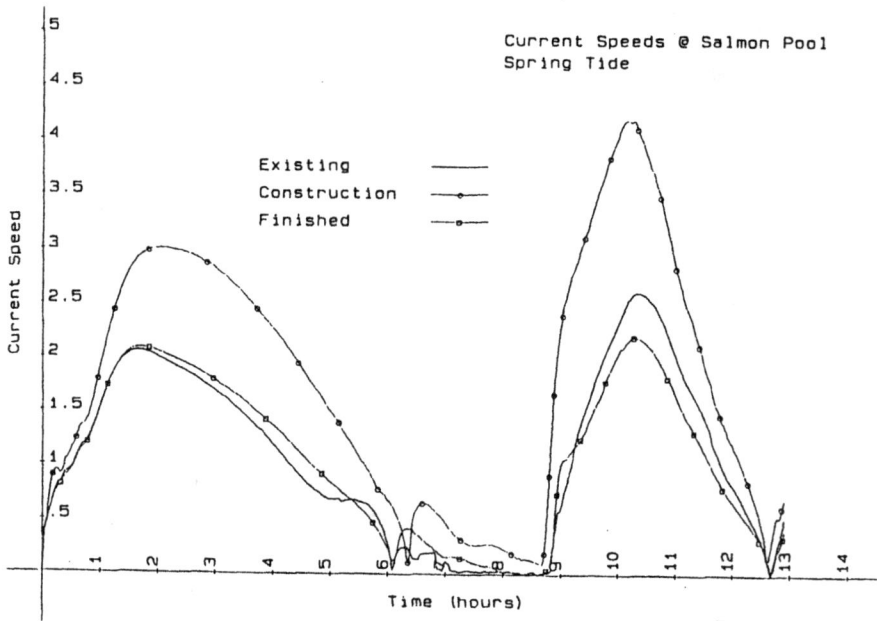

Figure 4

144

13 Problems associated with 2-D modelling of coastal waters

P. A. Mackinnon and D. C. Keiller

ABSTRACT

The implementation of the EC directive on bathing waters has required the promoters of sewage disposal schemes to satisfy the regulatory authorities that the standards specified in the directives will be met on beaches around the outfalls. In many circumstances numerical models are used to demonstrate that the proposed outfalls will be satisfactory.

The basis for the dispersion modelling of effluents from outfalls is frequently a two-dimensional numerical model of coastal waters in the locality. The location of the model boundaries and the choice of data used to drive the tidal flows in these models has been found by experience to be critical to their success.

The difficulties inherent in selecting appropriate boundary conditions are discussed using examples from recent models of coastal areas in the UK and overseas. The examples demonstrate that careful choice of the location of a model boundary is often necessary to ensure that the calibrated model can correctly reproduce the dominant circulation patterns in the bays, headlands or estuaries which form the subject of the model study. This choice must often rely on the modeller's understanding of the dominant flow patterns in the area prior to setting up the model.

Once the model boundary has been established, the decision to use level or flow boundaries along each of the open sea boundaries is often not straightforward, although it is important to the overall success of the calibration. Examples of some of the problems encountered and the solutions adopted will be presented.

Introduction

An increasing concern with environmental matters has led to the need for engineers to consider the environmental effects of proposed schemes along with the more traditional design criteria. In wastewater engineering, the practice of disposal of effluent at sea has come under recent scrutiny, with the introduction of the EC Bathing Water Directive and its equivalent in various overseas locations.

In order to perform an environmental study of a proposed scheme, the promoter is faced with the need to provide some quantified prediction of the future scenario, after commission of the scheme. Increasingly, promoters opt for mathematical modelling as a cost-effective means of indicating the changes likely to take place.

With the increased need to model coastal locations, the difficulties of modelling open sea conditions have been emphasized. Over the past few years, the demand for modelling in areas of complex bathymetry and tidal conditions has provided valuable experience in model development, particularly in respect of boundary conditions.

In the following pages, the development of models for various coastal locations in the UK and overseas is described. In each case, the features peculiar to the location and the influence of these features on the choice of boundary conditions are highlighted.

The initial examples deal with coastal and estuarine flows modelled during various studies of water quality in Hong Kong. The increasing complexity of modelling in open sea environments is illustrated with examples of modelling in the North Sea and off the coast of Northern Ireland.

In all of the studies described, the Bradford University DIVAST model [1] was used to simulate the tidal flow and dispersion characteristics.

Modelling of tidal patterns in sheltered coastal areas

Modelling of pollutant and heat dispersion at Lamma Island, Hong Kong

Lamma Island lies immediately south of Hong Kong island, in the South China sea. For most of the year, depth-averaged modelling is satisfactory to simulate the tidal conditions in the area. On neap tides, the tidal cycle is semi-diurnal. On spring tides, the tidal cycle exhibits a distinct diurnal pattern.

The general location of Lamma Island is illustrated in Figure 1 and in greater detail in Figure 2. The predominant flow patterns in the area are also shown. Prior to modelling, several important features relating to flow patterns in the area were known. Close to the shore, a deep channel has been dredged to allow access for bulk carriers unloading coal at the Lamma Island power station. Local knowledge of the area indicated that a large-scale eddy existed in Ha Mei Wan, the main bay which contains two popular bathing beaches. Apart from these short stretches of beach, the coastline is generally rocky, falling away steeply from the shore. In the area modelled, to the west of Lamma Island, the water is relatively shallow, with typical depths in the immediate area of the island of the order of 15 to 20 metres.

FIGURE 1   *Map of Hong Kong and surrounding area*

The tidal range in the area is relatively small, at approximately 2 metres on a spring tide and 1 metre on a neap. No significant phase lags between high water at different locations on the coast of Lamma Island are known to exist.

Based on an estimated tidal excursion, the required extent of the model was determined. The model covers an area of approximately 8 km by 10 km, with a cell size of 83.33m. In order to achieve numerical stability, a timestep of 24 seconds was adopted for calculations.

In the case of the Lamma Island model, modelled data on tidal flows and water surface elevations were available from the Hong Kong Government's Victoria Harbour (WAHMO) model [2]. The data were available at 250m intervals along the boundaries of the Lamma Island model.

In modelling the tidal characteristics of the area, initial efforts using the boundary combination shown in Figure 2a proved unsuccessful due to minor incompatibilities between the elevations specified on the northwestern and southeastern limits of the model. The level data along the boundaries were specified to 0.0001 m. Due to the

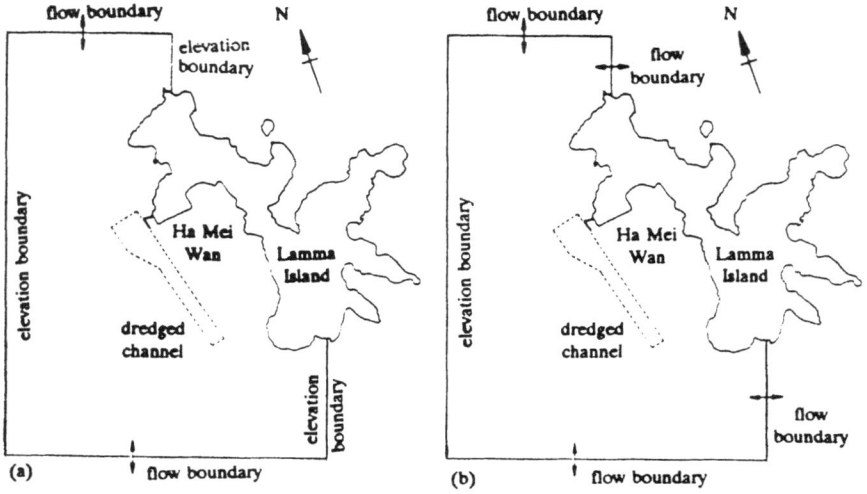

FIGURE 2   *Boundary conditions for the Lamma Island model*

relatively small distance (8 km) between these boundaries, small differences in elevation specified at opposite boundaries resulted in significant variations in flow across the model. This problem was overcome by adopting the alternative boundary combination shown in Figure 2b. By specifying flow on all except the northwestern boundary, the recorded flow patterns in the area were satisfactorily reproduced. In particular, the model was observed to reproduce the effects of refraction at the dredged channel and the formation of a gyre in Ha Mei Wan at certain states of the tide.

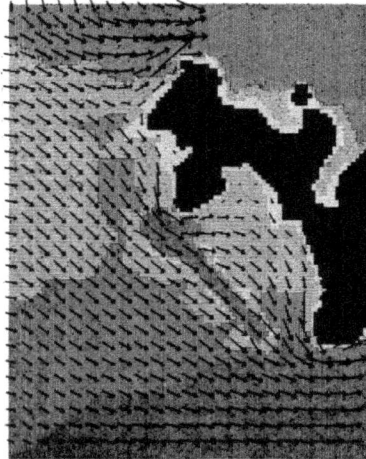

FIGURE 3   *Modelled flow patterns at Lamma Island*

After calibration, the model was used to predict the effect of a reclamation on sedimentation patterns in the areas and the dispersion of leachates from a coastal lagoon. In subsequent studies using the same model, the dispersion of effluent and thermal waste from the power station cooling water system was quantified, and the recirculation of heat discharged from the outfall through a series of nearby intakes was investigated. In addition to prediction of depth-averaged concentrations of pollutants, the model was used to predict plume behaviour after release of cooling water from the outfall. Figure 3 shows an example of the results obtained from the model during extensive studies on tidal patterns and the dispersion of effluent discharged from the power station.

Modelling of effluent dispersion near Peng Chau, Hong Kong

Increasing populations in Hong Kong have led to the development of dormitory suburbs in the outlying islands. Figure 4 shows north east Lantau and nearby islands, including Peng Chau, which have recently been developed as residential areas.

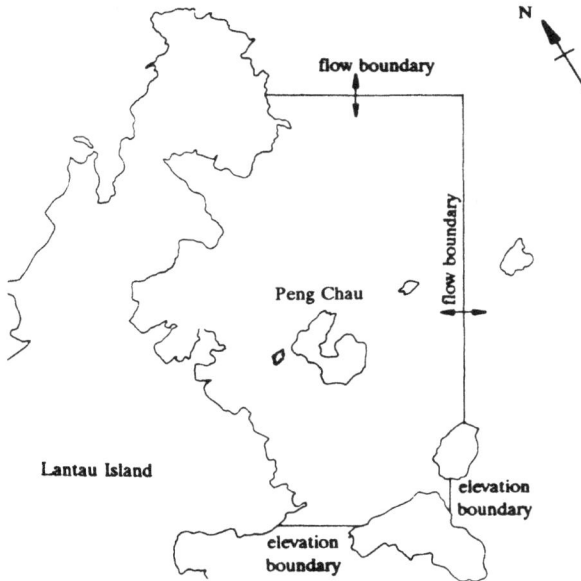

FIGURE 4   *Boundary conditions for the Peng Chau model*

The expanding population and the Hong Kong Government's stricter requirements on wastewater disposal require improved sewage disposal facilities. In order to satisfy the environment protection authorities, a study of the water quality in the area was undertaken, predicting the conditions likely to occur as a result of the improved discharge. The future quality of the bathing waters in the vicinity of local beaches was a focal point in the overall study.

The location of Peng Chau in Hong Kong is shown in Figure 1. The direction of flow of the main tidal current is approximately north-south, with islands, bays and inlets causing local recirculation and changes in overall flow direction.

The model covered an area of 6 km by 8 km. Maximum water depths of 20m are typical throughout the area. Spatial variations in depth are gradual in all areas.

As in the Lamma Island study, the DIVAST model was adopted for prediction of tidal patterns and effluent dispersion. Modifications to the software allowed modelling of interacting pollutants. Boundary conditions for the model were derived from modelled flows and water surface elevations from the Hong Kong Government's Victoria Harbour model.

In the case of the Peng Chau model, the boundaries were aligned with the direction of longshore tidal movement. They were positioned to minimise the length of open boundary whilst maintaining an adequate distance from the outfall to model the tidal excursion of effluent from the proposed discharge point.

In order to adequately define the predominant north-south flows, tidal movement at the two larger open boundaries (the northeastern and southeastern boundaries) was controlled using flows as opposed to water surface elevations. At each of these locations, significant flows are present across the boundaries throughout the tidal cycle. The southwestern boundary was used to define elevation throughout the model. As in the Lamma Island model, the limited extent of the area of interest and the small tidal range in the area resulted in little phase lag in tidal propagation across the area of the model. For this reason, definition of the water surface elevation at only one end of the model was sufficient. The conditions at the remaining boundary, a narrow strait between two islands at the southern end of the model, were controlled using water surface elevation. The boundary combination adopted is shown in Figure 4.

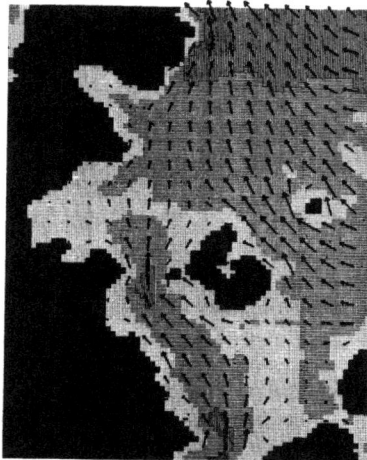

FIGURE 5   *Modelled flow patterns near Peng Chau*

The model was used to predict the dispersion and decay of pollutants discharged to the sea. The parameters investigated included dissolved oxygen, Biochemical Oxygen Demand, organic and inorganic nitrogen compounds, suspended solids and bacteria. Figure 5 illustrates the flow patterns predicted by the model.

Modelling of tidal characteristics in open sea areas

The models described in the preceding paragraphs provide an insight into model design, but represent relatively straightforward modelling conditions, particularly where a continuous input of high quality boundary data is available on a regular basis and at high resolution across the area to be modelled. Furthermore, modelling of small areas represents a straightforward scenario when compared with modelling of large expanses of tidal water, where phase differences in tidal propagation can result in significant effects on flow patterns.

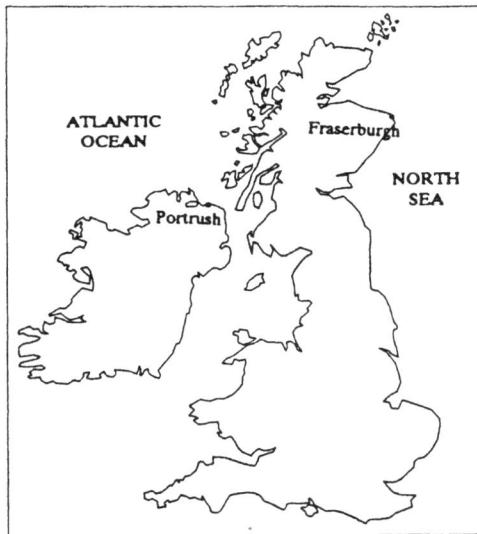

FIGURE 6   *Location of UK coastal models*

In the cases described earlier, the gradual variations in water depth and small tidal range combined to result in relatively straightforward tidal modelling. In the following cases, examples of more complex models of large areas of open sea are described, and the inherent difficulties in this type of model are discussed. In both of the cases mentioned, the areas modelled are located in the United Kingdom. Figure 6 shows the location of the areas described.

Modelling of bacterial dispersion at Fraserburgh, Scotland

In order to assess the environmental effects of a proposed long sea outfall, a tidal model for the coastal area at Fraserburgh in northeast Scotland was developed. The model was used to investigate the bacterial dispersion characteristics in the coastal waters surrounding the outfall. The study comprised an investigation of the likelihood of compliance of the proposed sewage discharge with the bacteriological standards detailed in the EC Bathing Water Directive.

The tidal characteristics in the Fraserburgh area are complicated by the interaction of tidal streams from the Moray Firth and the North Sea. In addition to this, several trenches of exceptionally deep water exist relatively close to the shore. In these zones, depths in excess of 200 metres of water are encountered, and rapid transitions between relatively shallow and deep water occur over small distances.

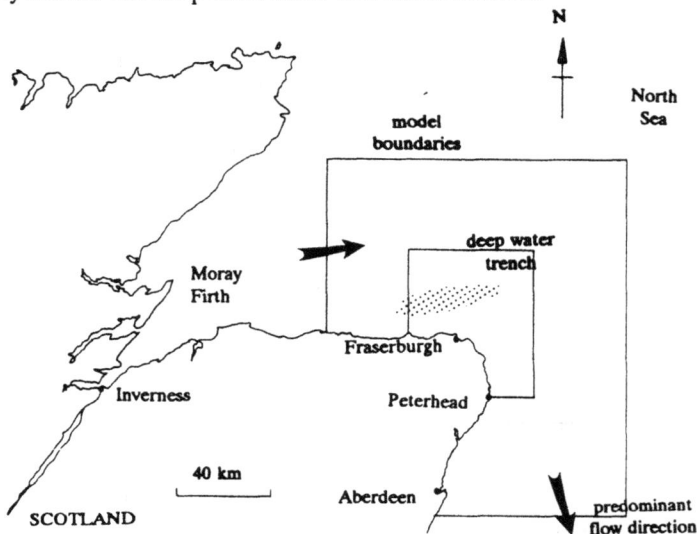

FIGURE 7   *Extent of the Fraserburgh model*

A mathematical model of the flows in the area was developed using the DIVAST model. Figure 7 shows the location of the model and the main features of the area. Boundary data for the model were available from the Proudman Oceanographic Laboratory model of UK coastal waters. However, at approximately 9 km spacing, the spatial resolution of data nodes was such that they were too sparse for direct use in the nearshore area around Fraserburgh. The spacing of data in this locality resulted in the need for a large-scale outer model which was used to produce boundary conditions for a detailed model of the area around the proposed outfall.

The outer model extended over a distance of 138 km from north to south, and 108 km from east to west. A cell size of 3 km was adopted for this model. The inner model covered an area of 39 km by 39 km, with a cell size of 333m.

For both models, it was decided to control the flows by using tidal elevations on all four of the open boundaries. This decision was based on the fact that no single predominant flow path could be identified throughout the tidal cycle. Due to the open sea environment at the headland, significant changes in flow direction occur during the tidal cycle.

In the Fraserburgh region, water depths close to the shore are often in excess of 40 m. Water depths greater than 200 m are found in trenches within 8 kilometres from the coast. During model development, problems were encountered in computing flow patterns in regions of sudden variations in water depth. Due to the rapid depth changes, the principles of depth-averaged flow are unlikely to apply fully within the deepest parts of these trenches. In order to model the flows across these regions, criteria to restrict the effective depth were applied.

The models were calibrated and verified by checking velocities against those recorded at coastal current meters sites. Figure 8 illustrates the flow patterns predicted by the model. Modelled tidal elevations at ports in the area were checked against data from Admiralty publications.

In the case of the Fraserburgh models, correlation between modelled and recorded data was achieved without defining flows on any of the boundaries. Specification of water surface elevation on all boundaries was seen to produce a satisfactory solution. The scale of the models was such that differences in tidal elevation between opposite boundaries were sufficient to permit relatively accurate definition of flows on this basis alone.

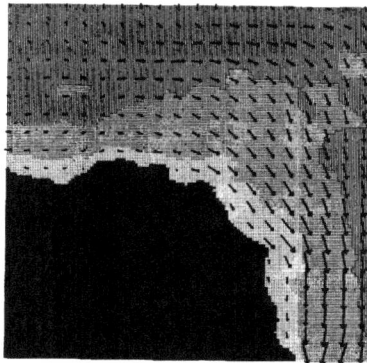

FIGURE 8   *Modelled flow patterns at Fraserburgh*

The models were used to predict bacterial dispersion from a proposed new long sea outfall. Float tracking and dye dispersion tests carried out subsequent to model calibration confirmed the validity of the model for predictions in the coastal area around Fraserburgh.

Tidal modelling for effluent discharge at Portrush, Northern Ireland

Perhaps the most complex of the models described in this paper is a model recently developed for the coastal area around Portrush, on the north coast of Northern Ireland. As in the Fraserburgh study, the model was primarily developed to assess compliance of effluent discharges with EC bacteriological standards.

The area modelled is shown in Figure 9. The tidal patterns in this area are dominated by several unusual features. Firstly, on each tidal cycle, a large volume of water flows through the constricted opening of Lough Foyle influencing flow patterns up to several kilometres offshore. In the nearshore region strong tidal streams flowing past islands and headlands produce eddies at various locations along the coast.

The second important feature in this area is the presence of an amphidromic point on the shores of Islay, off the west coast of Scotland, north of the area under investigation. The effect of this feature is to cause rapid changes in tidal range over relatively small distances and unusual characteristics in the phase of the tidal wave between neighbouring ports. The direction of progression of the tidal wave along the coast is seen to vary, depending on the tidal range. High water progresses along the coast from west to east on neap tides. Depending on the magnitude of spring tides, high water in the central region of the area of interest occurs simultaneously or after that at the eastern boundary.

FIGURE 9 *Extent of the Portrush model*

154

Water depths in the area vary dramatically, in some instances increasing by over 100m over a horizontal distance of 1 kilometre. The bed topography varies significantly within the model; the gradual changes in depth in the western area are in sharp contrast to the eastern area, where depths of the order of 50 m are encountered within 700m of the shore.

The tidal streams in this area are strong, with velocities in excess of 3 knots occurring in offshore areas. Close to the coastline, tidal currents accelerate through a narrow strait between a group of small islands and a rocky promontory at Portrush. Velocities of up to 2½ knots occur regularly in this area.

Boundary data for the Portrush model were available from the same source as those for the Fraserburgh model. At 9 km spacing, data nodes were relatively sparse for accurate definition of tidal conditions along the boundaries of the model.

Due to the sparsity of boundary data, a large scale outer model was developed, to generate adequate boundary conditions for more detailed models. The full extent of the outer model was 42 km along the coast, and 35 km perpendicular to the coast. The selected grid size was 600m.

For this model, a combination of elevation and flow boundaries was adopted. The inner area of Lough Foyle was not modelled, but the effect of its presence on the coastal zone was modelled by including a flow boundary at the narrowest point of the estuary. Using existing data on velocities at the estuary mouth and computed tidal volume changes within the estuary, boundary conditions were produced. The use of a flow boundary in this location allowed detailed specification of the process which has a major influence on pollutant transport in the nearby coastal zone.

The open sea boundaries were modelled by aligning the outer model with the predominant tidal stream and specifying the flow on the eastern and western limits of the model. Tidal elevations were controlled along the northern boundary of the model. The main reason for this was the need to specify elevation over the width of the model, in order to account for phase differences across the area. In the finer scale models, the flow parallel to the boundary was also specified in these cells.

Once the offshore tidal patterns were established, an inner model with finer resolution was developed. In order to achieve adequate definition of flows through straits between a group of islands close to the coast, a grid size of 100 metres was adopted for the inner model. This model was constructed at a different orientation to the outer model. This was due to the fact that tidal movement is predominantly parallel to the shore in this region. The modified orientation also resulted in improved definition of coastal features.

Calibration of the models was achieved by comparison with flow data recorded in the coastal region during the course of the study. Initial calibration tests showed that, while predicting the correct tidal range at Portrush, the model significantly underestimated flows in both coastal and offshore areas. Due to the deep water conditions which applied throughout most of the model, adjustment of friction had no significant effect in altering flows throughout the model. Considerable adjustment of the original boundary conditions was required to produce flows comparable to those recorded. The required increases in flow were achieved mainly by alteration of the phase and amplitude of tidal elevations across the model. The final results showed correlation between modelled and recorded flows, with satisfactory prediction of tidal range and phase at ports in the area. Figure 10 shows an example of the flow patterns predicted using the model.

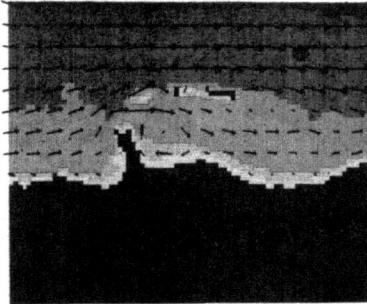

FIGURE 10   *Modelled flow patterns at Portrush*

Conclusions

The preceding paragraphs have illustrated the philosophy developed at Binnie & Partners for design of two-dimensional models of coastal and estuarine environments. The increasing demand for numerical modelling as a implicit part of engineering studies for both effluent disposal and reclamation works has resulted in the need for efficient methods for model production. Bearing in mind the complex nature of coastal modelling and the many variables involved, the ability to form a systematic approach to model development is a necessity if the methods are to become a widely accepted practice for solution of the problems and uncertainties encountered. As demonstrated in the examples quoted, the single most important factor in model design lies in the modeller's ability to recognise and interpret the salient tidal characteristics prior to model development.

REFERENCES

[1]   Falconer, R. A., (September 1984), A Mathematical Model Study of the Flushing Characteristics of a Shallow Tidal Bay. Proceedings of the Institution of Civil Engineers, part 2, Vol 77, pp 311-332.

[2]   Government of Hong Kong, Territory Development Department, Urban Area Development Office (April 1989), Hydraulic and Water Quality Studies in Victoria Harbour, Final Report - Part 1, Mathematical Modelling. Binnie & Partners (Hong Kong).

14 Sensitivity of a 2-dimensional hydrodynamic model to boundary conditions

A. M. Cawley and M. Hartnett

Abstract

Computer based numerical models are firmly established as practical tools for solving hydraulic-engineering problems. Finite difference and finite element techniques have been used to develop general purpose software to solve the equations of surface water flow and solute transport. [3,7] Such applications are now widely used throughout the engineering community for planning and design purposes. In order to achieve the best results from a model an understanding of the basis of the model is imperative, in particular it is necessary to understand how varying boundary conditions of the model affect the results.

This paper details the application of a two-dimensional hydrodynamic model, DIVAST, to a complex waterbody [4]. The boundary conditions of the model are discussed, and the sensitivity of the model results to changes in some of the boundary conditions is examined. The results from the sensitivity analyses are presented and discussed.

Introduction

Hydrodynamic modelling of bays and estuaries is used to predict the circulation patterns and water surface elevations of the waterbodies for various meteorological and astronomic conditions [8]. Typically a study is carried out in a number of stages: divide the domain into discrete grid points; define the bathymetry and topography of the domain; calibrate and validate the model; and finally use the model for predictive purposes [9]. All of these stages are important for the successful completion of the study, however, the calibration and validation stage is considered the most important aspect of the study.

Calibrating and validating a model are two distinct processes. Calibration means tuning a model such that it is forced to imitate some physical phenomena as closely as possible. With regards to hydrodynamic modelling this procedure is generally carried out by varying model parameters until model results compare favourably with measured values of current speeds and directions and water surface elevations. Validation means proving that the calibrated model accurately describes physical phenomena at locations in the domain and for environmental conditions other than these used to calibrate the model.

The assignment of correct boundary conditions plays an important role in calibrating a model, thus it is necessary to understand which boundary conditions most influence the results of the model. The following sections describe the theory and application of a two-dimensional finite difference hydrodynamic model, DIVAST (Depth Integrated Velocity and Solute Transport), to a complex waterbody, namely, Killary Harbour, Co. Galway, Ireland. The nature of Killary Harbour and the construction of the model are described in detail. The sensitivity of the model to the prescribed boundary conditions is considered by comparing model results derived for different sets of boundary conditions against measured physical oceanographic data.

A problem frequently encountered when carrying out hydrodynamic modelling studies for clients is: At what stage is the model considered validated?. A model that appears well validated to one assessor may seem unsuitable to another. To address this problem, a simple methodology for assessing the efficiency of the model is outlined. By using this methodology the hydraulic modeller can determine when the specified boundary conditions and empirical coefficients are generating sufficiently accurate hydrodynamic conditions within the domain.

Mathematical Formulation of Hydrodynamic Model

The equations of motion used in two-dimensional (depth integrated or depth averaged) hydrodynamic models are derived using two basic laws of physics [1], namely;

    i)   The law of conservation of mass.
    ii)   The law of conservation of momentum.

Consider a parallelpiped, as shown in Fig. 1, then the first law can be stated as:

The increase in fluid mass into a parallelpiped of dimensions Δx, Δy, $H(t)$ over an increment of time Δt is equal to the net transfer of fluid mass into the parallelpiped during time Δt:

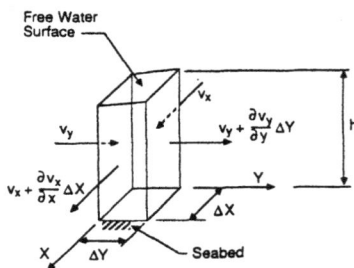

FIGURE 1. *Flow Through a Parallelpiped*

$$\rho \Delta x \Delta y \frac{\partial H}{\partial t} \Delta t = -\left(\frac{\partial V_x}{\partial x} \Delta x \Delta y + \frac{\partial V_y}{\partial y} \Delta y \Delta x \right) \rho H \Delta t + \rho Q \Delta t$$

(1)

Dividing across Eq. (1) by $\rho \, \Delta x \, \Delta y \, \Delta t$ and rearranging terms gives the mass continuity equation for the model as:

$$\frac{\partial H}{\partial t} + H \frac{\partial V_x}{\partial x} + H \frac{\partial V_y}{\partial y} = \frac{Q}{\Delta x \, \Delta y}$$

(2)

The second law applied to the parallelpiped may be stated as:

The increase in fluid momentum in the parallelpiped over an increment of time Δt is equal to the net transfer of impulse momentum into the parallelpiped during Δt. This can be expressed mathematically as:

$$\frac{d}{dt} \left[\int_{\Delta V} \rho \, \underset{\sim}{v} \, dV \right] \Delta t = \int_{\Delta S} \underset{\sim}{f}_s \, \Delta t \, dS + \int_{\Delta V} \underset{\sim}{f}_c \, \Delta t \, dV$$

(3)

Equation (3) can be written as two scalar differential equations in two mutually perpendicular directions by firstly depth integrating and then simply evaluating the integrals. For a constant density turbulent fluid flow on a rotating earth, the depth integrated momentum equation for flow in the x horizontal co-ordinate direction can be expressed as [5]

$$\frac{\partial V_x H}{\partial t} + \beta \left[\frac{\partial V_x^2 H}{\partial x} + \frac{\partial V_x V_y H}{\partial y} \right] - f V_y H + g H \frac{\partial \eta}{\partial x} - \frac{\rho_a C^* W_x (W_x^2 + W_y^2)^{1/2}}{\rho}$$
$$+ \frac{g n^2 V_x (V_x^2 + V_y^2)^{1/2}}{H^{1/3}} - \varepsilon H \left[2 \frac{\partial^2 V_x}{\partial x^2} + \frac{\partial^2 V_x}{\partial y^2} + \frac{\partial^2 V_x}{\partial x \partial y} \right] = 0$$

(4)

A similar expression can be derived for flow in the y horizontal co-ordinate direction.

Likewise, the depth integrated form of the conservation equation (2) can be written as:

$$\frac{\partial \eta}{\partial t} + \frac{\partial V_x H}{\partial x} + \frac{\partial V_y H}{\partial y} = \frac{Q}{\Delta x \, \Delta y}$$

(5)

These three differential equations must be solved throughout the fluid domain for the unknown variables Vx, Vy and H. The solution of these differential equations may be obtained using either finite difference or finite element solution schemes. In the case of this study an alternating direction fully centered implicit finite difference solution scheme [4] was used.

Killary Harbour Physical Oceanography

Killary Harbour is a fjord - like inlet on the west coast of Ireland, located approximately 50km due north of Galway Bay, see Fig. 2. Killary Harbour is about 13km long and 0.7km wide with a mean depth of 20m [13]. Outside the mouth of the harbour, the approaches to the harbour open to a wide embayment exposed to Atlantic swells. The water structure in the harbour varies between being semi-stratified and well mixed, depending on surface water runoff into the harbour and wind conditions [10]. However, the water structure at the mouth of the harbour exhibits little or no stratification and a two-dimensional model is considered adequate for hydrodynamic modelling of this area.

The mean spring tidal range in Killary Harbour is 3.6m which induces strong rectilinear currents of 0.5m/sec at the entrance to the harbour [12]. The water circulation in the approaches to the harbour is significantly influenced by southwesterly winds, which cause longshore currents to be generated, particularly on the flood tide. On the ebbing tide the strong currents generated at the mouth of the harbour behave as a jet of water discharging into the approaches to the harbour. This jet induces eddy currents to the north and south of the jet, resulting in areas of siltation.

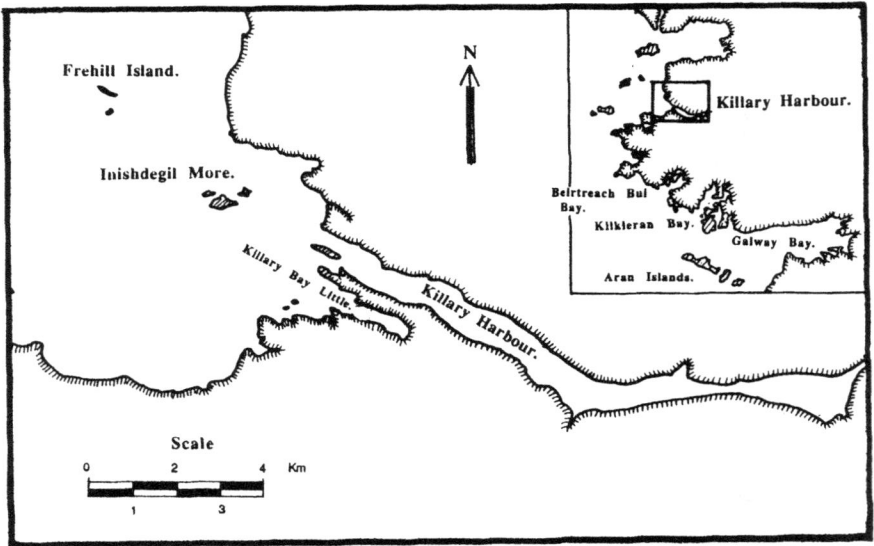

FIGURE 2   *Killary Harbour with Approaches*

Hydrodynamic Model

A finite difference hydrodynamic model of Killary Harbour with approaches was constructed by overlaying a finite difference mesh over Admiralty Chart No. 2706 and defining the bathymetry at the mesh grid points. The mesh grid points were equally spaced at 125m in two mutually perpendicular directions, see Fig. 3. The mesh consists of 131 grid points along the I-axis and 76 grid points along the J-axis (i.e. western boundary). Fig. 4 shows an isometric view of the bathymetry of the mouth of Killary Harbour as modelled in this study. The orientation of the view is 90° from true north.

Each grid point within the domain is specified as being either a land node, a wet node, a wet/dry node or a boundary node. A land node implies that this grid point never gets flooded and is not included in the computational domain. A wet node implies that the elevation of the bathymetry at that grid point is below the lowest astronomical tide and hence is always flooded. A wet/dry node implies that the elevation of the bathymetry at that grid point is

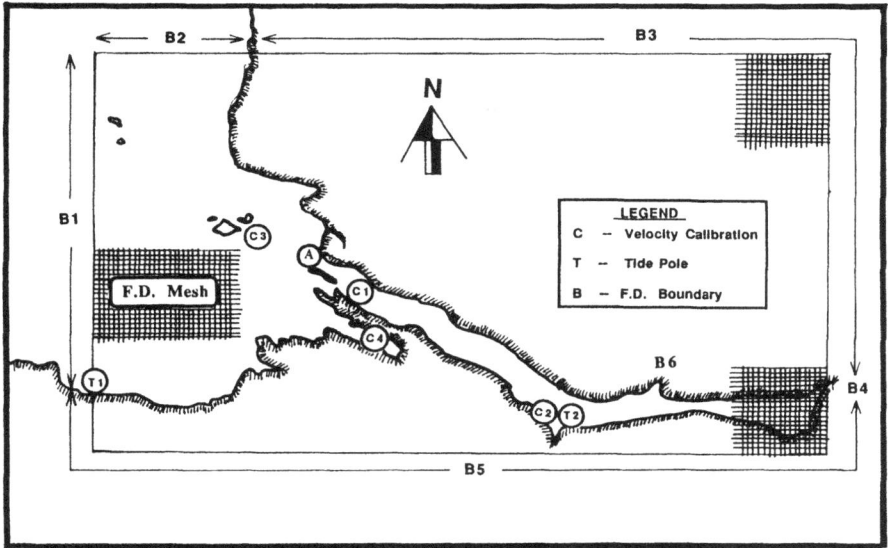

FIGURE 3   *Finite Difference Model of Killary Harbour*

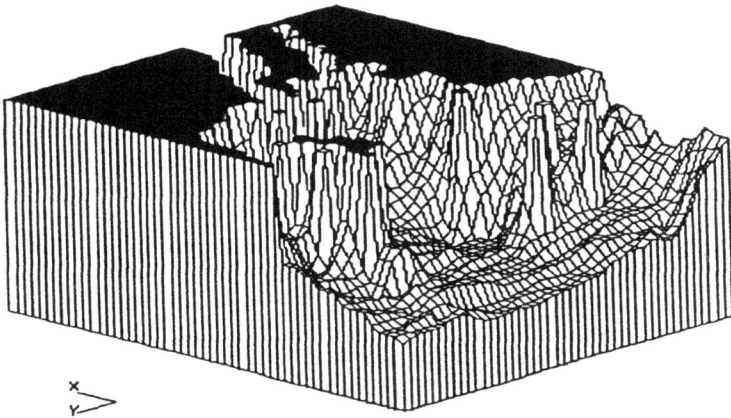

FIGURE 4   *Isometric View Of the Mouth of Killary Harbour*

between mean high water and the lowest astronomical tide and the grid point will alternate between being flooded and dry depending on the stage of the tide and on the roughness coefficient specified at that location [6]. A grid point is considered a boundary node when a constraint is specified at the node, in particular the tidal height or current velocity is specified over the duration of simulation at open-sea boundary nodes [2].

Sensitivity of Model to Boundary Conditions

Numerical hydrodynamic models such as DIVAST, are more sensitive to the choice of specified boundary conditions specified than to other model parameters such as bed friction coefficients or eddy viscosity. The sensitivity of the model to boundary conditions plays an important role in the choice of the boundary conditions and location of the boundaries.

The boundary conditions used in the hydrodynamic model include specifying zero normal velocities along the closed land boundaries with tidal heights or flux densities being specified along the open boundaries. The type and location of these boundaries are shown in Fig. 3. In all coastal hydrodynamic studies a water elevation boundary is required in order to simulate the tidal dynamics. However, the specification of a water elevation boundary requires that circulation occurs normal to the boundary and that the parallel component of flow is zero. Subsequently, given that the flow direction in the approaches to Killary Harbour is principally East-West, a tidal elevation boundary was chosen along the western boundary. The water elevation boundary condition requires that uniform hydrodynamic conditions prevail along the boundaries, to provide these it is recommended that bathymetry data near the open sea boundary are smoothed to reduce large variations in water depths between adjacent grid squares. Tidal elevations were specified at a single grid square on the open boundary at each time step over the tidal cycle. To ensure that recirculation along the open boundary does not occur at slack water (i.e. near high and low tide), the remaining grid squares on the boundary were adjusted for the Coriolis effect. The correction to the tidal height at the m'th boundary grid square on the water elevation boundary can be expressed in finite difference form as follows:

$$\eta_m = \eta_{m-1} + \frac{\Delta y f}{2g}(U_m + U_{m-1})$$

(6)

Flow boundaries were introduced along the northern open-sea boundary and at respective freshwater inflow locations, namely the Erriff (B4) and the Bundorragha (B6) rivers. It is important that the location of open boundaries, particularly water elevation boundaries, are chosen sufficiently distant from areas of interest to ensure that boundary effects do not influence the results. From experience the authors have found that this boundary should be at least 10 grid squares from areas of interest.

Tidal heights may be specified as either a discrete series of surface elevation heights over the tidal cycle or as a smooth continuous function, such as a sine curve or some high order polynomial. A number of model sensitivity analyses were carried out on the effects of inputting the tidal boundary condition as a series of discrete tidal heights and as a sine curve. Fig. 5(a) and (b) shows the hydrodynamic output of current speeds and directions at observation site A when the tidal heights at the open boundaries were specified as a series of discrete tidal heights. These tidal heights were extracted at half hourly intervals from measured tidal data at T1 and were filtered for spurious values. A linear interpolation procedure was applied to determine the tidal heights at every half time step from the series of half-hourly tidal inputs . However, this tidal input generated high intense numerical oscillations in the hydrodynamic output as is evident from Fig. 5(a). The cause of such

162

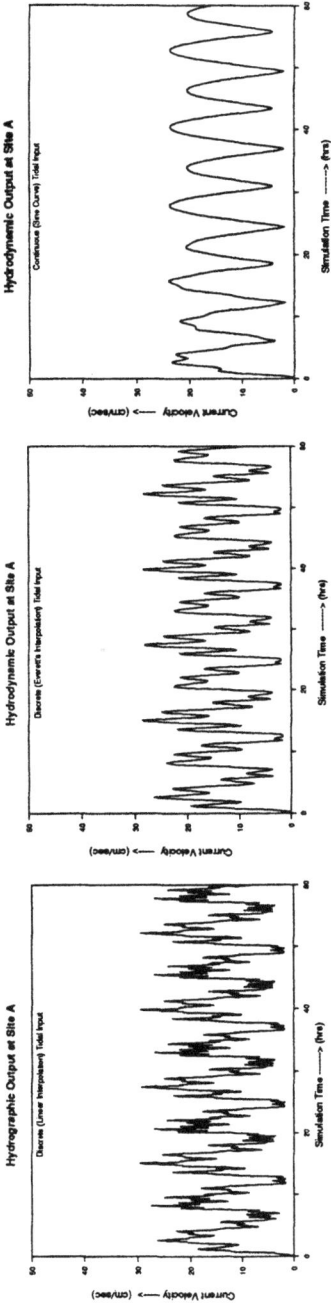

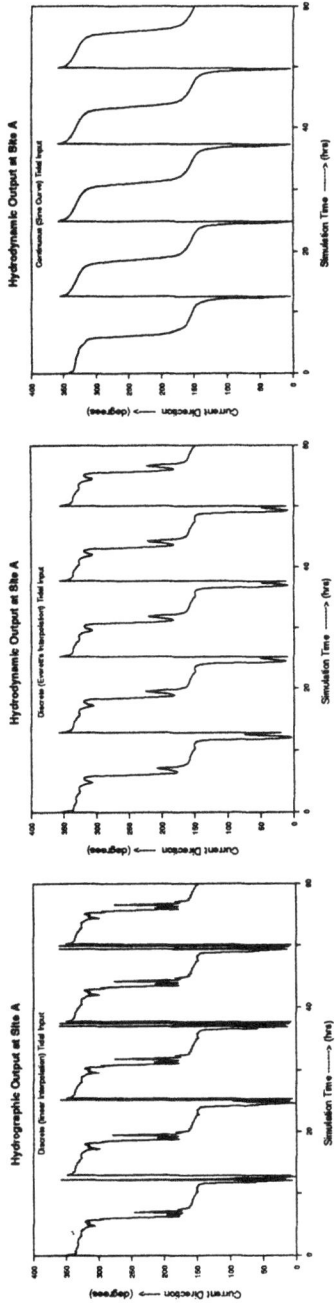

FIGURE 5 (a), (b) & (c)   Hydrodynamic Output at Site A for Discrete and Continuous Tidal Inputs.

163

intense oscillatory behaviour can be principally attributed to relatively large changes in the slope of the tangent to the tidal curve due to linear interpolation between data points. To overcome the above problem Everett's interpolation formula [15], which fits an interpolation polynomial based on even-order central difference to the data, was applied. The degree of oscillatory behaviour in the hydrodynamic output was reduced considerably as can be seen from Figure 5(b). However, some oscillations in the current velocity time series remained.

Fig. 6 presents multiple graphs of spring tidal curves for Killary Harbour reduced from measured data at T1, the standard port (Galway) tidal curve for Killary Harbour and a sine curve distribution. It is evident from Fig. 6 that a sine curve well represents both the measured and standard port tidal curve distributions. It is the experience of the authors that in general the tidal curves for open coastal embayments along the western and southern seaboards of Ireland are well represented by a sine curve. A hydrodynamic analysis was carried out using a sine curve with amplitude equal to the tidal range of 4.2m previously used in the discrete tidal input simulations and a period of 12.4 hours equivalent to the measured tidal cycle for Killary Harbour. The sine curve simulation produced smoother current velocity and water elevation time series than the discrete tidal input simulations (see Figs. 5 (a), (b) and (c)). Because water bodies are damped viscous systems the current velocity and water elevation time series distributions of an embayment should not generally exhibit oscillatory characteristics.

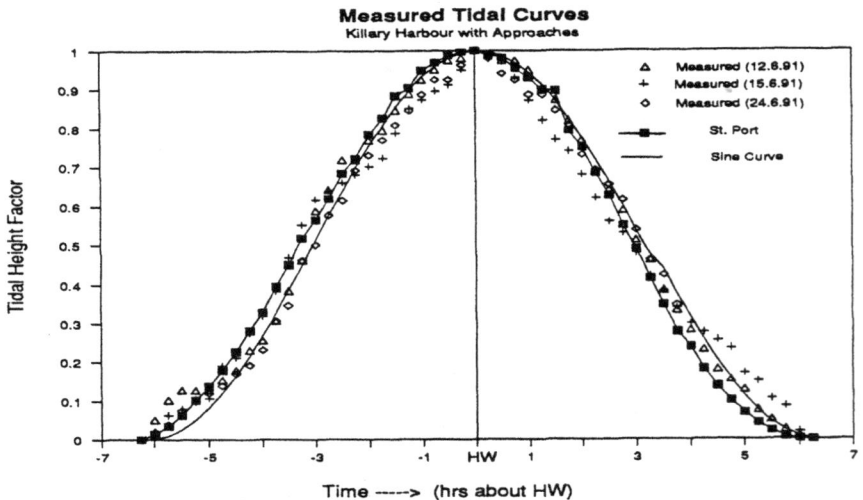

Measured Tidal Curves
Killary Harbour with Approaches

FIGURE 6   *Measrued and Modelled Tidal Curves for Killary Harbours*

From other hydrodynamic modelling work carried out by the authors it has been found that when a sine curve is used to specify tidal dynamics the model acts in a well behaved manner. Conversely, analyses using discrete tidal height inputs have shown that the hydrodynamic output can vary considerably for different size of the time steps. Figs. 7 and 8 present results of a time step sensitivity analysis for both a sine curve and discrete tidal height inputs for Casheen Bay, Co. Galway, Ireland. As can be seen from Fig. 7 little variation in current

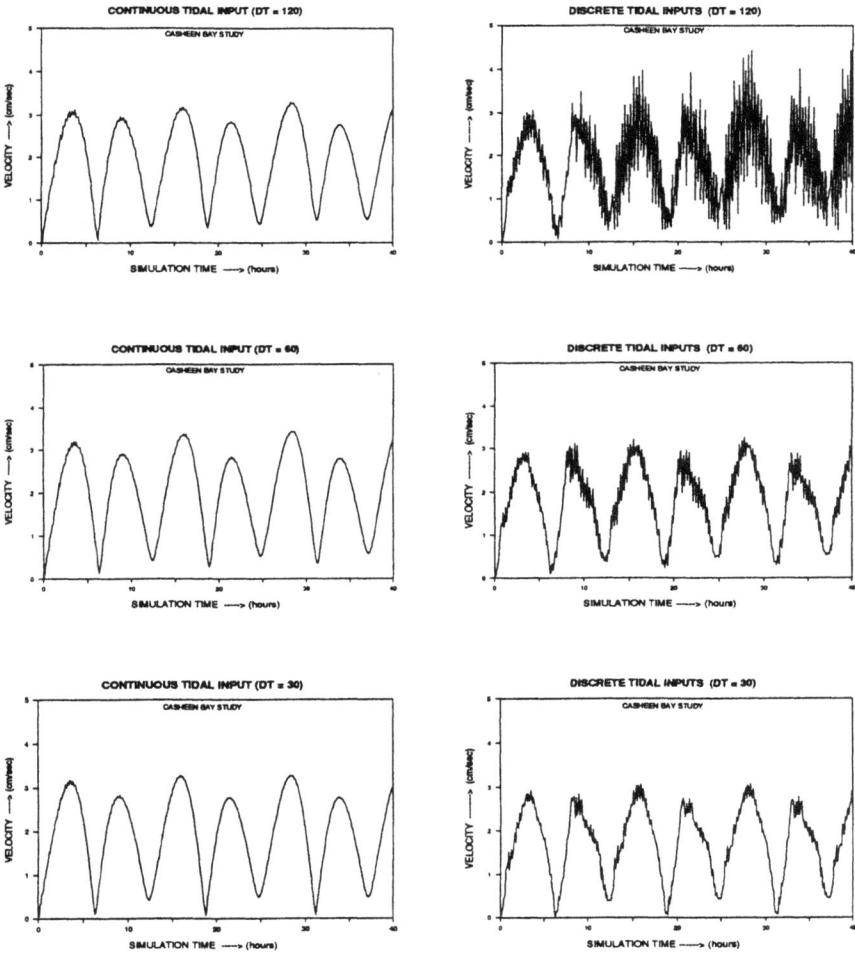

FIGURE 7 *Time Step Sensitivity Analysis - Casheen Bay, Co. Galway*

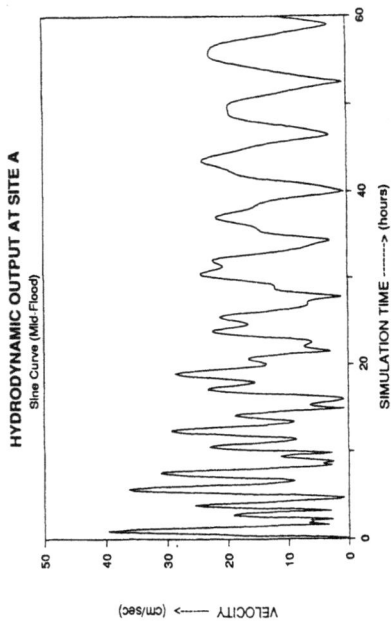

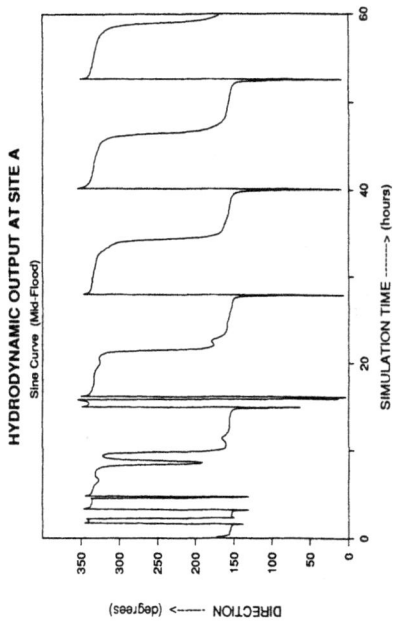

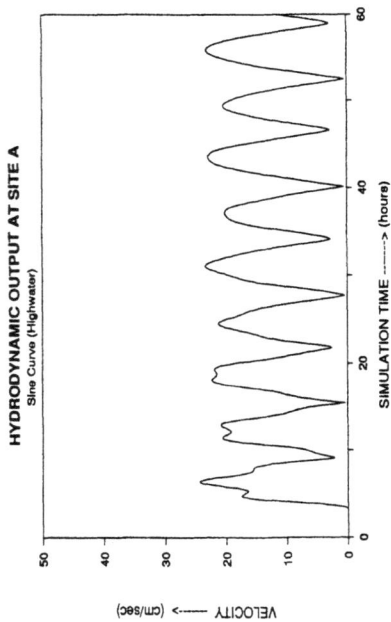

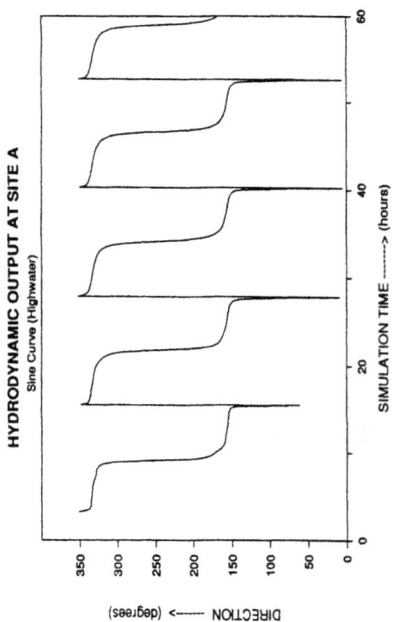

FIGURE 9 *Hydrodynamic Analysis (Mid-Flood)*

FIGURE 8 *Hydrodynamic Analysis (Highwater)*

166

velocities occurs for the sine curve simulation, while for the discrete tidal input simulation the current velocity results for the different time steps varies considerably. It may also be observed that as the time step is reduced the discrete tide simulation resembles more closely the sine curve simulation results. Normally the acceptable time step for such models are governed by the Courant number. It is recommended for the purpose of accuracy that the Courant number for implicit finite difference schemes be less than 8 [14]. However, when specifying discrete tidal inputs the analysis time step may need to be smaller than that specified by the Courant number constraint to produce realistic results as was found with the Casheen Bay analyses.

Inherent in most finite difference hydrodynamic models are cold start effects which cause numerical instabilities at the start of simulations. Analyses carried out on the Killary Harbour model have revealed that the stage of the tidal cycle at which the simulation commences influences the degree and duration of these numerical instabilities. Figs. 8 and 9 presents the hydrodynamic results at site A for two analyses commencing at highwater and mid-flood respectively. The effects of the cold start for the analysis commencing at highwater are less intense and dissipate out much faster than those of the analysis commencing at mid-flood as shown in Figs. 8 and 9. The reasons for this are that the slope of the tidal curve at highwater is zero while the the slope at mid-flood is at a maximum causing a greater shock to the model. As a result, it is recommended that simulations should be started at either high or low tides to reduce the degree and duration of the cold start effect.

Calibration of Hydrodynamic model

The previous section discusses the sensitivity of model results to changes in boundary conditions. The model was calibrated by selecting boundary conditions and empirical coefficients such that model results compare favourably with measured data. In this study the model was calibrated against measured values of current speeds, directions and water surface elevations.

One of the difficulties encountered when calibrating a hydrodynamic model is to determine when the model results are considered accurate enough. When calibrating a model it must be bourne in mind that physical measurements are also subject to error, particularly during low flows when boat movement can affect the readings. The authors suggest that the accuracy of the model can be determined by computing the mean percentage difference between recorded and predicted water surface elevations and current velocities. If we consider the graph in Fig. 10, the mean percentage difference between the measured and predicted current velocities is given as:

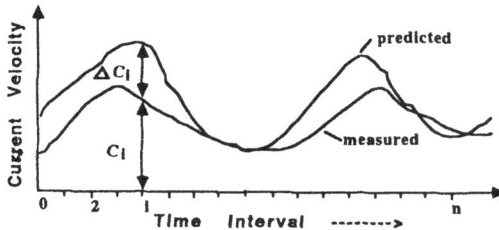

FIGURE 10. *Error Estimation*

$$E = \left[\frac{1}{N+1} \sum_{i=0}^{N} \varepsilon_i \right] \times 100$$

(7)

When comparing current velocities values of percentage difference (E) between 10% and 20% are considered practical. Whereas a value of 5% is considered practical when comparing water surface elevations. It is suggested also that an upper bound be placed on any single value of ε_i.

A field survey was carried out in June 1991 to establish the tidal regime at the open boundary of the model and to provide calibration data throughout the domain. The field survey included water surface elevation measurements at two locations for spring and neap tidal conditions and continuous recording of current speeds and directions over fifteen days at three locations. The locations where these measurements were recorded are shown in Figure 3. Further tidal flow data was made available from a previous field survey carried out by Hensey Glan Uisce Teo at two sites in November 1987. During both surveys current measurements were recorded at a number of depths below the water surface. Because the hydrodynamic model used by MCS was of the depth integrated type, a comparison between measured and computed values was carried out by averaging the measured data using the formula:

$$V_{av} = \frac{1}{(H + L)} \int_{-H}^{L} v_i \, dx_i \tag{8}$$

where V_{av} is the depth integrated current, H is the water depth from the still water line to the seabed, L is the tidal amplitude and V_i is the recorded current data at depth X_i.

A sinusoidally varying water surface elevation was applied at boundary B1 specifying the tidal height and period as measured at T1. At the early stages of calibration low flows were specified across boundary B2 using sine curve distributions with different amplitudes and phases. However best results were obtained by specifying streamline flow conditions along boundary B2. Wind conditions measured during the course of the fieldwork were applied over the surface of the model. Estimated mean annual river inflows determined using a water balance approach, were specified for both freshwater inputs as no actual flow records existed.

The following empirical coefficients were specified for all calibration and subsequent model runs:

(i)	Bed resistance	30 mm
(ii)	Eddy viscosity coefficient	1.000
(iii)	Wind energy transfer coefficient	0.0026

The model results were compared against hydrographic measurements recorded at the locations shown in Fig. 3. Good agreement was obtained in each case. Comparisons between measured and recorded current speeds and directions and water surface elevations at locations C1 and T2 are represented in Figs. 11 and 12 respectively.

Conclusions

The application of a finite difference based numerical hydrodynamic model to Killary Harbour, Co. Galway, Ireland, has been described in detail. The background mathematics to the model were detailed and the sensitivity of this type of model to certain boundary conditions were discussed.

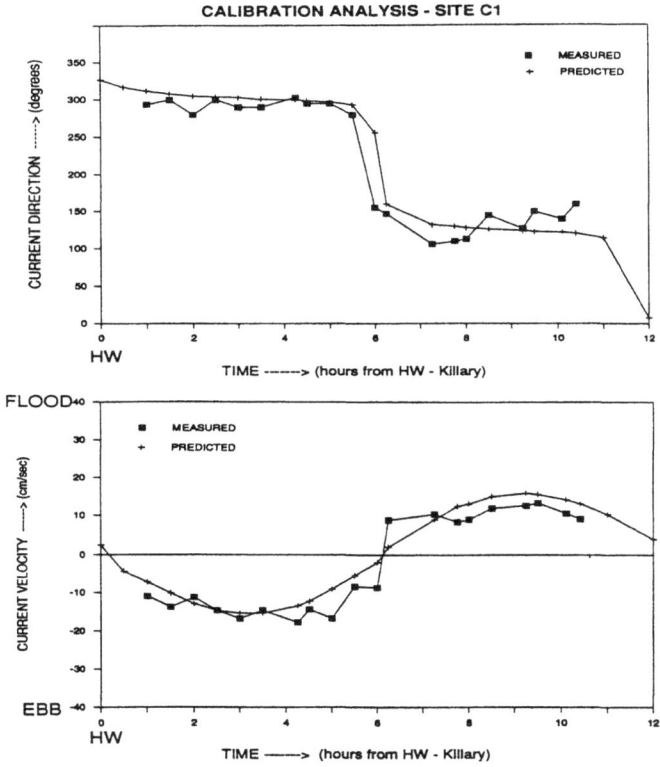

FIGURE 11   *Current Velocity Calibration Analysis - Site C1*

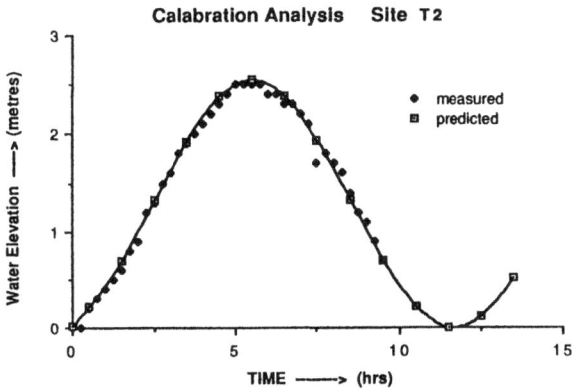

FIGURE 12   *Water Elevation Calibration Analysis - Site T2*

The main finding of the sensitivity analyses can be summarised as follows:

- When possible the tidal dynamics at an open sea boundary should be specified as a sine curve.

- When tidal dynamics cannot accurately be represented by a sine curve a higher order interpolation function should be fitted to the measured data and then specified at the relevant boundary.

- To achieve sufficient accuracy it may be necessary that the model be constructed so that the Courant number is less than 8 when using discrete tidal inputs. Sensitivity analyses should be carried out to determine the time step for such analyses.

- Variations in empirical coefficients of the model, such as eddy viscosity and bed roughness, do not have as significant an effect on model results as variations in boundary conditions.

- "Cold start" numerical instabilities are minimised when simulations are commenced at times of zero slope on the tidal curve.

- Hydrodynamic models are calibrated and validated against measured data, thus it is necessary to ensure that the accuracy of the measured data is within acceptable limits.

References

1) Abbott, M.B., (1979) "Computational Hydraulics - Elements of the Theory of Free Surface Flow", Pitman Publishing Ltd., ISBN 0-273-01140-5.
2) Blumberg, A.F., and Kantha, L.H., February (1985), "Open Boundary Condition for Circulation Models", Journal of Hydraulic Engineering, ASCE, Vol. 111, No. 2, pp. 237-255.
3) Brebbia, C.A. and Partridge, P., (1975) "Finite Element Models for Circulation Studies", Mathematical models for Environmental problems, Proceedings of International Conference held at University of Southampton, Pentech Press, London.
4) Falconer, R.A., (1986), "A two-dimensional mathematical model study of the nitrate levels in an inland natural basin". International Conference on Water Quality Modelling in the Inland Natural Environment, BHRA Fluid Engineering, Bournemouth, England. pp 325 - 344.
5) Falconer, R.A, (1991), "Review of Modelling Flow and Pollutant Transport Processes in Hydraulic Basins", Proceedings of the First International Conference on Water Pollution: modelling, Measuring and Prediction, Southampton, UK, Computational Mechanics Publications, pp. 3-23.
6) Falconer, R.A. and Chen, Y., (1991), "An improved Representation of Flooding and Drying and Wind Stress effects in a 2-D Tidal Numerical Model", Proceedings of the Institution of Civil Engineers, Part 2, Research and Theory, Vol. 91.
7) Fischer, H.B., List. E.J., Koh, R.C. Y., Imberger, J., and Brooks, N.H.., (1979) "Mixing in Inland and Coastal Waters". Academic Press, 1979, ISBN 0-12-258150-4.
8) Harlemann, F.A., (1986) "Estuary and Coastline Hydrodynamics" Ippen, A.T. (Ed.), McGrath - Hill Book Company Inc.
9) Hartnett, M., (1991) "Water Quality Modelling - A Case Study", Proceedings of the First Environmental Engineering Conference, University College Cork, Ireland, pp 377-

10) Keegan, F.B. and Mercer, J.P, (1986) "An Oceanographical Survey of Killary Harbour on The West Coast of Ireland" Royal Irish Academy, Vol. 86, B, No. 1.

11) Nihoul, J.C.J, (1975) " Modelling of Marine Systems" Elsevier Scientific Publishing Company, Amsterdam.

12) Macmillian, (1991) " The Macmillian and Silkcut Nautical Almanac", Macmillian Press.

13) MCS, (1991) "Solute and Particulate Waste Dispersion Analysis of Killary Harbour and its Approaches" Unpubl.

14) Smith, G.D., (1985)"Numerical Solution of Partial Differential Equations: Finite Difference Methods", 3rd Edition, Oxford University Press, ISBN 0-19-859650-2.

15) Spencer, A.J.M et Al., (1977) " Engineering Mathematics Volume 1" Van Nostrand Reinhold Company Limited.

Nomenclature

C^*	air-water interfacial resistance coefficient
f	coriolis parameter
f_c	Coriolis force vector per unit volume
f_s	surface force vector (f_{sx}, f_{sy}) per unit surface area along ΔS
g	graviational acceleration
H	the instantaneous water depth, parallelpiped
n	mannings bed roughness coefficient
Q	external flowrate into a parallelpiped
Q_{in}	net water inflow to waterbody
Q_{out}	net water outflow from waterbody
V	Volume of waterbody
$v = (v_x, v_y)$	two-dimensional velocity vector
V_x, V_x	depth mean velocity components in x, y directions
W_x, W_y	wind velocity components in x, y directions
β	correction factor for non-uniformity or vertical velocity profile
ΔS	surface enclosing the parallelpiped
Δt	time step
ΔV	volume of the parallelpiped
$\Delta x, \Delta y$	horizontal grid dimensions
ε	depth mean eddy viscosity
η	water surface elevation above or below chart datum
ρ_a	air density
ρ	water density
η_1	specified tidal height at 1st grid point on boundary
U_m	the normal velocity component at the m'th grid location
ΔC_i	the absolute difference
C_i	the i'th measured current velocity or water elevation
ε_i	the i'th fractional difference
E	the mean percentage difference

Part 2
WATER QUALITY MODELLING

15 Application of a 2-D depth integrated model for tidal flow and pollutant transport to the northern shore of the Firth of Forth

J. P. Hodder, N. J. Parkinson and R. A. Falconer

Abstract

A two dimensional model for tidal and wind driven flow and solute transport has been applied to simulate the dispersal of sewage effluent along the northern shore of the Firth of Forth, in the U.K. The model has been validated against measurements carried out in the present and previous studies and run to determine outfall lengths for three proposed outfall extensions in the area. Results have been provided for both primary and secondary treatment options and have been used as input to preliminary engineering design. This paper describes the application of the model to formulate design lengths for the proposed outfalls.

1.   Introduction

The Firth of Forth is a funnel shaped area of coastal waters on the eastern side of the Forth Bridges, in Scotland, some 50 km long and 20 km wide at its widest point (Figures 1 and 2). It has relatively high populations on both north and south shores which discharge their sewage effluent to the sea, for treatment and dispersal.

At present, treated and untreated sewage on the northern (Fife) shore discharges through a series of short sea outfalls between Inverkeithing and Kinghorn as shown in Figure 3. The effluent consists of non-toxic supernatent liquor from domestic sewage generally subject to primary settlement treatment. The outfalls are presently in insufficient depth of water to allow adequate dispersion of sewage effluent. This has resulted in failure of water quality to meet the European Community (EC) standards on two registered bathing beaches in the area, namely Aberdour Silversands and Pettycur.

175

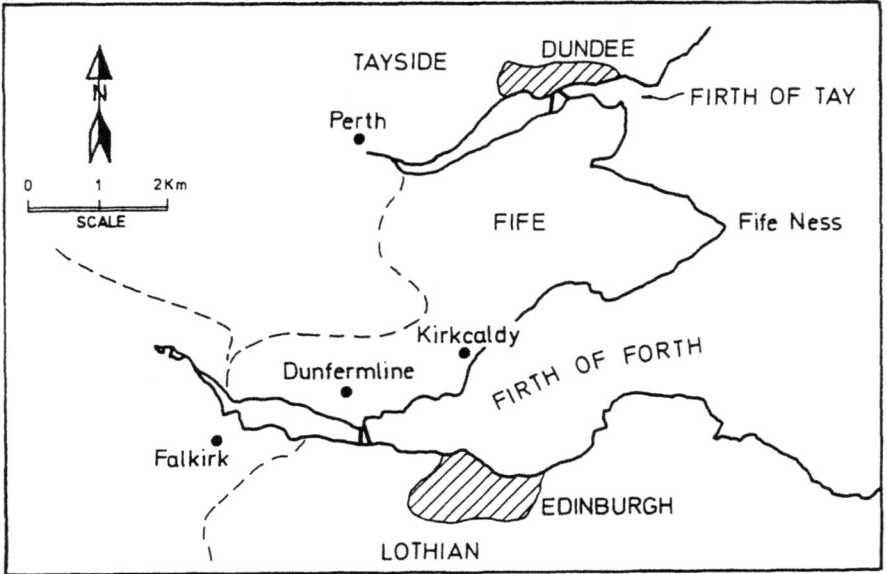

Figure 1   *Map Location*

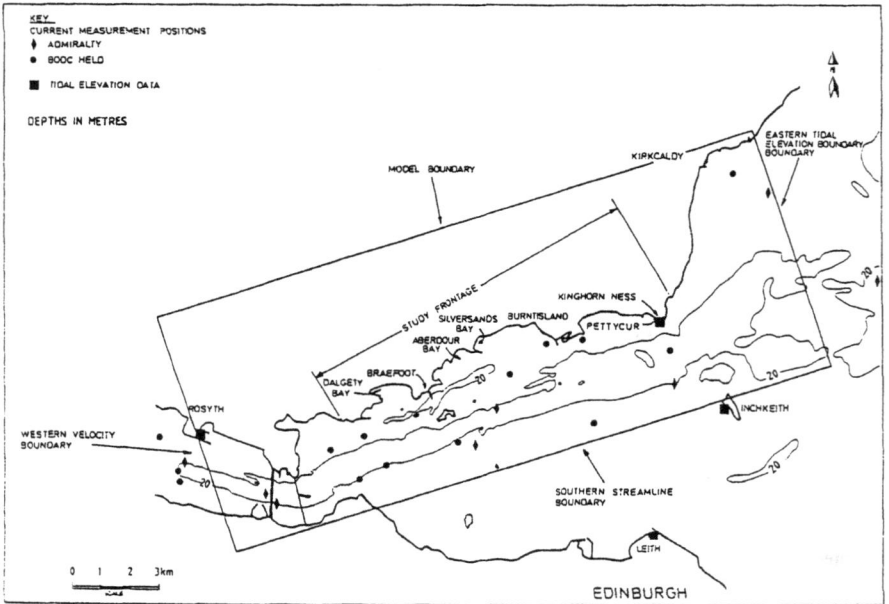

Figure 2   *Study area and extent of model*

As part of a remedial procedure Fife Regional Council (FRC) have lengthened an outfall at Pettycur and plan to combine and lengthen many of the other discharges to give adequate dispersion so that the quality standards can be met. The study looked at proposals to lengthen existing outfalls at Aberdour West and Silversands and to construct a new sewage treatment works and outfall at Burntisland (Figure 3). FRC commissioned a modelling study with field measurements to evaluate the appropriate treatment and optimise outfall lengths at the three sites.

Since a large quantity of hydrodynamic and dispersion data already existed in the area, a review of data sources was carried out before conducting measurements specific to the present exercise. This showed that there have been extensive measurements of tidal currents and tracer dispersion within the Firth which could be used for model boundary and validation data within the main channel area. The recording current information was held largely at the British Oceanographic Data Centre (BODC) at Bidston. However further site measurements were required to validate the model within the study area. The extent of historical data is shown in Figure 2.

2.   Environmental Quality Standards

The standards applicable to the study area are those defined by the EC Bathing Waters Directive (1976) and Municipal Waste Waters Directive (1991). Although the area contains only two bathing beaches, the regulatory authority, the Forth River Purification Board (FRPB), have indicated that the guideline ("G") limit of 100 <u>Escherichia</u> Coliforms (E Coli)

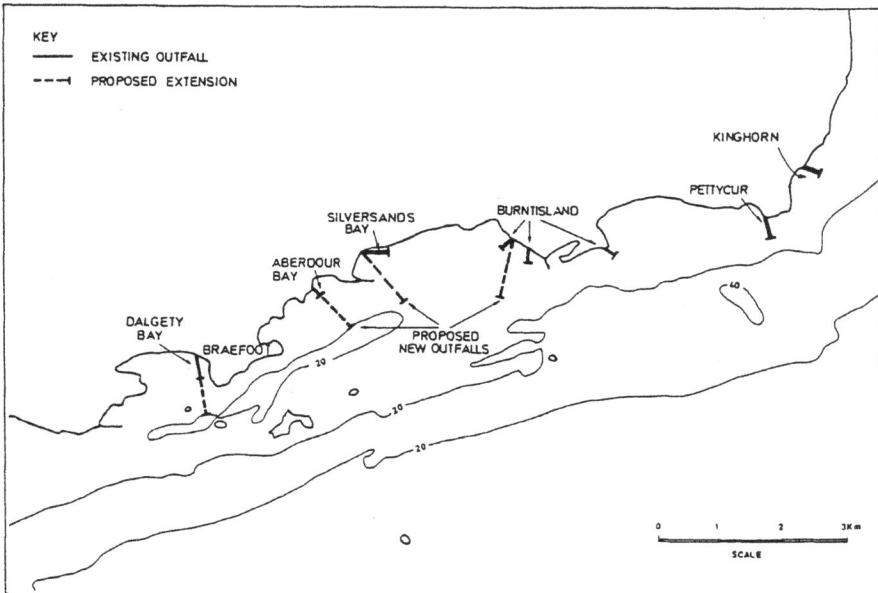

Figure 3   *Existing and proposed outfalls*

177

per 100 ml water for 80% of samples should apply to the whole coastline apart from the industrialised frontages at Braefoot and Burntisland. At the latter locations the mandatory ("M") limit of 2000 E Coli per 100 ml for 95% of samples applies.

Initial dilution standards to be met at the outfall locations are 120 x for areas subject to "G" values and 50 x for areas subject to "M" values, for primary treated sewage effluent. For secondary treatment 50 x applies.

3. Numerical model

3.1 Theory

The model was required to replicate the water movements within the Firth and dispersion of effluent from the proposed outfall scenario within the study area shown in Figure 2. The chosen model for the study simulates depth mean currents and dispersion of effluent in two dimensions, with the flows considered to be well mixed in depth. The hydrodynamic module is based on the depth integrated equations of motion and includes all shallow water processes of importance, including:

-       Local and advective accelerations
-       Rotational effects
-       Surface wind stress
-       Bed friction
-       Turbulence induced lateral shear.

The dispersion module is based on the solute transport equation, written for E Coli, and includes:

-       Advection
-       Longitudinal dispersion
-       Lateral diffusion
-       Decay
-       Source inputs

The hydrodynamic and solute transport equations are solved of each half timestep to ensure mass continuity of effluent within the model.

A more detailed description of both modules can be found in References 1 and 2.

3.2 Set up

The model was set up over the area shown in Figure 2, covering an area of 23 km by 9.5 km. The model was aligned approximately 17 degrees north of east, parallel as far as possible with the free streamline flow within the estuary. It was first necessary to check that the extent of the model covered the full excursion of effluent discharged from any outfall within the study area.

The western (velocity) boundary was located near Rosyth, where three historic measuring stations gave a good description of the velocity distribution across the channel. The eastern boundary was located to the east of Kirkcaldy and was described by tidal elevations. Tidal elevation data were sought from Leith and Rosyth and also two other measuring sites during the project as shown in Figure 2. All boundary data were corrected to depth average and the model set up to run for two tidal conditions, including:

i)   27-28 August 1991, Mean Spring Range (4.8m)
ii)   3-4 September 1991, Mean Neap Range (2.4m)

The model was set up for two grid mesh sizes, namely a coarse grid of 240 m and a fine grid of 90m. The purpose of this was to enable a quick turn around time for initial outfall configurations, whilst retaining a greater accuracy for refinement of outfall lengths during later stages of the project.

Bacterial input to the model was determined by taking random samples from the outfalls during the field survey in August 1991. The effluent concentrations were considered as representative of an outfall discharge of 2 x Dry Weather Flow (DWF) and these values have been used as input to the study as design conditions for the summer season. The discharges used in the study were 2 x DWF as related to design populations for the area.

The bacterial flux was checked as being representative of similar discharges elsewhere in the Firth of Forth (Reference 3), and relates to a primary level of treatment. For secondary treatment the bacterial flux was reduced by factor of 10 (Reference 5).

4.   Field measurements

In addition to the offshore tidal measurements a series of current and dispersion measurements were undertaken during the August to September 1991 period, to provide the necessary calibration and validation data for the modelling exercise. These included the following measurements in the vicinity of the two proposed Aberdour outfalls and supplemented earlier measurements carried out at Burntisland:

i)   Recording current measurements at two locations for a 15 day period.
ii)   Float tracking exercises over Spring and Neap tides.
iii)   Dye dispersion tests over Spring and Neap tides.
iv)   Initial dilution tests over Spring tides.
v)   Bathymetric and geophysical surveys.
vi)   Wind speed and direction.

5.   Model validation

5.1   Hydrodynamic module

Calibration of the model was carried out for a mean spring tide for both coarse and fine models. Model results were compared initially with tidal levels at three locations, and showed good agreement. Depth averaged velocities were available from the BODC

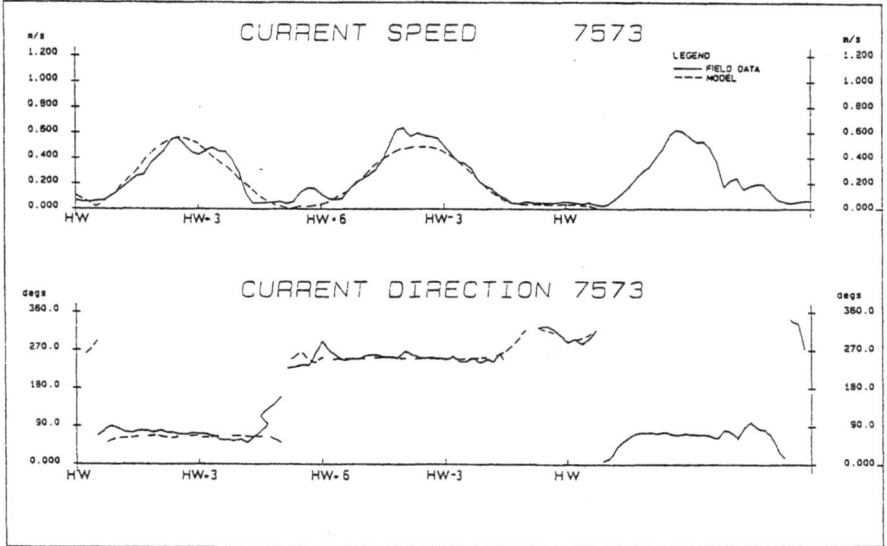

Figure 4   *Comparison between measured and predicted currents at Aberdour East station*

database of information at 18 locations. Comparisons showed very good agreement both along the main channel and also at nearshore locations. An example comparison at the Aberdour Silversands current meter location is shown in Figure 4.

Output from the model was also produced in a vector map format for each hour of the tide and compared with similar presentations of site data, both around the outfall sites and further offshore. This showed encouraging agreement between model predictions and measurements, with the model accurately reproducing the observed gyre formations to the west and south east of Burntisland.

During this process good agreement between model predictions and measurements was obtained without the need for fine tuning any model parameters.

5.2   Dispersion module

All coefficients used, ie diffusion, dispersion and decay time, vary with site and hydrodynamic and meteorological conditions. The approach was to take established values and test their validity to the individual sites.

The results from field tests showed reasonable agreement with published values in the literature. Sensitivity tests were carried out in addition, to examine the effect of varying these parameters. These tests showed that predicted effluent concentrations were relatively insensitive to changes in the diffusion and dispersion coefficients. Variations of the decay (T_{90}) time, showed that the model results were slightly sensitive to changes of T_{90} within the range of 4 - 10 hours (i.e. typical day time values see References 4 and 6), but greater sensitivity occurred at much higher values. A T_{90} of 10 hours was adopted as a typical mean during summer conditions.

6.   Run methodology

The model was run for two tidal conditions (i.e. Mean Spring and Mean Neap) and a variety of wind conditions representative of the local area, as shown in Table 1. As can be seen from the Table, the summer season is particularly sensitive to south westerly and north easterly winds and calm conditions.

Table 1   *Wind probability matrix*

DIRECTION								TOTAL
NE		SE		SW		NW		
S	P	S	P	S	P	S	P	
0		0		0		0		20%
4m/s	24%	4m/s	6%	4m/s	39%	4m/s	9%	78%
				9m/s	2%			2%

Probability expressed as percentage occurrence of the summer season.

S   -   Speed
P   -   Probability

Output from the model was expressed in terms of exceedence of the "G" and "M" effluent concentration levels at 35 locations as shown in Figure 5. These points were chosen to give representative results over the area relating to water quality standards. Model results could then be obtained for a variety of conditions (representative of a summer season when sampling is carried out) and then summed to compare directly with "G", or "M", limits, as appropriate.

The model was run for a variety of conditions for both the coarse and fine grid scales to determine suitable outfall lengths for the three proposed discharges. Initially the existing discharges elsewhere in the model domain remained "on", but it was found that these tended to mask the effect of altering the proposed outfall lengths, and so the majority of tests were undertaken with the existing discharges turned "off".

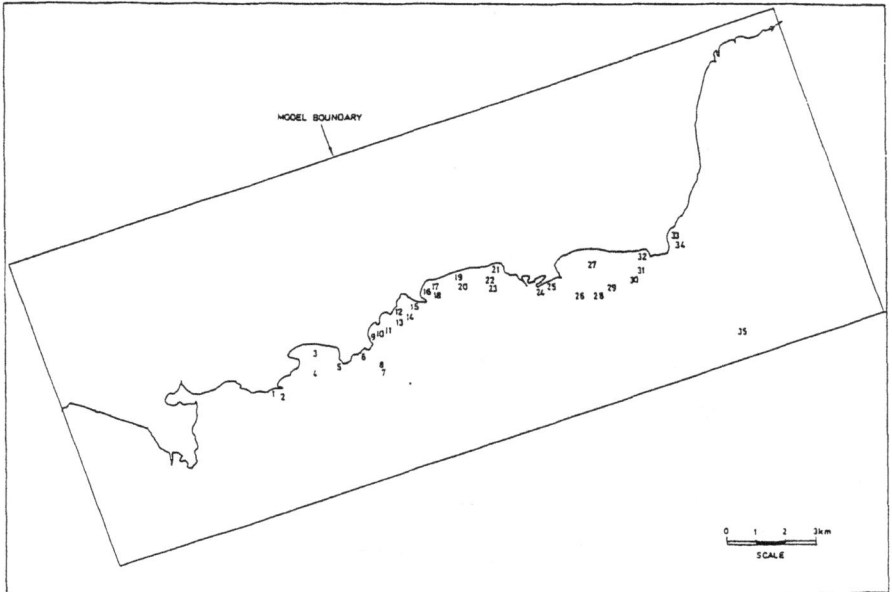

Figure 5   *Locations in model examined for solute concentrations*

7.   Model results

7.1   Primary treatment option

The coarse grid scale model tests were carried out with all three outfall lengths equal to 750 m and then 1000m. It was soon found that the 750m long outfalls would be too short for the required standards to be met along the coastline.

The fine grid scale model tests were restricted largely to the three oufalls being 1000m long. Figure 6 shows the model output for a high water spring tide with a no wind condition. The white vectors indicate currents whilst the colour shadings represent bacterial concentrations as shown in the attached key. The plot shows the effect of the flood tide in causing high concentrations to the west of Aberdour. It also shows an amount of "ponding" at the outfall locations at slack water.

Conditions were improved on the shoreline on the ebb tide as shown in Figure 7, which illustrates conditions at three hours after high water. Although the area of the slick is relatively long, effluent in concentration has been taken away from the beaches and the inter-tidal banks which are shown in light blue in the Figure.

Exceedences at the 35 sample points were plotted in the form shown in Figure 8 for various outfall lengths, and summed to represent summer seasonal conditions. Figure 8 shows general compliance with the G limit, apart from the area offshore between Braefoot and Aberdour Bay.

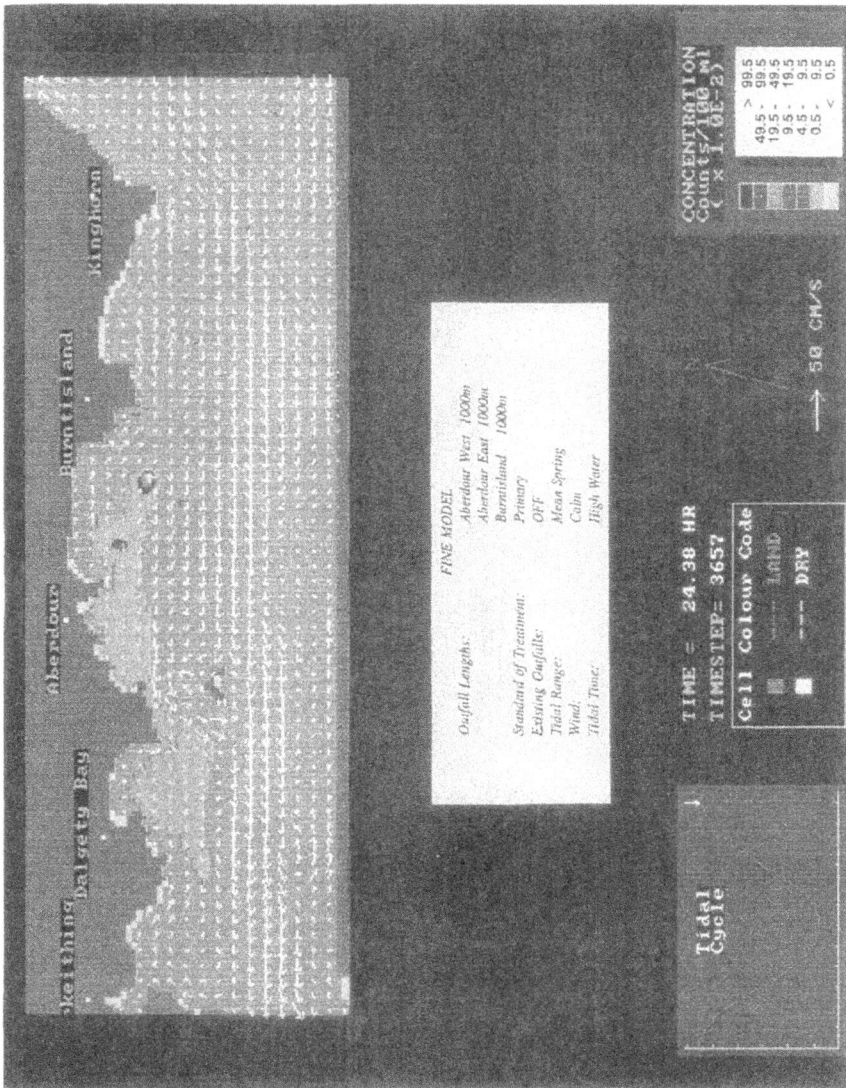

Figure 6   *Fine scale model results, 3 x 1000m outfalls, primary treatment, spring tide, no wind, HW*

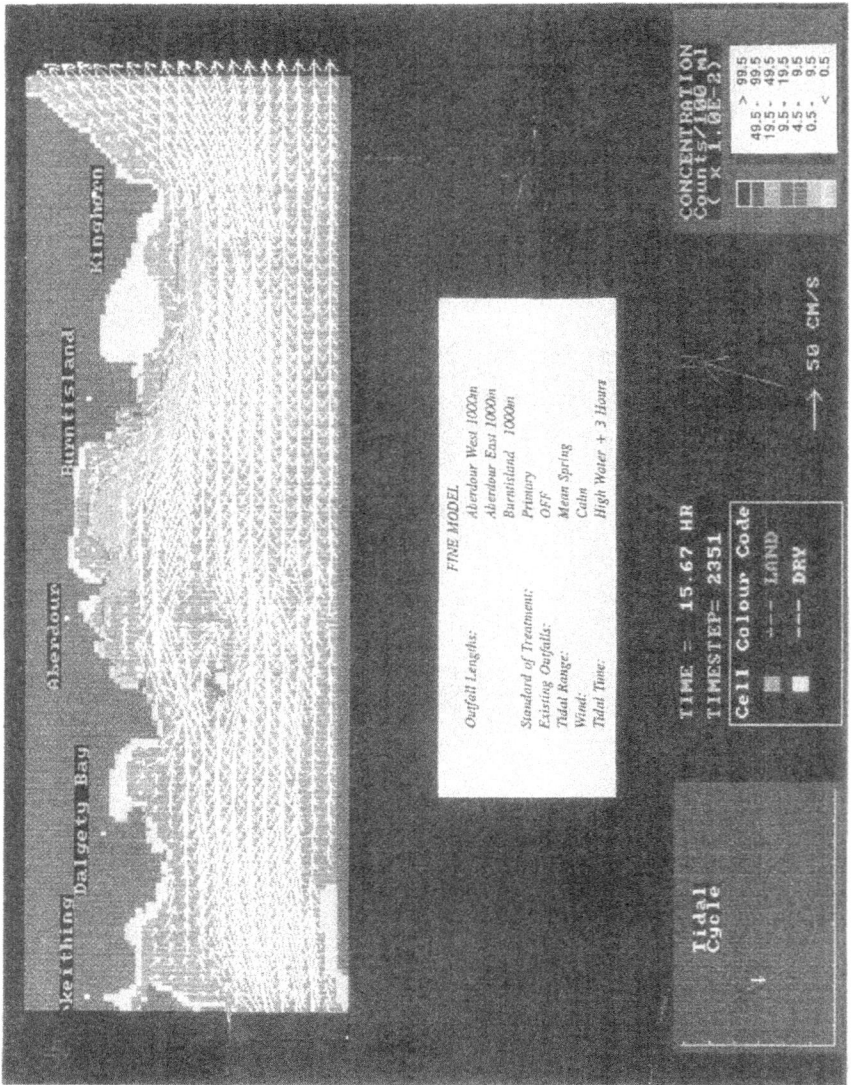

Figure 7   *Fine scale model results, 3 x 1000m outfalls, primary treatment, spring tide, no wind, HW + 3 hours*

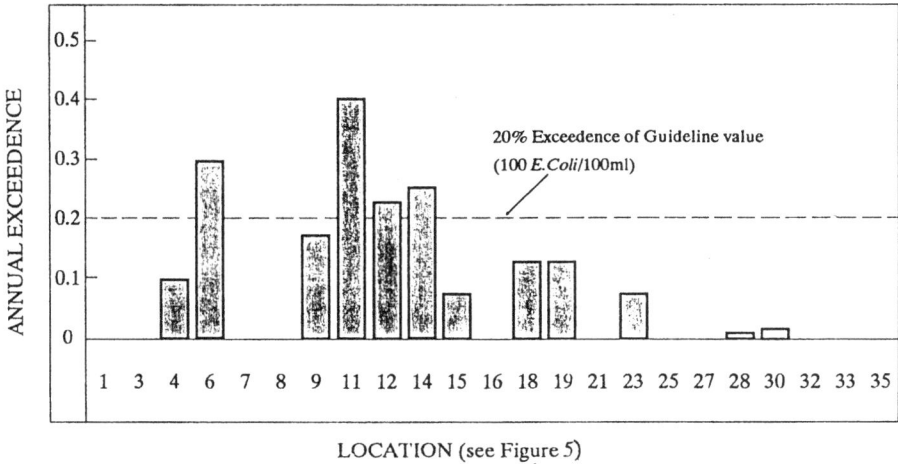

Figure 8 *Seasonal exceedences of "G" limit fine grid model, 1000m outfalls. primary treatment*

7.2 Secondary Treatment Option

The above procedure was repeated for three outfalls lengths with secondary treatment applied to the effluent. Results indicated that outfall lengths of 600 m at each proposed location would satisfy the required water quality standards.

7.3 Sensitivity of Results to Wind Direction

Since the beach at Silversands tends to "fail" during prolonged south easterly wind conditions the FRC were interested in the effect of such winds on model predictions. Results showed that with the proposed long outfalls, Aberdour Bay would be detrimentally affected during spring tides, but elsewhere the bathing waters would not be unduly sensitive to winds from this sector.

Model tests showed that high winds from the south west would result in particulary high concentrations along the whole frontage, but that the distribution was dependent on tidal range. At lower wind speeds the distribution of concentration is broadly similar to the seasonal mean.

8. Conclusions

The present study has investigated the improvement of water quality standards along the northern shore of the Firth of Forth by the provision of long sea outfalls and improved treatment of sewage effluent. The modelling study has investigated the extension of three outfalls at Aberdour Bay, Silversands Bay and Burntisland. Results of the study have shown that if primary treatment of sewage is adopted, outfall lengths should be at least 1000m at each site. If secondary treatment is adopted, then much shorter lengths of 600m may be used. These results are being used as input to the detailed engineering design. Final model runs are being undertaken to confirm the precise outfall lengths.

Acknowledgements

The Authors would like to thank Fife Regional Council for their permission to publish the results of this study. Thanks are also due to Forth Rivers Purification Board and Forth Ports Authority for their assistance during the study and BP for their release of current meter data held by BODC. The model study was carried out by Howard Humphreys and BMT Ceemaid Limited with specialist advice from Bradford University. Survey work was undertaken by Brown and Root Survey.

References

1.   Falconer, R.A., (1976), "Mathematical modelling of jet - forced circulation in reservoirs and harbours", PhD Thesis, University of London.
2.   Falconer, R.A., (1986), "A two-dimensional mathematical model study of the nitrate levels in an inland natural basin", *Proceedings of the International Conference of Water Quality Modelling in the Inland Natural Environment*, BHRA Fluid Engineering, Bournemouth, Paper J1, pp.325-344.
3.   Forth River Purification Board, (1988), *A Numerical Model of Faecal Coliform Concentrations Arising from the Proposed Sewage Oufalls for East and West Wemyss*, Report TW12/88, Tidal Waters Section, FRPB, Heriot Watt Research Park, Edinburgh.
4.   Gameson, A.L.H., (1986), "Investigations of sewage discharges to some British coastal waters", Chapter 8, *Bacterial Mortality*, Part 3, Report TR 239, WRC, Medmenham, Bucks, U.K.
5.   Gould, D.J.and Fraser, J.A.L., (1991), "Operational use of Oxymaster in sewage effluent disinfection and an appraisal of its environmental impact", *Presented in the Scientific Section of Symposium on Technology Options for Producing Higher Quality Effluent, IWEM, London.*
6.   Gould, D.J. and Munro, D., (1981), "Relevance of microbial mortality to outfall design", *Coastal Discharges - engineering aspects and experience*, ICE, London, pp 45-50.

16 Some aspects of tidal flow and water quality modelling of the North Devon coastal waters

Y. Kaya

ABSTRACT

As part of a comprehensive strategy for environmental protection, South West Water commissioned the Babtie Group to construct a mathematical model to simulate tidal flow, bacterial dispersion, and water quality in the North Devon coastal waters.

The model is based on a 2-dimensional, depth averaged, alternating direction implicit, finite difference scheme originally developed by Professor R.A. Falconer. Application of the model to this particular area necessitated a special form of radiation conditions on the open boundaries, which renders the boundary transparent to outgoing transients, yet permits the background tidal and mean elevations to be maintained. Representation of wind stress effects was improved by assuming a parabolic vertical velocity profile, which, in addition to the inclusion of the surface stress components, enables correction of advective accelerations for wind effects.

The bacterial dispersion and water quality model was constructed to allow simultaneous modelling of up to 9 pollutants. To improve numerical simulation of buoyant plumes in the near field, pollutants are assumed to be confined to within a predefined top layer and advected with near surface velocities obtained from the assumed parabolic vertical velocity profile.

1   Introduction

The model area covers about 52 km of the North Devon coast in the Bristol Channel from Porlock in the east to a point about 6 km west of Baggy Point, including Woolacombe, Ilfracombe, Combe Martin and Lynmouth, as shown in Figure 1. The Bristol Channel has been a focus of attention for engineers in recent years. Much of the recent work has been concerned with the possible effects of a barrage on the tidal regime, see Owen (1980), Uncles (1983) and Stephens (1986).

Within the model area there are several important bathing beaches, a National Park and a designated Area of Outstanding Natural Beauty. A number of small outfalls discharge effluent into the coastal waters mainly in the form of sewage and storm water run-off. There are also a number of water courses which discharge at the coast. The largest river within the model catchment area is the River Lyn which meets the sea at Lynmouth. Ilfracombe and Lynmouth are popular medium-sized holiday resorts.

Tides in the Bristol Channel are predominantly semi-diurnal, the largest components being the M2 (principal lunar), followed by S2 (principal solar), see Uncles (1983). Mean tidal ranges are about 8.5 m for spring tide and 3.5 m for neap tide. Tidal current speeds near the coast of the model area are about 0.75 m/s at spring tide and stronger further offshore.

The mathematical modelling system is based on the DIVAST (Depth Integrated Velocity and Solute Transport) developed by Falconer, see Falconer (1986). The 2-dimensional depth averaged equations of continuity and momentum are solved using an Alternating Direction Implicit (ADI) finite difference numerical scheme.

The original model was split into two basic components:- i) a hydrodynamic model referred to as FLOFIELD and ii) a water quality model, DISPOL. The modelling suite has been structured in a modular way which allows execution of both of these models separately or simultaneously.

A number of modifications and additions have been made to the original model for this specific application. The main additions to the hydrodynamic model included implementation of a special radiation boundary condition which renders the boundary transparent to outgoing waves, and an assumption of parabolic vertical velocity distribution which essentially allows prediction of a quasi-three dimensional velocity field. The former was used for treating numerical oscillations which originated from reflection of transient waves between a water elevation open boundary and the coastline geometry in this particular case. The latter is used for advecting buoyant plumes with near surface velocities instead of the conventional depth - averaged velocities.

The North Devon model was constructed on a space staggered grid having a regular mesh size of 400 m. The hydrodynamic model simulates the water surface elevation and depth averaged current velocities for each wet grid square. It was calibrated and verified using tidal data which was made available from a comprehensive environmental survey. The flooding and drying procedure employed in the present scheme essentially follows that of Falconer (1986), and Falconer and Chen (1991).

The water quality model is based on the solution of the 2-dimensional depth averaged advection-diffusion equation. The advective transport of concentration is represented with a third-order spatial finite difference scheme, QUICK as given in Falconer and Liu (1987). It was shown by Falconer and Liu (1987) that this scheme has a higher accuracy in modelling high concentration gradients than the conventional second-order central-difference representation.

To assist in data input, model execution and interpretation of model output a versatile and easy to use menu-driven User Interface has been incorporated. All modelling functions within the North Devon suite as well as file handling requirements and batch queue job submissions are controlled via the user interface.

2   Hydrodynamic model

2.1   Basic equations

The hydrodynamic model is based on the solution of the two-dimensional depth integrated equations of motion and continuity. For a constant density turbulent flow described in a cartesian co-ordinate system these equations can be expressed as (Falconer 1986):-

$$\frac{\partial UH}{\partial t} + \beta \left[\frac{\partial U^2 H}{\partial x} + \frac{\partial UVH}{\partial y} \right] - fVH + gH \frac{\partial \eta}{\partial x} - \tau_{sx} + \tau_{bx}$$
$$- \varepsilon H \left[2 \frac{\partial^2 U}{\partial x^2} + \frac{\partial^2 U}{\partial y^2} + \frac{\partial^2 V}{\partial x \partial y} \right] = 0 \tag{1}$$

$$\frac{\partial VH}{\partial t} + \beta \left[\frac{\partial UVH}{\partial x} + \frac{\partial V^2 H}{\partial y} \right] + fUH + gH \frac{\partial \eta}{\partial y} - \tau_{sy} + \tau_{by}$$
$$- \varepsilon H \left[\frac{\partial^2 V}{\partial x^2} + 2 \frac{\partial^2 V}{\partial y^2} + \frac{\partial^2 U}{\partial x \partial y} \right] = 0 \tag{2}$$

$$\frac{\partial \eta}{\partial t} + \frac{\partial UH}{\partial x} + \frac{\partial VH}{\partial y} = 0 \tag{3}$$

Eqs. (1) and (2) are the momentum equations in the x- and y- directions respectively and Eq.(3) is the continuity equation. In the above equations U,V = depth mean velocity components in the x, y directions respectively, H = total depth of flow, t = time, β = correction factor for non-uniform vertical velocity profile, f = coriolis parameter, g = gravitational acceleration, η = water surface elevation with respect to datum, ε = depth mean eddy viscosity, τ_{sx} and τ_{sy} = water surface shear stress components in the x and y directions given by

$$\tau_{sx,y} = \frac{\rho_a C^* W_{x,y} \bar{W}}{\rho} \tag{4}$$

where ρ_a = air density, C^* = air - water interfacial resistance coefficient, $W_{x,y}$ = wind velocity components in x and y directions, $\bar{w}$ = absolute wind speed, ρ = water density, and τ_{bx} and τ_{by} = sea bed shear stress components expressed as

$$\tau_{bx,y} = \frac{g q_{x,y} \, \bar{q}}{H^2 \, C_z^2}$$

(5)

where $q_{x,y}$ = depth integrated discharge per unit width, q = absolute discharge per unit width and C_z = Chezy roughness coefficient.

The above equations were solved by an alternating direction implicit finite difference scheme on a space staggered grid. All partial differential terms were approximated by fully centred difference equations in both space and time by iteration. The scheme is basically second order accurate with no stability constraints. Full details of the finite difference scheme are given in Falconer (1986).

2.2   Open boundary conditions

The model incorporates three types of open boundaries. The eastern open boundary is driven by current velocities while the western and south-western open boundaries are driven by water surface elevations. The northern open boundary was assumed to have zero normal velocities (no flow is allowed across the boundary) and free slip tangential velocities. This assumes that the tidal flow is mainly in the east - west direction along the northern open boundary.

This assumption was later justified by the field survey data which indicated a predominantly east - west flow direction except for a short distance at the western end of the boundary where some cross-flow is likely to occur. However, the effect of this relatively small amount of cross-flow on model predictions are unlikely to be significant.

Some difficulties were encountered in trying to calibrate the hydrodynamic model when a water elevation boundary was used along the eastern boundary instead of a velocity boundary. The model predictions in this case were poorer on the eastern side of the model near the open boundary which indicated that imposing level boundary conditions along the eastern open boundary could not produce accurate enough current velocities which effectively control the tidal flow in and out of the model area. Using a velocity/flow boundary at this specific location greatly assisted in resolving the calibration difficulties encountered.

Along the western and south-western open boundaries the water surface elevation was predefined as a function of time. Such boundaries do not allow transients generated inside the region to be transmitted outwards. Model instability which appeared after several tidal cycles was investigated by numerical experiments which showed waves being reflected back and forth between the south side of the western boundary and the opposite coastline. The numerical oscillations were first generated in the south west region and propagated towards the north and east directions. The amplitude of these oscillations gradually increased and the model became unstable after two consecutive tidal cycles. This phenomenon had not previously been encountered.

To overcome these numerical oscillations an appropriate form of a radiation boundary condition was introduced along the western open boundary. The method employed is fully described by Blumberg and Kantha (1985) which uses a simplified radiation boundary condition given as

$$\frac{\partial \eta}{\partial t} + gH^{\frac{1}{2}} \frac{\partial \eta}{\partial n} = - \left(\frac{\eta - \eta_k}{T_f} \right) \tag{6}$$

in which n is the direction normal to the planar boundary. The term on the right hand side represents damping which tends to force the value of η at the boundary to some known value η_k with a time scale of the order of T_f. $T_f = 0$ corresponds to the original level boundary conditions where no disturbances are allowed to pass out through the boundary; and $T_f = \infty$ represents the pure radiation condition (see Blumberg et al 1985) and renders the boundary transparent to waves travelling in the positive n direction with phase speed ($gH^{\frac{1}{2}}$). $T_f = 2r/(gH^{\frac{1}{2}})$ where r is the average distance between the western boundary and the opposite coastline.

Numerical experiments carried out with the above boundary conditions showed no oscillations in both water levels and current velocities for simulations of several consecutive tide cycles.

2.3   Surface wind effects

In most depth averaged tidal numerical models the effects of surface wind have been included as an additional shear stress term at the free surface, see Falconer (1986). Also an assumption is generally made for the shape of the vertical velocity profile taken, in most cases, either as a logarithmic or a seventh power law velocity distribution, Falconer (1986).

Koutitas (1988) and Koutitas et al (1986) stated that wind generated velocity profiles may vary considerably over the depth and suggested using a second order parabolic velocity distribution Falconer (1991). For the x - direction this can be expressed as

$$u = \left[\frac{3C_x}{4} - \frac{3U}{2} \right] \left[\left(\frac{z}{h} \right)^2 - 1 \right] + C_x \left[\left(\frac{z}{h} \right) + 1 \right] \tag{7}$$

where u = velocity at elevation z, and where z = 0 at the surface and z = -H at the bed,

$$C_x = \frac{H}{\varepsilon} \rho_a C_f W_x \frac{\sqrt{W_x^2 + W_y^2}}{\rho} \tag{8}$$

in which C_f is the air-fluid resistance coefficient.

The depth integrated advective acceleration given in Equation (1) can then be determined using Equation (7). This gives the value of β as 1.2 and a correction term in the form:-

191

$$\frac{\partial}{\partial x}\left[\frac{C_x\ UH}{20} + \frac{C_x^2\ H}{120}\right] + \frac{\partial}{\partial y}\left[\frac{C_y\ UH}{40} + \frac{C_x\ VH}{40} + \frac{C_x\ C_y\ H}{120}\right] \qquad (9)$$

A similar correction term can be derived for the y - direction advective acceleration for Equation (2). Further details are given in Falconer and Chen (1991).

Once the depth averaged velocity field is determined the values of U and C_x can be substituted directly into Equation (7) to give the vertical velocity profile which permits the computation of the current pattern at any depth. If, for example, the free surface velocity components or the averaged velocity components for a predefined top layer are required for subsequent use in a surface/near surface advective diffusive pollutant transport model they can be computed from Equation (7).

Figure 2 shows comparison of depth - averaged velocities with near free surface and near sea bed velocities. The near surface velocities were obtained by averaging vertically varying horizontal velocities over a top layer of one quarter of the total water depth. Similarly the near bed velocities correspond to averaged velocities within a bottom layer of one quarter of the total depth. It is interesting to note that the differences between near bed and near surface velocities are greater in deep water, towards the top left corner in Figure 2, and smaller in the shallower region, towards the bottom right corner.

In the North Devon model facilities are provided for advection calculations using averaged velocities from a top layer of 2 m, 3 m, 4 m or 5 m as well as the depth - averaged velocities.

2.4   Cold and Hot Start

At the beginning of each hydrogynamic simulation the model checks if an initial conditions data file exists in the data base corresponding to the tide cycle to be simulated. If such a file exists the model uses that data as the starting flow conditions. If, on the other hand, no such data exists the model automatically calculates appropriate initial conditions prior to the actual simulation.

The common practice for setting up initial conditions from scratch is to start from a zero velocity field and a horizontal free surface and let the model run until appropriate flow conditions are established everywhere within the model area. In general, these starting conditions are closely satisfied at or near high and low water where velocities approach zero as flow changes direction. Starting from high or low water and running the model for half to one tide cycle is usually adequate to establish appropriate starting conditions for the actual simulation. For the present model, however, to derive such conditions requires simulation of at least five consecutive tide cycles.

The flow conditions within the North Devon model area are strongly influenced by the funnelling effects of the Bristol Channel. At high water, for instance, there is a significant slope in water surface which is inclined towards the eastern end of the model, yet the general flow direction is still from west to east, running against the average surface slope. To establish such conditions from scratch requires considerable computational effort as aiming to satisfy correct surface elevation differences between the west and east end of the model generates an initial flow field in exactly the opposite direction of the actual flow. This incompatibility between the surface slope and the flow direction creates oscillations which dissipate over a relatively long period of simulation as the flow gradually changes direction.

To reduce the computational effort required to set up the initial conditions as described above an alternative method has been employed. By trial and error, it was found that if the simulation to establish initial conditions is started from about 1.5 hours after the high water then running the model until the following high tide is adequate to generate appropriate initial flow conditions for the subsequent tide cycle. It seems that starting from such specific flow conditions allows much quicker transision to the actual flow conditions where the tide can flow against the direction of surface slope. The difference in water surface elevation between the west and east boundaries could be as much as 0.8 m.

At the end of the simulation of each tidal cycle the flow conditions are written out to a specific file in the data base to be used for a subsequent simulation of the next tidal cycle.

3   Water quality model

The water quality model is based on the solution of the 2-dimensional depth averaged advection - diffusion equation expressed as, see Falconer (1986)

$$
\frac{\partial SH}{\partial t} + \frac{\partial SUH}{\partial x} + \frac{\partial SVH}{\partial y} - \frac{\partial}{\partial x}\left[HD_{xx}\frac{\partial S}{\partial x} + HD_{xy}\frac{\partial S}{\partial y}\right]
$$
$$
+ \frac{\partial}{\partial y}\left[HD_{yx}\frac{\partial S}{\partial x} + HD_{yy}\frac{\partial S}{\partial y}\right] \tag{10}
$$

in which S is the depth mean solute concentration and D_{xx}, D_{xy}, D_{yx} and D_{yy} are the depth mean dispersion - diffusion coefficients in the x and y directions respectively, Falconer 1986,

The dispersion coefficients are given by Falconer (1986) as:-

$$
D_{xx} = \frac{(k_l U^2 + k_t V^2) H\sqrt{g}}{\sqrt{(U^2 + V^2)} C_z}
$$
$$
D_{yx} = \frac{(k_l V^2 + k_t U^2) H\sqrt{g}}{\sqrt{(U^2 + V^2)} C_z} \tag{11}
$$
$$
D_{xy} = D_{yx} = \frac{(k_l - k_t) UV\sqrt{g}}{\sqrt{(U^2 + V^2)} C_z}
$$

where k_l is the longitudinal depth mean dispersion coefficient (= 5.93) and k_t is the lateral depth mean turbulent diffusion coefficient (= 0.15).

The solution of the above equations can be accomplished similar to the hydrodynamic equations by conventional second-order finite difference methods. However to increase the accuracy of predictions, particularly in the regions of high concentration gradients, the advective transport terms are represented with a third-order spatial finite difference scheme known as QUICK (Quadratic Upstream Interpolation for Convective Kinematics). The application of the method to the present model essentially follows that of Falconer and Liu (1987) and Lui and Falconer (1989).

For this particular application the water quality model is set up to simulate:

- Total Coliforms
- Faecal Coliforms
- Biochemical Oxygen Demand (BOD)
- Total Organic Nitrogen
- Ammonical Nitrogen
- Nitrate Nitrogen
- Dissolved Oxygen (DO)
- 2 user defined parameter

Equations (10) and (11) are adequate to describe the conservative pollutants. For non-conservative pollutants additional reaction equations must be solved simultaneously with Eqs. (10) and (11). For the coliforms and user defined parameters a first order reaction equation is used.

$$dS/dt = -KS$$

where S is the concentration and K is a decay rate (constant). A conservative pollutant such as salinity or dye can be modelled by setting the K coefficient to zero.

The decay rate constant is usually defined at a certain temperature which is modified by the model for the water temperature being simulated.

The nitrogens and dissolved oxygen must be modelled simultaneously as the nitrogens form a chain of reactions and they both influence and are influenced by the dissolved oxygen levels.

The model allows pollutant inputs from 10 rivers and streams, 7 existing outfalls and up to 16 additional user defined outfalls. Interaction of similar types of pollutants discharged from several locations is automatically dealt with.

4   Conclusions

A system of mathematical models has been developed to enable South West Water Services Limited to gain a thorough appreciation of the existing conditions within the model area and assess the potential effects of future engineering solutions on the local marine environment.

A special radiation boundary condition was introduced along the western water elevation open boundary which successfully allowed outgoing transient waves through the boundary yet maintained the background mean tidal elevation.

Representation of surface wind effects has been improved by introducing a parabolic vertical velocity distribution which permits advection of buoyant plumes with near surface velocities.

The water quality model allows simultaneous computation of up to 9 water quality parameters.

The correlation between the predicted and observed water surface elevations and current velocities was generally very good (Figs. 3 and 4) in an area which experiences a high tidal range.

5 Acknowledgements

The author wishes to express his thanks to South West Water Services Limited for their permission to publish this work and Metocean Consultancy Limited for their assistance during the preparation of this paper. The encouragement provided by colleagues and in particular by Professor R.A. Falconer at various stages is greatly valued by the author.

The views expressed in this paper are those of the author, and should not be assumed to be those of either South West Water Services Limited or Metocean Consultancy Limited.

6 References

1. Blumberg, F.A. and Kantha, L.H., (1985), "Open Boundary Conditions for Circulation Models", *Journal of Hydraulics Engineering, ASCE,* Vol. 111, No. 2, pp. 237-255.

2. Falconer, R.A., (1986), "A Two-Dimensional Mathematical Model Study of the Nitrate Levels in an Inland Natural Basin", *Proc. Int. Conf. on Water Quality Modelling in the Inland Natural Environment, BHRA Fluid Engineering,* Bournemouth, Paper J1, pp. 325-344.

3. Falconer, R.A. and Liu S., (1987), "Modelling Solute Transport Using QUICK Scheme", *Journal of Environmental Engineering, ASCE,* Vol. 114, No. 1, pp. 3-20.

4. Falconer, R.A. and Chen, Y., (1991), "An Improved Representation of Flooding and Drying and Wind Stress Effects in a 2-D Tidal Numerical Model", *Proc. Inst. Civil Engineers, Part 2, Research and Theory,* Vol. 91, pp. 659-678.

5. Koutitas, C.G., (1988), *Mathematical Models in Coastal Engineering*, Pentech Press Limited, London.

6. Koutitas, C. and Gousidou-Koutita, M., (1986), "A Comparative Study of Three Mathematical Models for Wind-Generated Circulation in Coastal Areas", *Coastal Engineering,* Vol. 10, pp. 127-138.

7. Lui, S. and Falconer, R.A., (1989), "Application of the QUICK difference Scheme for Two-Dimensional Water Quality Modelling", *Proc. Int. Conf. Hydraulic and Environmental Modelling of Coastal, Estuarine and River Waters,* University of Bradford, pp. 360-370.

8. Owen, A., (1980), "The Tidal Regime of the Bristol Channel: A Numerical Modelling Approach", *Geophys. J.R. Astr. Soc.* Vol. 62, pp. 59-75.

9. Stephens, C.V., (1986), "A Three-Dimensional Model for Tides and Salinity in the Bristol Channel", *Continental Shelf Research,* Vol. 6, No. 4, pp. 531-560.

10. Uncles, R.J., (1983), "Hydrodynamics of the Bristol Channel", *Marine Pollution Bulletin,* Vol. 15, No. 2, pp. 47-53.

Model Boundary

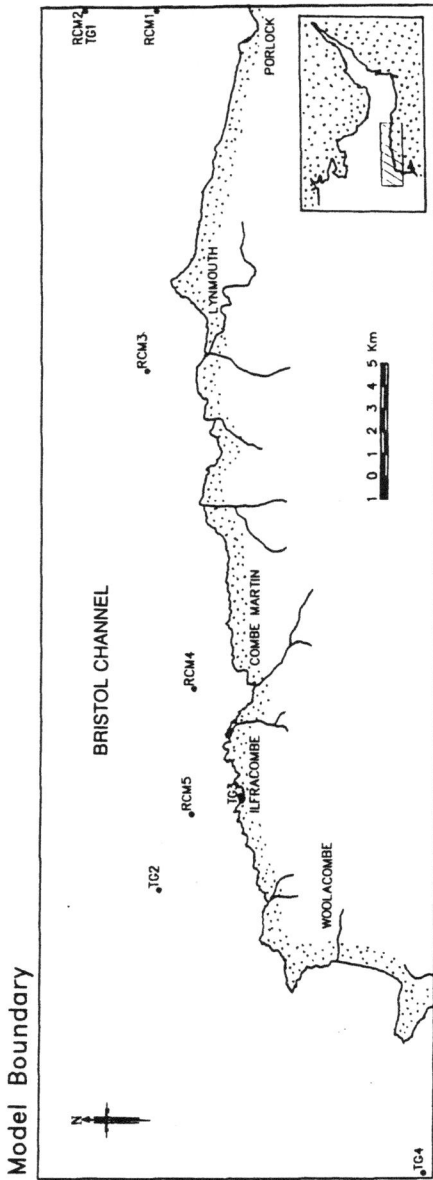

FIGURE 1 – Model Boundaries and Location of Recording Instruments

196

CURRENT SPEED AND DIRECTIONS AT TIME :9.00 hrs
(244800E, 149100N)

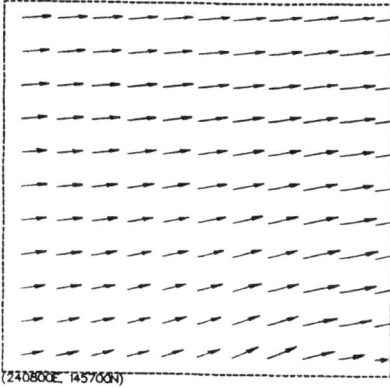

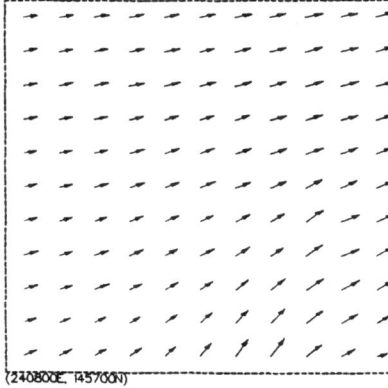

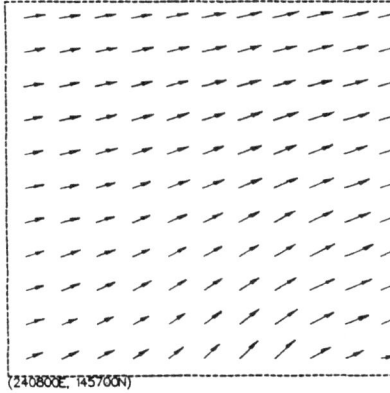

(240800E, 145700N)

Near Surface

CURRENT SPEED AND DIRECTIONS AT TIME :9.00 hrs
(244800E, 149100N)

(240800E, 145700N)

Near Bed

CURRENT SPEED AND DIRECTIONS AT TIME :9.00 hrs
(244800E, 149100N)

(240800E, 145700N)

Depth Averaged

FIGURE 2   -   Predicted velocities at mid-flood of a spring tide

197

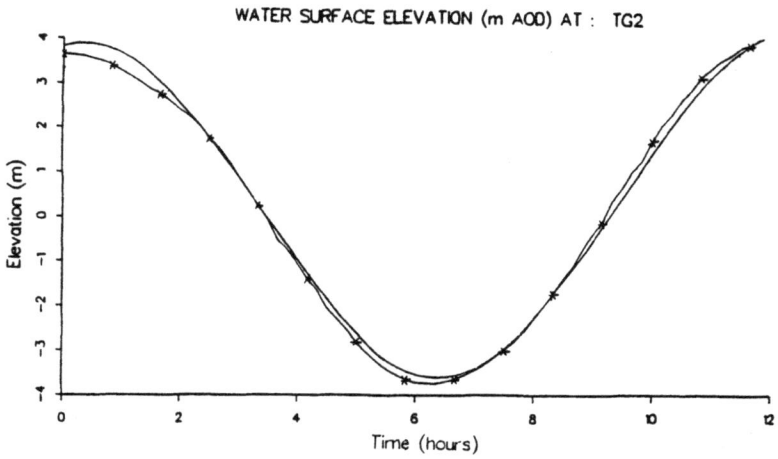

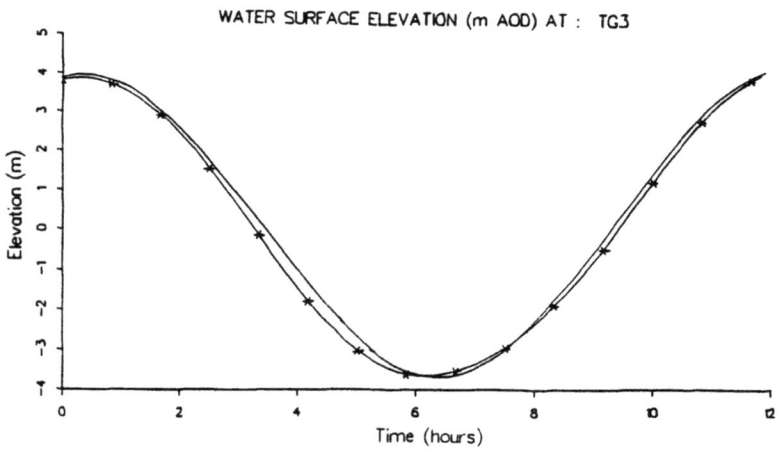

FIGURE 3   -   Comparison of predicted and measured water surface elevations at TG2 and TG3

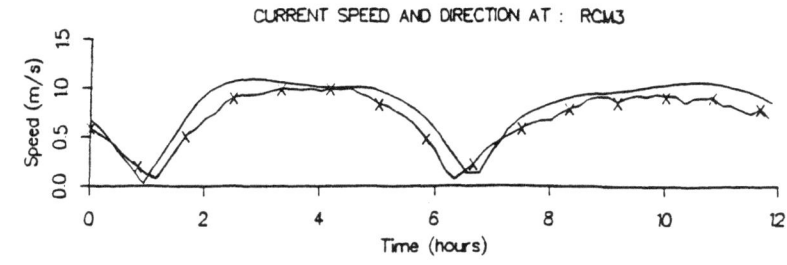

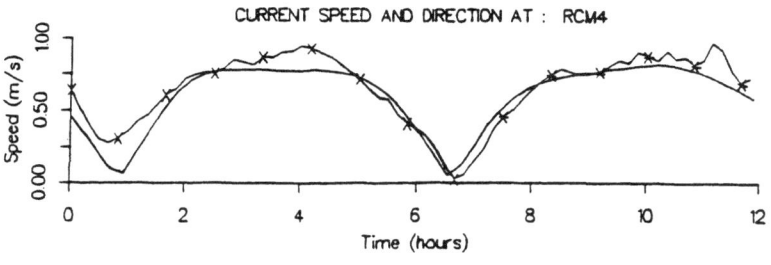

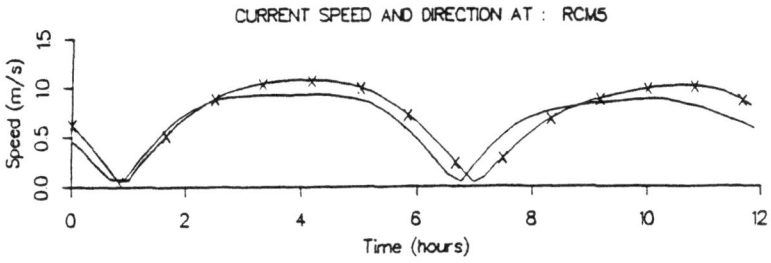

FIGURE 4 - Comparison of predicted and measured velocities at RCM3, RCM4 and RCM5

17 OIL-RW: A mathematical model for predicting oil spills trajectory and weathering

L. Mansur and D. M. Price

ABSTRACT

A random walk dispersion model has been developed at HR Wallingford for predicting the trajectory of oil slicks.

Quasi 3D flow fields are calculated using a two-dimensional, depth averaged model with applied profiles to account for the vertical structure of the flow and wind driven currents. The oil slick is assumed to contain a given distribution of particle sizes which in its turn determines the buoyancy of the oil droplets. The oil particles move in three-dimensions, making use of the vertical structure to simulate the effect of shear on the oil slick.

The oil weathering predictions are made using mathematical algorithms which simulate the chemical and physical changes of the oil following its spillage. A data file of the properties of several crude oils is accessed by the model to determine the original oil characteristics, the environmental conditions are also required to calculate the resulting oil properties. The following properties are predicted by the model: evaporative loss, water content, density, viscosity and pour point.

Results from the weathering model showed good agreement with data obtained from field measurements and the oil trajectory predictions displayed physical characteristcs observed in real oil spills.

1. Introduction

The accurate prediction of the trajectory and properties of oil slicks is an essential prerequisite for the control and prevention of oil pollution. Computer models for predicting the track and fate of oil when it is spilled are the primary tool for contingency planning operations, training of personnel involved with oil incidents and the deployment of adequate measures for the control of oil pollution.

A sophisticated random walk contaminant dispersion model has been developed at HR Wallingford to determine the movement and fate of oil slicks. The advection and diffusion of discrete oil particles is simulated in both coastal and offshore waters. The hydrodynamic processes implemented in the model allow accurate predictions of the slick movements. The model also accounts for the physical and chemical changes of the oil which affect both the behaviour of a slick and the type of control methods required under the given circumstances.

2. Prerequisites for modelling oil weathering

Following an oil spill, several physical, chemical and biological processes take place in the oil at various time spans and magnitudes; resulting in major changes in the properties of the oil.

Weathering - or degradation - of oil spills is a very complex process affected by both the properties of the oil and the environmental conditions prevailing during an oil spill.

Changes in the chemical composition of the oil as well as the effects of the environmental conditions result in a change of state of the slick from an oil-on-water to an oil-in-water dispersion to a water-in-oil emulsion. The formation of stable emulsions and the cumulative effects of the weathering processes lead to subsequent alterations in the composition of the oil and thereafter the formation of tarry residues which are the end results of an oil slick.

In order to simulate the properties of the weathered oils, a prepared file on the properties of 58 crude oils is accessed directly by the model and provides the following information (CONCAWE, 1983):

- oil name,
- distilled properties,
- density,
- pour point,
- asphaltene content, and
- viscosity.

3. Dispersion mechanism of oil slicks

In order to simulate the spreading of a slick, several environmental parameters are required in the model.

These include the water temperature, the tidal and wind-driven currents, and the turbulent diffusion of the oil particles in the water.

The oil slick is assumed to be split into oil droplets due to the effect of breaking waves; these droplets are mixed instantaneously into the water column following a size distribution contant by volume in the range 10-1000 μm (Forrester, 1971).

3.1 Advection of the oil slick

The depth-averaged tidal flows used in the model allow the simulation of buoyant plumes in coastal waters by including a representation of the depth structure of tidal currents in the model. This is expressed by a logarithmic velocity profile shown in Figure 1 and derived as follows:

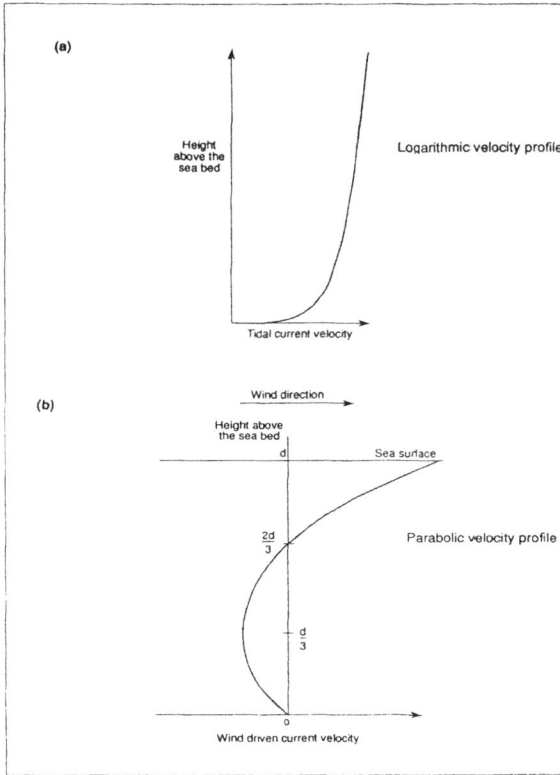

Figure 1 Depth structure of (a) tidal (b) wind-driven currents in OIL-RW

$$U(z) \quad = (U_*)_T/k_0 \log_e(30.1 \ z/k_s) \tag{1}$$

where

U = current velocity (ms^{-1})
$(U_*)_T$ = friction velocity for a tidal current (ms^{-1})
k_0 = von Karman's constant
z = distance above the sea bed (m)
k_s = roughness length (m)

In addition to advection by mean tidal currents, the oil particles move in response to wind-driven currents. The wind-driven current speed at any depth in the water column is computed from:

$$U_W(z) = \quad S(3(1-z/d)^2 - 4(1-z/d) + 1) \tag{2}$$

203

where

U_w = wind-driven current velocity (ms^{-1})
S = surface wind-driven current speed (ms^{-1})
d = water depth (m)

The surface wind-driven current is assumed to be parallel to the wind vector with a speed given by:

$$S = \alpha w \tag{3}$$

where

α = 0.03
w = wind speed at 10m above the sea surface.

Equation 3 gives rise to a parabolic wind-driven velocity profile as shown in Figure 1.

3.2 Turbulent diffusion of the oil particles

The lateral and vertical diffusion displacements are incorporated in the model to simulate the effects of turbulent eddies. Particles in OIL-RW are subjected to random displacements in addition to the ordered movements which represent advection by mean currents. The motion of simulated plumes is therefore a random walk, being the resultant of ordered and random movements.
An additional term is added to calculate the vertical velocity of the oil droplets. For the two particle sizes studied in this work, the rise velocity follows Stokes law and is expressed as:

$$U_B = g\,d^2(1 - D_0/D_w) / 18\,V \tag{4}$$

where

U_B = vertical rise velocity of the oil droplets
d = droplets diameter (m)
D_0 = density of the oil (Kg/m^3)
D_w = density of the water (Kg/m^3)
V = viscosity of sea water (m^2/s)

The variation in rise velocity therefore varies with the particle size and density as shown in Figure 2, which in its turn affects the distribution of the oil particles throughout the water column.

3.3 Shoreline stranding of spills

The model also accounts for the beaching of the oil which depends on the shoreline type (Gundlach, 1985).
The oil stranding percentages considered are as follows:

- moderate-to-high-energy rocky shores   0-1%
- coarse grained beaches   20-35%
- fine-grained beaches   40-60%

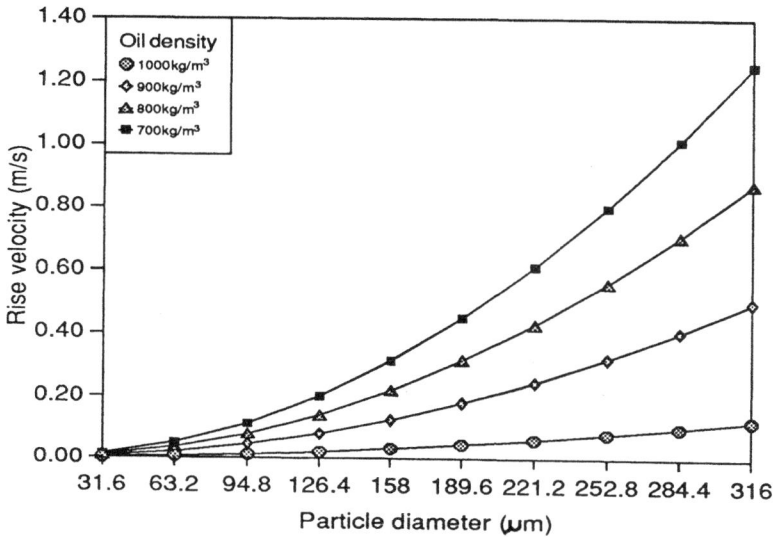

Figure 2 Rise velocity of oil droplets of different sizes

4. Oil weathering algorithms

4.1 Evaporative loss

Evaporation can remove between 20-50% of spilt crude oils, more than 75% of refined petroleum fuel and 10% or less of residual fuel oils, such as bunker C.
  Evaporative loss is controlled by several parameters and can be expressed as follows:

$$N = kaP/RT \tag{5}$$

where N = molar flux of the oil (mol/s)
  k = mass transfer coeficient (m/s)
  a = surface area (m^2)
  P = vapour pressure (Pa)
  R = gas constant (= 8.314 Pa m^3/mol°K)
  T = environmental temperature (°K)

Equation 5 is solved following the assumption that P/RT can be estimated from the oil distillation curve. Buchanan (1987) showed that correct predictions were obtained using:

$$\ln P = 21.4 - T_B/15 \tag{6}$$

where T_B = boiling temperature of the crude oil, which is respectively related to the oil distillation curve.
  The surface area is calculated from the slick thickness which ranges between 0.04-0.2 mm (NAS, 1989) and the volume of oil spilled.

205

Finally, the value of k for hydrocarbons, is derived from wind tunnel studies as:

$$k = 2.5 \times 10^{-3} \, U^{0.78} \tag{7}$$

where U is the wind speed (m/s).

4.2 Emulsification rate

Water-in-oil emulsions are formed in sea conditions rapidly and as a result of natural forces. Huang and Elliot (1977) stated that the turbulent energy provided by a gently rolling sea state is sufficient for emulsification of spilt oils.

Crude oils can take up to 90% of their volume in water thus changing into a heavy brown mixture often refered to as 'chocolate mousse'.

Factors controlling the formation of water-in-oil emulsions are:
- wind speed,
- presence of asphaltenes and parafins, and
- pour point.

4.3 Viscosity changes

As well as affecting the motion of the oil, viscosity affects oil spills control measures. The application of dispersants for instance ceases to be effective when the viscosity is higher than 10^4 cP. The applicability of other clean-up methods such as sorbents and pumping is also influenced by the viscosity rates of spilt oils.

Viscosity is calculated using the Mooney equation:

$$V_{emul} = V_a \exp(2.5 \, W/1\text{-}KW) \tag{8}$$

where

V_{emul} = viscosity of the emulsion (cSt)
V_a = adjusted oil viscosity (cSt)
W = fraction of water content
K = emulsification constant = 0.65

The original oil viscosity is first related to the fraction of oil evaporated and then adjusted to the environmental temperature using:

$$\log \log (V_a) = \log \log (V_T) + CF \tag{9}$$

$$\log (V_T + 0.7) = \log (V_{40} + 0.7) - 0.01 \, (40\text{-}T) \tag{10}$$

where

V_T = viscosity at environmental temperature (cSt)
C = viscosity constant = 1
F = volume fraction distilled
V_{40} = viscosity at 40°C (cSt)
T = environmental temperature (°K)

4.4 Density changes

The original density of crude oils is affected by various weathering processes, which increase the oil density and cause an oil slick to submerge beneath the surface layer of the water and eventually sink, thus affecting the detection of the slick.

Once emulsions are formed, the density of the oil becomes that of the original crude oil and of the water incorporated in the emulsions. This is expressed as:

$$D_{emul} = Y_W D_W + (1 - Y_W) D_{oil} \tag{11}$$

where D_{emul} = density of the emulsion (kg/m^3)
           Y_W = fractional water content
           D_W = density of sea water (kg/m^3)
           D_{oil} = density of the oil (kg/m^3)

Corrections are also made to relate the densities of weathered oils to the prevailing environmental temperature and the resulting evaporative loss.

4.5 Pour point

The pour point of an oil is the lowest temperature at which an oil will still flow. The temperature of sea water and the increase in the pour point will therefore affect the behaviour of spilt oils.

Pour point of oil emulsions is a function only of the fraction of oil evaporated, and is derived using:

$$PP_{emul} = PP_i + C_5 F \tag{12}$$

where

PP_{emul} =   pour point of emulsion (°C)
PP_i =   initial pour point of oil (°C)
C_5 =   slope of pour point distillation curve

Further details on the derivation of the equations discussed in this paper can be found in HR Wallingford (1991).

5. Model results and validation

Tests on the OIL-RW random walk model were run using a 1500m grid calibrated 2d-depth averaged HR TIDEFLOW model of the Irish Sea with U and V velocity components stored explicitly and not as harmonics. The area chosen was close to the shoreline which ran approximately from SE to NW and the main flows were alongshore. Simulations were done for 13 hours with a release of 720kg of oil over a period of 1 hour from an outfall situated 2 Km from the shore.

Two sizes of oil droplets were considered in the simulations, the droplet diameters were 31.6 and 316 μm according to the constant volume distribution of the droplets. A wind speed of 5 m/s was applied in two different directions (alongshore and inshore) with a wind drift factor of 0.03.

The oil particles demonstrate different rise velocities for the different droplet sizes. Due to the logarithmic tidal current and parabolic wind profiles, particles closer to the water surface move faster than those residing deeper in the water column. As smaller particles exhibit lower rise velocities, they therefore tend to remain for a longer time at a greater depth than larger particles. This causes increased shear in patches of small particles compared with that of larger ones, accounting for the smaller concentrations and larger patch size.

Figures 3 and 4 show the location of the oil patches with onshore and alongshore winds. It is evident from figure 3 that the larger particles are affected to a greater extent by the wind than the smaller ones. This can also be seen in figure 4. Figure 3a is more dispersed than 4a

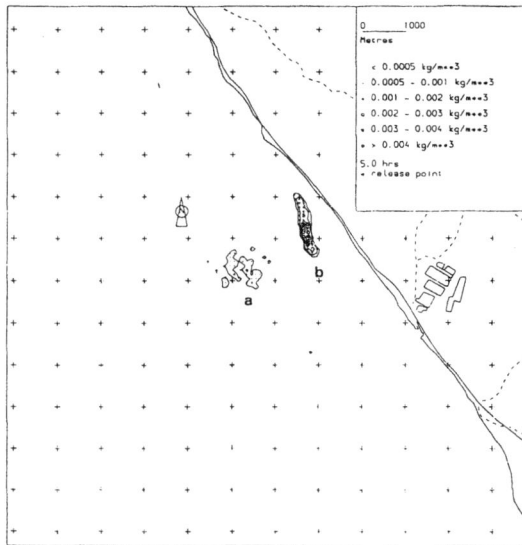

Figure 3 Simulation of oil particles trajectory of (a) 31.6 μm and (b) 316μm diameter with wind blowing from 225° from grid north (onshore wind).

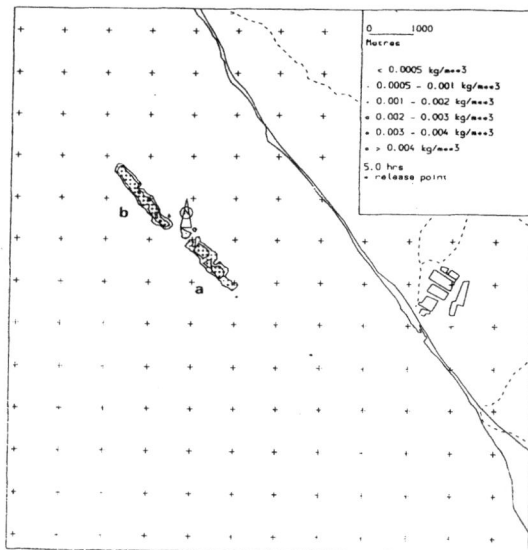

Figure 4 Simulation of oil particles trajectory of (a) 31.6 μm and (b) 316 μm diameter with wind blowing from 135° from grid north (alongshore wind).

because the patch is being advected in two perpendicular directions by the wind and tidal currents, thus increasing the shear effect in two directions instead of one. The simulations highlight the variation in the distance between the oil patches and the extent of mixing for different droplets sizes, current speed and direction. This phenomenon also explains the presence of a small oil slick lagging behind the major slick in real spills.

Another major development in OIL-RW is that coastal predictions are made more realistic than those in earlier models through the specification of a wind-driven current profile which varies over the water depth. This prevents the accumulation of particles in model cells adjacent to the coastline when the specified wind is directed onshore.

The weathering predictions were compared with results from field experiments carried out in the North Sea which involved continuous monitoring of two oil spills (Buchanan et al, 1988 and Hurford, 1989). The model's predictions showed good agreement with the data obtained from field measurements; results are listed in Figures 5 and 6. The model allows the simulation of spills with different time scales and environmental conditions.

The model correctly accounted for the variation in the oil behaviour in the two studied cases. While Flotta crude viscosity reached 10^4 cP in less than 24 hours, Forties crude viscosity remained below the limit for dispersant application. The fluctuation in the viscosity data from the field results can be related to measurement errors or collection of debris during sampling. Another point of special interest to the weathering trends of the oil is the increase of the pour point above water temperature (15°C) which causes the emulsion to destabilise and thus water shedding from the emulsion to occur.

6. Summary

The OIL-RW random walk model correctly predicts the lag of a patch of small particles behind that of larger ones. This TIDEFLOW-2D based method is more applicable in inshore areas where harmonic analysis does not work. If needed, fine resolution flows can be produced enabling one to resolve physical structures, intertidal areas, eddies etc. It is envisaged that the random walk and weathering model be coupled together to run simultaneously, allowing the user to make predictions regarding the fate of the spill and the best clean up method to use. Simulations containing a realistic distribution of particle sizes will be run in the future so as to produce a more realistic slick.

7. Acknowledgments

The authors would like to thank Dr C T Mead for his developement of the original random walk contaminant dispersion model, PLUME-RW and Dr G V Miles and B R Wild for their efforts during the development of the model.

8. References

1. Buchanan, I., 1987. *Methods for Predicting the Physical Changes in Oil Spilt at Sea.* Warren Spring Laboratory, Report No. LR 609 (OP) M.
2. Buchanan, I. and N. Hurford, 1988. *Results of the Forties Fate Trial, July 1987.* Warren Spring Laboratory, Report No. LR 671 (OP) M.
3. CONCAWE, 1983. *Characteristics of Petroleum and its Behavior at Sea.* Report No. 8/83, The Hague.
4. Forrester, W.D., 1971. "Distribution of suspended oil particles following the grounding of the tanker Arrow". *Journal of Marine Research* 29(1):151.
5. Gundalch, E.R., T.W. Kana and P.D. Boehem, 1985. "Modelling spilled partitioning in nearshore and surfzone areas". *Proceedings of the 1985 Oil Spill Conference.* API. Washington D.C. pp379-383.

6. HR Wallingford, 1991. OIL-RW : *A mathematical model for oil slicks dispersion and weathering*. HR Wallingford Ltd. Report IT 364, August 1991.
7. Huang, C.P. and H.A. Elliot, 1977. "The stability of emulsified crude oils as affected by suspended particles". D.A. Wolfe (ed.), *Fates and Effects of Petroleum Hydrocarbons in Marine Organisms and Ecosystems*. Pergamon Press.
8. Hurford, N., 1989. *The Large-scale Fate of Oil Trial in the North Sea*, August 1988. Warren Spring Laboratory, Report No. CR 3138 (MPBM).
9. NAS, 1989. *Using Oil Spill Dispersants on the Sea*. Committee on Effectiveness of Oil Spill Dispersants, National Research Council, National Academy Press, Washington D.C.

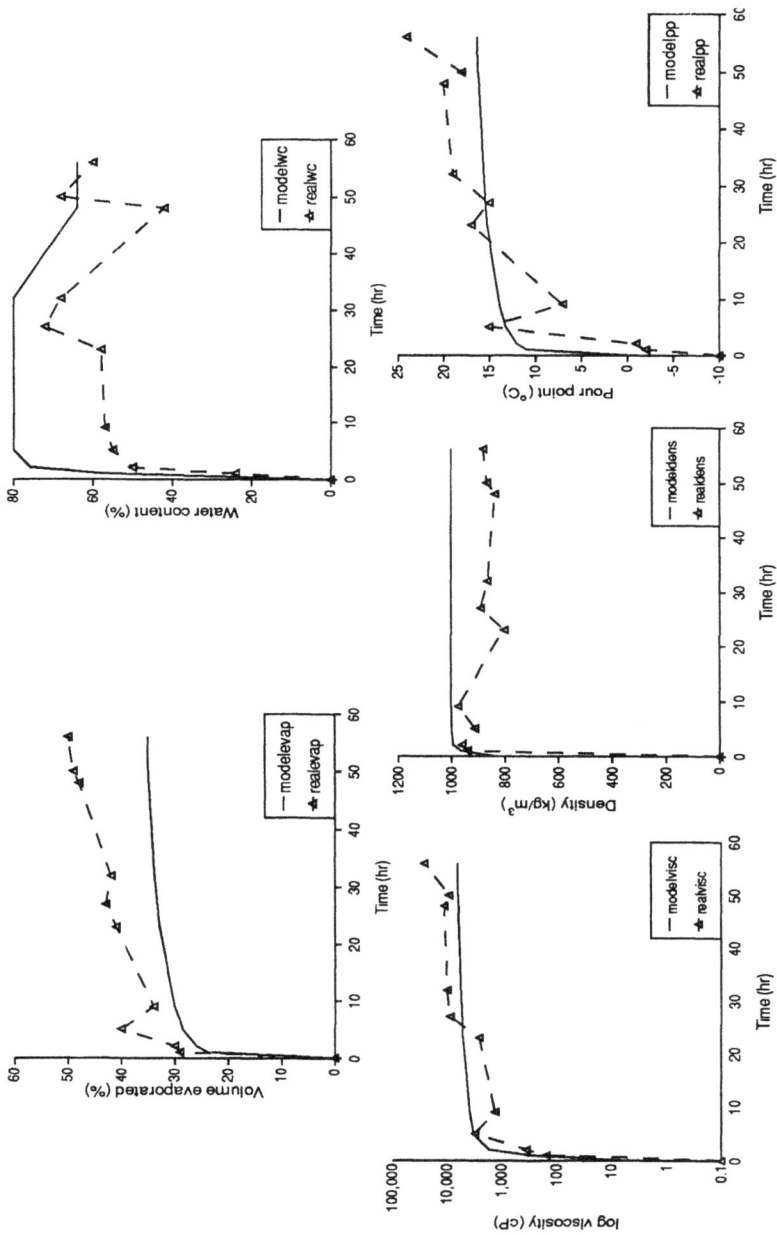

Figure 5. Model predictions of Forties crude trial compared to field results.

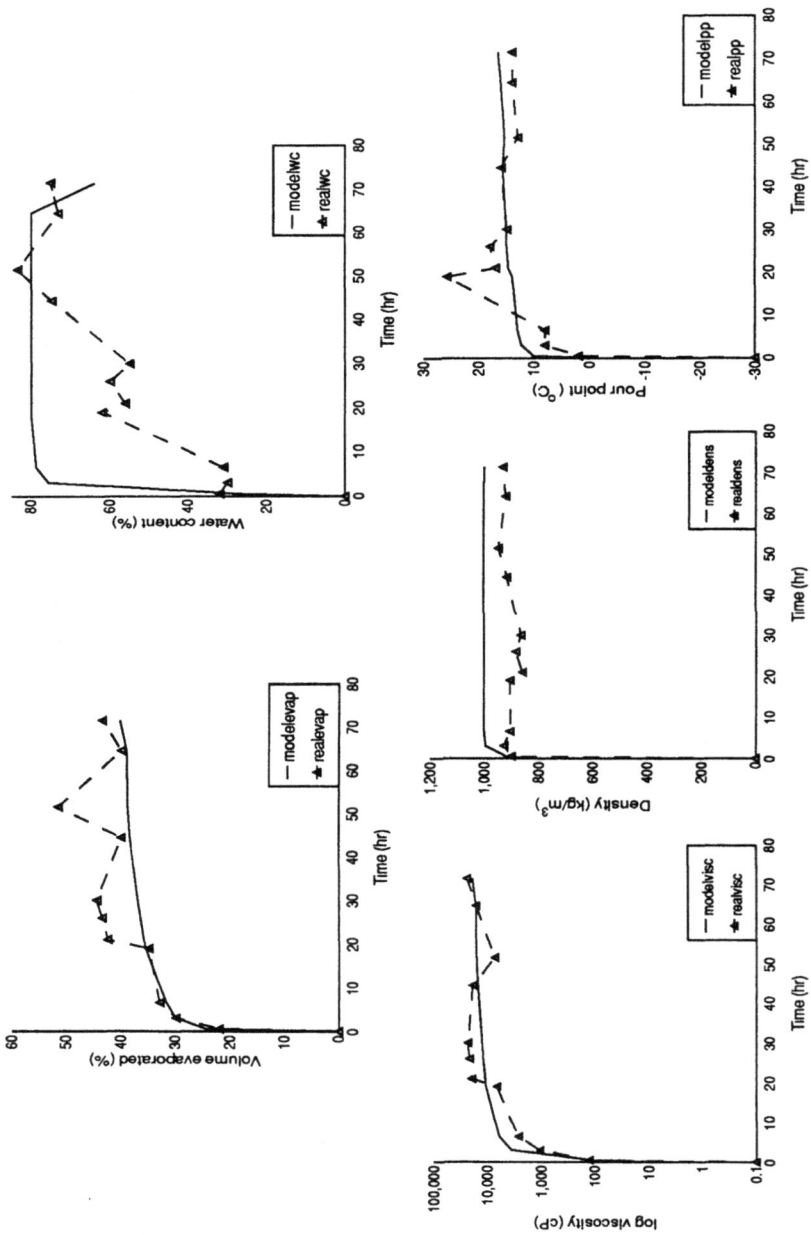

Figure 6. Model predictions of Flotta crude trial compared to field results.

18 Ecological modelling of the complex of primary producers in a coastal system, The Sound, Denmark; response to decrease in nutrient loadings

H. K. Bach

Abstract

The variety of primary producers (microalgae, macroalgae and rooted macrophytes) is different from area to area. In ecological modelling of the potential impact on an area, it is important to include in the model the actually present primary producers. First, this will provide the impact on all the present primary producers and not just one (e.g. phytoplankton), and secondly, it will assure that the ecological model depicts the fundamental processes and interactions taking place in the considered water system.

A study for The Sound between Denmark and Sweden called for an ecological model including the three groups of primary producers: phytoplankton, macroalgae (filamentous algae) and rooted macrophyte (Eelgrass). The study concerned the impact from a planned link across The Sound between Denmark and Sweden, both under the construction work and after the completion of the bridge/tunnel system.

The calibration of the model was based on measurements for the year 1989. The nutrient loadings to the area were provided from this period as well. The evaluation of the effect of the standing bridge was based on the expected nutrient loading condition for the year 2000. A side effect of this study was then a prediction of the effect of a $\approx 50\%$ reduction in nutrient loadings to the area.

The calculation for the year 2000 showed as expected a decrease in phytoplankton production and also in macroalgae biomass. Expected since these organisms take up the nutrients necessary for their growth from the water. Eelgrass getting nutrients from the bed gained biomass at the decrease in nutrient loadings due to reduced shading from phytoplankton and macroalgae.

1. Introduction

Many marine ecosystems are founded on or contain more than one main primary producer. The dynamic behaviour of systems with a compound of primary producers depends on the behaviour of each of these as well as their interrelations. Especially coastal communities can consist of several groups of primary producers: phytoplankton, macroalgae (like *Ulva lactula, Cladophora sp, Ectocarpus siliculosus,* and *Pilayella littoralis*) and rooted macrophytes (like *Zostera marina L* and *Ruppia sp*). The low water depths of the coastal zone allow the plants and macroalgae to attach to the bottom and grow in spite of the limited light conditions at the bottom. Some macrophytes and macroalgae can even grow at deep waters, e.g. *Laminaria saccharina* and *Ceramium sp* and *Polysiphonia sp*. Characterization of a biological community includes the determination of the important primary producers and their interrelationships. Some of the species compete for nutrients (e.g. phytoplankton versus macroalgae like *Ulva* or *Ectocarpus*), and others compete for light (e.g. phytoplankton versus macrophytes like *Zostera*; macroalgae like *Ulva* or *Ectocarpus* versus macrophytes like *Zostera*). The competition between macroalgae and rooted plants often results in well known belts of vegetation changing along a depth gradient.

Efforts in evaluating the effects of eutrophication and the changes which could be obtained by reducing eutrophication have, during the recent years, often involved modelling of the eutrophication process. In order to make predictions by modelling, it is realized that the model must include all the important types of primary producers and their interactions. This includes multi species phytoplankton/zooplankton modelling [7], modelling of macroalgae/phytoplankton competition for nutrients [5], phytoplankton/macrophyte interaction concerning light conditions [2, 8, 13] and phytoplankton/macroalgae/macrophyte competition for light and/or nutrients [6].

The study to be presented here includes modelling of two species of phytoplankton (spring diatoms and summer algae), the filamentous macroalgae *Ectocarpus* and *Pilayella*, and the rooted macrophyte *Zostera marina L* (eelgrass). The study area, i.e. the Sound between Denmark and Sweden, comprises areas of deep waters (up to 15 m) dominated by phytoplankton, and very shallow areas where eelgrass, filamentous algae etc. can grow. The area is considered eutrophic, hence a eutrophication model was applied. Two large cities (Copenhagen and Malmö) discharge waste water to it, and the area is recipient for agricultural runoff from both countries. The effects of reducing something like 50% of the nutrient input to the area (the future plans for the two countries) were modelled.

The model concept, the calibration, and the predicted results are presented. The basic assumptions of the eutrophication modelling are that macroalgae and the phytoplankton compete for the nutrients and that both have a shading effect on rooted macrophytes like eelgrass. The presentation concentrates on the interactions between these three primary producers and the effects of light (water transparency) and of eutrophication (nutrients).

2.   Materials and Methods

Study Site

The study site is a part of the Sound between Denmark and Sweden from the south of Copenhagen to the north of the island of Saltholm (see figure 1).

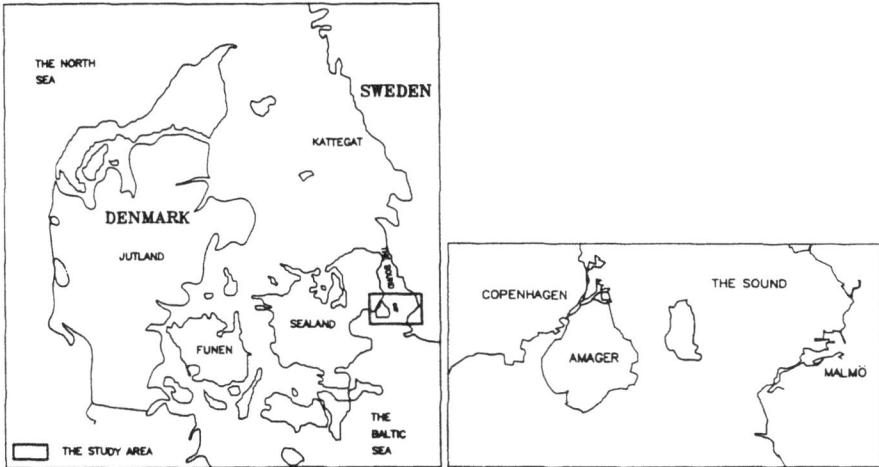

Fig. 1   The study site: a part of the Sound between Denmark and Sweden

The topography of the area is shown in figure 2. The figure shows the bathymetry used in the computation.

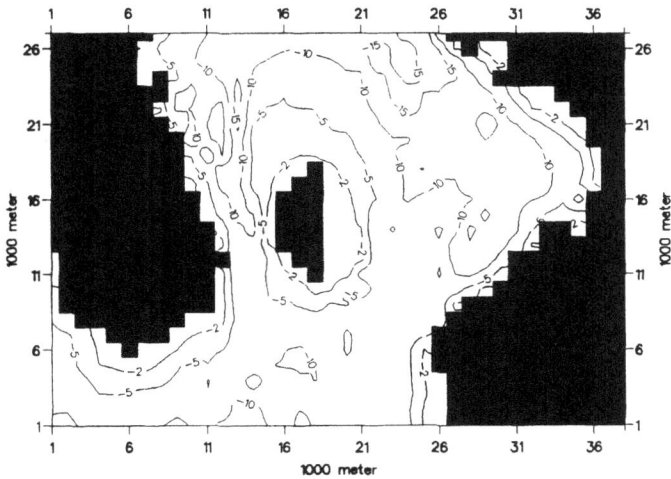

Fig. 2   Topography of the considered area. Model bathymetry is shown in meters

215

The Sound is one of the three water ways for exchange of water between the Baltic Sea and the Skagerak (North Sea). The salinity of the water from the Skagerak is around 30 ‰ whereas the water of the Baltic Sea is brackish ($\approx$ 8-15 ‰). Two layered flow in the Sound is therefore observed especially in the northern part of the Sound. In the southern part of the Sound a barrier (Drogden) with relative low water depth ($\approx$ 8 m) separates the Skagerak and Baltic waters. The net flow through the Sound is from south to north corresponding to the fresh water flow into the Baltic Sea. The actual flow is determined by climatological changes over Scandinavia and by the wind conditions resulting in both northerly and southerly flow. The wind and current direction changes usually within less than five days and often at a higher frequency. The condition of the area thus changes in a very dynamic way.

Biological Structure (Primary Producers); measured Values

The hydraulic conditions of the Sound influence the ecological system including the phytoplankton and the distribution and density of the bottom vegetation. The phytoplankton community is dominated by diatoms ($\approx$ 75 species) and dinoflagellates ($\approx$ 50 species) with a geographical gradient from north to south, as the richness in species decreases from north to south coincident with the salinity gradient. The phytoplankton primary production has been increasing over a span of years. Measurements from the thirties and fifties in the northern part of the Sound show a doubling [14]. The mean values of the phytoplankton primary production measured during the period of 1975-1989 for the growth season are in the range 0.2-1.2 gC/m^2/day. Corresponding values of total nitrogen and total phosphorus are in the ranges 300-450 $\mu g/m^3$ and 20-40 $\mu g/m^3$ respectively. The phytoplankton primary production is moderate compared to estuarine conditions, which reflects the close connection to water of more open areas (e.g. the Skagerak).

The distribution of vegetation and the various species were mapped based on a field survey in August 1990 and an overview is shown in figure 3.

Eelgrass is the most widely distributed and quantitatively the dominating macrophyte in the area. Eelgrass grows in the range 1 - 6 m's with maximum biomass at 2 - 4 m's. The eelgrass meadows are overgrown with filamentous algae (*Ectocarpus* and *Pilayella*) to a varying degree.

The belts of rooted vegetation are important biotopes as they have a fauna, which is varied and rich in species. They are important habitats for fish either permanently or as spawning areas and for a range of small creatures like *Crustaceans*, snails, *Bryozoans*, etc. Secondly, the rooted plants like eelgrass can stabilize the sediment and thereby protect the area against erosion.

The estimated biomasses and distribution of eelgrass and filamentous algae based on measurements from August 1990 are shown in figures 4 and 5.

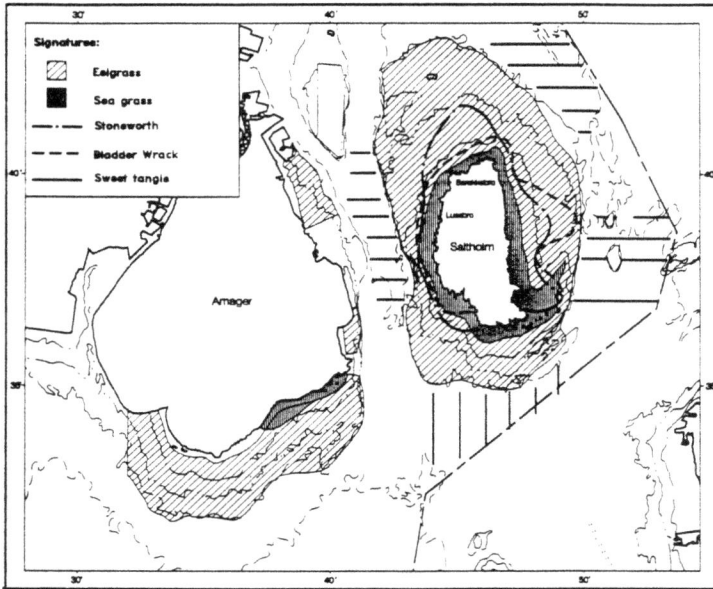

Fig. 3   The most important types of vegetation

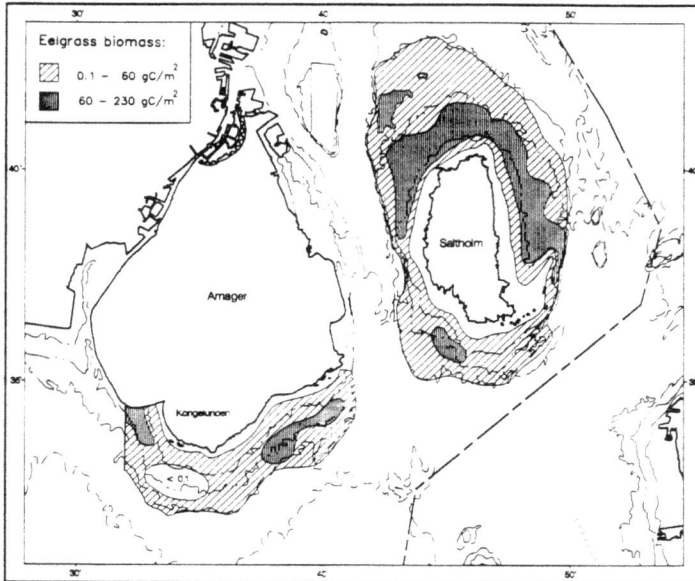

Fig. 4   Estimated biomass of eelgrass from measurements, August 1990

217

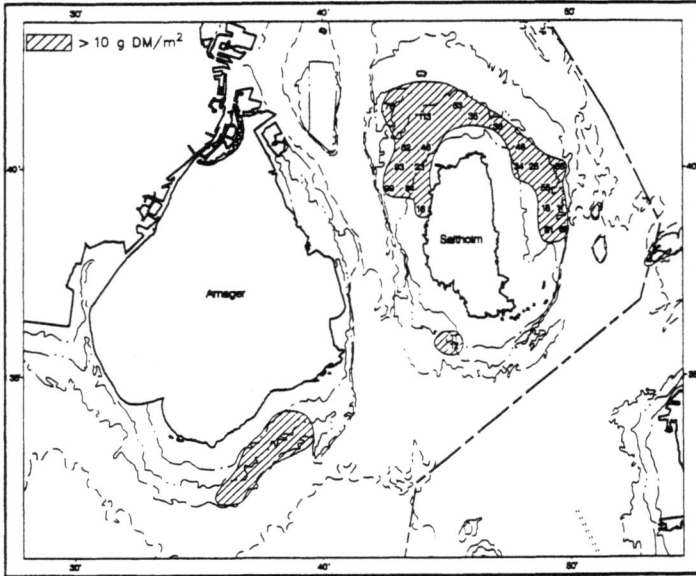

Fig. 5   *Biomass of macroalgae (Ectocarpus and Pilayella) measured August 1990. The numbers within the hatched area show the actual measured biomass (g DM/m²)*

The Eutrophication Model

The applied model includes a description of the pelagic system and the benthic vegetation representing eelgrass and filamentous algae. The model includes 16 state variables representing the three primary producers: phytoplankton (carbon, nitrogen, and phosphorus), macroalgae (carbon, nitrogen, and phosphorus) and rooted macrophytes (shoot biomass and shoot density), organic matter in the water (detritus carbon, nitrogen, and phosphorus), the inorganic nutrients (nitrogen and phosphorus) and finally oxygen. Thus 11 of the 16 included state variables constitute the pelagic system and 5 the benthic vegetation. The model describes the seasonal and spatial variations of these variables. The seasonal variations depend on a number of forcing functions: water exchange, influx of light, water temperature, nutrient loadings, and the conditions in the surrounding areas (boundary conditions).

The carbon, nitrogen and phosphorus cyclus as regarded in the model is shown in figure 6, exemplified by the carbon cyclus (nitrogen and phosphorus are in principle the same). An oxygen balance is included based on the processes of the carbon cyclus.

218

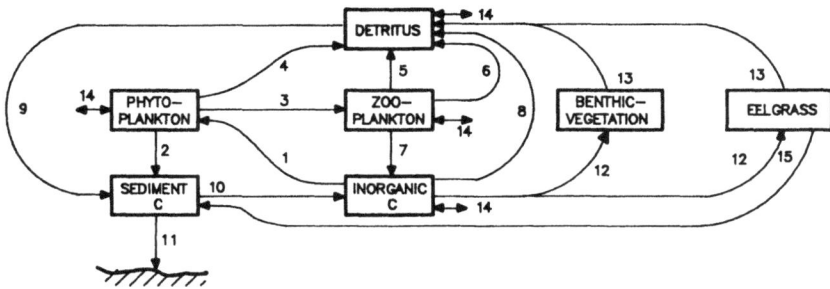

Fig. 6   Processes and state variables in the eutrophication model, exemplified by
the carbon cyclus

1. PRODUCTION, PHYTOPLANKTON
2. SEDIMENTATION, PHYTOPLANKTON
3. GRAZING
4. EXTINCTION, PHYTOPLANKTON
5. EXCRETION, ZOOPLANKTON
6. EXTINCTION, ZOOPLANKTON
7. RESPIRATION, ZOOPLANKTON
8. MINERALIZATION OF SUSPENDED DETRITUS

9. SEDIMENTATION OF DETRITUS
10. MINERALIZATION OF DETRITUS
11. ACCUMULATION IN SEDIMENT
12. PRODUCTION, BENTHIC/EELGRASS VEGETATION
13. EXTINCTION, BENTHIC/EELGRASS VEGETATION
14. EXCHANGE WITH SURROUNDING WATERS.
15. SEDIMENTATION OF EELGRASS

The pelagic system is described by the growth of phytoplankton, grazing of phytoplankton, and the transformation of living phytoplankton and grazers to dead organic material (detritus) (process No. 1, 3, 4, 5, and 6). In addition phytoplankton
and detritus are subject to sedimentation (process No. 2 and 9). The release of
nutrients from the degradation of organic matter in the water and sediment and the
corresponding oxygen demand are also included (process No. 10). The growth of
phytoplankton depends on the light conditions, water temperature, and nutrient
concentrations. The nutrient dynamics in phytoplankton are expressed by internal
pools of nitrogen and phosphorus. This means, that algae growth can take place
even though the nutrient concentrations in the water are very low.

The growth of macroalgae is dependent on light, temperature, and nutrients like
phytoplankton. The nutrient dynamics of the macroalgae are described by internal
pools of nitrogen and phosphorus similar to phytoplankton. The spatial distribution
of macroalgae biomass will be dependent on the topographical conditions (the
water depth) since the light decreases with increasing depth and since the wave
exposure increases with decreasing water depth. At high biomasses of macroalgae,
a selfshading effect will influence the growth.

In the description of the eelgrass growth, two types of growth are distinguished
in the model: one is leaf elongation (i.e. increase in biomass of the individual
shoot) and the other is the development of new shoots. The eelgrass biomass is
then the product of the biomass of each shoot and the number of shoots per m^2.
Applying this approach in modelling of eelgrass, enables the model to reflect a
situation where the eelgrass biomass at two different places is equal, but the shoot
density is different. And on the other hand a situation where the shoot density can
be the same at two different places but the total biomasses are different due to

differences in the biomass of the individual shoots. The growth of eelgrass is light and temperature dependent and dependent on the wave exposure. Nutrient limited growth for eelgrass is not included in the model. Eelgrass can take up nutrients from the water and from the interstitial water of the sediment and the nutrients can be translocated within the plant itself. Nutrient limitation is therefore not expected to be an important factor for eelgrass growth. The seasonal variations in eelgrass biomass is a direct result of the seasonal variation in light and temperature and more indirectly influenced by the variation in macroalgae biomass via a shading effect. Filamentous algae will at significant biomasses have a severe shading effect on rooted vegetation. The spatial distribution is as well as for macroalgae depending on the light penetration and the wind and wave exposure (which is a function of water depth).

The important factors concerning competition between phytoplankton and macroalgae are the availability of nutrients and the light dampening effect of phytoplankton (chlorophyll-a) on macroalgae. The shading effect of filamentous algae on eelgrass is the key point in the competition between macroalgae and rooted vegetation.

Eelgrass and the macroalgae have different affinities for optimum depth. Eelgrass prefers water depths of 2 - 6 m's, wehreas *Ectocarpus* and *Pilayella* prefer deeper waters, e.g. 3 - 7 m's. One of the reasons is assumed to be a higher impact from wind and waves on macroalgae compared to eelgrass. The two species will though interfere in some areas/depths (3 - 6 m's).

The model description for the eelgrass part can be found in [2], for macroalgae in [10] and an overview is given in [11].

Model Setup

To describe the very dynamic transport of water through the Sound, a two-dimensional fully dynamic, hydrodynamic, and advection/dispersion model, the model system MIKE 21, has been chosen [1, 4] This model is based on the finite difference discretization in space with rectangular elements in the computational mesh. In this case a mesh of 1000x1000 meter elements was chosen. This means that the waters within these 1000x1000 meters are homogeneous. This refinement is sufficient to describe the variation in the biological parameters, e.g. phytoplankton and vegetation found by measurements. The hydrodynamics are based on time series of wind and of currents/water levels at the model boundaries representing a mean-to-minimum situation, including high and low current speeds and northerly as well as southerly wind and current directions.

The boundary conditions for the variables describing the pelagic phase are found from long term measurements at sampling stations south and north of the model area. Seasonally varying intensities of the photosynthetic active light at the surface are taken from measurements at a monitoring station outside Copenhagen. Water temperature within the area and data on which the model calibration was based are provided by a field survey covering the year 1989 [12]. Nutrient loadings from sewage outlets, agricultural runoff, and atmospheric deposition from the Swedish and the Danish side have been estimated for the 1990 situation (status) and for the projected situation "year 2000" according to the future plans for the two countries [3]. The inputs of nitrogen and phosphorus are shown in table 1.

Table 1 *Nutrient loadings in the present situation (status) and for the projected situation "year 2000"*

Nutrient loadings (tonnes/year)	Status Denmark	"year 2000"	Status Sweden	"year 2000"
Total nitrogen	7975	3564	7988	5525
Total phosphorus	1160	434	487	140

The eutrophication model was calibrated preliminary using a very coarse "box-model" for the area. This calibration was verified afterwards in the more detailed model.

The simulation period in the detailed model was six months from April to September in order to cover the intensive growth period for macroalgae and rooted vegetation. The computational time step was 15 min. for the advection-dispersion model and 3 hrs. for the eutrophication model. The coupling of these models is explained in [9].

3.   Results

Present Situation

The calculated results for the present situation are compared with measured data of phytoplankton primary production, total nitrogen and total phosphorus for the pelagic system and with the measured biomasses and distribution of filamentous algae and eelgrass for August 1990 (the time of the bottom vegetation survey).

Phytoplankton primary production, total nitrogen and total phosphorus for two mesh points are shown in figure 7. The nutrient concentrations are within the measured ranges (see section 2). The agreement with measured data is satisfying.

The areal distribution of phytoplankton production for August is shown in figure 8.

The calculated distribution of eelgrass and macroalgae for August are shown in the figures 9 and 10 and can be compared with the measured data of figure 4 and 5 respectively. There it shows (in measurements and computations) how the maximum biomass of macroalgae will be found at deeper waters compared to eelgrass. This is stressed by the results in figure 11, where biomasses at three different water depths are shown for the whole computational period. The maximum biomass of macroalgae is seen at the deepest point in contradiction to eelgrass and where the highest eelgrass biomass is calculated is the minimum macroalgae biomass. In the intermediate zone, the macroalgae have a shadowing effect on the eelgrass as it is entangled with the eelgrass plants.

221

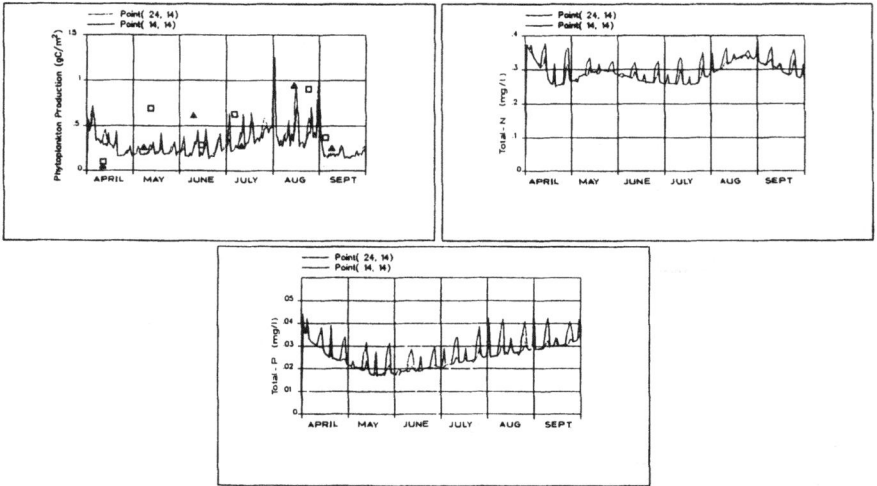

Fig. 7 Computed values of phytoplankton production, total nitrogen and total phosphorus for two positions

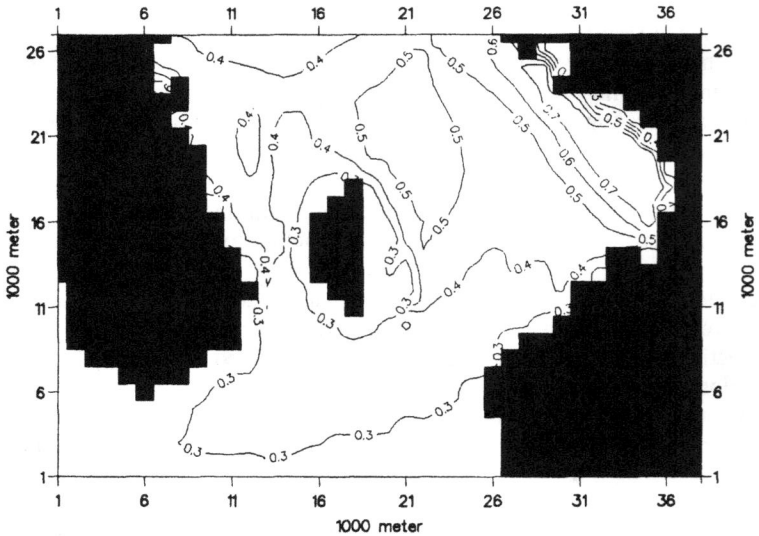

Fig. 8 Areal distribution of phytoplankton production (August)

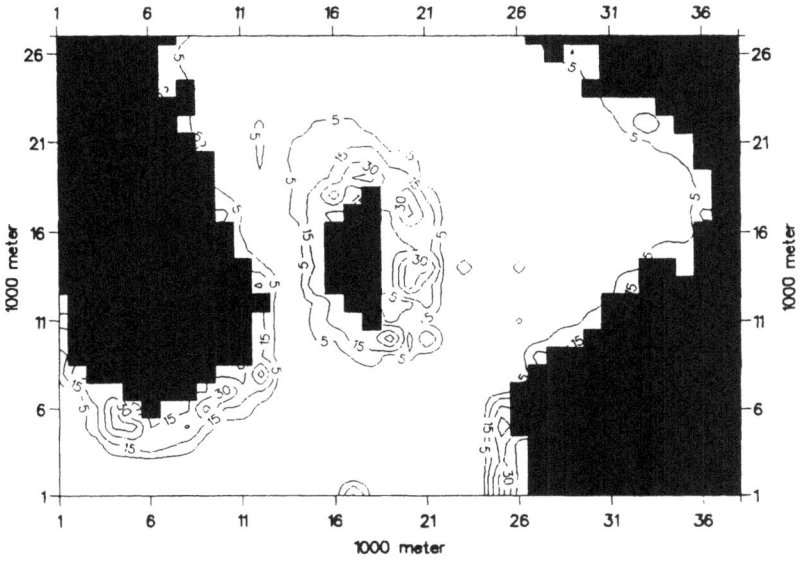

Fig. 9   Areal distribution of eelgrass biomass (August)

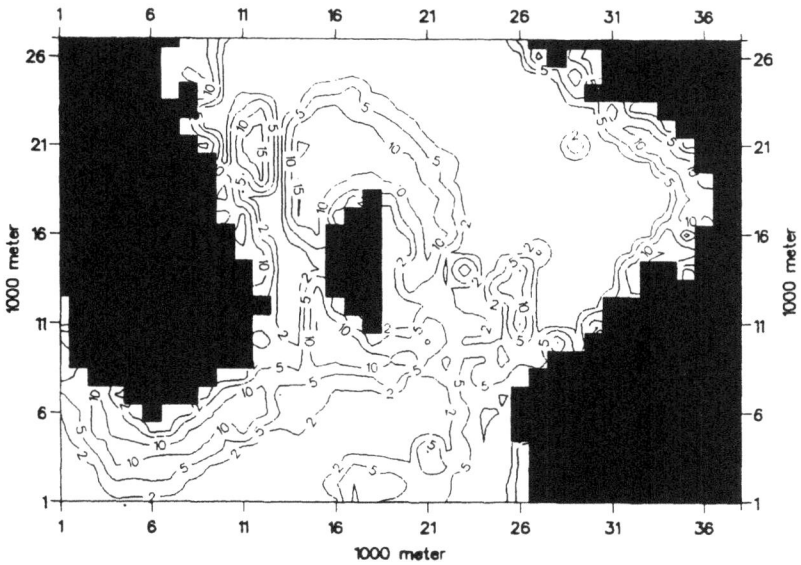

Fig. 10   Areal distribution of filamentous algae biomass (August)

The shading effect of eelgrass is triggered at a certain macroalgae biomass level, by calibration found to be 15 gC/m². This means that the eelgrass biomass is depressed especially in the north-westerly and south-westerly side of the island Saltholm. Macroalgae grow more at the western side of the island than on the eastern side, mainly due to the nutrient inputs from the city of Copenhagen. The phytoplankton biomass is influenced by regional conditions (the Baltic Sea and the Skagerak) more than by the local conditions of the area which of course contribute to the regional conditions. This explains why the phytoplankton production varies very little within the area.

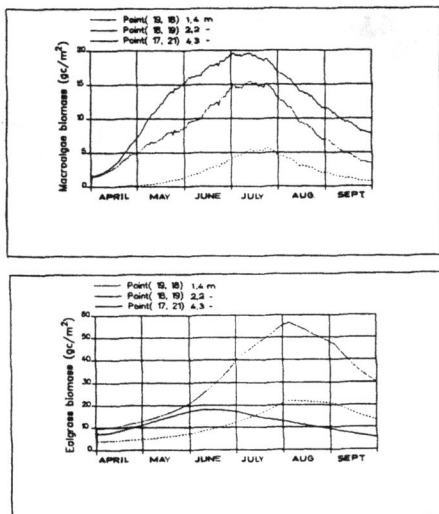

Fig. 11   Biomasses of macroalgae and eelgrass at three different water depths at a position north of the island Saltholm

"Year 2000"

The predicted phytoplankton production and changes in biomasses of eelgrass and macroalgae for August are shown in figures 12, 13, and 14 respectively.

The phytoplankton production and the macroalgae biomasses both decrease due to the decrease in nutrient inputs. Eelgrass biomass increases as expected due to the decreased shading from phytoplankton and macroalgae. The area covered with eelgrass does not change very much, however. This is probably due to the fact that the depth limit of eelgrass is determined by the general extinction of the water (including phytoplankton) rather than shading from macroalgae at these depths (6 - 7 meters).

The decrease in macroalgae biomass is up to 20% and the increase in eelgrass biomass up to 70%.

The predicted decrease in macroalgae biomass is in the area with significant biomasses 10-20%. The decrease is above 20% in some areas north of Saltholm, which is influenced by the waste water outlet from the city of Copenhagen.

The main decreases in phytoplankton production are seen north of Saltholm again in the area receiving waste water from Copenhagen. Only very small differences are predicted south of Saltholm.

The highest increases in eelgrass biomass are not coincident with the highest decreases in macroalgae biomass. They are predicted to be in areas where the macroalgae biomass decreases below the shading limit of 15 gC/m^2. This is north- and south-west of Saltholm. The increase in eelgrass biomass north of Saltholm is also due to the decrease in phytoplankton. East of Saltholm, where high biomasses of eelgrass are found, almost no increase in biomass is predicted. This is due to an unchanged phytoplankton production. The changes in eelgrass biomass is then a combination of shading from macroalgae and from chlorophyll-a.

The decrease in macroalge is also influenced by the decrease in phytoplankton. Leaving phytoplankton out would have lead to a larger decrease than predicted with the complex model.

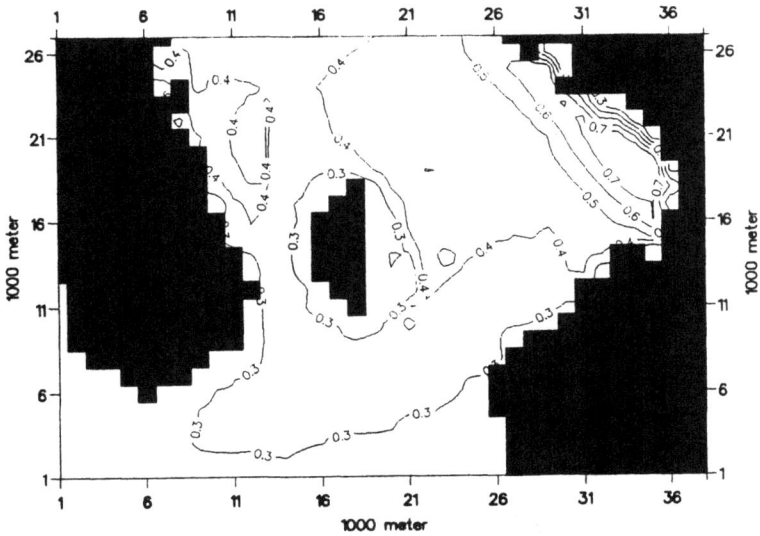

Fig. 12   Phytoplankton production "year 2000"

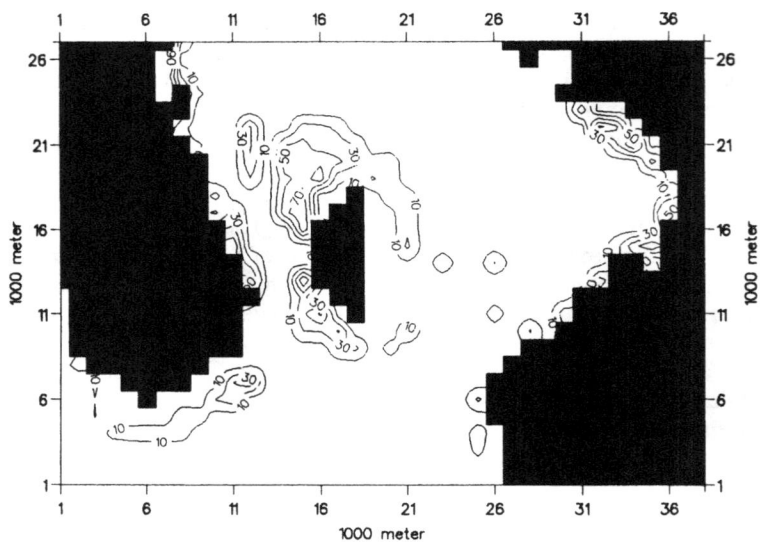

Fig. 13 Eelgrass biomass "year 2000" substracted year 1990 (%)

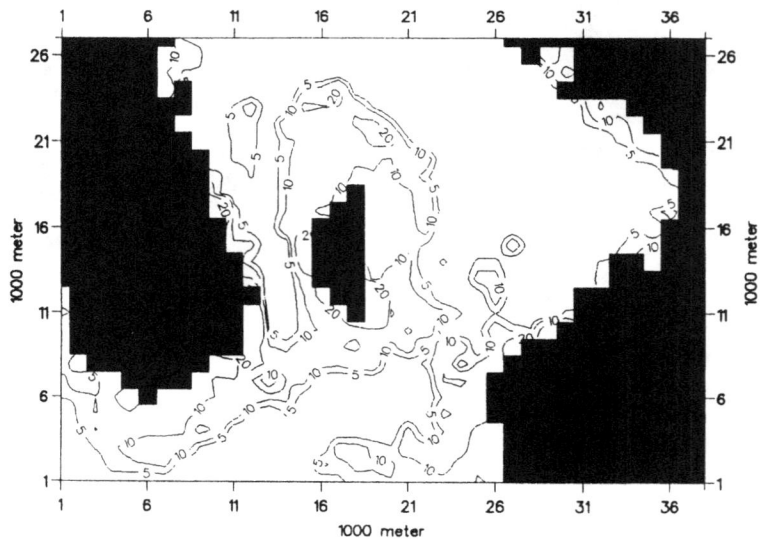

Fig. 14 Macroalgae biomass year 1990 substracted "year 2000" (%)

4. Conclusions

The presented results emphasize the importance of including all the important primary producers in an ecosystem. Changes in growth and biomass of one will influence the others. Here phytoplankton was found to influence both macroalgae and eelgrass, but differently in different areas. Macroalgae was predicted to influence eelgrass biomass. The changes in macroalgae biomass had different effects due to the way the filamentous algae shade the eelgrass plants.

The decrease in nutrient loadings had the qualitatively expected effects on the biomass, but the ditributed picture of decrease (macroalgae) or increase (eelgrass) could not have been predicted without a model like the one applied. The changes would quantitatively also be very difficult to predict without this tool.

5. References

[1]   Abbott, M.B., McCowan, A. and Warren, I.R., (1981), Numerical Modelling of Free-surface Flows that are Two-dimensional in Plan, *Transport Models for Inland and Coastal Waters*, Academic Press, pp 222-283.

[2]   Bach, H.K., (1992), A Dynamic Model Describing the Seasonal Variation in Growth and the Distribution of Eelgrass (Zostera marina L.). I. Model Theory, *Ecological Modelling*, in press.

[3]   Cowi/VKI Joint Venture, 1990, (in Danish), Investigations of the Consequences for the Marine Environment of a Link across the Sound, KM 4.2 Mapping of Loadings, Technical Report, No. 0.1. *Report to DSB/Vejdirektoratet.*

[4]   Ekebjærg, L. and Justesen, P., (1991), An Explicit Scheme for Advection-Diffusion Modelling in Two Dimensions, *Computer Methods in Applied Mechanics and Engineering*, 88, pp 287-297.

[5]   Malmgren-Hansen, A. and Warren, R., (1992), Management of Urban, Industrial and Agricultural Pollution to Control Coastal Eutrophication, *Science of the Total Environment* (in press).

[6]   Rasmussen, E.K., Bach, H.K. and Riber, H., (1992), Use of an Ecological Model in Assessing the Impact from Sediment Spill on Benthic Vegetation, *Proceedings of the 12th Baltic Marine Biologists Symposium*, August 25-30, 1991, Helsinore, Denmark.

[7]   Rose, K.A., Swartzman, G.L. Kinding, A.C. and Taub, F.B., (1988), Stepwise Iterative Calibration of a Multi-species Phytoplankton-Zooplankton Simulation Model Using Laboratory Data, *Ecological Modelling*, 42, pp 1-32.

[8]   Verhagen, J.H.G. and Nienhuis, P.H., (1983), A Simulation Model of Production, Seasonal Changes in Biomass and Distribution of Eelgrass (Zostera marina) in Lake Grevelingen, *Mar. Ecol. Prog. Ser.*, 10 pp 187-195.

[9]   Vested, H.J., Jensen, O.K., Ellegaard, A.C., Bach, H.K. and Rasmussen, E.K., (1991), Circulation Modelling and Water Quality Prediction, *Proceedings of 2nd International Conference on Estuarine and Coastal Model-*

ling, 13-15 November 1991, Tampa, Florida, U.S.A.

[10] Water Quality Institute, (1989), Macroalgae Model. Description of a Model for Implementation in the North Lake of Tunis. *Prepared for Sir William Halcrow and Partners, U.K.*

[11] Water Quality Institute, (1990), General Description of the Mathematical Model for Eutrophicated Coastal Areas.

[12] Water Quality Institute, (1990), (in Danish), Measurements of Primary Production and Bioassays in the Sound and Køge Bay 1989. *Report to the Copenhagen Council.*

[13] Wetzel, R.L. and Neckler, (1986), A Model of Zostera Marina L. Photosynthesis and Growth: Simulated Effects of Selected Physical-Chemical Variables and Biological Interactions, *Aquatic Botany*, 26, pp 307-323.

[14] Øresundskommissionen, (1984), (in Swedish), The Sound, Present State - Effects of Nutrients, *Statens Naturvårdsverk*, Report No. 3008.

19 Experience and recommended practice for the construction and objective calibration of coastal pollution models

N. V. M. Odd and D. G. Murphy

Abstract

The paper describes the authors experience and ideas for a recommended practice for the selection, construction and objective calibration of coastal pollution models.

1 Introduction

In recent years, there has been a large expansion in the application of mathematical coastal pollution models for UK water companies, developers and NRA regional organisations. Many models, developed as part of contracts won by competitive tender, have not performed as well as expected because they were specified incorrectly or set-up and calibrated using inadequate data and poor methodologies. In many cases, there has been an emphasis on user-friendly aspects of the models rather than their technical quality. At the same time the NRA regional organisations have raised their standards for accepting the calibration of models used to support planning applications.

The purpose of this paper is to describe the authors experience and ideas for the recommended practice for the selection, specification, construction and calibration of coastal pollution models with an emphasis on the physical aspects of the processes causing the dispersal of effluents.

1.1 Definitions

The coastal environment is the zone where oceanic phenomena overlap with tidal and fluvial phenomena. To pollute the coastal zone is to degrade the natural purity of the water and to make it offensive and harmful to human, animal and plant life. Water quality standards are set to control pollution and meet water quality objectives. Water quality variables include temperature, turbidity, salinity, dissolved oxygen, BOD, ammoniacal nitrogen, nutrients, coliforms, chlorophyll, heavy metals (dissolved and adsorbed), hydrocarbons, PCB's, toxic chemicals, floating debris, oil and grease.

A mathematical pollution model is a systematic calculation of the movement of water, solutes and suspended matter, element by element, throughout the modelled area at successive time intervals. Models may be categorised as 'research', 'engineering', 'policy' or 'commercial'. Research models tend to concentrate on the details of particular physical, chemical or biological processes and are not generally used to assist in decision making. Engineering models are usually designed to predict the effect of particular engineering works such as an individual outfall. Policy or planning models have to be able to predict the cumulative effect of a large number of point and diffuse sources of pollution over a relatively large area and perhaps throughout a whole annual seasonal cycle. A commercial model is one which is sold or licensed and may be used by non-specialist modellers.

1.2 The Use of Coastal Pollution Models

The Hong Kong Environmental Protection Department, who are heavy users of coastal pollution models have described (D.Choi Personal Communication) their use of such models in assisting water quality, management and planning as follows;

(a) Strategy development

Models providing realistic representation of the system concerned, are used as tools in strategy development and to examine responses of the environment to the various proposed changes, such as extensive reclamation, increased effluent discharges arising from population and urban growth and implementation of water pollution control measures. Since mathematical models can simulate various scenarios in a short time, it is possible to assess a wide range of alternative strategies which would occur large sums of capital expenditure if implemented. Model simulations help to reduce the risk of failure in achieving the goals.

(b) Setting effluent quality requirements and treatment levels

Effluent quality requirements are set to safeguard the beneficial uses of the receiving water. Usually, there are several discharge points into a single water body; mathematical models can be used to assess its assimilative capacity and determine the maximum acceptable pollution loads, which indicates the levels of treatment required for the discharges.

(c) Refinement of monitoring programme

Environmental monitoring is expensive, but information obtained is essential for better understanding of the system of interest and enables more rational decision making when necessary. Modelling results can help to identify the inadequacy or redundancy of the existing monitoring programme, hence, improve the e
ffectiveness of the information capturing exercise with the same level of available resources.

The UK water companies and the NRA regional organisations probably have very similar uses for coastal pollution models.

2   2D and 3D models

The ideal model is three-dimensional with an infinitely variable grid able to resolve detail in-plan and in the water column. The model should include the interaction of waves and tidal currents. The wind-field should take into account the topography of the land. The model should be able to simulate transient phenomena such as storm water overflows, diurnal on-off shore breezes as well as seasonal variations and the ultimate fate of pollutants in the long term.

In reality; for practical applications to engineering problems it is necessary to use simpler models which have inherent approximations. It requires experience and skill to, firstly decide which processes may be ignored or simplified, and secondly to use the model to the best possible advantage. This usually means selecting a series of representative combinations of environmental conditions.

Waste water disposal models have to date, often been 2D depth-averaged models. However, these models are basically unsatisfactory for simulating wind driven circulations in coastal waters especially if they are stratified by salinity or temperature or if there are significant lateral density gradients at depth, which generate density currents. Even in vertically well-mixed conditions in the absence of density effects it is important to take into account the shearing effect of the velocity profile by predicting the 3D movement of effluent particles.

2D models fail to simulate :-
      (i) 3D wind driven circulations
      (ii) Buoyancy and stratification effects.
      (iii) Gravitational and geostrophic currents.
      (iv) Light penetration as a function of depth and turbidity.

Generally speaking it is easier to calibrate and interpret the predictions from the more sophisticated 3D models provided they do not have an excessive number of ill defined empirical coefficients. HR have used 2D layered and 3D models to study coastal pollution for ten years.

2.1 Model grids

There are three main model grid systems which are appropriate to coastal models:-
      (i)   rectangular (patched,variable)
      (ii) curvilinear
      (iii) finite element (quadrilateral or triangular)
The main advantages and disadvantages of the three methods are shown in Table 1.

Table 1 *Comparison of model grid systems*

	Patched rectangular	Finite element	Curvilinear
Bathymetric resolution	Variable	Good	Best in rivers
Ease of construction	Easy	Complex	Very complex
Code	Simple	Very Complex	Complex
Plotting	Simple	Complex	Complex
Computing	Medium	May be high	May be low
Transport models	Easy	May not resolve pollution fields	More complex
Numerical effects	Poor if 45° to narrow channel	Conservation not exact	Flow tends to follow long axis

The model grid must be able to resolve the essential details of the bathymetry, tidal currents, eddies, salinity or thermal fronts and pollutant distributions, which are usually not co-incidental thus favouring a fairly regular grid. It is not always possible to determine in advance where a coarse grid may be adequate.

2.2 Bathymetry

A basic requirement for a satisfactory model is high quality bathymetric survey data, which should contain sufficient detail in shallow inter-tidal areas, which are sometimes difficult to survey. Ideally, the survey should include sections orthogonal to channels at a spacing not exceeding more than one or two channel widths and there should also be two repeated sections along the channels to pick up deep holes or bars missed by the cross-sections. The spacing of the surroundings might vary between 0.3 and 2 times the proposed model grid size. The survey should not leave any area with large gaps. Ideally, it is also very useful to have a sounding line following the HW mark as closely as possible by using a shallow drafted survey vessel. The survey should be presented to the modeller as x,y,z triples, where x and y are the co-ordinates of a point in a specified grid system and z is the depth which should be reduced to a common datum.

Considerable care should be taken in interpolating the soundings onto a model grid. Often in the case of a poorly resolved survey it is necessary to add contours manually. The authors favour an interpolation procedure in which the bed level at a grid point is interpolated from the nearest sounding in each of four quadrants within a specified search radius. The bed level is evaluated as a weighted mean, based on the reciprocal of the distance, of the nearest soundings found within the search radius. The search radius may be increased in areas where the bed is very flat and reduced in the vicinity of steep gradients. A poor survey and a poor interpolation procedure will produce a poor model. At this stage, the model bathymetry should be compared with the survey by overlaying contours. It may be necessary to adjust the depths to resolve poorly resolved features of the bathymetry such as deep holes. It may also be necessary to remove artificially uneven bathymetry in the model, especially in drying areas exposed to strong tidal currents.

2.3 Boundary conditions

Many coastal models have failed to live up to their expectations as a result of poorly posed offshore boundary conditions. The basic requirement is that the model boundaries should contain pollutants discharged in the study area in a buffer zone for long enough to avoid any significant artificial flushing effects. In tidal waters, the boundaries should be well aligned with co-tidal lines, peak flood and ebb currents and they should be located well away from tidal eddies and headlands.

Results from a coarse grid regional model are a great help in selecting the optimum boundaries of more detailed local models. It is also important to specify the correct mix of boundary conditions in terms of water level, velocity or discharge. In coastal waters, the tidal currents are often relatively weak, and inertia effects are more important than frictional effects. As a result, peak currents are not in phase with the maximum water surface slope and may occur at any phase of the tidal cycle. In this case, it is essential to specify a tidal discharge or velocity along at least one open boundary. It is also usually essential to employ at least one soft boundary to avoid unrealistic seiching within, or concentration of flows at, the boundary of the model. It is usually necessary to allow for wind driven longshore currents.

The authors do not favour boundary tides defined directly from harmonic constants, allowing the user to simulate any tide, because it can set up unrealistic conditions. Instead, the authors prefer to set up the model to simulate a moderate number (2-7) of representative tides covering the full range of amplitudes, which can be validated and if necessary run with different wind or fluvial conditions.

There is usually little point in attempting to measure very small changes in the phase and level of the tide at either end of a small study area. Instead it is much better to use results from a regional model or to measure the tidal velocities at several points along the boundary carrying a significant tidal discharge. 2D or 3D regional models should also be used to define the effect of winds on the boundaries of the more local model.

3 Objective calibration of a coastal pollution model

The acid test of any mathematical model is its ability to reproduce reality. A model may be deficient for a number of reasons ranging from over-simplification of the physical processes simulated to the accuracy of the numerical method.

In order to assess the diagnostic capability of a mathematical model the simulated results must be compared to observations made in the region of interest. Formulating an objective method of comparison with observations is essential to a successful calibration of any model.

The calibration of a model implies the adjustment of inexactly known factors, such as bed roughness, boundary conditions or horizontal turbulent diffusivity to achieve an acceptable level of agreement with field observations. However, every effort should be made to use theory and observations to predetermine the values of all empirical coefficients used in a model. It is the definition of this acceptable level of agreement which we shall consider further. Once a model has been calibrated it is usually validated against another set of data to ensure the model can be used to represent a variety of different conditions. The validation process is carried out with no further adjustment to the model calibration parameters.

A model can only be calibrated effectively with a representative data set of field measurements. Naturally, the collection of data on field surveys is financially constrained and it is therefore essential to optimise the collection of data to give a clear picture of the processes occurring in the field. It is considered that the field measurements provide a synoptic window on reality. A project may be fortunate to have access to previous field studies which often provide useful data; but to simulate for example, the dispersion of pollutant from a proposed outfall site may require a specific site survey. Ideally, this should occur before the collection of further data.

It is recommended that an integral part of the study should be the assessment of available field data. This can have a two-fold benefit; as well as giving a guide to the need for more data it will also give an initial estimate of the suitability of a 2D or 3D model.

On the basis of the assessment of field data the calibration procedure should be clearly defined, indicating the variables to be considered and importantly, an estimate of the expected accuracy of the model. The goodness of fit of the model in comparison with the observations cannot be exactly predicted in advance due to the fact that it is dependent on the accuracy of the field observations which is not always very high. A fundamental difficulty in the comparison of a model with observations is due to their sparseness both spatially and temporally and the resolution of the model. However one should be able to specify statistical calibration targets which would be expected to be attained after the removal of spurious data in the field observations. It is then the responsibility of the modeller to undertake to explain any significant deviation from the expected performance of the model. The usage of statistical calibration targets is a necessary step towards a common framework for coastal pollution modelling. It enables water companies and regulatory bodies to assess rapidly the accuracy of a model, this assessment will inevitably influence the confidence the regulator has in predictions made by the model.

The standard practice is for coastal pollution models to consist of a hydrodynamic model, which should represent a variety of tidal and meteorological conditions; and a pollutant dispersion model which is driven by the flows from the hydrodynamic model. In practice, the random walk models are calibrated on the coefficient of horizontal diffusivity which has a narrow range of realistic values, typically about 0.2-0.5 m^2/s for British coastal waters. The accuracy of the dispersion model will be strongly correlated to the accuracy of the flow model. It is therefore essential for the hydrodynamic model

233

to achieve the required level of accuracy.

3.1 Calibration of tidal flow models

In determining a suitable calibration procedure for a hydrodynamic model one must take into account the variability of tidal and meteorological conditions prevailing during the field observations. Usually the data will be non-simultaneous and be spaced over a large number of different tides. It is then advisable to batch the observations according to fairly narrow tidal ranges and high water levels at a reference port. The model can then be used to simulate a number of representative repeating tides including combinations of tides with different wind and seasonal conditions. Wind and density driven residual currents play a vital role in the dispersion of pollutants in coastal areas.

The tidal levels and tidal currents in a coastal model are governed to a very large extent by the seaward boundary conditions. Within the model they are modified by resonance, inertia, shallow water and frictional effects.

The first aim of a calibration procedure should be to check the reproduction of the tidal levels and velocities on the open sea boundaries and the outermost parts of the model. Ideally, the observations should relate to exactly the same tide as that hindcast in the model. The main concern is the models capability to reproduce correctly the amplitude and phase of tidal levels and tidal currents. This can be done by carrying out a species analysis of model results and observations and comparing the in-plan spatial variation of the tidal constants. Alternatively, a least squares error analysis can be applied to model and observations at selected points. The former method is useful for checking the structure of the tidal regime including the pattern and magnitude of tide-averaged residual currents which are an important feature of coastal waters. A typical calibration target would be that 80% of the amplitude and phase of the boundary tides and currents should be within ± 5% and ±7° of the main observed harmonics. Residual currents should be within ±20% and ±20° of the observed amplitude and direction, respectively.

The next stage is to consider tidal propagation in the inner parts of the modelled area including any estuaries. This can be done in the same manner as the outer part of the model. However, the method of specific analysis tends to breakdown if the tide drys out at low water. Even in the case of least squares error analysis it is advisable to treat a long period of low water as a single value and give more weighting to the period near high tide. It is important that the model reproduces the correct tidal prism which is governed by HW levels and the accuracy of the bathymetry in the model. A typical calibration target would be that 80% of the tidal levels including a single LW level should be within 5% of the largest mean spring tide range in the modelled region. It is important that the model reproduces the correct pattern of tidal and residual currents because they govern the pattern and rate of dispersal of effluent. The results from recording current meters may be batched according to tidal range and several tidal cycles of observations can be compared to the model results with the same tidal range and wind and fluvial conditions.

There is a problem comparing measurements made at a point with a depth-averaged flow model. The model results can be modified by a correction factor based on the relative depth of the observations and the estimated roughness of the bed.

A typical calibration target for a single site would be that 80% of the observed values were within ±20% and ±20° of the simulated speeds and directions, respectively. It must be recognised that the target for the accuracy of tidal currents is much less than for tidal levels on account of the effects of large-scale turbulence.

If the velocity meter is situated in an area where the currents change rapidly in space, one may compare the observations with the simulated values within a search radius equal to the size of a model element. In addition one can also compare the observations and simulated currents in terms of the amplitude ratio (the gradient in linear regression analysis which should be unity) and the timing of the turn of the tide. In the case of a species analysis one has to quantify and compare the amplitude and phase of the northerlyand easterly components of each tidal constituent.

Examples of calibration statistics for a study of the Loughor Estuary are shown in Tables 2 and 3. The statistics are presented for a number of tide gauges and current meters for five representative tides. A graphical comparison of simulated model velocities with non-simultaneous observations is shown in Figure 1.

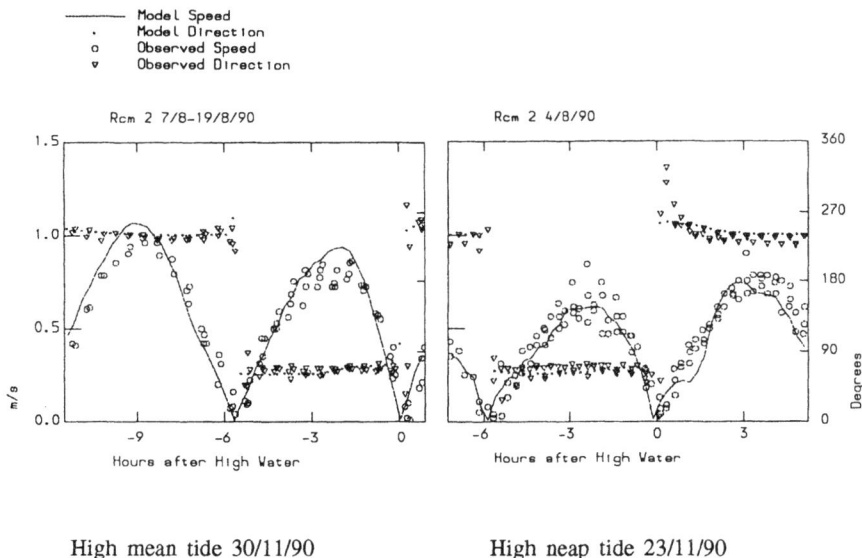

High mean tide 30/11/90   High neap tide 23/11/90

Figure 1 *Comparison of observed and simulated tidal velocities*

Table 2 *Loughor Estuary tidal levels - calibration statistics*

Correlation coefficient

Site	Burry	Llanelli	Loughor(1)
Low spring	0.992	0.984	0.989
High mean	0.990	0.964	0.980
Mean	0.993	0.963	0.983
High neap	0.999	0.969	0.994
Mean neap	0.999	0.987	0.991

Percentage of values with an accuracy of better than 5% (10%) of the tidal range at Burry

Low spring	68	100	78
High mean	69	73	47(87)
Mean	64	70	58(89)
High neap	92	67	67(89)
Mean neap	100	89	29(88)

Note 1. Low water treated as a single value

Table 3 *Loughor Estuary tidal velocities - calibration statistics*

Correlation coefficient - speed

Station	RCM1	RCM2	RCM3	Stn3	Stn4
Low spring	0.626	0.823	0.936	0.854	0.807
High mean	0.944	0.954	0.909	*	0.954
Mean	0.932	0.914	*	*	*
High neap	0.845	0.924	0.567	*	0.703
Mean neap	0.716	0.720	*	0.853	*

Amplitude ratio (observations/model)

Low spring	1.00	0.87	0.85	1.06	1.00
High mean	0.85	0.94	0.88	*	1.12
Mean	0.88	0.97	*	*	*
High neap	0.99	1.06	0.56	*	1.20
Mean neap	0.89	0.89	*	0.73	*

Percentage of values of speed with an accuracy better than 10%, (20%) or [30%] of the maximum observed speed.

Low spring	35(77)	39(82)	56(85)	46(79)	45(75)
High mean	64(92)	71	47(85)	*	72
Mean	65(95)	62(92)	*	*	*
High neap	60(88)	63(94)	25(46)[60]	*	50(70)
Mean neap	38(71)	31(58)[83]	*	30(56)[67]	*

Percentage of value of direction within 15° (25°) of the observed value

Low spring	76	89	67(85)	79	40(75)
High mean	78	90	48(76)	*	60(92)
Mean	78	87	*	*	*
High neap	79	82	83	*	57(71)
Mean neap	85	79	*	89	*

* no data available

One should then consider the overall 2D in-plan pattern of currents and tidal eddies. This may be facilitated by using a device such as OSCR which measures surface currents simultaneously on a grid over an area of more than 100 km². The results can be analysed in terms of tidal constituents and compared with the model on a similar basis. Alternatively, the results for the model may be compared directly with a series of instantaneous vector plots. However it is necessary to correct for the difference between surface and depth-mean velocities and to allow for the fact that a 2D model can not simulate 3D wind-driven circulations.

Besides checking that the model reproduces tide induced depth-averaged residual circulations in bays, it is also important to check that the model simulates the rate of growth, speed of rotation and decay of tidal eddies on each phase of the tidal cycle. This can sometimes be done by comparing model results with Admiralty flow charts.

The unique response of a volume of water in a coastal zone to an applied wind-stress in the form of a three-dimensional pattern of surface and subsurface currents can not be simulated in a 2D depth-averaged model. However, it is often the case that prevailing winds generate residual currents along the coast which are fairly uniformly distributed through the depth. This type of wind-driven current has to be imposed at one of the seaward boundaries of the model. It then appears as a steady residual depth-mean current in the model. One may be able to able to analyse a current meter record for day to day variations in tide-mean residuals and correlate them with the speed and direction of the prevailing wind. A similar analysis can be applied to a 3D model with an applied wind-speed. OSCR data can also be analysed to extract the surface wind-driven current which can be compared with 3D-model results.

It is particularly difficult to devise an objective method of comparing simulated and observed float/drogue tracks, because in reality each float or drogue is subject to the influence of a unique sequence of large, medium and small random turbulent motions.

In contrast, the track of a particle predicted directly from results of a flow model is smooth and repeatable without the effects of turbulent fluctuations. Furthermore, 2D models cannot reproduce the complex unsteady 3D patterns of wind-driven currents and they do not simulate any vertical structure of the flow or the effect of secondary currents in channels. In the authors experience, the best method of comparing a model with float tracks is to use a random walk model to release a number of particles within a small circle at the site and time of the release of the float or drogue, and to compare the observed position of the drogue with the cluster of simulated drogues. In the case of a 2D-model it is also necessary to apply a factor to increase the speed of the simulated float compared to the model depth-averaged velocity and to apply another factor to define the surface drift. One may then compare the simulated tracks and observed tracks by reference to the distance between the centre of gravity of the model floats and the position of the real float at successive phases of the tidal cycle.

3.2 Calibration of plume dispersion models

The most common use of a coastal pollution model is to predict the physical dispersal and decay of faecal coliforms in a weakly buoyant effluent from sewage treatment works or a more strongly buoyant and polluted discharge from storm water outflows. Coastal pollution models may also be used to predict the dispersal and utilization of nutrients or the dispersal and fate of dissolved or adsorbed toxic waste.

Plumes of decaying pollutants tend to form in long narrow sinuous zones with relatively high concentrations and are therefore most readily resolved using a random walk model based on a finer grid than the flow model. More conservative pollutants or nutrients generate higher background levels and produce a more diffuse concentration field. They are most readily simulated using an advective-dispersion model, which is usually based on the same grid as the flow model.

The calibration of pollution dispersion models in the past have been hindered by a lack of data. In the absence of calibration data one only has the accuracy of the hydrodynamic model to justify any predictions made from a pollution model. This implicitly assumes that the correct representation of physical processes are included in a dispersion model along with faith in the formulation and implementation of the numerical method. As a consequence, there is a requirement for an increase in the amount of synoptic data available which can be used for the calibration procedure.

Synoptic field data for use in calibrating a 'plume' model is generally obtained from a survey of the dispersal of a inert tracer released from either a known or proposed source of pollution. Suitable tracers include rhodamine-B dye or bacterial spores. The tracer will be spread over a tidal excursion which may be as much as 20km in UK coastal waters. An instantaneous snap-shot of the area is clearly not feasible and therefore allowances must be made when positioning the observations in time with respect to the simulated values.

It is recommended that an objective calibration procedure is formulated using a direct spatial and temporal comparison with field observations, allowing for the averaging of the mathematical model. A time series analysis of the simulation of the dispersion of a pollutant can also be compared with observations to illustrate behaviour at a point or within a region in a model area. This process can also be applied to a series of nearly instantaneous observations on a traverse across a pollutant plume. A goodness of fit can be applied to all the above methods but experience has shown that statistical analysis of direct spatial and temporal comparisons tends to give a clear objective assessment of the accuracy of a model.

Given a set of observed data which are positioned in space and time, one must firstly take into account the non-simultaneous nature of the observations. The model will give instantaneous results at discrete time intervals. The typical time step of a coastal pollution model will be about 10 minutes although this figure is dependent on the length of the simulation, the computer storage available and the numerical timestep. Hence one must batch the observations according to the temporal resolution of the model. The point observed values must then be located within a spatially averaged model grid cell. It is also important to take into account the possible distance travelled by the pollutant between the time of observation and the comparable simulation time. This implies a need to compare the observed pollution concentration with model results over a region, which can be defined by a suitable search radius or the distance travelled by the effluent particles can be estimated using simulated current velocities.

Graphical representation of the simulated concentrations of pollutant can incorporate observed concentrations. This presents a useful subjective comparison which can aid the calibration process. An objective analysis of the accuracy of the model in simulating the observed conditions can be formulated by finding the upper and lower bounds of simulated values over a region, defined by a search radius, surrounding an observation. A large search radius will increase the simulated range of values and it is recommended in the first instance that the width of a grid cell should be chosen.

Further difficulties arise from the fact that there exists a limit of detection of the tracer, additionally there will be a degree of experimental error associated with the actual measurements. For example, care must be taken to ensure the observed values are adjusted for background values. The presence of natural variability in the observed pollution concentrations caused by medium and large scale turbulence, the accuracy of the sampling and analysis techniques lead the modeller to decide upon a target expressed as an acceptable level of agreement between the simulated and observed values. A statistic relating the percentage of the observed values within a pre-determined factor of the simulated range of values will assess the performance of the model. For example one might say that 80% of the predicted dye concentrations within a search radius of 100m of the sample position should be within a factor of 2 of the observed concentration or within $\pm$ 2 x 10^{-9}v/v. Such a method does require a significant number of positive samples, typically 40-200. The method is independent of the pattern of sampling. The method of analysis tests the capability of the model to predict the position of the effluent as well as its dilution.

A second method of testing the accuracy of the model is to determine the percentage of negative samples (below the level of detection) which are also negative in the model. This analysis tests the extent to which the model has predicted tracer when none was found.

Similar techniques can be applied to advective-dispersion models and 3D model covering areas as large as the whole of the Southern North Sea.

An example of objective calibration statistics is shown in Table 4 for the comparison of simulations and observations of dye releases made during two different tidal conditions.

Table 4 *Loughor Estuary dye release simulation statistics (PLUME-RW with wind drift)*

	Low spring tide 18/10/90	High neap tide 28/7/89
No. of samples	490	272
Positive dye concentrations (> 2.5 x 10^{-10} v/v)	103	27
Negligible dye concentrations (< 2.5 x 10^{-10} v/v)	387	245
Percentage of positive dye samples within a factor of 2 of the simulated range within a 75(225)m search radius of the sample sites	53(84)	44(78)
Percentage of negative samples lying outside the simulated plume model < 2.5 x 10^{-10} v/v	90	97

Similar techniques can be applied to advective-dispersion models and 3D models covering areas as large as the whole of the Southern North Sea.

4 Summary and Conclusions

The authors have attempted to describe a guide for good practice in the setting-up and calibration of coastal pollution models. The authors propose that;
(i)   Historic data and the results from pilot models are used to optimise the collection of calibration data.
(ii)   Bathymetric surveys are devised to obtain the required degree of detail.
(iii)   Boundary conditions are setup to allow for inertia effects and residual currents.
(iv)   Objective calibration targets are set in advance in respect of flow and pollution models.
(v)   Ideally, three-dimensional models are required to simulate wind and density driven currents.

5 Acknowledgements

The authors would like to thank Wallace Evans for giving permission to show results from a model of the Loughor Estuary.

20 Application of higher order accurate schemes for advective transport in a 2-D water quality model

R. A. Falconer, S. Q. Liu and Y. Chen

ABSTRACT

Details are given of the application of various higher order accurate difference schemes for modelling the fate of faecal coliforms from a long sea outfall at Bridlington Bay, located along the east coast of Yorkshire, in the UK. Comparisons were made for the various schemes of the measured and predicted faecal coliform distributions, both along the outfall plume centreline and across the plume at various locations. The resulting comparisons showed that, although all of the schemes considered gave close agreement with the peak plume centreline faecal coliform counts, the transverse coliform distributions differed noticeably, with the third order upwind scheme giving the closest agreement with the measured data.

Introduction

With increasing public awareness and concern for hydro-ecological and environmental issues relating to coastal and estuarine waters, and with the increasing higher standards of EC directives etc. regarding water quality, numerical hydraulic models are becoming more widely used within the water industry for environmental impact assessment and feasibility studies. Also, as organisations such as the National Rivers Authority in the UK acquire and make more use of increased powers to prosecute polluters, and as computer capacity continues to increase for ever reducing costs, then the numerical model users - including consulting engineering firms and water companies - will continue to require increasingly accurate and more complex and sophisticated numerical models. In improving the accuracy of existing numerical models, one of the main areas of current research interest is in the numerical treatment of the advective terms of the solute transport equation. When a time centred second order accurate representation is used for these terms, then pronounced grid scale oscillations, or undershoot and overshoot, are generally produced in regions of relatively high concentration gradients. As time progresses these oscillations can propagate across the domain, particularly when the physical diffusion is relatively small, and the mass conservative solution can become meaningless physically as a result of the occurrence of negative concentrations or solute levels.

In recent years the authors have been investigating the properties of, and testing and applying, various higher order accurate schemes, see Falconer and Liu (1988) and Falconer and Chen (1991) and Chen and Falconer (1992), including in particular the QUICK, QUICKEST and Third Order Upwind Difference Schemes. However, to-date compared in the literature against analytical or idealised test cases, such as the one-dimensional advection of a plug or Gaussian tracer distribution in a steady flow, or the two-dimensional circular advection of a cone or Gaussian tracer distribution around a square. Although the QUICK scheme has been applied to Poole Harbour, in Dorset, UK, and compared with the central difference scheme for predicting nitrate levels in the basin, see Falconer and Chen (1991), the schemes referred to above have not previously been tested against field data from an existing and operational long sea outfall where relatively high concentrations gradients occur in open coastal waters.

In the current study these higher order difference schemes have been compared against field data for a long sea outfall at Bridlington Bay, on the North Yorkshire coast of the UK. Extensive field data have been provided by Yorkshire Water plc of the faecal coliform levels for the existing plume from the Bridlington long sea outfall. As can be seen from Figure 1, the outfall at Bridlington lies approximately 1500m from the coastline and is located in open coastal waters. These data provide an excellent opportunity to investigate the behaviour and accuracy of these schemes against prototype field data.

Differential Equations

The governing differential equations for the hydrodynamic model were based on the solution of the depth integrated Navier-Stokes equations, including the effects of the earth's rotation, bottom friction, wind shear and turbulence. Full details of these equations are given in Falconer (1991).

The governing solute transport equation was expressed in the following depth integrated form (see Falconer and Chen 1991):-

$$\frac{\partial \phi H}{\partial t} + \frac{\partial \phi U H}{\partial x} + \frac{\partial \phi V H}{\partial y} - \frac{\partial}{\partial x}\left[HD_{xx}\frac{\partial \phi}{\partial x} + HD_{xy}\frac{\partial \phi}{\partial y}\right]$$

$$- \frac{\partial}{\partial y}\left[HD_{yx}\frac{\partial \phi}{\partial x} + HD_{yy}\frac{\partial \phi}{\partial y}\right] - H\left[S_L + S_B + S_k \right] = 0 \qquad (1)$$

where ϕ = depth average solute concentration, H = total depth of flow, t = time, D_{xx}, D_{xy}, D_{yx}, D_{yy} = depth average dispersion-diffusion coefficients in x,y directions (see Falconer, 1991), S_L = direct and diffuse loading rate, S_B = boundary loading rate (including upstream, downstream, benthic and surface inputs or outputs) and S_k = total kinetic transformation rate.

Finite Difference Scheme

In solving the governing differential equations for the hydrodynamic model, an alternating direction implicit finite difference scheme was used, including a traditional space staggered grid scheme. The difference equations were fully centred in both time and space, with the advective accelerations and the turbulent diffusion terms being centred by iteration. The difference scheme has no stability constraints, although it was established that the accuracy of the scheme deteriorated rapidly when the Courant number (defined as $\Delta t \sqrt{(gH)} \Delta x^{-1}$, where Δt = time step, H = depth of water column and Δx = grid spacing) exceeded about 8. Full details of the finite difference scheme for the hydrodynamic equations are given in Falconer (1986).

For the finite difference representation of the solute transport equation, the scheme adopted was basically of the ADI type with a semi-implicit representation for the higher order terms. The general scheme for the first half timestep from time level n to $n + \frac{1}{2}$ can be expressed in the form:-

$$(\phi H)_{i,j}^{n+\frac{1}{2}} - (\phi H)_{i,j}^{n} + \frac{\Delta t}{4\Delta x} \left[(UH)_{i+\frac{1}{2},j}^{n+\frac{1}{2}} \left(\phi_{i+1,j}^{n+\frac{1}{2}} + \phi_{i,j}^{n+\frac{1}{2}} \right) - (UH)_{i-\frac{1}{2},j}^{n+\frac{1}{2}} \right.$$

$$\left. \left(\phi_{i,j}^{n+\frac{1}{2}} + \phi_{i-1,j}^{n+\frac{1}{2}} \right) \right] + \frac{\Delta t}{4\Delta y} \left[(VH)_{i,j+\frac{1}{2}}^{n} \left(\phi_{i,j+1}^{n} + \phi_{i,j}^{n} \right) - (VH)_{i,j-\frac{1}{2}}^{n} \right.$$

$$\left. \left(\phi_{i,j}^{n} + \phi_{i,j-1}^{n} \right) \right] - \alpha \frac{\Delta t}{\Delta x} \left[(UH)_{i+\frac{1}{2},j}^{n+\frac{1}{2}} \Delta_{xx} \overline{\phi}_{i+\frac{1}{2},j}^{n+\frac{1}{2}} - (UH)_{i-\frac{1}{2},j}^{n+\frac{1}{2}} \Delta_{xx} \overline{\phi}_{i-\frac{1}{2},j}^{n+\frac{1}{2}} \right]$$

$$- \alpha \frac{\Delta t}{\Delta y} \left[(VH)_{i,j+\frac{1}{2}}^{n} \Delta_{yy} \phi_{i,j+\frac{1}{2}}^{n} - (VH)_{i,j-\frac{1}{2}}^{n} \Delta_{yy} \phi_{i+\frac{1}{2},j}^{n} \right] + \beta \frac{\Delta t}{\Delta x} \left[(UH)_{i+\frac{1}{2},j}^{n+\frac{1}{2}} \right.$$

$$\Delta_{xy} \overline{\phi}_{i+\frac{1}{2},j}^{n+\frac{1}{2}} - (UH)_{i-\frac{1}{2},j}^{n+\frac{1}{2}} \Delta_{xy} \overline{\phi}_{i-\frac{1}{2},j}^{n+\frac{1}{2}} \right] + \beta \frac{\Delta t}{\Delta x} \left[(VH)_{i,j+\frac{1}{2}}^{n} \Delta_{yx} \phi_{i,j+\frac{1}{2}}^{n} \right.$$

$$\left. - (VH)_{i,j-\frac{1}{2}}^{n} \Delta_{yx} \phi_{i,j-\frac{1}{2}}^{n} \right] = \textit{Dispersion \& Diffusion terms}$$

$$+ \textit{ Source or sink terms } + \textit{ decay and interaction terms} \qquad (2)$$

where i,j = grid square locations in x,y directions and:-

$$\Delta_{xx} \phi_{i+\frac{1}{2},j} = \begin{cases} \phi_{i+1,j} - 2\phi_{i,j} + \phi_{i-1,j} & U_{i+\frac{1}{2},j} > 0 \\ \phi_{i+2,j} - 2\phi_{i+1,j} + \phi_{i,j} & U_{i+\frac{1}{2},j} < 0 \end{cases}$$

$$\Delta_{xy} \phi_{i+\frac{1}{2},j} = \begin{cases} \phi_{i,j+1} - 2\phi_{i,j} + \phi_{i,j-1} & U_{i+\frac{1}{2},j} > 0 \\ \phi_{i+1,j+1} - 2\phi_{i+1,j} + \phi_{i+1,j-1} & U_{i+\frac{1}{2},j} < 0 \end{cases}$$

and similarly for $\Delta_{yx} \phi_{i,j+\frac{1}{2}}$ and $\Delta_{yy} \phi_{i,j+\frac{1}{2}}$. The values of α and β depend upon the difference scheme being used and vary accordingly:-

(i)   For the central difference scheme $\alpha = \beta = 0$

(ii)   For the QUICK difference scheme $\alpha = 1/16$ $\beta = 1/48$

(iii)   For the third order upwind scheme $\alpha = 1/12$ $\beta = 1/36$

The general scheme is time centred by iteration, with the entire equation first being solved by setting:-

$$\overline{\phi}_{i+\frac{1}{2},j}^{\,n+\frac{1}{2}}\Bigg|_1 = \phi_{i+\frac{1}{2},j}^{\,n}$$

and for the second (and any subsequent) iterations the corresponding terms are evaluated from:-

$$\overline{\phi}_{i+\frac{1}{2},j}^{\,n+\frac{1}{2}}\Bigg|_{k+1} = \phi_{i+\frac{1}{2},j}^{\,n+\frac{1}{2}}\Bigg|_k$$

where $\overline{\phi}_{i+\frac{1}{2},j}^{\,n+\frac{1}{2}}\Big|_k$ = kth estimate of $\overline{\phi}_{i+\frac{1}{2},j}^{\,n+\frac{1}{2}}$ etc.

The iteration can be stopped at this point or continued to (K) iterations until:-

$$\overline{\phi}_{i+\frac{1}{2},j}^{\,n+\frac{1}{2}}\Bigg|_K \simeq \overline{\phi}_{i+\frac{1}{2},j}^{\,n+\frac{1}{2}}\Bigg|_{K-1}$$

For either case the error is $0(\Delta t^2)$, with little improvement in the accuracy being observed for $k > 1$. The dispersion-diffusion terms were expressed in a fully time centred form as given by Falconer (1986).

Model Application

In applying the numerical model including the various representations of the advection terms of the solute transport equation to a practical long sea outfall study, the region around Bridlington Bay (see Figure 1) was represented in the model using a nested coarse and fine grid. Firstly, a coarse grid model of 57 x 50 grid points, each of grid size $\Delta X = 1080$ metres, was set up covering the region from Scarborough to just south of Withernsea. The model was run for average tide conditions using a time step of 360s and a bed roughness height of 80mm, with time varying water elevations and velocity components being prescribed along the northern and southern boundaries respectively from the HR Wallingford North Sea Model. This coarse grid model then provided boundary conditions for the fine grid model around Bridlington Bay, which consisted of 60 x 58 grid points, each of grid size $\Delta x = 180$ metres. The fine grid model was run using a time step of 40s and the same bed roughness height, i.e. 80mm, and with the discharge from the Bridlington outfall being set to a mean daily value of 0.26 $m^3 s^{-1}$.

The faecal coliform level at the outlet was set to a daily mean value of 2.7×10^7 counts per 100ml, with the initial coliform level across the domain being set to zero for both models. For the solute level open boundary conditions in the coarse grid model, the faecal coliform level was evaluated by extrapolation from the adjacent grid squares

244

for outflow conditions (see Falconer, 1991). However, for inflow conditions the boundary was assumed to be far enough away from Bridlington long sea outfall for the boundary concentration to be set to zero. For the fine grid model the solute level for outflow conditions was obtained by linear extrapolation as before, with the inflow concentration now being prescribed from the coarse grid model. From extensive field data (see O'Neill, 1989) and calibration tests the decay rate for faecal coliform was set to 7day^{-1}, corresponding to a T_{90} value of just under 8hr.

The corresponding mean spring tide flow patterns for a tidal range of 3.6m were compared with drogue results commissioned by Yorkshire Water plc. The resulting comparisons between the observed and predicted velocity fields agreed closely and typical coarse and fine grid velocity distributions are given in Figures 2 and 3 respectively, with the velocities only being reproduced at every other grid square. The corresponding results showed that a strong tidal eddy occurred in the lee of Flamborough Head around high tide and with higher velocities generally occurring in the shallow submerged sandbanks off Bridlington Bay and just beyond the outfall.

For the corresponding faecal (or F) coliform distributions, a typical predicted plan distribution is given in Figure 4 just before high water. However, in order to test more accurately the numerical model predictions of the faecal coliform levels, and compare the accuracy of the various schemes for high concentration gradients, the model predictions of the peak faecal coliform levels along the outfall plume were compared directly with those measured for the Bridlington Outfall by Yorkshire Water plc. The resulting comparisons are shown in Figure 5 for the central, QUICK and third order upwind difference schemes. Comparisons undertaken for the QUICKEST difference scheme, without filtering etc, indicated that this scheme was inherently unstable when extended directly to two-dimensions from the one-dimensional form. This result was subsequently confirmed in a detailed stability analysis by Chen and Falconer (1992), where the QUICKEST scheme was shown only to be stable for two-dimensions when all the terms of the Taylor series were included for advective dominated flows. The resulting comparisons in Figure 4 show a particularly close agreement between all three numerical schemes and the field data, with the QUICK and third order upwind schemes producing almost identical longitudinal peak concentration distributions.

To investigate further the properties of these schemes, comparisons were also made of the lateral faecal coliform concentration distributions across the plume at two transects, namely at 80m and 200m respectively along the plume. The corresponding results in figures 6a and 6b show that the QUICK and third order upwind difference schemes agree relatively closely with the field data - although the lateral faecal coliform distribution data are relatively sparse. For these two schemes some undershoot and overshoot (or grid scale oscillations) occur across the plume, although this effect is much less significant than the central difference scheme where peak negative concentrations near the outfall are in excess of 10,000 coliform counts per 100ml. In addition to the high negative concentrations for the central difference scheme, this scheme also less accurately predicts the peak concentration near the outfall. A detailed numerical analysis of the properties of the QUICK and third order upwind schemes by Chen and Falconer (1992) has shown that the third order upwind scheme is slightly more accurate than the QUICK scheme and therefore leads to lower negative concentrations. This result is further confirmed in Figure 6 where, although both schemes give similar predictions for the coliform distributions, the third order scheme exhibits lower negative concentrations - a result which is particularly noticeable at a distance of approximately 400m from the peak value in figure 6a.

CONCLUSIONS

The authors have recently undertaken an extensive analysis of the numerical properties of various higher order accurate schemes for modelling the advection terms of the solute transport equation. In this analysis the third order upwind difference scheme was shown to be more accurate than the QUICK scheme and the QUICKEST scheme was shown to be numerically unstable when extended directly from the one-dimensional form and without diffusion. All three schemes were computationally efficient and relatively straightforward to implement for complex boundary conditions, including flooding and drying.

The schemes have been tested and compared against field data in the form of faecal coliform concentrations from an offshore long sea outfall in Bridlington Bay. Comparisons between the measured and predicted peak concentration along the plume centreline showed good agreement between the field data and the predicted results. The QUICK and third order upwind difference schemes gave almost identical results, with the second order central difference scheme also agreeing reasonably closely with the field data. Comparisons were also undertaken between the predicted and field measured coliform distributions at two transects across the plume, with the results again agreeing closely with the field data for the QUICK and third order upwind difference schemes. For both transects the central difference scheme exhibited relatively large negative concentrations, with the peak concentration also being over predicted. In confirming earlier studies on the relative accuracy of the QUICK and third order upwind schemes, the schemes gave similar predictions although the third order upwind difference scheme gave slightly reduced negative concentration predictions.

ACKNOWLEDGMENTS

The authors are grateful to the following organisations for funding and supporting this research project:- the European Commission (contract no. CL1*0390-UK), the Science and Engineering Research Council (grant no. GR/F/02892), Yorkshire Water plc and UNIRAS A/S. The computer simulations were undertaken on IBM PS/2 machines using a Tektronix 4693D colour printer, provided respectively by IBM and Tektronix study grants.

The authors are also grateful to the following contacts for their assistance in connection with this project:- Dr J G O'Neill and Mr T Toole (Yorkshire Water plc), Dr C Carle and Dr M Jern of (UNIRAS A/S), Ms S Hooper (IBM UK Ltd) and Mr S Greenup (Tektronix UK Ltd).

REFERENCES

[1] Falconer, R A and Liu, S Q, (1988), Modelling Solute Transport Using QUICK Scheme, *Journal of Environmental Engineering*, ASCE, Vol.114, No.1, pp.3-20.
[2] Falconer, R A and Chen, Y P, (1991), Water Quality Modelling in Coastal Waters Using an Implicit QUICK Difference Scheme, *Proceedings of International Symposium on Environmental Hydraulics*, Hong Kong (Ed. Lee, J W and Cheung, Y K), A A Balkema Publishers, Vol.1, December, pp.741-746.
[3] Chen, Y and Falconer, R A, (1992), Modified Forms of the Third-Order Convection Second-Order Diffusion Scheme for the Advection Diffusion Equation, *submitted for publication to Applied Mathematical Modelling*, pp.1-27.

[4] Falconer, R A, (1991), Review of Modelling Flow and Pollutant Transport Processes in Hydraulic Basins, *Proceedings of First International Conference on Water Pollution: Modelling, Measuring and Prediction,* Southampton, Computational Mechanics Publications, September, pp.3-23.

[5] Falconer, R A, (1986), A Two-Dimensional Mathematical Model Study of the Nitrate Levels in an Inland Natural Basin, *Proceedings of the International Conference on Water Quality Modelling in the Inland Natural Environment,* BHRA Fluid Engineering, Bournemouth, Paper Jl, June, pp.325-344.

[6] O'Neill, J G, (1989), Field Techniques for Validation of Bacterial Decay and Dispersion Models, *Proceedings of the International Conference on Hydraulic and Environmental Modelling of Coastal, Estuarine and River Waters,* Bradford (Ed. Falconer, R A, Goodwin, P & Matthew R G S), Gower publishing Co Ltd, September, Paper 57, pp.625-636.

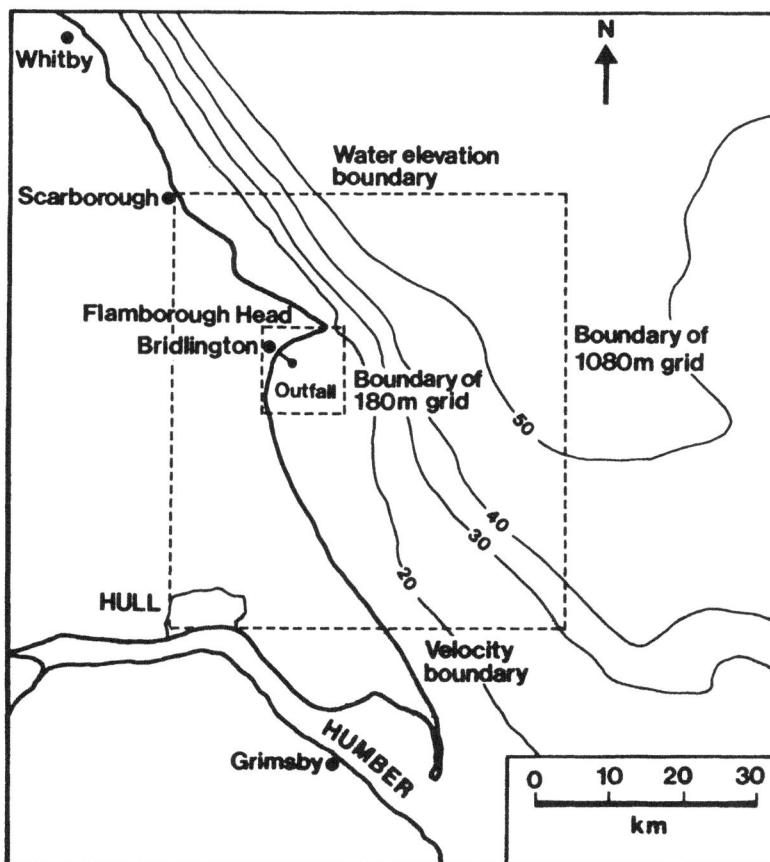

Fig.1. Illustration of Bridlington Bay, Showing the Outfall Site and the Coarse and Fine Grid Boundaries

247

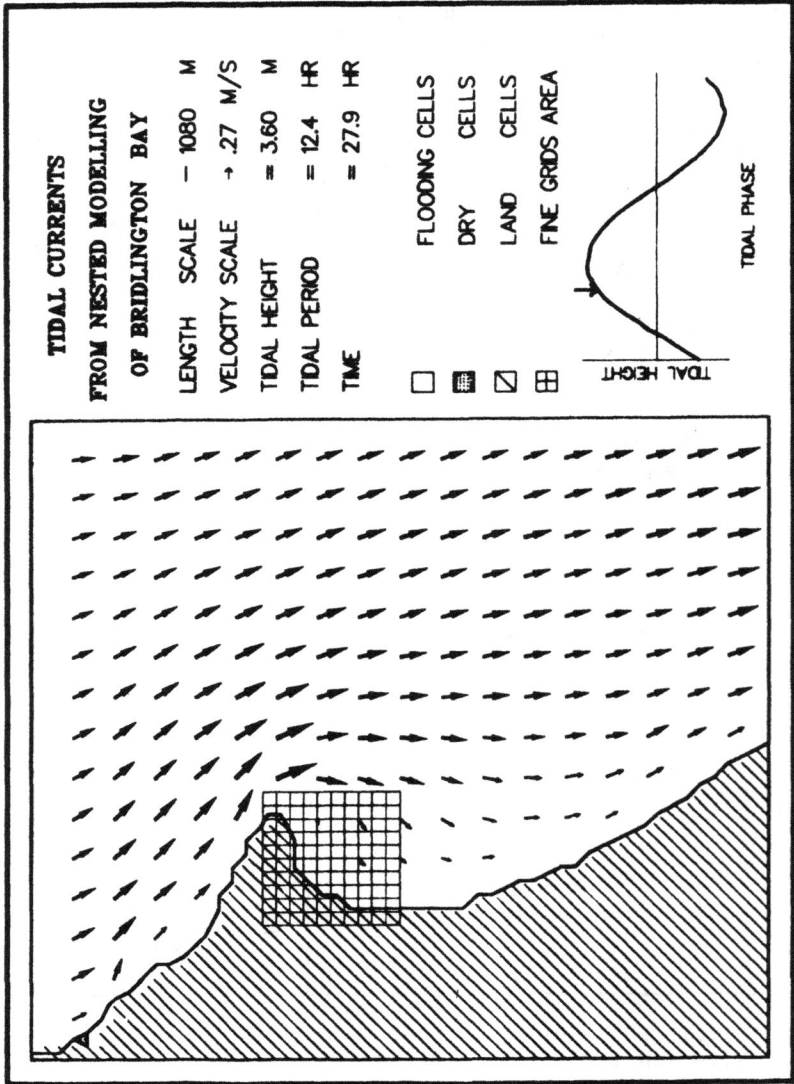

Fig.2. Predicted Coarse Grid Velocity Distribution Just Before High Water

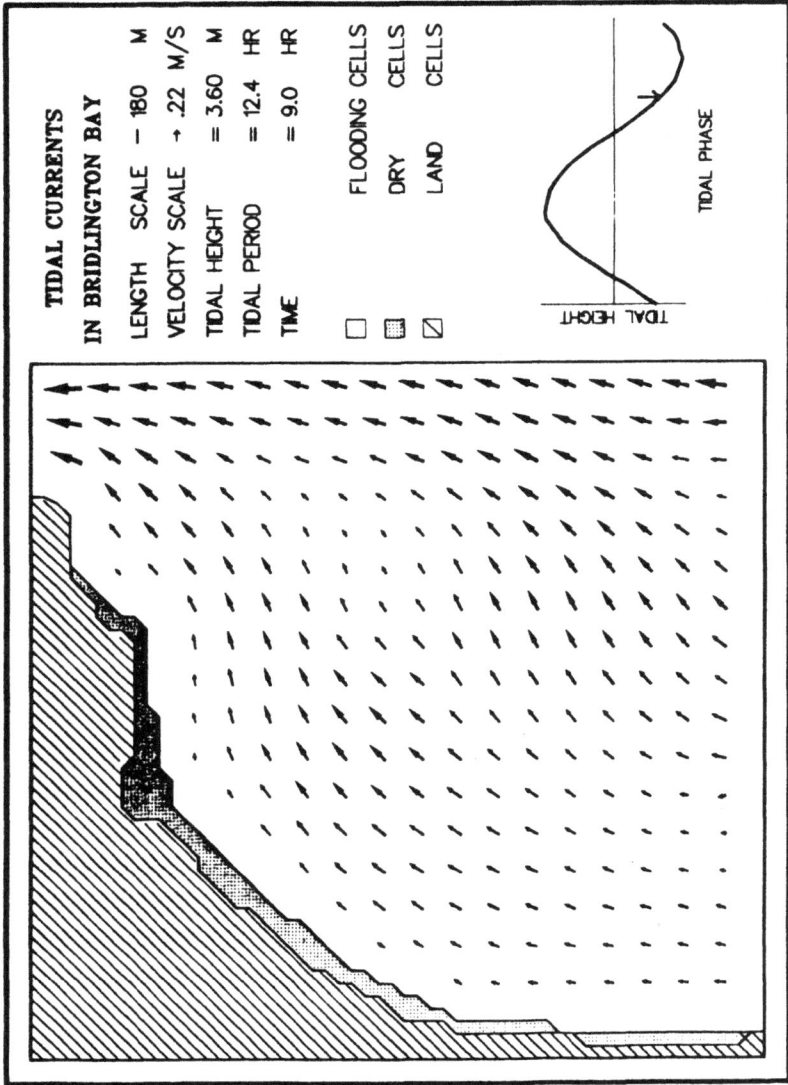

Fig.3. Predicted Fine Grid Velocity Distribution Just Before Low Water

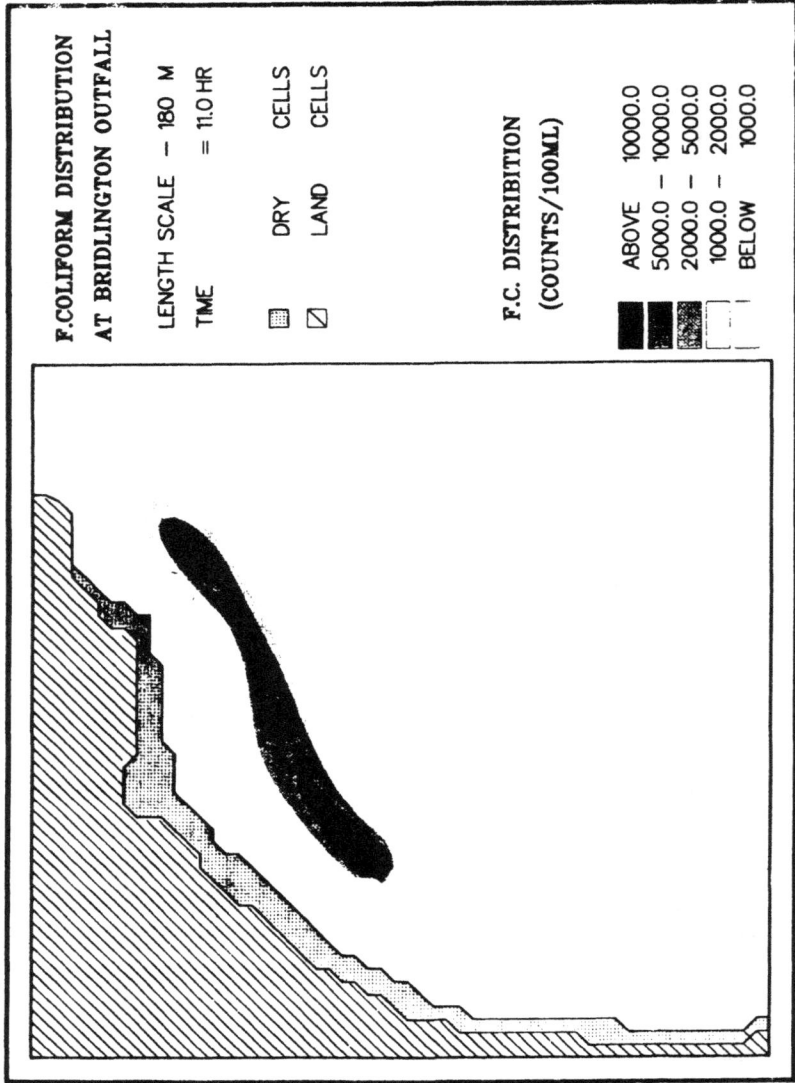

F. COLIFORM DISTRIBUTION AT BRIDLINGTON OUTFALL

LENGTH SCALE — 180 M

TIME = 11.0 HR

DRY CELLS

LAND CELLS

F.C. DISTRIBITION (COUNTS/100ML)

ABOVE 10000.0

5000.0 — 10000.0

2000.0 — 5000.0

1000.0 — 2000.0

BELOW 1000.0

Fig.4. Predicted Faecal Coliform Distributions From Bridlington Sea Outfall Just Before High Water

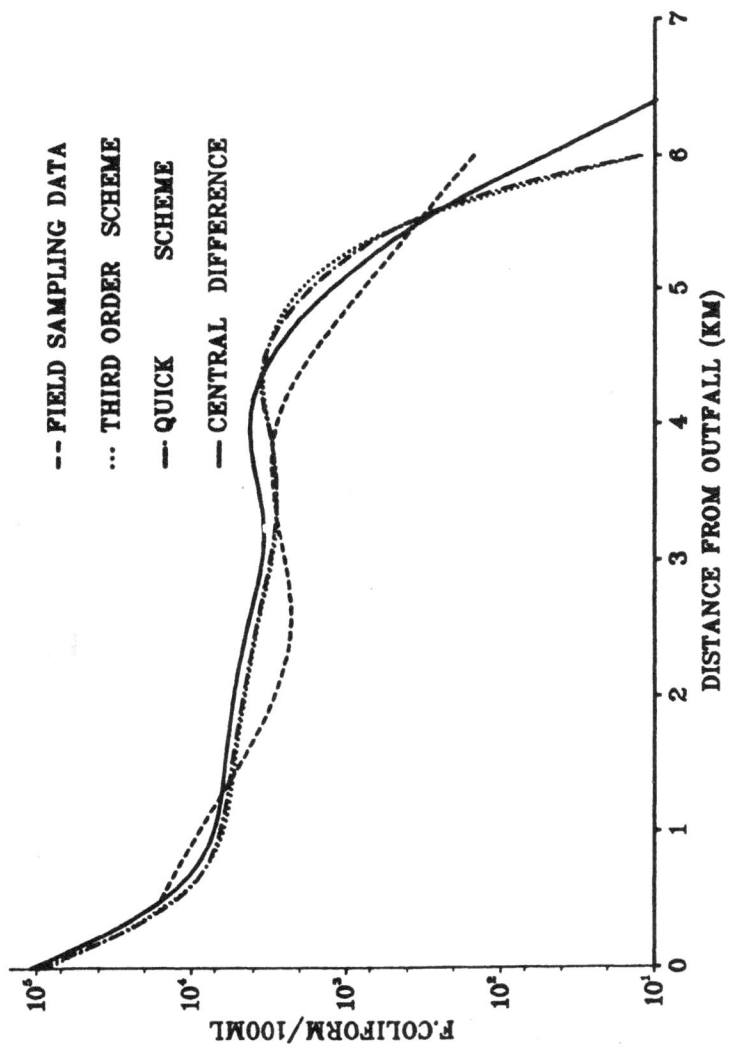

Fig.5. Comparison of Predicted and Measured Faecal Coliform Distributions Along the Outfall Plume Centreline for Various Difference Schemes

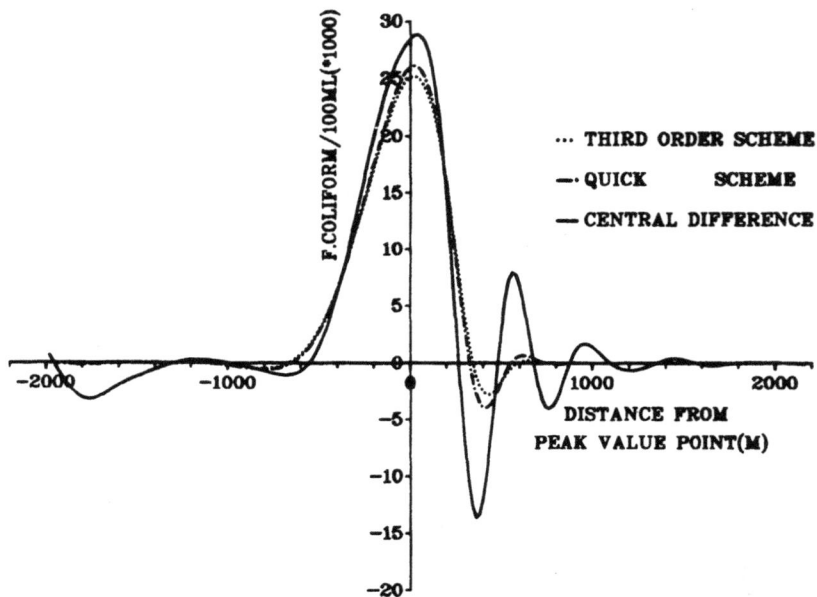

(a) 200m From Outfall

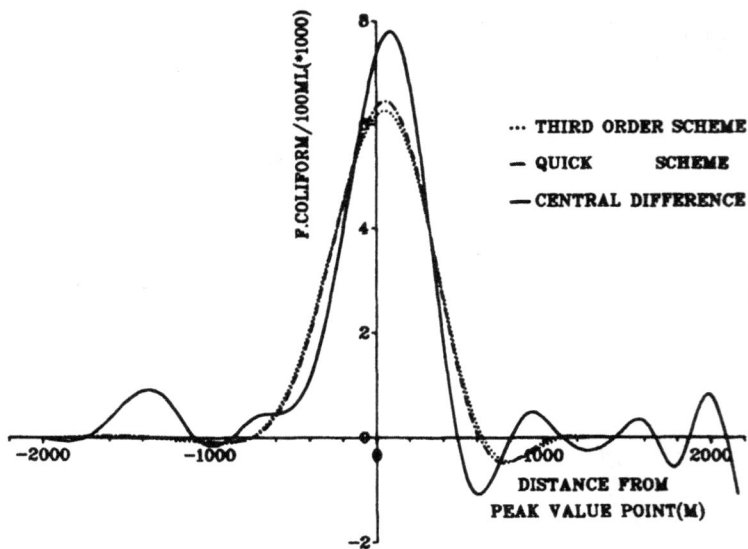

(b) 800m From Outfall

Fig.6. Comparison of Predicted Faecal Coliform Distributions for Various Difference Schemes Across the Outfall Plume

21 Evaluation of dispersion coefficient in a bay

T. Komatsu, M. Sagara and S. Yano

ABSTRACT

One dimensional dispersion coefficient should be varied with each location in a natural bay and originally determined by flow characteristics. In this study, the idea is introduced that the dispersion coefficient is proportional to the product of local representative velocity and local representative length scale. One dimensional numerical calculation of contaminant diffusion were carried out for some actual bays. When local M2 tidal maximum velocity and either local width of a bay or local tidal excursion are considered as the representative velocity and the length scale, the proportional constant can be determined so that the computational results of contaminant diffusion agree well with the observational ones. This proportional constant was found to be a function of complexity of topography of a bay. As the result of this study, one dimensional simulations of contaminant diffusion in a bay can be carried out very easily with a little knowledge on characteristics of tidal flow.

1. Introduction

In order to predict environmental pollution and water purification in a bay or a gulf, two or three dimensional numerical simulations for pollutant diffusion are often adopted. Although it is necessary to estimate eddy diffusivity for such calculations on pollutant diffusion, there has been no way available to estimate easily even one dimensional dispersion coefficient without using observational results. In most cases, diffusion coefficient was assumed to be constant by approximate estimation or determined so as to coincide calculated results with observational ones. Therefore, it has been required to estimate one dimensional dispersion coefficient with high accuracy not only to carry out accurate one dimensional simulation but also to enable us to give boundary conditions for two and three dimensional simulations.

The dispersion coefficient should be varied with each location in a natural bay and originally determined by flow characteristics only. It is natural to assume that the dispersion coefficient is proportional to the product of local representative velocity and local representative length scale.

One dimensional numerical simulations of contaminant diffusion were applied to Seto Inland Sea, Ariake Bay, Hakata Bay, Kagoshima Bay, Beppu Bay and Bungo Channel. When the local M2 tidal maximum velocity and either the local width of the bay or the local tidal excursion are considered as the representative velocity and the length scale, the proportional constant can be determined so that the computational results agree well with the observational ones. It is found that the proportional constant is not universally constant but a function of complexity of topography of a bay. With the result the dispersion coefficient in a bay can be easily estimated. Therefore, simulations of contaminant diffusion could be carried out with a little information on characteristics of tidal current and topography of a bay.

2. Estimation of dispersion coefficient

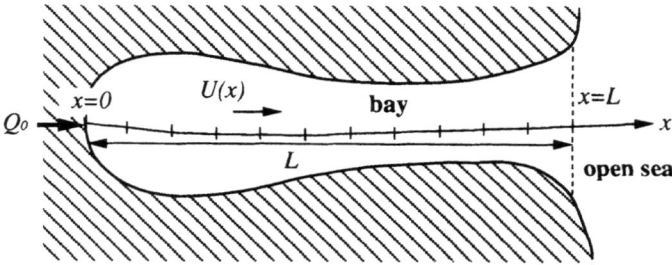

FIGURE 1 *Definition of x-axis in a bay*

When the x-axis is taken along the centerline of a bay like FIG. 1, the governing equations for one dimensional contaminant diffusion are written as follows:

$$U(x) = \frac{1}{A(x)} \left\{ Q_0 + \int_0^x q(\xi)\, d\xi \right\} \qquad \text{(continuity equation)} \quad (1)$$

$$\frac{\partial C}{\partial t} + U\frac{\partial C}{\partial x} = \frac{1}{A}\frac{\partial}{\partial x}\left(A\, D\, \frac{\partial C}{\partial x} \right) - \frac{1}{A} q\, C + \frac{m}{\rho A} \qquad \text{(advection-diffusion equation)} \quad (2)$$

for steady case,

$$UAC - AD\frac{\partial C}{\partial x} = \left[UAC - AD\,\frac{\partial C}{\partial x} \right]_{boundary} + \frac{R(x)}{\rho} \qquad R(x) = \int_0^x m(\xi)d\xi \qquad (3)$$

where $C(x,t)$ is the concentration of conservative contaminant, $A(x)$ is the sectional area of bay normal to x-axis, $U(x)$ is the cross-sectional averaged velocity, $D(x)$ is the one dimensional dispersion coefficient including all factors which are related to mixing and dispersion, Q_0 is the inflow across the section at $x=0$, $q(x)$ and $m(x)$ are the water quantity and the mass of contaminant supplied per unit time, unit length. Split Operator Approach was adopted to calculate

Eq.(2). The 6-point scheme developed by Komatsu et al.[1] was used for the advection of left side and the 2nd order central finite difference scheme for the dispersion of right side.

Expressing the representative length of one dimensional mixing by local tidal excursion or width of bay, it is assumed that the dispersion coefficient can be estimated as follows:

$$D(x) = \alpha V_M(x)^2 T \tag{4}$$

$$D(x) = \beta V_M(x) B(x) \tag{5}$$

where $V_M(x)$ is the local M2 tidal maximum velocity at each cross-section, $B(x)$ is the local width of bay normal to x-axis, $V_M(x)T$ is the length which is proportional to the tidal excursion, T is the period of M2 tide (12.42 hr.), α and β are the proportional constants.

When there is an island in a bay, $B(x)$ is defined as B_1 which is the largest width (FIG. 2). The proportional constants α, β are determined by fitting so as to coincide the calculated results of one dimensional simulation with the observational results.

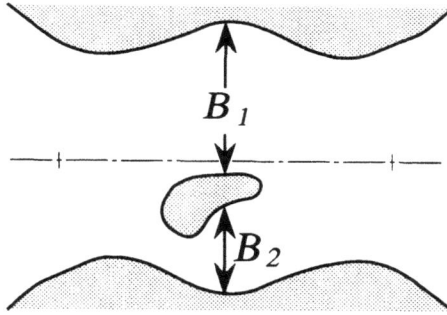

FIGURE 2 *Definition of bay width*

3. Determination of values of α and β

In six bays, namely, Seto Inland Sea, Ariake Bay, Hakata Bay, Kagoshima Bay, Beppu Bay and Bungo Channel, the salinity distributions were calculated by using Eq.(1) and either Eq.(2) or Eq.(3) as the basic equations. Then dispersion coefficients were estimated by Eq.(4) and Eq.(5). The one dimensional coordinate for numerical simulations in Seto Inland Sea is shown in FIG. 3.

There are 21 computational grid points from $x=9$ to $x=29$ and the space interval between two grids is 20km. Time increment Δt is 3.0 hours. Under those conditions the numerical simulation of contaminant diffusion was carried out. Various quantities at each section are given in TABLE 1. M2 tidal maximum velocity on each grid point was obtained from linear interpolation using data obtained by Wada and Kadoyu[2]. The values of width of a bay were read from a marine chart. The flow rate of class A rivers were read from the discharge chronological table, and ones of rivers under class B rivers were estimated from the product of the specific discharge presumed from those of class A rivers and the catchment area. The sectional area and the chlorinity were obtained from Hayami and Unoki[3]. As boundary conditions of both sides $C_9=18.55‰$ and $C_{29}=19.06‰$ were used, and then $Q_9=14.6\times10^3 m^3/year$ (Komatsu et al.[4]).

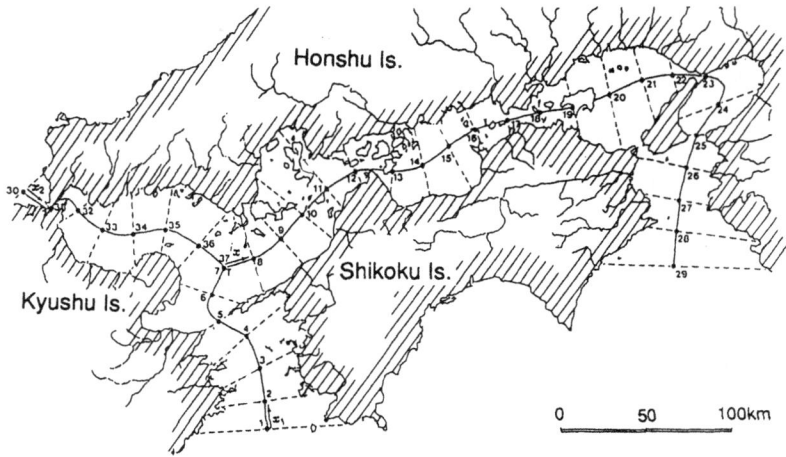

FIGURE 3 *Coordinate for numerical simulations in Seto Inland Sea*

TABLE 1 *Various quantities at each section in Seto Inland Sea*

Coordinate	Sectional Area A (km²)	Width of the Bay B (km)	Inflow Rate q (*)	M₂ Tidal Maximum Velocity (cm/sec)	Cl(Upper Layer) (‰)	Cl(Lower Layer) (‰)
9	1.899	36.0	11.8	20.5	18.47	18.55
10	1.788	30.0	93.1	15.0	18.29	18.40
11	1.239	25.5	175.3	78.5	18.18	18.33
12	0.705	12.5	50.6	76.5	18.29	18.35
13	0.215	15.5	22.2	67.0	18.22	18.31
14	0.633	32.5	13.2	52.0	18.15	18.26
15	0.810	42.5	27.3	36.5	18.09	18.21
16	0.252	14.5	109.7	37.5	18.14	18.23
17	0.204	8.5	29.2	62.0	17.90	18.01
18	0.195	8.5	183.7	69.0	17.74	17.86
19	0.639	16.5	10.5	48.5	17.59	17.71
20	1.788	56.0	75.4	27.5	17.69	17.73
21	1.131	45.0	96.9	19.0	17.56	17.58
22	0.264	18.5	9.6	108.5	17.61	17.58
23	0.244	6.0	446.5	64.0	17.46	17.44
24	1.377	53.0	21.0	30.5	17.44	17.86
25	0.370	4.5	63.7	20.5	18.30	18.51
26	1.680	26.0	114.0	13.0	18.45	18.66
27	1.884	38.0	122.5	16.5	18.49	18.77
28	3.981	57.5	45.6	13.5	18.83	19.07
29	27.810	99.0		31.0	18.85	19.06

* Unit $10^6 \cdot$ m³/km/year

256

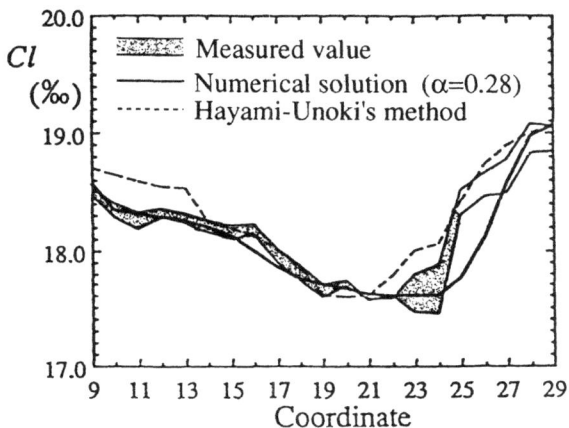

FIGURE 4 *Calculated steady salinity distribution in Seto Inland Sea with α=0.28*

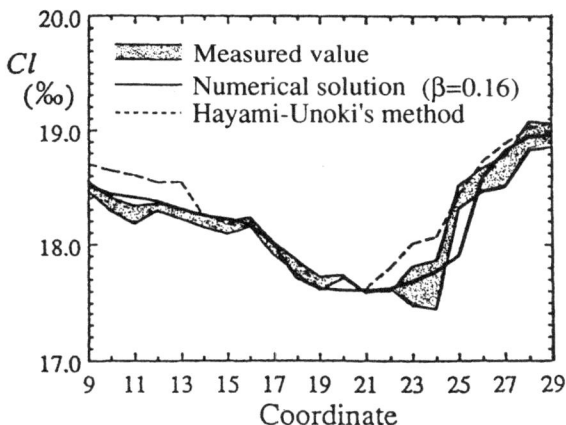

FIGURE 5 *Calculated steady salinity distribution in Seto Inland Sea with β=0.16*

The comparisons of the calculated steady salinity distribution with the measured value and the calculated results by Hayami and Unoki[3] are shown in FIG. 4 and FIG. 5. It was found that the calculated results agree better with the observation result than the result by Hayami-Unoki's method.

The coordinate for numerical simulations in Hakata Bay is shown in FIG. 6. Steady calculation was carried out under the conditions as follows:the space interval between two grids is 2.0 km, $m(x)=0$, and the salinity flux at the innermost part of Hakata Bay is equal to 0. Various quantities at each section are given in TABLE 2. The sectional area and the width of the bay were read from Fukuoka City's data. M2 tidal maximum velocities were obtained from the result of two dimensional numerical simulation of tidal current. The comparisons of the calculated salinity distributions with the measured values are shown in FIG. 7 and FIG. 8. Although both results agree well with the observational result, it seems that FIG. 8 shows better agreement.

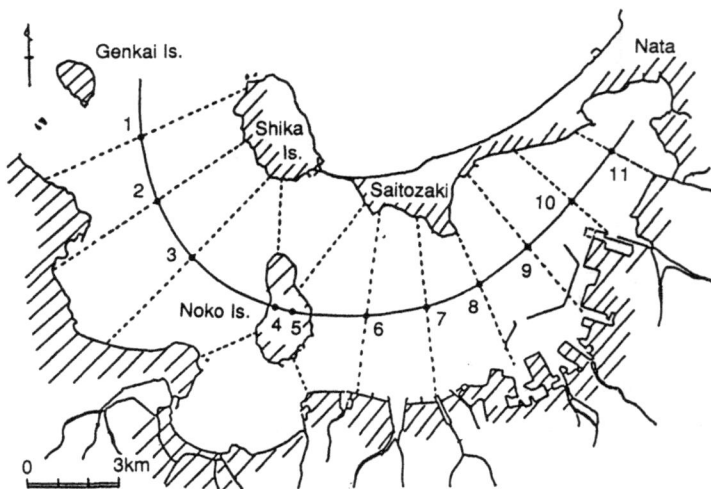

FIGURE 6 *Coordinate for numerical simulations in Hakata Bay*

TABLE 2 *Various quantities at each section in Hakata Bay*

Coordinate	Sectional Area A (km²)	Width of the Bay B (km)	Inflow Rate q (*)	M₂ Tidal Maximum Velocity (cm/sec)	Cl (Upper Layer) (⁰/₀₀)	Cl (Lower Layer) (⁰/₀₀)
1	99.06	6.88	3.39	19.4	18.84	18.64
2	115.67	7.33	10.22	14.1	18.75	18.52
3	118.22	7.38	0.0	16.2	18.62	18.37
4	52.54	2.35	50.21	21.6	18.26	18.22
5	46.43	3.28	11.90	17.4	18.14	17.96
6	52.46	5.70	66.48	16.3	17.92	17.70
7	46.09	5.75	44.27	14.1	17.92	17.63
8	35.32	4.68	107.29	14.4	17.60	17.48
9	34.84	5.70	7.26	13.3	17.41	17.40
10	23.06	3.90	111.49	11.5	17.33	17.17
11	12.00	2.78		6.5	16.83	16.80

*Unit $10^6 \cdot m^3/km/year$

258

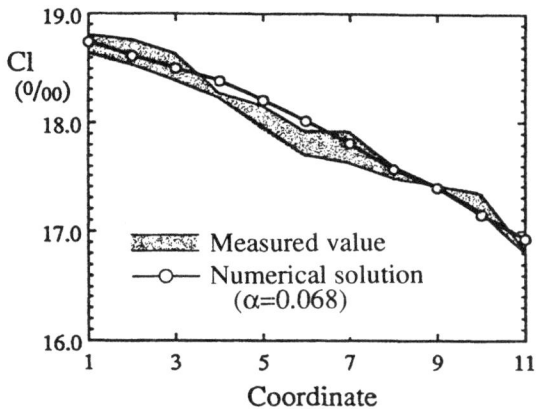

FIGURE 7 *Calculated steady salinity distribution in Hakata Bay with α=0.068*

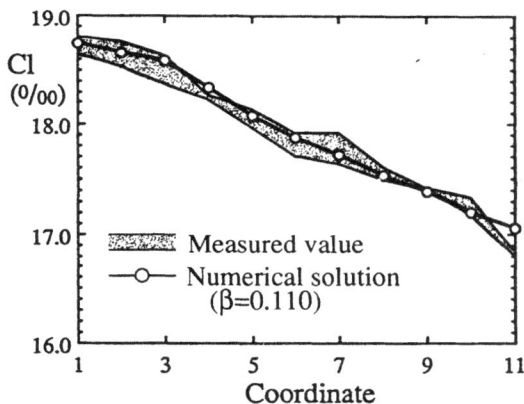

FIGURE 8 *Calculated steady salinity distribution in Hakata Bay with β=0.110*

The best fitting value of the proportional constant in other four bays are shown in TABLE 3 with the averaged tidal excursion and the averaged width of the bay for each bay. In Ariake Bay, no value of α could calculate salinity distribution fitted well to the observational one. There were some bays where the dispersion coefficient expressed by Eq.(4) could give better computational result than that by Eq.(5). Therefore, the following parameter is introduced to look into the factor making this discrimination.

$$\lambda = \frac{\overline{V_M\,T}}{B} \tag{6}$$

259

	Averaged Tidal Excursion $V_M T$	Averaged Width of the Bay B	$\dfrac{\overline{V_M T}}{\overline{B}}$	Proportional Coefficent	
				α	β
Seto Inland Sea	16.3 km	30.8 km	0.53	0.28	0.16
Hakata Bay	6.7 km	5.1 km	1.32	0.068	0.11
Kagoshima Bay	6.7 km	11.8 km	0.57	0.04	0.06
Ariake Bay	55.6 km	13.8 km	4.03	—	0.044
Beppu Bay	6.8 km	14.0 km	0.49	0.035	0.020
Bungo Channel	21.6 km	42.4 km	0.51	0.37	0.19

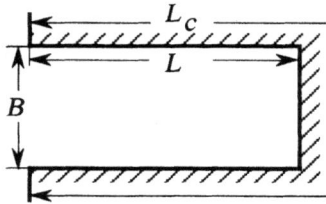

FIGURE 9 *Definition of L_c and L*

As a result of comparing the salinity distributions calculated by Eq.(4) and Eq.(5) with the observed ones in detail, if $\lambda < 1$, it seems that Eq.(4) give better results. On the other hand, if $\lambda \geq 1$, Eq.(5) is better. Since the proportional constant α and β may be affected by complexity of topography of a bay, relations between the values of α and β and topography on six bays have been investigated. As shown in FIG. 9, define L_c as a length of shoreline of a bay and L as a length of a bay along centerline. In addition, define ψ as a parameter representing complexity of topography of a bay as follows:

$$\Psi = \frac{L_c}{L_e} - 1 \tag{7}$$

where $L_e = 2L$ in case both sides of a domain are open, or $L_e = 2L + B$ in case only one side is open. As the bay is rectangular as shown in FIG. 9, the parameter of complexity ψ takes 0. With the increase of the length of shoreline, namely, the increase of complexity of topography of the bay, the value of ψ has a tendency to increase. FIG. 10 and FIG. 11 show the relation between ψ and α and that between ψ and β. It is confirmed that both α and β are unique functions of ψ. Therefore, the values of them can be determined by ψ.

TABLE 4 *Length of shoreline Lc and parameter of complexity of topography of bay ψ*

	Length of Shoreline Lc (km)	Length of Bay L (km)	Parameter of Complexity Ψ
Seto Inland Sea	2514.0	400.0	2.14
Hakata Bay	92.4	22.0	0.88
Kagoshima Bay	254.1	67.5	0.73
Ariake Bay	194.3	60.0	0.62
Beppu Bay	106.4	27.5	0.54
Bungo Channel	937.0	70.0	5.70

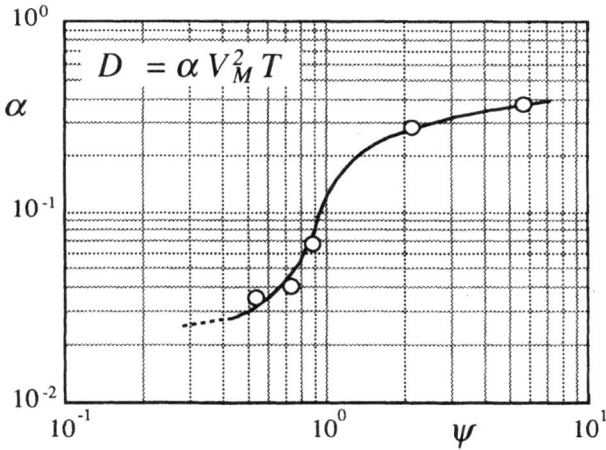

$$D = \alpha V_M^2 T$$

FIGURE 10 *The relation between α and ψ (λ<1)*

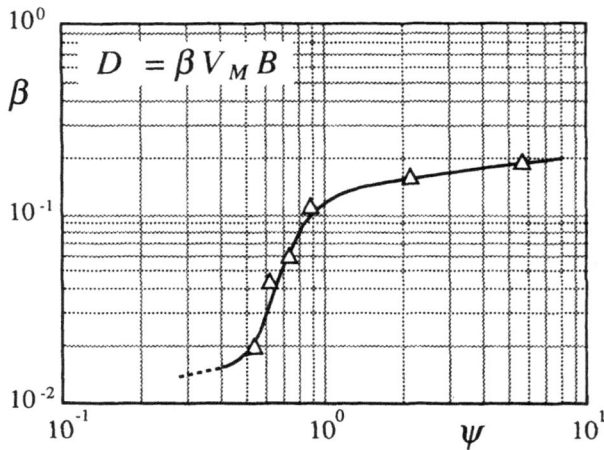

FIGURE 11 *The relation between β and ψ (λ≧1)*

4. Conclusion

We attempted to estimate the local one dimensional dispersion coefficient in a bay by considering local representative velocity and local representative length scale. It was assumed that the dispersion coefficient is proportional to the product of local representative velocity and local representative length scale. In this study, the M2 tidal maximum velocity and either the local width of the bay or the local tidal excursion at the cross-section were introduced as the representative velocity and length scale. One dimensional numerical simulations of contaminant diffusion were carried out on six bays in Japan. As a result of this study, it was found that the values of the proportional constant α and β are functions of the parameter of complexity of topography of a bay ψ, and the selection between α and β depend on the discriminating parameter λ. If $\lambda <$ 1, α should be adopted, while if $\lambda \geqq 1$, β should be used. By applications of this result, it is possible to estimate easily the one dimensional local dispersion coefficient including the effect of tidal current in a bay.

REFERENCES

[1] Komatsu, T., Holly, F. M., Nakashiki, N. and Ohgushi, K., (1985), Numerical Calculation of Pollutant Transport in One and Two Dimensions, *Journal of Hydroscience and Hydraulic Engineering*, **Vol. 3**, No. 2, pp 15-30.
[2] Wada, A. and Kadoyu, M., (1974), Flow Regime and Characteristic of Dispersion in Seto Inland Sea, *Proceedings of the 21st Japanese Conference on Coastal Engineering*, JSCE, pp 297-302 (in Japanese).
[3] Hayami, K. and Unoki, S., (1970), Cross Flow of Sea Water and Contaminant Diffusion in Seto Inland Sea , *Proceedings of the 17th Japanese Conference on Coastal Engineering*, JSCE, pp 385-393 (in Japanese).
[4] Komatsu, T., Sagara, M., Asai, K. and Ohgushi, K., (1989), Estimation of Contaminant Dispersion Coefficient in Seto Inland Sea, *Proceedings of Coastal Engineering*, JSCE, **Vol.36**,pp 804-808 (in Japanese).

262

22 Water renovation rate in Bouzas basins

M. Montero, A. Lloret and A. Ruiz-Mateo

ABSTRACT

Velocity field of a harbour placed at an intermediate stretch of an estuary is simulated, assuming a standing wave behavior. In order to establish renovation rate for different harbour layouts and tidal ranges, behavior of a tracer is also simulated.

Figure 1. Layout of Vigo Ria.

1. Introduction.

Bouzas basins are located at Vigo's Ria left margin, in the North-West coast of Spain (Fig.1). The shape of the harbour as well as its relatively small entrance (Fig.2) gives rise to the accumulation of pollutants at the most internal basins. In addition most of the sewage and some of the industrial waste water generated in the area are finally discharged to Bouzas basins. The logical result of these two factors is a very deplorable image of dirtiness.

The authorities of the harbour entrusted to CEPyC (CEDEX), a research center of the Ministry of Public Works and Transportation, the study of possible solutions -if any- for their pollution problem, pointed to accomplish a more extensive exchange between basins

263

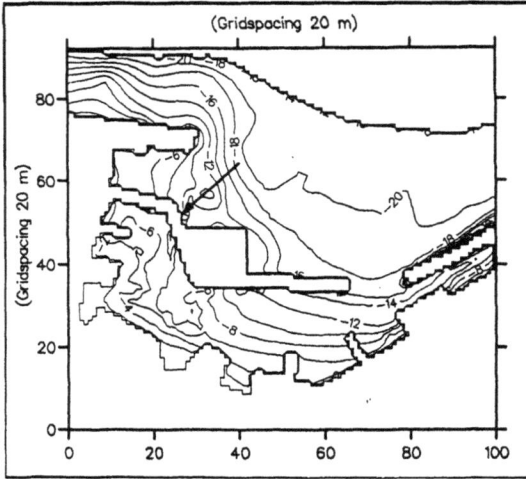

(Gridspacing 20 m)

Figure 2. Grid of Bouzas basin.

polluted water and outside cleanner water.

The first proposed alternative regarded the opening of a tunnel through the protection dock at the western end of the basins (Fig.2), whilst the second one was based on a continuous clean water input at the same point. Both possibilities had to be considered under different harbour layouts because of some new works which were about to be undertaken in short or medium-term.

The analysis of both alternatives was carried out using the numerical model MIKE21 [2] in order to compare one to another and to evaluate their efficiency.

2. Hydrodynamical analysis.

As it is well known [1,4], estuarine tide behavior corresponds to a standing wave arising from the total reflexion of the tidal progressive wave at the closed end of the estuary. For the unidimensional standing wave, surface elevation and horizontal velocity are determined as follows:

$$\eta(x, t) = A\cos\left(\frac{2\pi t}{T}\right)\cos\left(\frac{2\pi x}{L}\right) \tag{1}$$

$$v(x, t) = A\left(\frac{c}{h}\right)\sin\left(\frac{2\pi t}{T}\right)\sin\left(\frac{2\pi x}{L}\right) \tag{2}$$

where A is the maximum amplitude, i.e. the wave amplitude at the reflexion point, c is the celerity and x is the distance from the end of the ria.

The order of magnitude of wavelength ($\approx 10^5$m) is big enough to ensure that any nodes of the wave would be outside the study area, and hence every point inside would be in phase.

Once the physical system is defined, a brief description of the numerical model is needed.

MIKE21 model undertakes a bidimensional hydrodynamic calculation over a rectangular discrete grid by finite differences method. Starting from some imput data, the model provides water level and velocity for every time step and grid point. The input data must define the initial conditions over the whole grid as well as the evolution of open boundaries over the whole simulation period. Calibration parameters like bed ressistance or eddy viscossity will not be discussed here.

The essential points for a successful application of the model are: the correct choice of the grid (area, orientation, dimensions and gridspacing), the appropriate determination of the system state at the beginning of simulation, and the establishment of accurate boundary conditions for all the period covered by simulation.

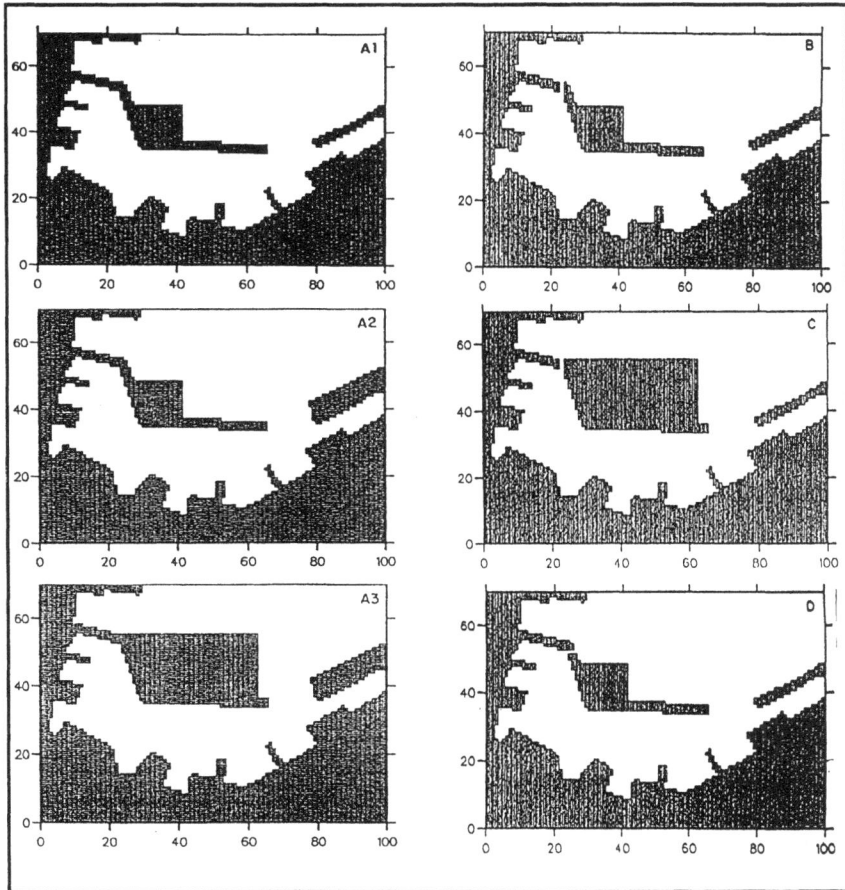

Figure 3. Grids used for simulation of alternatives A1, A2, A30, A31, A32, B, C and D.

2.1. Boundary Conditions.

The usual way of proceeding is the input of a water level senoidal variation around a mean value whether at a single open boundary, or at two opposite boundaries with an empirical or calculated phase difference.

Using the first option for input on a grid covering the whole estuary from the western boundary chosen to the closed end, the steady state solution would have finally been a

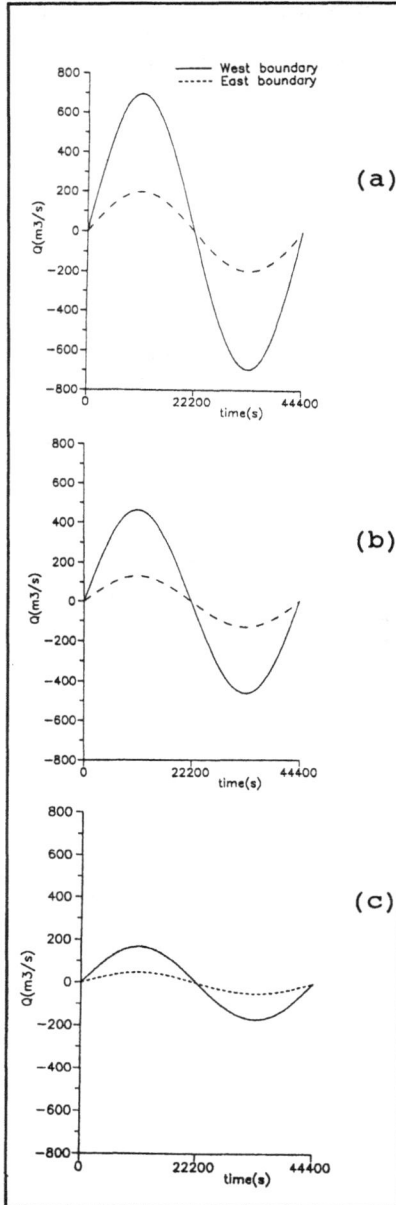

Figure 4. Time evolution of boundary conditions for cases (a) OPTIMUM, (b) MEDIUM and (c) WORST.

standing wave. But the dimensions of the area would have required an extremely large grid, which implies a large number of calculation points, or else a loss in area definition with the consequent loss of accuracy in results.

The choice of a smaller grid (Fig.2) leads to the second option, that is to the opening of two boundaries. Since a standing wave was being considered, determination of time evolution at both boundaries did not consist in the calculation of a phase difference but an oscilation amplitude difference.

Calculation of a difference between west and east boundary oscillation amplitudes was achievable. Surface elevations were taken from statistic tidal forecast and horizontal peak velocities from [5]. Those data where supossed to correspond to the same point (namely, any point along west boundary), which was at an unknown distance from the origin, and the amplitude at this point was as well unknown.

With the available data, it was possible to calculate the standing wave amplitude and therefore velocity and surface elevation at any point and time using equations 1 and 2 (see Table 1).

Several tests were carried out, both over discrete Bouzas area bathymetry and a simpler flat channel grid. Surprisingly, the calculated current pattern had nothing to do with what was expected.

However, much more realistic behavior was observed when flux density instead of water level boundary conditions were impossed, despite of using the same initial data.

Therefore, it was decided to use flux density as boundary condition, wich was equivalent to establish the filling and emptying system of the grid.

Flux density calculation. Once west boundary cross section was "measured", peak flux density could be calculated starting from peak velocity. The time variation of it, as it was for velocity, should correspond to an expression such as:

$$Q(t) = Q_0 + Q_{max} \sin\left(\frac{2\pi t}{T}\right) \qquad (3)$$

Since only the velocity at west boundary was known, only peak flux across this boundary could be calculated as the product of peak velocity and cross section. Flux trough the opposite boundary had to be estimated as follows: the grid has a surface S, measured at LLW, and should be filled from low water up to h (high water level). The volume $V = h \cdot S$ had to be reached by means of water entering through west boundary and despite of water flowing out through east boundary:

$$V = \int_0^{T/2} [Q^W(t) - Q^E(t)]\, dt = (Q_{max}^W - Q_{max}^E)\, \frac{T}{2\pi} \tag{4}$$

This estimation was carried out for three different tidal ranges (OPTIMUM, MEDIUM and WORST cases; Fig.4). The same process was also used to calculate the boundary evolution for a 44h period from the forecast data (REAL case). Variables for all cases are shown in Table 1.

Concerning to initial conditions, it was necessary to consider the surface shape at low water since the model automatically set zero velocity at its time origin.This was made using (1), which means that possible variations along de north-south direction were neglected.

The different harbour layouts which have been simulated can be seen in Figure 3, and the complete set of simulations is summarized in Table 2. For all of them, the application of the conditions mentioned above gives rise to the espected circulation pattern (maximum velocity at medium water and zero at low and high waters), and a time evolution of water surface as shown in Figure 5.

Table 1. Tide parameters used for calculations.

CASE			$2\eta_{max}$[*] (m)	A(m)	DESCRIPTION	
OPTIMUM[**]			3.386	1.707	25 largest tidal ranges average.	
MEDIUM[**]			2.248	1.133	Average of the whole year.	
WORST[**]			0.826	0.416	25 shortest tidal ranges average.	

REAL	date	time	η_0[***] (m)	T/2(s)	$2\eta_{max}$[****] (m)	A(m)
1.	07/07	17:08	1.00			
				22560	1.99	1.003
2.		23:24	2.99			
				22740	2.09	1.053
3.	08/07	05:43	0.90			
				22320	2.15	1.083
4.		11:55	3.05			
				22740	2.22	1.119
5.		18:14	0.83			
				22140	2.25	1.134
6.	09/07	00:23	3.08			
				22620	2.31	1.164
7.		06:40	0.77			
				22500	2.44	1.230
8.		12:55	3.21			

[*]Tidal range at the closed end of the ria.
[**]The period (T) used is 44400 s.
[***]Surface elevation above LLW at Vigo's harbour.
[****]Tidal range at Vigo's harbour.

267

3. Renovation analysis.

Once the hydrodynamical problem was solved, renovation rate r for each case and alternative had to be estimated. The calculation of this variable, defined as the fraction of the initial polluted volume replaced by clean water during a tidal cycle, was carried out by means of a relatively simple aproximation to the problem that allows to study more complex situations.

The simplest way to estimate r was to simulate the behavior of a conservative tracer by impossing the same tracer concentration (namely 1.7 mg/l) to every grid point inside the basins, and setting to zero the concentration outside.

The aim of using a tracer is just to distinguish the basins water from the outside water. In general, if V_p is the volume which go out from the basin in a tidal cycle, and V_p' is the volume of basin water that comes in again, the renovated volume will be V_p-V_p', and renovation rate

$$r = \frac{V_P - V_P'}{V_0} \tag{5}$$

where V_0 is the volume of the basin itself. The described situation is equivalent to an accidental "spill" as long as it is punctual in time and homogeneous in space. One tidal cycle was simulated, beginning at high water, and mean concentration inside the basins was calculated at the end of the cycle.

The expression for calculating renovation rate per tidal cycle was

$$r = \frac{C_0 - C_f}{C_0} \tag{6}$$

which is completely equivalent to (5). For REAL case, renovation rate per tidal cycle was calculated as

$$r = 1 - \sqrt[3]{\frac{C_3}{C_0}} \tag{7}$$

Table 2. Summary of simulations.

CASE ALTERNATIVE	OPTIMUM	MEDIUM	WORST	REAL
A1	X	X	X	X
A2		X		
A30, A31, A32		X		
B	X	X	X	X
C	X	X	X	X
D	X			

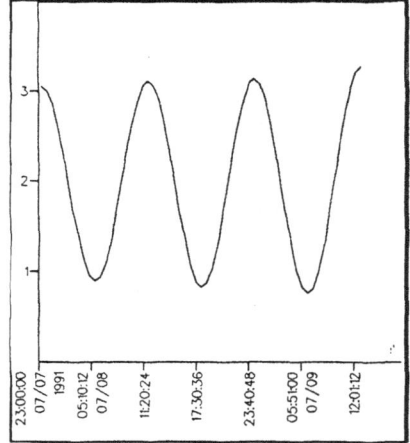

Figure 5. Time evolution at grid point (60,60); REAL case, A1 alternative.

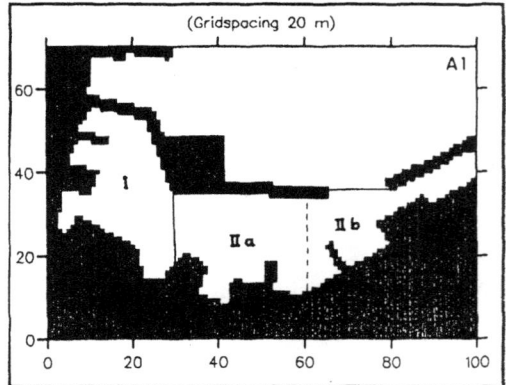

Figure 6. Basins division into zones. (Z-II covers Z-IIa and Z-IIb).

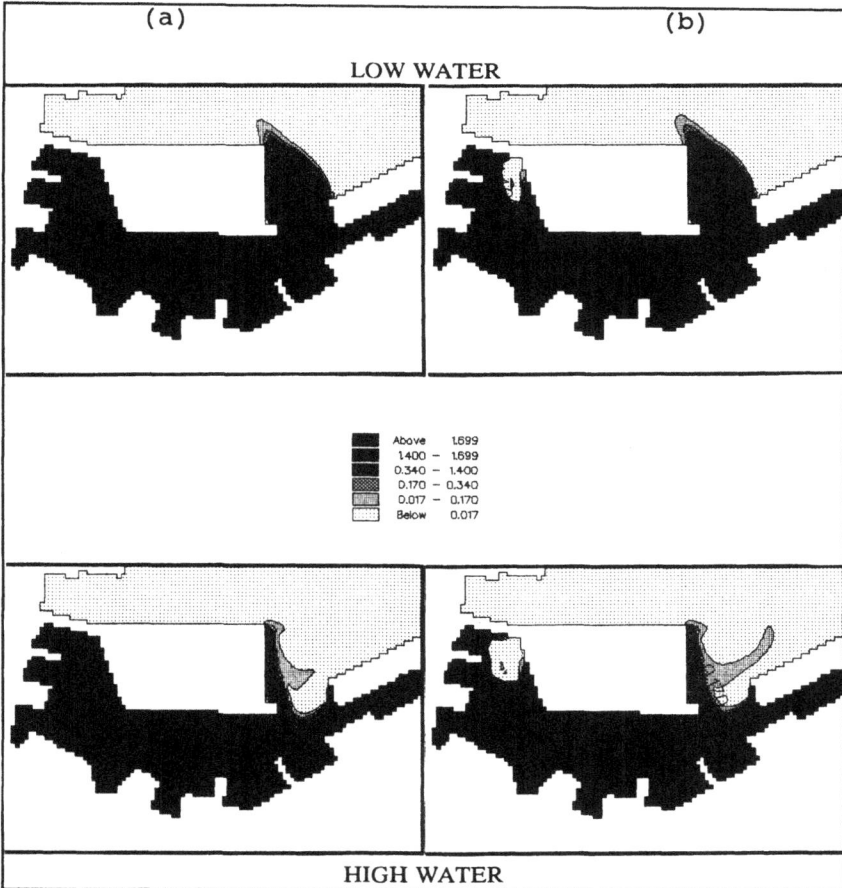

Figure 7. Concentration contours (mg/l) from simulations A30 (a) and A32 (b) at low and high water.

Some results from the complete set of simulations are shown in figures 7 and 8 and renovation rates can be seen in tables 3 and 4. It should be said that **r** has been calculated for the whole basin (TOTAL) and for four subareas inside. The first one (Z-I), is supposed to behave independently from the second (Z-II). At the same time, Z-II is divided into two different zones, Z-IIa and Z-IIb (Fig.6).

Table 3. Renovation rate for OPTIMUM (a), MEDIUM (b) and WORST (c) case.

(a)

Alternative	TOTAL	Z-I	Z-II	Z-IIa	Z-IIb
A1	.149	.005	.207	.173	.249
B	.172	.149	.181	.127	.246
C	.113	.186	.084	.009	.175
D	.162	.028	.213	.183	.250

(b)

Alternative	TOTAL	Z-I	Z-II	Z-IIa	Z-IIb
A1	.068	.000	.094	.036	.164
A2	.068	.000	.093	.038	.160
A30	.044	.000	.061	.009	.124
A31	.048	.021	.058	.009	.118
A32	.082	.202	.036	.007	.070
B	.078	.087	.075	.017	.145
C	.054	.121	.028	.004	.057

(c)

Alternative	TOTAL	Z-I	Z-II	Z-IIa	Z-IIb
A1	.012	.000	.016	.000	.035
B	.013	.013	.013	.000	.029
C	.014	.022	.011	.000	.023

Table 4. Renovation rate per tidal cycle for REAL case.

Alternative	TOTAL	Z-I	Z-II	Z-IIa	Z-IIb
A1	.061	.000	.086	.041	.148
B	.071	.080	.067	.024	.125
C	.042	.105	.020	.006	.038

4. DISCUSSION

4.1. About boundary conditions.

The set of undertaken simulations confirms some well known aspects of numerical modelling.

First of all, the correct choice of the grid is of great importance for a satisfactory response of the model. At the present case, the suitable orientation because of harbour geometry was not appropriate for the simulation of external velocity field: differences between east and west boundaries cross-sections to reproduce similar flows. Therefore it was necessary to introduce an island at the upper left corner to achive an external flow according to the expected estuarine flow.

On the other hand, the model shows to be very sensible to small variations of surface elevation, but not so much for flow variations. In order to reach soon a regular steady state of velocity field, flux density boundary conditions have been found more useful than water level boundary conditions.

4.1. About renovation rate.

As can be inferred from results presented above, renovation rate will depend on tidal range, on harbour geometry and also on the process employed to accomplish an increase of water exchange.

Dependance on tidal range is monotonous but not linear. The data obtained from simulations seem to follow a potential law with exponent close to 2, but it would be necessary a larger number of simulations to establish an specific function.

Concerning to geometry, two different aspects can be distinguished. In one hand we have the entrance relative position and size, that will not be discussed here because those are the same for all the simulations. On the other hand, we have the influence of external geometry on the external velocity field.

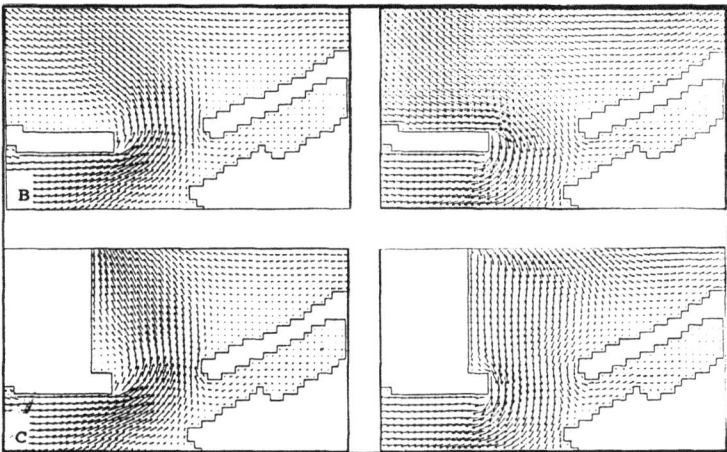

Figure 8. Velocity field at the entrance for B and C alternatives.

271

There is a clear difference between simulations A1 and A30, and between B and C. The filling work at left side of the entrance (Figure 3) has an important negative effect on renovation rate in the area Z-II, essentially because a change in flow direction at the external area closer to the entrance (Figure 8).

In simulations A1 and B, inflow and outflow directions differ in some degrees, and the nearer is the entrance, the larger is the difference. On the contrary, in C and A30 simulations, the direction is almost the same for inflow and outflow in the nearest 400 meters, which means that most of polluted water comes back to the basins.

In relation to the method used for renovation enhance, renovation rate has been represented versus injected flow (Figure 9) for simulations A30, A31 and A32. In all the areas described in figure 7, a linear dependance is observed that was expected for the area Z-I, but unexpected for the remaning areas. Renovation rate corresponding to C alternative have been also plotted. The equivalent flow assigned to this alternative has been calculated assuming that Z-I area renovated volume V_{REN} is due to a continuous and constant injection Q_{eq} of clean water throughout the tidal cycle:

$$Q_{eq} = \frac{V_{REN}}{T} \tag{8}$$

where T is the tidal period.

As can be seen in Figure 9, C alternative is always less efficient than the real injection of Q_{eq} except for Z-I area. Therefore, water outflow through the tunnel has a more negative influence on renovation at the entrance than clean water injection as long as this is under 12 m^3/s.

Figure 9. Renovation rate versus injected flow (m^3/s). C alternative is represented at $Q = 5.6 \ m^3/s$.

4.3. About possible uses of parameter r.

Parameter r allows for estimation of the number of tidal cycles necessary to reduce concentration to any fraction of its initial value C_0:

$$\frac{C_n}{C_0} = (1-r)^n \tag{9}$$

where C_n is concentration after n tidal cycles (Figure 10).

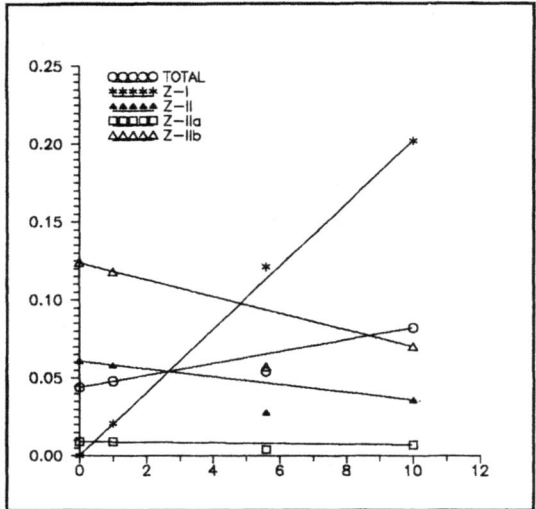

Also it is possible to calculate an equilibrium value C_{eq} for concentration inside the basin for the more usual case of a continuous discharge of polluting substances (i.e. sewage). In the case of a constant discharge Q_v of concentration C,

$$C_{eq} = \frac{CQ_vT}{V_0 r}$$

(10)

Evolution of concentration for this case is shown in Figure 11.

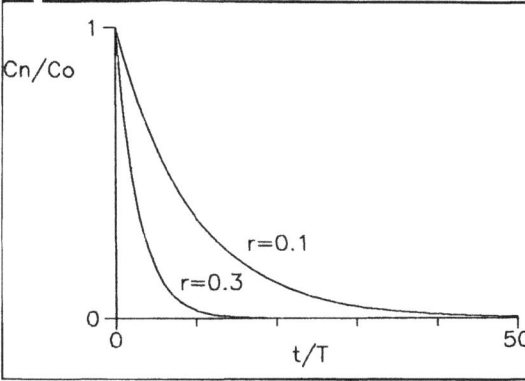

Figure 10. Concentration evolution for an accidental spill.

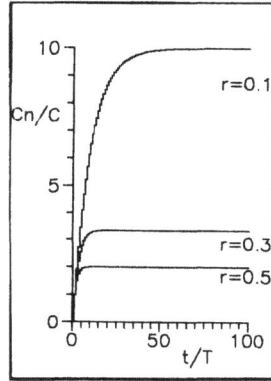

Figure 11. Concentration evolution for a continuous discharge.

5. REFERENCES.

[1] Bowden,K.F. (1983) *Physical Oceanography of Coastal Waters*, Ellis Horwood Ltd., England.
[2] Danish Hydraulic Institute (1991) *User's Guide for MIKE21.*
[3] Fischer,H.B. et al (1979) *Mixing in Inland and Coastal Waters*, Academic Press, New York.
[4] Lloret, A. (1991) Report: "Limpieza de las Darsenas de Bouzas". CEPYC (CEDEX), Madrid.
[5] Ruiz-Mateo, A. (1983) Report: "Dinamica Marina de la Ria de Pontevedra". CEPYC (CEDEX), Madrid.
[6] Ruiz-Mateo, A. (1987) Report: "Predimensionamiento de 37 Emisarios Submarinos Situados en las Rias Gallegas". CEPYC (CEDEX), Madrid.

Part 3
SEDIMENT TRANSPORT MODELLING

23 Numerical modelling of sedimentation around Johor Port

G. Li and P. Wright

ABSTRACT

A mathematical model is used to model the tidal currents and the sedimentation in Johor Strait around Johor Port area, Malaysia. A special nested grid system is adopted to enhance the numerical modelling accuracy. while the required computation effort is maintained to a reasonable level. After calibration against the extensive field measured data, the model is applied to predict tidal currents, sediment transport and long-term bathymetric changes for various recommended Phase 4 expansion schemes of Johor Port in order to obtain the optimal expansion scheme.

Introduction

The port of Johor is located at Pasir Gudang at the southern tip of the Peninsular Malaysia. It lies on the north shore of the Johor Strait, separating Malaysia from Singapore, which provides the port with sheltered deep water, less than 3 hours sailing time from the main shipping lanes which pass through the Straits of Singapore.

The port is very young compared with long-established ports of the region, having only started operation in 1977 but, despite its close proximity to one of the world's largest ports on Singapore, its growth has been dramatic. So much so that, after only 14 years, a third phase of development is already under construction and a fourth phase is currently being prepared for implementation.

The first phase of the port has been sited close to the widest part of this section of the Johor Strait and opposite the end of the deep water Serangoon Channel (see figure 1). The subsequent second phase formed an easterly extension while Phase 3 will extend the port westward to the practical development limit in this direction. This meant that, when the time came to consider the fourth phase, the only remaining site for expansion in the present port

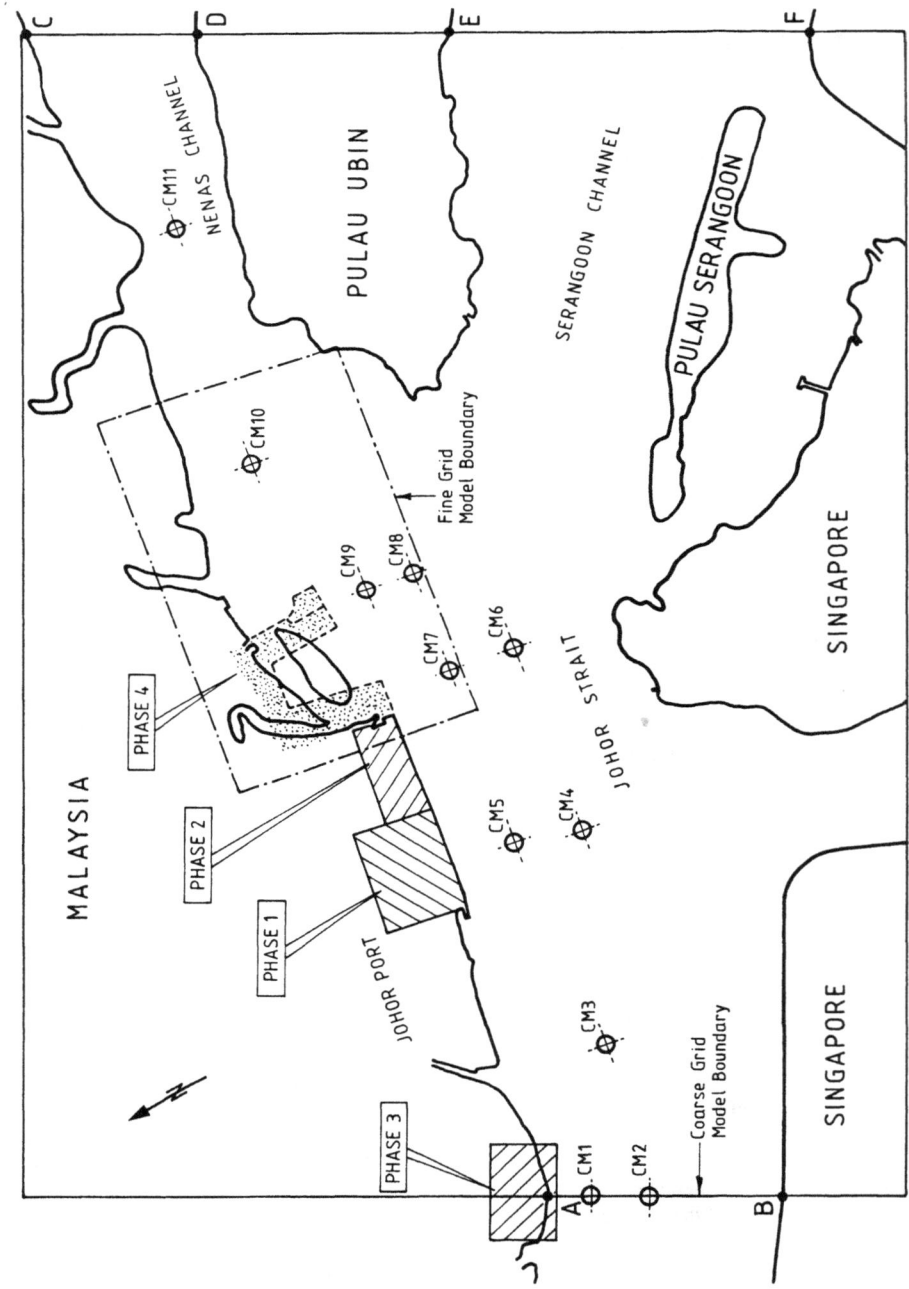

Figure 1 – Plan of Johor Port and the mathematical modelling area

location was a somewhat restricted area at the eastern end of the existing berths.

The site for Phase 4 is, as can be seen from figure 1, bounded to the east by a fabrication yard for large modules for offshore oil rigs. It also lies to the east of the natural line of the Serangoon Channel at the entrance to the narrower and more shallow Nenas Channel. This channel runs between the Malaysian mainland and the island of Pulau Ubin, part of Singapore, so the international boundary runs down the middle of the channel, further restricting the potential for dredging out a wide shipping channel (without prior international agreement). The site itself extends little more than 600 metres east of the existing berths but by the time consultants came to be appointed for the project in early 1991, the port's requirements for extra berths was significantly in excess of this. It was therefore necessary to consider a radically different solution to a simple straight-line extension.

The consultants, a consortium of Engineering and Environmental Consultants (EEC) of Kuala Lumper and Posford Duvivier, UK based international consulting engineers, were invited by the Johor Port Authority to consider a range of options, including the possibility of forming a harbour with berths running north-south as well as east-west (see figure 2). This plainly provided a much greater berth length but also presented a number of potential problems.

Some of these problems were associated with basic port planning issues such as the balance between berth length and land area but the issues which form the basis of this paper are concerned with the effects on the existing tidal and sediment transport regime of such a radical alteration in land form and water depth. The existing site includes a small mangrove-covered island and shallow tidal channels with considerable inter-tidal areas. The new layout would involve extensive reclamation covering the channels and half the island while the remainder of the island and other land would be removed to dredge a harbour some 13m deep extending inshore 500m.

It was decided from the outset that it was very important to know as much as possible about the existing hydrodynamic conditions and sediment transport and their changes due to the port expansion. Mathematical modelling techniques were exploited to investigate: (a). siltation and/or scour effects for both the new berths and existing berths, (b). tidal current behaviour to assess its effect on ship handling, and (c). the effects of deepening the end of the Nenas Channel.

In this paper, a mathematical model used for the study is presented. The model consists of a hydrodynamic model, a suspended sediment transport model, a bed-load sediment transport model and a morphologic evolution model. The calibration of the model against the extensive field measured data is detailed and its application to one of the recommended Phase 4 expansion schemes of Johor Port is also demonstrated.

2. The mathematical model

The study area is under the influence of a semi-diurnal tide. Therefore, the model must be able to simulate the unsteady flow. One-dimensional model could not reflect the cross-sectional change which would have significant effect on the sedimentation in the area, and three-dimensional models demand too much computation effort for any realistic application to such a large scale modelling project. An unsteady depth-integrated two-dimensional model was therefore chosen for the study.

2.1 The hydrodynamic model

The hydrodynamic model consists of flow continuity equation and momentum equations in two horizontal directions.

279

Flow continuity equation

$$\frac{\partial\,(\,uH\,)}{\partial\,x} + \frac{\partial\,(\,vH\,)}{\partial\,y} + \frac{\partial\,\eta}{\partial\,t} = 0 \qquad (1)$$

Momentum equations

$$\frac{\partial\,(\,uH\,)}{\partial\,t} + \beta\left\{\frac{\partial\,(\,uuH\,)}{\partial\,x} + \frac{\partial\,(\,uvH\,)}{\partial\,y}\right\} = fvH - gH\frac{\partial\,\eta}{\partial\,x} + \frac{1}{\rho}\,(\,\tau_{xw} - \tau_{xf}\,)$$
$$+ \frac{\partial}{\partial\,x}\left(\frac{\sigma_{xx}H}{\rho}\right) + \frac{\partial}{\partial\,y}\left(\frac{\tau_{xy}H}{\rho}\right) \qquad (2)$$

$$\frac{\partial\,(\,vH\,)}{\partial\,t} + \beta\left\{\frac{\partial\,(\,uvH\,)}{\partial\,x} + \frac{\partial\,(\,vvH\,)}{\partial\,y}\right\} = - fuH - gH\frac{\partial\,\eta}{\partial\,y} + \frac{1}{\rho}\,(\,\tau_{yw} - \tau_{yf}\,)$$
$$+ \frac{\partial}{\partial\,x}\left(\frac{\tau_{yx}H}{\rho}\right) + \frac{\partial}{\partial\,y}\left(\frac{\sigma_{yy}H}{\rho}\right) \qquad (3)$$

Where η = water elevation above datum; h = water depth below datum; $H = \eta + h$, total water depth; u = depth-mean water velocity in X-direction; v = depth-mean water velocity in Y-direction; t = time; $f = 2\omega\sin\phi$, Coriolis parameter; ω = angular velocity of the Earth's rotation; ϕ = geographical latitude; β = correction coefficient for the vertically non-uniform velocity distribution; ρ = water density; τ_{xw} = wind-induced surface shear stress in X-direction; τ_{yw} = wind-induced surface shear stress in Y-direction; τ_{xf} = bed shear stress in X-direction; τ_{yf} = bed shear stress in Y-direction; σ_{xx}, τ_{xy}, τ_{yx}, σ_{yy} = turbulence stress; g = gravitational acceleration.

The coupling of the turbulence enclosure, the formulae used for the wind and bed shear stress and the numerical schemes etc. are described in detail by Falconer [3] and Falconer and Li [4].

2.2 The sediment transport model

2.2.1 Suspended sediment transport The suspended sediment transport model is based on the mass conservation by taking into account advection, diffusion, resuspension and deposition.

$$\frac{\partial\,(\,SH\,)}{\partial\,t} + \alpha\left\{\frac{\partial\,(\,SQ_x\,)}{\partial\,x} + \frac{\partial\,(\,SQ_y\,)}{\partial\,y}\right\} = \frac{\partial}{\partial\,x}\left(\varepsilon_x H\frac{\partial\,S}{\partial\,x}\right) + \frac{\partial}{\partial\,y}\left(\varepsilon_y H\frac{\partial\,S}{\partial\,y}\right)$$
$$+ E - D \qquad (4)$$

$$D = w_f\,S \qquad (5)$$

where $Q_x = uH$; $Q_y = vH$; α = correction coefficient for the vertically non-uniform distribution of suspended sediment concentration; S = depth-averaged suspended sediment concentration; $\varepsilon_x = \varepsilon_y = \nu_t / \sigma$, turbulent eddy diffusivity; σ = Prandtl-Schmidt number (0.5 ~ 0.9); E = resuspension rate; D = deposition rate.

Resuspension rate is determined by applying mass conservation law to a steady and uniform flow, i.e.

$$w_f \, s = - \, \epsilon_z \frac{\partial \, s}{\partial \, z} \tag{6}$$

Its analytical solution is

$$s(z) = s_a \left(\frac{H-z}{z} \frac{a}{H-a} \right)^{z_*} \qquad a \leq z \leq H \tag{7}$$

$$z_* = \frac{w_f}{\gamma \kappa u_*} \tag{8}$$

where a = reference level; s = sediment concentration at equilibrium status; s_a = suspended sediment concentration at the reference level; z = vertical coordinate; z_* = suspension parameter; ϵ_z = vertical diffusion coefficient.

The following parabolic distribution of the vertical diffusion coefficient is made use of in deriving Eq.(7):

$$\epsilon_z = \frac{z}{H} \left(1 - \frac{z}{H} \right) \kappa u_* H \tag{9}$$

where κ = Von Karman coefficient.

Bearing in mind that we are dealing with depth-integrated model, depth-averaged concentration S used in Eq.(4) and Eq.(5) is different from the s(z) in Eq.(6) & Eq.(7). To get the depth-averaged equilibrium concentration, Eq.(7) is integrated vertically over the water depth from z = a to z = H, and divided by H - a :

$$S_e = \frac{s_a}{H - a} \int_a^H \left(\frac{H - z}{z} \frac{a}{H - a} \right)^{z_*} dz \tag{10}$$

where Se = depth-averaged equilibrium concentration level. Therefore the resuspension rate E

$$E = w_f \, S_e \tag{11}$$

Bijker's formula is used for the calculation of reference concentration level s_a

$$s_a = \frac{q_b}{6.34 \; U_* \, K_s} \tag{12}$$

where q_b = bed-load sediment transport rate, i.e. transport amount at unit time across unit width.

2.2.2 *Boundary conditions* At open boundaries, suspended sediment concentration is provided and uniform normal flux condition applies.

281

$$\frac{\partial^2 S}{\partial n^2} = 0 \qquad (13)$$

At solid boundaries, zero normal flux condition applies

$$\frac{\partial S}{\partial n} = 0 \qquad (14)$$

2.2.3 Bed-load sediment transport Kalinske-Frijlink formula is used in the calculation of bed-load sediment transport rate.

$$q_b = 5D_{50} \, \mu \, u_* \exp\left\{\frac{-0.27 \, (R - 1) \, C^2 \, D_{50}}{\mu \, (u^2 + v^2)}\right\} \qquad (15)$$

where

$$\mu = \left(\frac{C}{C_{90}}\right)^{\frac{3}{2}} \qquad (16)$$

$$C = 18 \, \log\left(\frac{12H}{K_s}\right) \qquad (17)$$

$$C_{90} = 18 \, \log\left(\frac{12H}{D_{90}}\right) \qquad (18)$$

μ = coefficient of sand ripple influence; u_* = friction velocity; $R = \rho_s / \rho$, relative density; ρ_s = sand density; C = overall Chezy coefficient; C_{90} = Chezy coefficient related to grains.

2.3 Morphologic evolution model

According to mass conservation law

$$(1-m) \, \frac{\partial h}{\partial t} = \frac{\partial \, (q_{sx} + q_{bx})}{\partial x} + \frac{\partial \, (q_{sy} + q_{by})}{\partial y} + E - D \qquad (19)$$

where q_{sx} = SuH, suspended sediment flux in X-direction; q_{sy} = SvH, suspended sediment flux in Y-direction; q_{bx} = bed-load sediment transport rate in X-direction; q_{by} = bed-load sediment transport rate in Y-direction; m = porosity of the bed material; E = resuspension rate; D = deposition rate.

3. Model calibration

3.1 Field data

Before the model can be applied to predict the impact of future port expansion, it is necessary to select and adjust the values of model parameters and coefficients so that the model can actually reproduce the available observed or field measured conditions. 11 currents meters were installed around the port area as shown in Figure 1, during the period of 27 June to 30 June 1991 for the modelling purpose. The tidal condition of that period was basically a spring tide condition during which the best accuracy of the field data collection could be expected since the tidal currents in the area is generally weak (less than 0.5 m/s). Velocity, salinity and turbidity were measured at about every 15 minutes at the 11 current metering locations.

3.2 Selection of finite difference grid system

A nested finite difference grid system was used in order to obtain more detailed information around the Phase 4 expansion area of Johor Port. The axes of the coarse grid system was such selected that Y-axis was parallel to the main longitudinal flow direction and X-axis was in the main cross-section direction. With this specially chosen coordinate system, zero velocity component of the cross-section direction at the open boundaries can be reasonably assumed to simplify the open boundary conditions without losing modelling accuracy. While in the fine grid model, Y-axis was parallel to the existing wharf line of Phase 1 and Phase 2 so that the most accurate representation of the solid boundaries in the Phase 4 area can be achieved. In view of the fact that velocity changes less significantly in the longitudinal flow direction than in the cross-section direction in the majority of the coarse grid modelling area, larger grid space (100m) for the longitudinal direction than for the cross-section direction (50m) was used so that higher accuracy than that of using a uniform grid system can be achieved if computation effort was kept at the same level or less comutaiton effort would be needed if required accuracy was fixed. 25m × 25m mesh size was used for the fine grid system.

3.3 Open boundary conditions

As shown in Figure 1, the modelling area lies in the border between Malaysia and Singapore. This feature complicated the field data collection. Therefore, all the 11 current meters were located on the Malaysia side, and no data was available for the open boundary E-F. To remedy the data shortage, the open boundary E-F was treated as a water level boundary and the water level and suspended sediment concentration were extrapolated from open boundary C-D. Open boundaries A-B and C-D were treated as velocity boundaries and the velocity and suspended sediment concentration were interpolated from CM1 and CM2 and extrapolated from CM11 respectively.

3.4 Calibration of hydrodynamic model

The model calibration was carried out by using the coarse grid model to adjust the bed roughness conditions and the phase lag between open boundaries A-B and C-D. The modelling area covered by the coarse grid model is shown in Figure 1. Having used the data at CM1, CM2 and CM11 in deriving the open boundary input, the rest of the current metering data was used to calibrate the model. The modelled results agree well with the measured data according to the comparison as shown in Figure 2. The calibrated bed roughness length was 50 mm and the tidal phase at open boundary A-B was 0.6 hour behind that at open boundaries C-D and E-F.

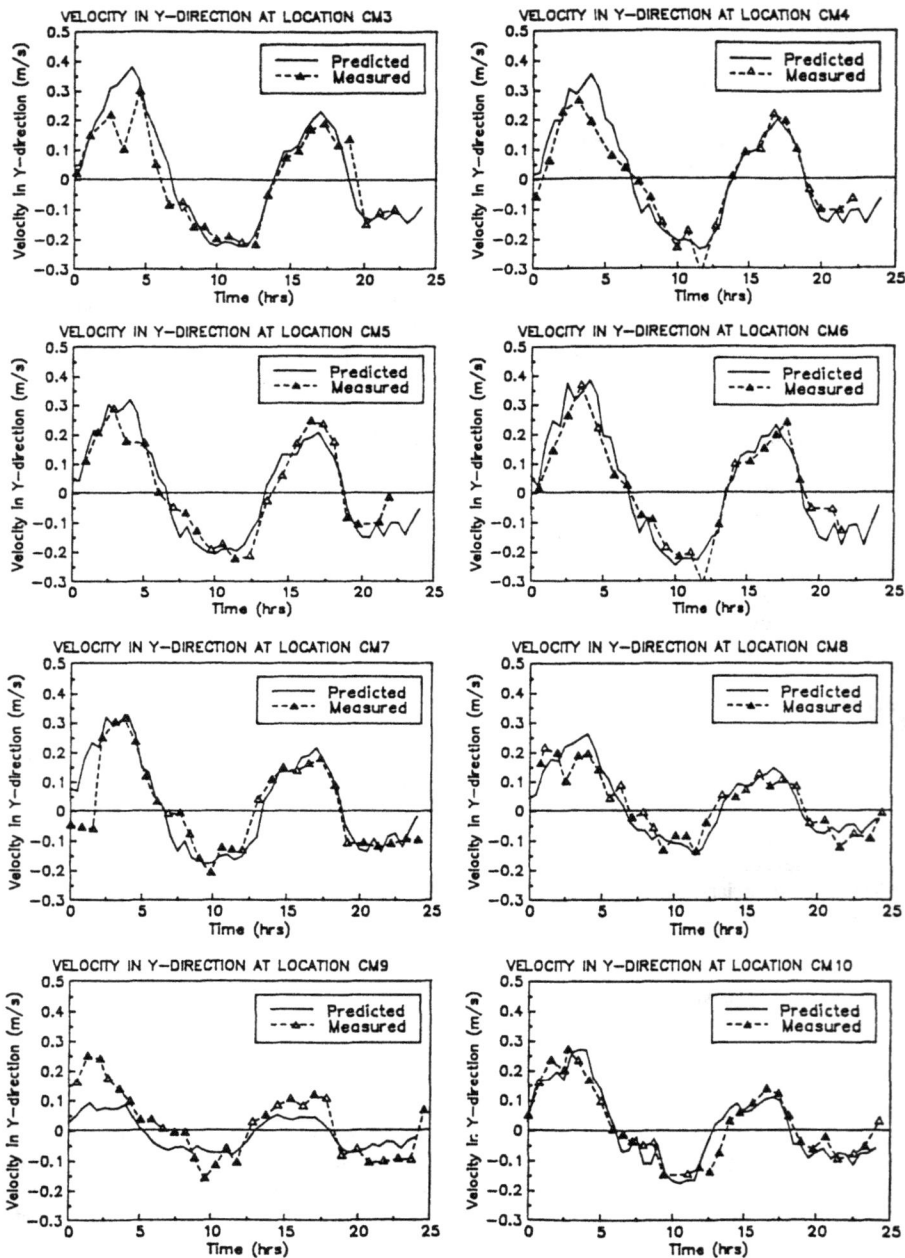

Figure 2 – Comparison between the modelled and the measured velocity in Y-direction

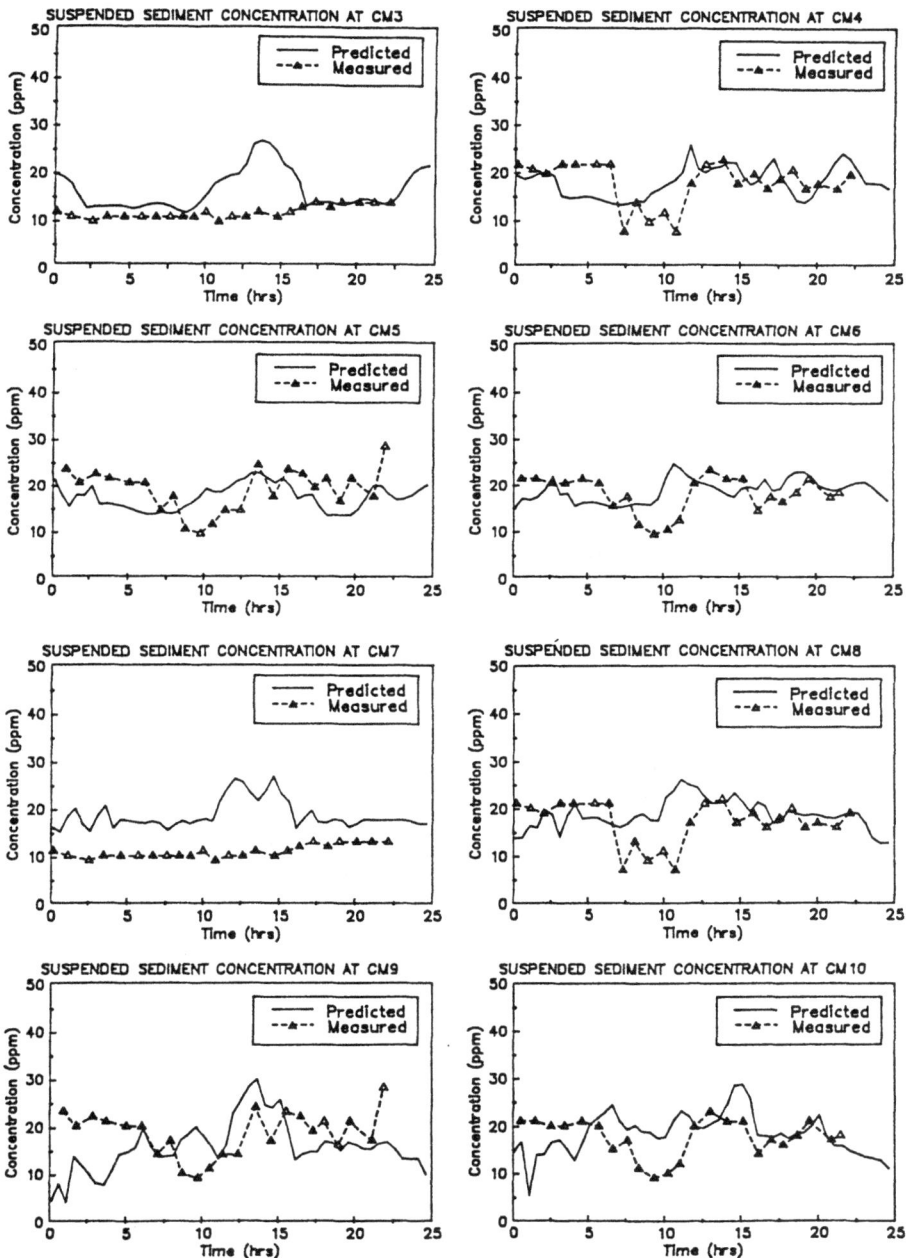

Figure 3 – Comparison between the modelled and the measured suspended sediment concentration

3.5 Calibration of suspended sediment transport model

The main calibration parameters of the suspended sediment transport model were particle size and distribution. Although 27 bed samples were taken in the port expansion area, the laboratory analysis gave a very wide range of particle size distribution. D_{50} ranges from 10 μm to 800 μm, and D_{90} from 75 μm to 3500 μm. The best calibration result was obtained when $D_{50} = 60$ μm and $D_{90} = 200$ μm. In general, the modelled results agree reasonably well with the measured data as shown in Figure 3. However, some discrepancies can be seen. This is possibly due to the existence of some cohesive silt in the bed material, and the cohesive silt behaves quite differently from the sandy material which the model is designed for.

4. Model application

After calibration, the model was applied to predict the tidal currents and to assess the long-term trend of morphologic evolution for various proposed Phase 4 expansion schemes of Johor Port. The application to one of the preferred schemes is detailed in this section. Figure 4 shows the bathymetry around the expansion area covered by the fine grid model.

Morphologic evolution is an integrated result under all tidal and fluvial conditions during the period concerned. A reliable quantitative prediction can only be achieved by running the numerical model through all the contributing input conditions. Unfortunately, it is impracticable to do so in today's technology and equipment. The way out in future perhaps lies in the development of a new kind of models such as a parametric model [1]. In this study, a qualitative prediction of bathymetric changes was carried out by making the following simplificaitons: (a). same tidal conditions applied to each year of the period considered, (b). all year round tidal conditions were represented by mean spring tide condition, mean neap tide condition and mean tide condition, (c). fluvial flow and its carried sediment discharge was ignored, and (d). the three representative tidal conditions were of equal weighing factor.

The whole set of the mathematical model, including the hydrodynamic model, the sediment transport model and the morphologic evolution model was run through those simplified hydrodynamic input conditions for both the coarse and fine grid system. Figure 5 shows the velocity field predicted by the fine grid model at the mean water of the mean spring flood tide when the maximum tidal velocity occurs. As expected, the tidal currents in the fine grid modelling area are generally low and the maximum velocity is only 0.4 m/s occurring at about 150 metres outside the new berth J-K. A clockwise eddy during the period of the mean water of flood tide to the high water, and an anti-clockwise eddy at the low water can be seen both near the entrance of the harbour and in the front of the river dicharge area. But those eddies are too weak to cause any significant effect on the long-term bathymetric changes as will be shown later. The maximum velocity inside the harbour is 0.16 m/s occurring at the entrance near the corner J of the new berths at 1.22 hours after the high water of the mean spring tide.

Predicted suspended sediment concentration level is also low and relatively uniform over the whole modelling area, and fluctuates narrowly throughout a tidal cycle, as can be seen in Figure 6. This is because the suspended sediment paricles are very fine in such a low velocity hydrodynamic environment. Not surprisingly, bed-load sediment transport rate is small as well as shown in Figure 7, which shows the highest bed-load sediment transport rate at all time considered.

Predicted yearly bathymetric changes are shown in Figure 8. Deposition occurs inside the entire harbour and in the area near shore, which agrees with the low flow trapping effect, while erosion occurs in the offshore area. The most severe scour occurs in the area near the berthing corner K where currents converge and the existing bathymetry changes sharply as seen in Figure 4 and Figure 5. In a long run, sediment transport would smooth the bathymetry

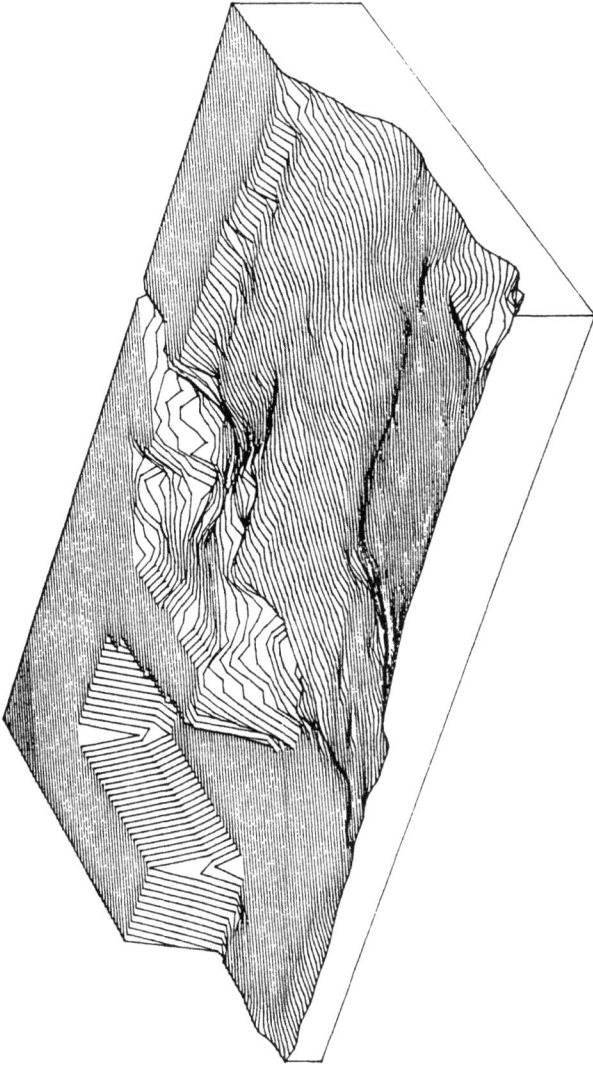

Figure 4 – Bathymetry of planned Phase 4 expansion area of Johor Port

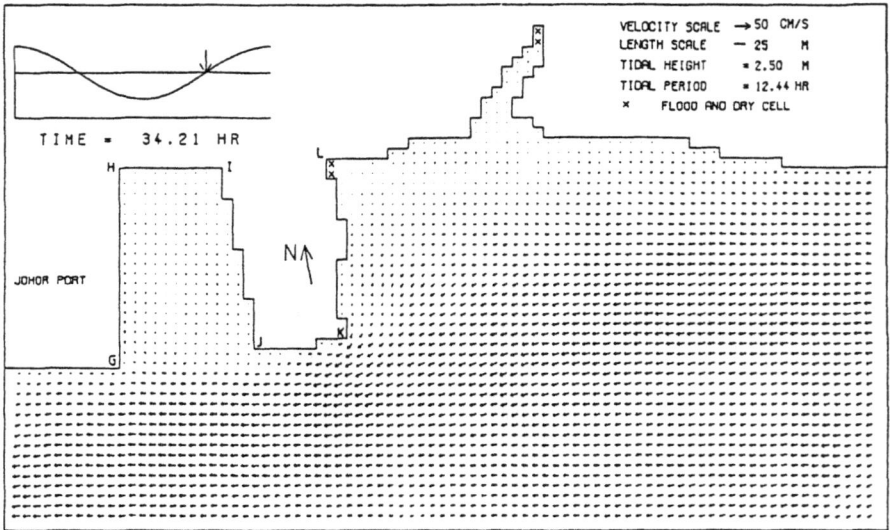

Figure 5 – Predicted velocity field at mean water of mean spring flood tide. Fine grid model

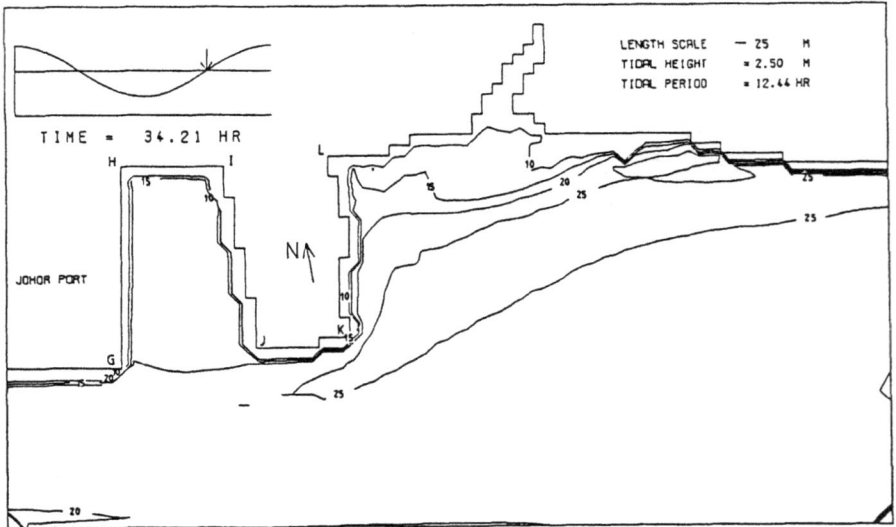

Figure 6 – Predicted suspended sediment concentration (ppm) at mean water of mean spring flood tide. Fine grid model

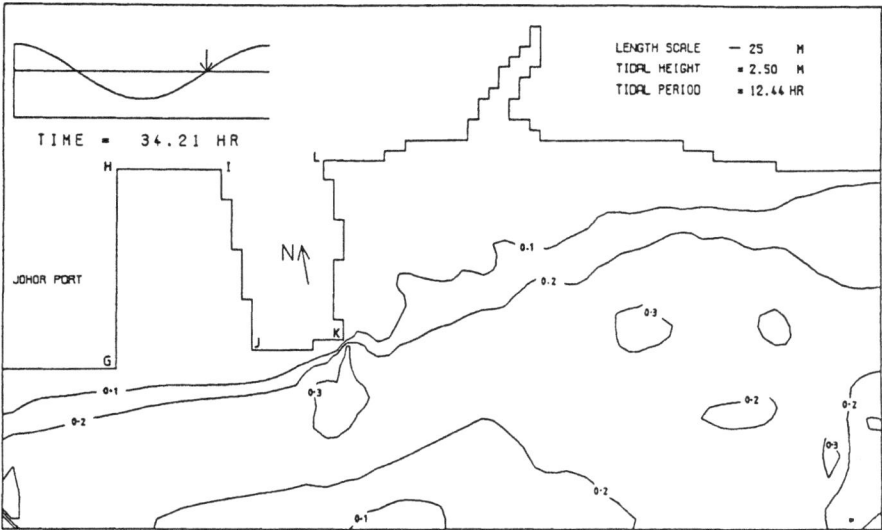

Figure 7 – Predicted bed-load sediment transport rate (mm^2/s) at mean water of mean spring flood tide. Fine grid model

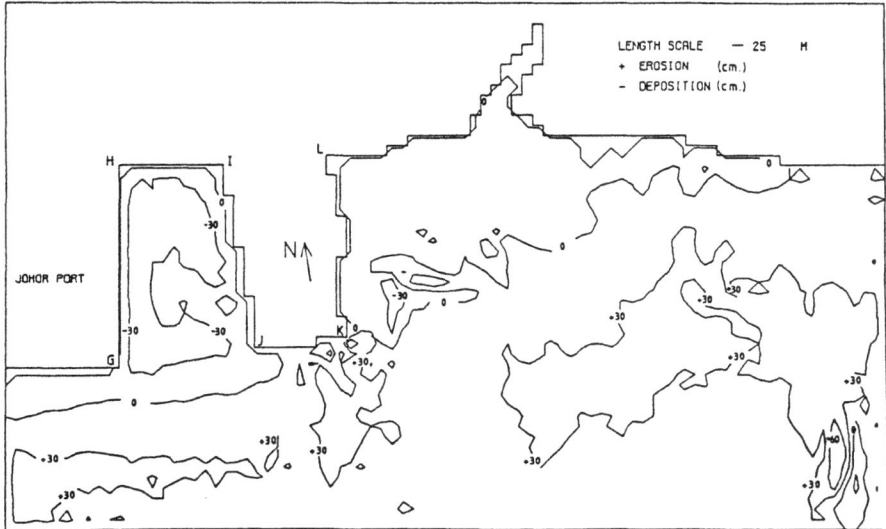

Figure 8 – Predicted yearly bathymetric change around the Phase 4 expansion area of Johor Port

if there exist no strong eddies and sharp velocity changes, so that shallow area would become deeper and deep area shallower. This phenomenon is clearly demonstrated in the area near the berth corner K and at the bottom right area of Figure 8.

5. Conclusions

A mathematical model consisting of a hydrodynamic model, a sediment transport model and a morphologic evolution model is presented. The special finite difference grid systems used in the study, i.e. different grid sizes for longitudinal and cross-section directions in the coarse grid system and the best alignment between the fine grid system and the planned berth line, can enhance the modelling accuracy while the required computation effort is maintained to a reasonable level. The simulated tidal currents agree well with the extensive field measured data. Good agreement has also been achieved between the modelled and the measured suspended sediment concentration although some discrepancies exist. The discrepancies may be caused mainly by the existence of some cohesive sediment material which behaves rather differently from the sandy material which the sediment transport model is designed for. The application of the model to one of the recommended Phase 4 expansion schemes of Johor Port shows that it can also give a reliable qualitative prediction of the long-term trend of bathymetric changes. However, the magnitude of the predicted bathymetric changes should not be treated as a precise prediction since the representation of the yearly tidal conditions is highly schematised and the modern sedimentation theories used in the model are still far from being perfect. Any possible fluvial discharge and seasonal climate changes should also be taken into account in the interpretation of the modelling results.

REFERENCES

1. De Vriend, H.J., (1991), Mathematical modelling and large-scale coastal behaviour. Part 1: Physical processes and Part 2: Predictive models, *Journal of Hydraulic Research*, No.6, Vol. 29, pp 727-752.
2. Deguchi, I. and Sawaragi, T., (1988), Effects of structure on deposition of discharged sediment around rivermouth, *Proceedings of Coastal Engineering*, pp 1573-1587
3. Falconer, R.A., (1986), A two-dimensional mathematical model study of the nitrate levels in an inland natural basin. *Proceedings of Inter. Conf. on Water Quality Modelling in Inland Natural Water Environment*, British Hydraulic Research Association, Bournemouth, England.
4. Falconer, R.A. and Li, G., (1992), Modelling tidal flows in an island's wake using a two-equation turbulence model, *Water, Maritime and Energy, Proceedings of ICE*, to be published.
5. Koutitas, G.C., (1989), *Mathematical Models in Coastal Engineering*, Pentech Press Ltd., London.
6. Van Rijn, L.C., (1984), Sediment transport, Part I: Bed load transport, *Journal of Hydraulic Engineering, ASCE*, N0.10, Vol. 110, pp1431-1456.
7. Van Rijn, L.C., (1984), Sediment transport, Part II: Suspended load transport. *Journal of Hydraulic Engineering, ASCE*, N0.11, Vol.110, pp1613-1641.
8. Yalin, M.S., (1972), *Mechanics of Sediment Transport*, Pergamon Press.

24 Causes of siltation in the port of Montevideo

J. C. I. Piedra Cueva and H. Rodriguez Borelli

SYNOPSIS

The purpose of this paper is to propose sediment transport and a silting process for the port of Montevideo. It describes the hydraulic factors influencing the distribution of sediments within the Montevideo Bay and studies their responsibility in the siltation of the port of Montevideo.

Sediment concentration measurements together with the results of a tidal currents numerical model, reveals that the Port siltation rates are not only due to fine suspended sediment deposition but to other mechanisms as well. The record of a bi-frequency (33 and 210 Kz) DESSO-10 echosounder shows the presence of a fluid mud layer, 0.10 m thick outside the navigation channels and 2.0 m thick inside them.

Waves produce the turbid, near-bed layer which is transported by currents into port approach channels and port. Storm weather produces a great sediment resuspension. The principal roll of storms is to redistribute the sediment contributing to channel siltation.

1. Introduction

At present, about 1:200.000 m³/year of fine sediments are dredged for the maintenance of the outer harbour and docks of the port of Montevideo. For the sake of minimizing dredging costs it is of great importance to know the mechanics of sediment transport in the Montevideo harbour.

The sedimentological process in the bay and port of Montevideo was thought to be due to the erosion -suspension transport - sedimentation and consolidation, caused by the astronomical tide. This theory gave good results in locations with macrotides, where the system's hydrodynamics is mainly controlled by the astronomical tides and the maximum current velocities reach values of 2 m/s, showing clearly defined cycles which agree with tide predictions. However, in the area which is studied here the system's hydrodynamics is different. The effects of the astronomical tides, winds, river flows and Coriolis force are comparable and their interaction is very complex. For this reason, it is not correct to relate the cause of the port's siltation to only one of the existing dynamic factors.

The purpose of this paper is to propose a sediment transport and silting process to the port of Montevideo.

2. Description of the site

2.1 *Location*

The bay and port of Montevideo are located in the Southern coast of Uruguay, in the intermediate area of River Plate (Figures 1 and 2). The bed materials of the bay and its surroundings are silts and clays with little consolidation. The average depth is 3,5 m, and the deepest areas are the inlet channel and docks with a depth of about 10 m.

Soil studies made in the area show layers of mud with a thickness of 4 m with no resistance in penetration tests.

The area of the bay outside the navigation channels are relatively balanced, with an annual sedimentation rate of less than 5 cm.

2.2 *Sediments Characteristics*

The principal components of the bed sediments of the bay are montmorillonite and illite and in less degree caolinite and organic material, with nitid differences in the granulometric composition in the different areas of the bay (1). On the west extreme of the bay fine sands are found, at the central sector there is a wide area of silts and in the east, in the port area, there are fine silts and clays. There is a clear gradation from fine sands to clays in a W-E direction. These areas were determined from mud samples taken from the surface layers of the bay bed.

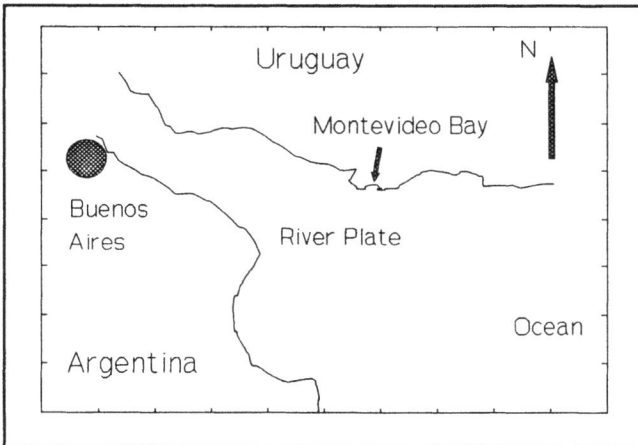

Figure 1. *Location of Montevideo Bay*

Sedimentation velocity of 0.14 mm/s for 50% of the sedimentation was obtained with laboratory tests, but there are no in situ values. The initial suspended concentration was 1.7 g/l, which is a value seldom reach in the bay.

2.3 *Salinity*

The port is located in the variation area of the salt intrusion limit. The saline range 0-33 % is located between Punta Brava and Piriápolis, and the movable salt intrusion limit between Punta del Tigre and Punta Brava, varying its position with minimum and maximum river discharges respectively (2). The salinity longitudinal variation is mainly determined by the river flows, and also affected by tides and winds.

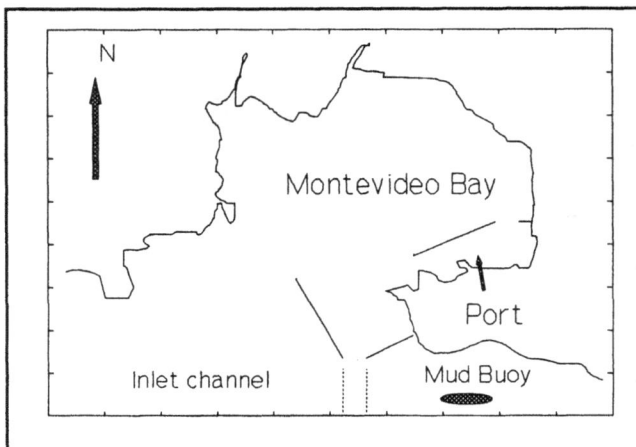

Figure 2. *Montevideo Bay*

According to Silvester's criterium, the River Plate is a well mixed estuary. The volumetric rate between oceanic and fluvial waters which enter the estuary in a tidal cycle is 0.01 (3). Salinity stratification is limited to the deepest areas of the channels, where the presence of a saline wedge is occasionally detected.

2.4 *River discharges and sediments*

The mean fresh water discharge of the River Plate is approximately 22.000 m3/s. The Coriolis force pushes the main fluvial currents to the Uruguayan coast. The solid discharge brought by Uruguay and Paraná rivers is 8:000.000 tons/year, with a suspended sediment concentration varying from 50 to 250 mg/l (4), in the interior area of the River Plate. The River Plate's wash load has appropriate conditions for sedimentation within the movable salt intrusion limit. The maximum turbidity layer which is formed by the meeting of river discharge with marine water is located in this area.

Measurements of suspended sediments made in the bay area during 6 months show that the average concentrations are approximately 60 mg/l (Figure 3)

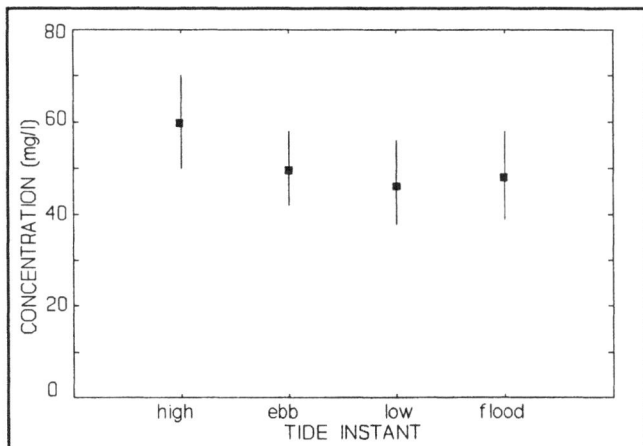

Figure 3. *Average suspended sediment concentration*

2.5 *Tides*

The tidal regime is of microtides; there are astronomical tides on the Uruguayan coast of 0.40 m amplitude, while the meteorological tides reach 1.5 m. The mean velocities caused by the astronomical tides seldom exceed 0.5 m/s.

The meteorological tides are more important than the astronomical tides. The mean weight of the meteorological tide compared to the astronomical tides, taking into account the mean volumes of water displaced for each one in the same period of time, is of 3:1. Figure 4 shows the levels' spectrum of the tide register with the mareograph located in the port of Montevideo. In this figure the importance of the low frequency components, related to meteorological events is clearly shown, in opposition to the importance of the diurnal and semidiurnal frequencies corresponding to the principal astronomical components.

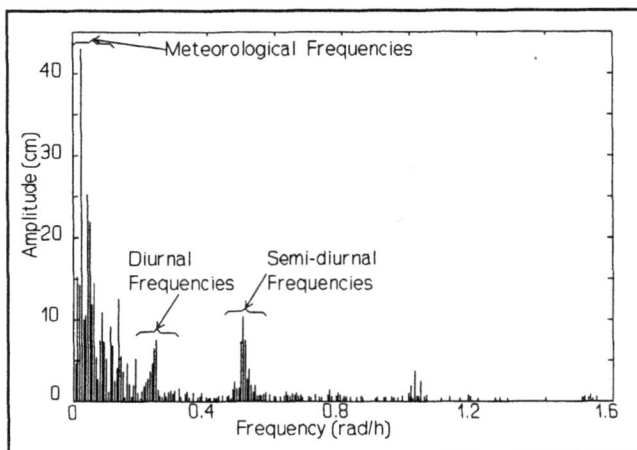

Figure 4. Spectrum of tide levels

2.6 *Winds and waves*

A significant wave climate, of little height by persistent throughout time, is favoured by the great extension of the River Plate. In the port area, waves of a mean height of 40 cm are present a 60% of the year. Storms with winds greater than 80 Km/hr and lasting from 6 to 25 hours have a frequency of 4.2 storms/year. The measurements of currents made in the surroundings of Montevideo during a period of 15 days, show the existence of a prevailing flow direction. Likewise from 28 flow inversion that must be present in that period, 12 do not happen due to the meteorological effects (3).

2.7 *Dredging and discharge*

The annual volume dredged in the docks, outer port and part of the inlet channel is nearly 1.2 10^6 m3/year. These dredged sediments are discharged at present at the area named "Mud Buoy", which is shown in Figure 2.

3. Transport processes and deposition

The classical theory on sediment transport in estuaries consider that in a tidal cycle sediments are eroded from the bed and suspended during ebb and flood, and deposited during high and low waters.

Figure 3 shows the mean values and 95% confidence intervals of the average concentration values for each tidal condition. With this data processing, it is evident that there is no significant difference in suspended sediment concentration during an astronomic tidal cycle.

This same fact can also be seen considering that during a tidal cycle there is no friction velocity u_*, due to the astronomical tide, which would be sufficiently high to produce cyclical bed erosion and resuspension of recently deposited material. The mean friction velocity obtained in the bay is about 5 mm/s, which is substantially smaller than the u_{*d} (critical sedimentation friction velocity) and u_{*e} (of erosion) used by different authors (5,6,7,8,9). They respond to stresses produced by unidirectional currents on little consolidated soils. Also, according to Migniot (10), newly deposited sediments are suspended with u_* approximately 2 cm/s, a value that is greater than those existing in the site too. This evidence shows the little erosive capacity of the astronomical tidal currents present in the area which is studied.

A volume of 3 10^9 m3/year flow crosses the area of the outer harbour and docks. This value was determined with a hydrodynamical numerical model. Using the mean concentration previously referred to (60 mg/l), the volume of sediment that crosses the area is 5.5 10^5 tons/year. Comparing this value with the volume dredged (4 10^5 m3/year) practically all the suspended sediment that reaches the area must settle to explain the dredged volume. Therefore and taking into account the sediment concentration measurements the port siltation is not due only to the settlement of suspended sediment.

In the area of the port of Montevideo there is a fluid mud layer of high sediment concentration, which moves partially or totally disconnected from the rest of the water column. This layer may transport great quantities of sediments, even if moved by modest currents, due to its high sediment concentration. Figure 5 shows an echosounder registration made over the entrance channel of the port, where the evidence of the existence of this fluid mud layer can be seen. The echosounder (DESSO 10) reflections for different frequencies (33 y 210 Kz) responds to rough changes in the sediment concentration profile in the range 15-200 g/l, quoted in literature as characteristic of the presence of a fluid mud layer (11). With echosounder registration the thickness of the layer is obtained, which is 0.10 m outside of the channel and around 2 m inside it.

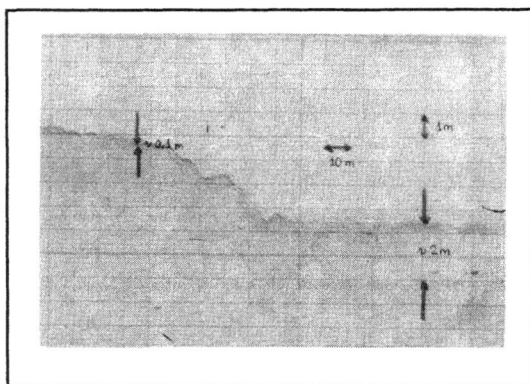

Figure 5. *Echosounder registration across the navigation channel*

As was previously quoted, the meteorological tide in the River Plate has an important weight. During stormy weather, a great sediment resuspension is produced, with a surface concentration around 300 mg/l. These sediments commonly settle down a few hours later after the storm has ceased. So it can be said that the principal roll of these storms is to redistribute the sediment, helping the navigation channels' siltation. In-situ sedimentation measurements made recently with mud traps show that the principal sediment deposition is produced immediately after the end of the storms and that the settlement produced with astronomical tides only is negligible than that produced by the storms.

In the last 40 years the dredged material from the bay and entrance channel has been discharged in the area named "mud buoy". Bathymetrical registrations made in that area show that the loss in depth is only a few centimeters after all that period of time. This fact, and the existence of shallow waters with the slope towards the inlet channel, suggests that part of the dredged and discharged material in that area returns to the channel and port.

4. Conclusions

Siltation in the bay and port of Montevideo is produced by the combined effects of currents, waves and winds and not solely by tidal currents as was thought until now.

Echosounder registrations (Figure 5) and laboratory measurements show the existence of a fluid mud layer that moves over the bed without mixing with the water column.

Its high sediment concentration is capable of producing great rates of siltation, even with weak currents.

The action of the storms produce a rapid sediment suspension and a later settlement immediately after the end of the storm which produces a sediment redistribution in the area, and as a consequence the port and channels siltation.

ACKNOWLEDGEMENTS

We thank the "Administración Nacional de Puertos" for its assistance, which made this paper possible. We are equally grateful to Dr. Rafael Guarga and A/P Elias Kaplan from IMFIA for their kind help.

REFERENCES

[1] Ayoup Zouain,R., (1981), Caracteristicas del Comportamiento Sedimentologico de la Bahia de Montevideo, *Publicacion 81-04*, SOHMA. Uruguay.

[2]   Nagy,G.,; Lopez Laborde,J. y Anastasia,L.,
        (1987), Caracterizacion de Ambientes del Río
        de la Plata Exterior (salinidad y turbiedad).
        Investigaciones Oceanológicas, **vol 1**.

[3]   Vinzon,S., (1991), Modelacao de Transporte de
        Substancias passivas em Corpos de Agua Rasos,
        Tesis de Maestria, Universidad de Rio de
        Janeiro, Brasil.

[4]   Ottmann,F. y Urien,C.M., (1966), Sur Quelques
        Problemes Sedimentologiques dans le Río de la
        Plata, *Revue de Geographie Physique et de
        Geologie Dynamique*, **vol. VIII, Fasc. 3**, pag.
        209-224.

[5]   Mehta,A.J., (1986), Characterization of
        Cohesive Sediment Properties and Transport
        Processes in Estuaries, *Estuarine Cohesive
        Sediment Dynamics*, pag. 290-325.

[6]   Harrison,A.J. & Owen,M.W., (1971), Siltation
        of Fine Sediments in Estuaries, *IAHR*, pp
        D1-1/-8.

[7]   Krone,R.B., (1962), Flume Studies of the
        Transport of Sediment in Estuarial Shoaling
        Processes, Final Report, *University of
        California, Berkeley*.

[8]   Nicholson,J. and O'Connor,B., (1986), Cohesive
        Sediment Transport Model, *J. of Hydraulic
        Engineeing*, ASCE, **Vol. 112**, No 7, pp 621.

[9]   Cole,P. and Miles,G., (1983), Two-Dimensional
        Model of Mud Transport. *J. of Hydraulic
        Engineering*, **Vol. 109**, No 1, pp. 11.

[10]   Migniot,C., (1981), Dynamique Sedimentaire
        Estuarienne: Materiaux Cohesifs et Non
        Cohesifs, *Oceanis*, **vol. 6**, fasc. 4, pp.
        359-432.

[11]   Ross,M.; Chung,L. and Mehta,A.J, (1987), On
        the Definition of Fluid Mud, *Proceeding
        Hydraulic Engineering Conference*, pp. 231-236.

25 A morphodynamic modelling system

Z. B. Wang, B. Karssen, A. Roelfzema and
J. C. Winterwerp

ABSTRACT

A morphodynamic modelling system has been developed by DELFT HYDRAULICS. In this paper the background of the system and two examples of its application are described.

1 Introduction

Morphological developments of estuarine and coastal areas change the geometry and bathymetry and thereby influence the flow pattern and all other physical processes in the region. Knowledge about the morphological development is important for hydraulic engineering as well as for studying other (e.g. ecological) processes. Therefore mathematical models for predicting the morphological changes due to natural influences and human interference in estuarine and coastal regions are urgently needed.

There are various mathematical models for morphological development in coastal and estuarine areas (Gerritsen, 1990, Steijn, 1991, De Vriend, 1991). These models can be classified into two categories, viz. models based on empirical relations and dynamic models. The equations in the first category of models are based on the observations in nature and they are purely empirical (O'Connor et al, 1990, Di Silvio, 1989). On the other hand, the dynamic models are based on the mathematical description of the motion of water and sediment (Holly and Rahuel, 1990, Teisson, 1991).

Morphological changes are the result of interactions between the water movement and the sediment on the bed. A straightforward reasoning leads to the conclusion that a mathematical model for the morphological development should be based on the description of the water- and sediment-movement. Therefore dynamic models have a better theoretical basis than empirical models. However, empirical models have already proven their value for solving practical problems, whereas experience with the application of dynamic models is still very limited.

Despite the considerable efforts by many scientists in the whole world, morphodynamic modelling is still in its infancy (Tetzlaff and Harbaugh, 1989). This may be explained by several reasons. First, the computer capacity had been too limited for long-term morphodynamic simulations which require large computational times. It is only since recently that such computations have become possible. Second, the knowledge on the morphological processes, as well as the knowledge on the morphodynamic models is still too limited. Although the basic (sub-) processes, like water flow and sediment transport, are quite well understood, much less is known about the long term morphological development due to the interaction of these sub-processes.

A morphodynamic modelling system is presently under development at DELFT HYDRAULICS. Although it also can deal with steady flow conditions (non-tidal river problems), the system is mainly meant for tidal flow conditions such as in estuaries and tidal inlets. The system can deal with cohesive and non-cohesive sediments. In this paper a brief description of the system in the present version is given and two applications are described, followed by a discussion on the required research work to make this kind of systems more predictive.

2. The modelling system

2.1 General outline of the system

The modelling system consists of the following basic modules:

- A quasi-3D hydrodynamic module, which is based upon TRISULA (Stelling and Leendertse, 1991) solving the complete shallow water equations. A secondary flow model is applied to account for the vertical structure of the flow.

- A quasi-3D sediment transport module describing bed load and suspended load transport due to current and waves. At this moment, sediment transport due to waves is incorporated by modifying the bed shear stress.

- A sediment balance module yielding the sedimentation and erosion rate and the new bed level.

The modules communicate through a 'quasi-steady' morphodynamic time-stepping mechanism as indicated in fig.1. During the flow computation the bed level is assumed to remain invariant and during the computation of the bed level (with a time step of a number of tidal periods) the flow and sediment transport are assumed invariant to the bed level changes. This approach was first described by De Vries (1959) for morphological computations in rivers. More complicated procedures for morphological computations in tidal regions have been studied by Haugel(1978) and Latteux (1987). However, the experience in the application of these procedures is still very limited. Therefore this system is restricted to the straightforward procedure shown in fig.1, which implies the following basic assumptions:

- the time scale of the bed level change is much larger than the time scales of the flow and of the sediment transport,

- the sediment concentration is so small that the presence of sediment in the water has a negligible influence on the flow.

The modules in the system are linked together on program level. However, the system is organised in such a way that various modules can be run after each other, either on-line or off-line. The off-line option is particular useful in the case that more morphological runs are needed based on the same flow conditions for instance in the case of model calibration or for sensitivity analyses.

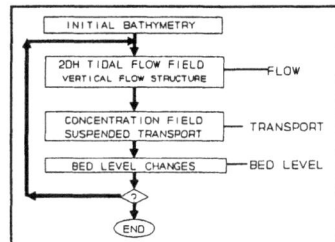

Fig.1 structure of the system

2.2 The Current Module

The flow computation is based on the so-called quasi-3D approach. First the depth-averaged flow field (the main flow) is computed by solving the shallow water equations with TRISULA, a program package of DELFT HYDRAULICS.

The velocity profile of the main flow is assumed to be logarithmic. In the cross-stream direction the secondary flow model of Kalkwijk and Booij (1986, also see De Vriend, 1981) is applied.

The secondary flow model takes into account the influence of the curvature of the main flow stream-line and of the geostrophic acceleration (assuming Ekman depth >> water depth). The vertical profile of the secondary velocity component is schematised into a straight line (fig.2).

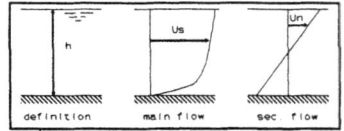

fig.2   flow   velocity profiles

2.4   The sediment transport module

2.4.1 Sand transport   In case of sand, the sediment transport is divided into bed load and suspended load. The bed load transport is calculated with a transport formula, e.g. van Rijn's (1984a). Further, the down-slope gravitational effect on the magnitude of the transport is taken into account via

$$S_b = S_{be} \left(1 - \alpha_b \frac{\partial z_b}{\partial s}\right) \tag{1}$$

Herein S_b =   effective bed load transport rate,
          S_{be} =   bed load transport rate according to the transport formula,
          α_b =   constant coefficient,
          z_b =   bed level,
          s =   horizontal coordinate in the main flow direction.

The down-slope effect on the direction of the bed load transport is included via

$$tg(\phi) = tg(\delta) - \gamma \frac{\partial z_b}{\partial n} \tag{2}$$

Herein δ =   angle between bed shear stress and s-direction,
       ϕ =   angle between the bed load transport and s-direction,
       n =   horizontal coordinate in the lateral direction,
       γ =   coefficient depending on the Shields parameter.

The suspended load transport is modelled with a quasi-3D approach based on an asymptotic solution of the 3D convection-diffusion equation for suspended sediment. This approach was first presented by Galappatti (1983, see also Galappatti and Vreugdenhil, 1985) for the width-averaged situation and later extended by Wang (1989) to the 3D case.

First the depth averaged sediment concentration C is solved from the following equation

$$\frac{\partial C}{\partial t} + U\frac{\partial C}{\partial x} + V\frac{\partial C}{\partial x} - \frac{\partial}{\partial x}(D_x\frac{\partial C}{\partial x}) + - \frac{\partial}{\partial y}(D_y\frac{\partial C}{\partial y}) = \frac{C_e - C}{T_a} \qquad (3)$$

in which t = time
      U, V = depth-averaged flow velocity in x- and y-direction
      x, y = horizontal Cartesian coordinates
      D_x, D_y = horizontal diffusion coefficients

In the right-hand part of equation (3) C_e is the equilibrium depth-averaged concentration which is calculated from a transport formula for suspended load, e.g. Van Rijn's (1984b). T_a is the adaptation time derived from the quasi-3D model, which expresses the relaxation effect of the suspended sediment concentration.

$$T_a = \eta\frac{h}{w_s} \qquad (4)$$

In which w_s is the fall velocity and η is a coefficient depending on w_s/u_*, where u_* is the bed shear stress velocity. Similar formulation of the suspended sediment transport model is also used by Lin et al (1981) and Falconer and Owens (1990).

The quasi-3D model also provides information on the vertical structure of the sediment concentration, which depends on the parameter w_s/u_* and on the defect-concentration C_e-C. The resulting 3D concentration field $c(t,x,y,z)$ can be used to determine the transport rates, using

$$S_x = \int_{z_b}^{z_b+h} (uc - D\frac{\partial c}{\partial x})\,dz \qquad (5)$$

$$S_y = \int_{z_b}^{z_b+h} (vc - D\frac{\partial c}{\partial y})\,dz \qquad (6)$$

Where S_x, S_y = suspended transport rate in x- and y-direction,
     u, v = flow velocity in x- and y-direction.

Since the flow velocity profile as well as the sediment concentration profile are constructed from parameterized profile functions, which can be determined a priori, equations (5) and (6) can be parameterized.

2.4.2 *Mud transport*   In case of mud, only the suspended load transport is taken into account. It is modelled in a similar way to the suspended sand transport. However, the source term on the right-hand side of the concentration equation (3) is then replaced by the sedimentation-erosion formulation of Partheniades and Krone (see f.i. Winterwerp, 1989).

$$\frac{\partial hC}{\partial t} + \frac{\partial UhC}{\partial x} + \frac{\partial VhC}{\partial x} - \frac{\partial}{\partial x}(hD_x\frac{\partial C}{\partial x}) + - \frac{\partial}{\partial y}(hD_y\frac{\partial C}{\partial y}) = E - D \quad (7)$$

with

$$E = M\left(\frac{\tau_b}{\tau_e} - 1\right) \qquad for \ \ \tau_b > \tau_e \qquad (8)$$

and

$$D = w_sC\left(1 - \frac{\tau_b}{\tau_d}\right) \qquad for \ \ \tau_b < \tau_d \qquad (9)$$

In these equations

E = erosion rate,
D = deposition rate,
M = erosion parameter,
τ_b = bed shear stress,
τ_e = critical shear stress for erosion,
τ_d = critical shear stress for deposition.

2.5   *The Bed Level Module*

The bed level is calculated from the mass balance equation for sediment:

$$\frac{\Delta Z_b}{NT} + \frac{\partial T_x}{\partial x} + \frac{\partial T_y}{\partial y} = 0 \qquad (10)$$

where ΔZ_b = bed level change in N tidal period,
N = an integer number,
T = tidal period,
T_x, T_y = residual sediment transport computed over one tidal period.

In case of mud, the bed level change may also be calculated by integrating the source term E-D in equation (7).

3. Applications

3.1. General

The system has been applied in a number of projects. The short-term simulations have already proven their values in practical applications. Up to now the long-term morphological simulations are restricted to research projects. This is due to two reasons. First, the computational costs are still prohibitive for large scale practical applications. Second, the morphological processes in estuarine systems are still not fully understood.

In the following two examples of the application of the system are given, one concerns short term simulations and the other concerns long-term simulations.

3.2. The Cochin Harbour problem

Cochin Harbour is located along the South-west coast of India. As an extension of the existing harbour, the construction of the Vallarpadam Container Transshipment Terminal is planned, involving the deepening and widening of the navigation channels. The modelling system was applied to predict the amount of siltation after the extension of the harbour in order to study the feasibility of the Terminal.

The bed in the area consists of fine grained sediment. No long-term morphological simulation is needed, because the channels will be dredged regularly to allow navigation. Hence, the bathymetry will not change much.

The sediment concentration equation (7) is linear. Its solution is linearly dependent on the concentrations at the model boundary and the parameter M. Therefore the solution may be split up into three partial solutions, yielding the contributions of sediment originating from the sea, originating from the river, and originating from erosion in the model area. These sources are designated as sea born sediments, river born sediments and local born sediments. The advantage of this approach is that the relevant processes are split up, which facilitates the analysis of the computational results, and thus the calibration of the results.

An example of the results is given in fig.3 showing the siltation rates due to river born sediment in the area of interest calculated for the situation after the extension. The right hand side of the figure shows a part of the Cochin backwater and the left part, beyond the gut, the access channel and a stretch of the surrounding sea at each side.

307

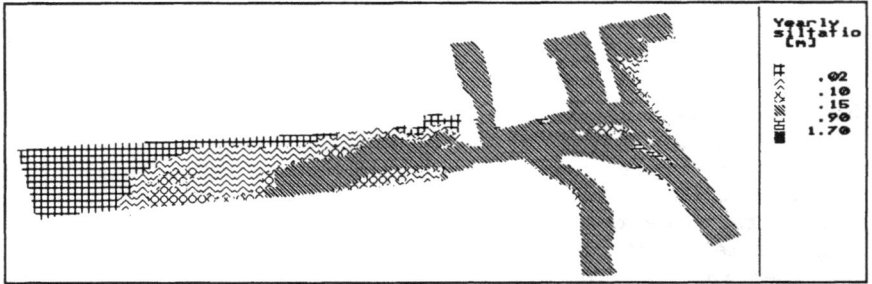

fig.3   The computed siltation rate due to river born sediment

3.3   "Het Friesche Zeegat" Model

"Het Friesche Zeegat" is a tidal inlet system in the Dutch Wadden Sea with an area of about 450 km². Since 1969, it has undergone significant morphological changes due to the closure of the Lauwerszee. As a consequence the in- and out-going tidal volume has been reduced by about 34%.

The morphological evolution of this inlet system has been the subject of an integrated research project, the COASTAL GENESIS project and the WASP-project, part of the EC-MAST programme. The morphodynamic modelling system described here is used to construct a pilot model for this inlet system. The study is mainly aimed to obtain a better understanding of the morphological processes in the tidal inlet rather than to make specific predictions of the morphological development.

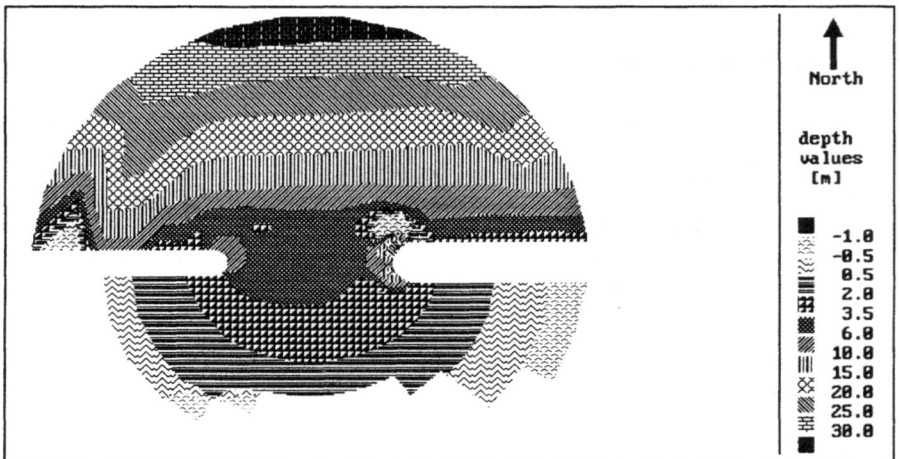

Fig.4   simulation without central bank

Though the study is presently still in progress, useful results have already been produced (Wang, 1991a, 1991b, Wang et al 1991). Despite many shortcomings, e.g. that the influence of waves is not included, the model appears to be a useful research tool. A lot of insight into the morphological processes in the tidal inlet system has been obtained from the simulations. Especially the long-term simulations with the model also give important information about the behaviour of the modelling system. Some results are given in fig.4 and fig.5 showing the calculated bathymetry at the end of two simulations. Both simulations started with a schematised initial bathymetry. The simulations are meant to investigate the ability of the model to develop a shoal-channel system starting from a "plane bed". The first simulation became unstable after 16 time steps as can be seen from the fluctuations in the left part of fig.4. In the second simulation, a bank in the gorge is added to the initial bathymetry. All other conditions remain unchanged. Such a central bank, "Engelsmanplaat", is present in the real situation. In contrast to the first simulation, this simulation remains stable. The results after 84 time steps (fig.5) shows the development of the outer delta and a shoal-channel system.

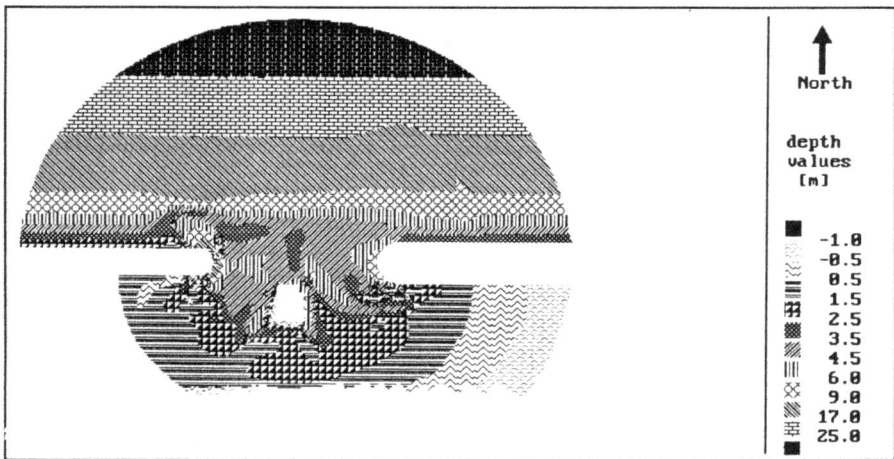

Fig.5   Simulation with central bank

The striking effect of a relatively minor change in the initial bathymetry suggests that the computation is quite sensitive to the planform. Up till now, no definite physical ex-planation could be found for this behaviour. Therefore more research is needed to make the modelling system a useful tool for practical applications.

4. Conclusions

A morphodynamic modelling system is under development at DELFT HYDRAULICS. It is aimed at modelling two-dimensional morphological changes under tidal flow situations. The system consists of a water flow module, a sediment transport module and a bed level change module. The quasi-3D approach is applied for the flow module and the sediment transport module. The bed level change module is based on the mass-balance for sediment. The system can deal with sand and mud.

Two examples of applications are discussed. One concerns short-term simulation with mud in Cochin Harbour (India) and the other concerns long-term morphological simulation for a tidal inlet in the Wadden Sea.

The modelling system appears to be a useful tool for short-term simulations in practical cases. Long-term simulations with the modelling system appear to be very helpful to study morphodynamic processes in tidal flow regions.

The modelling system at present state still needs to be improved. Especially the effect of wind, short waves and density current should be included. However, first a better understanding of the behaviour of the modelling system, especially concerning long-term simulations, is needed. For this purpose theoretical analyses as well as numerical experiments with the modelling system are required. At the same time more reliable field data have to be collected and analyzed in order to obtain better understanding of the morphological processes in nature and to test the modelling system.

Acknowledgement

The writers wish to thank Frederic R. Harris Limited, The Hague, The Netherlands, for giving the permission to publish the results from the Cochin project. We also would like to acknowledge the EC and Rijkswaterstaat for funding the research done with "Het Friesche Zeegat" model.

References

[1] Di Silvio, G. (1989), Modelling of the morphological evolution of tidal lagoons and their equilibrium configurations, XXII Congress of the IHAR, Ottwa, Canada.
[2] Eysink, W.D., (1991), Simple morphologic relationships for estuaries and tidal channels, handy tools for engineering, Proc. COPEDEC'91, Mombasa (in press).

310

[3] Falconer, R.A. and Owens, P.H. (1990), Numerical modelling of suspended sediment fluxes in estuarine waters, Estuarine Coastal and Shelf Science, 31, 745-762.

[4] Galappatti, R. (1983), A depth-integrated model for suspended transport, Delft University of Technology, Dept. of Civil Eng., Report nr. 83-7.

[5] Gallappati, R. and Vreugdenhil, C.B. (1985), A depth-integrated model for suspended sediment transport, J. of Hydr. Res., vol. 23, nr. 4.

[6] Gerritsen, F. (1990), Morphological stability of inlets and channels of the Western Wadden sea, Rijkswaterstaat, report nr. GWAO-90-019.

[7] Hauguel, A. (1978), Utilization des modeles mathematiques pour L' etude du transport solide sous L' action des courants de Maree, Report E42/78.41, EDF, Direction des Etudes at Recherches.

[8] Holly Jr., F.M. and Rahuel, J-L. (1990), New numerical/physical framework for mobile-bed modelling, J. of Hydr. Res., Vol. 28, No. 4.

[9] Kalkwijk, J.P.Th. and Booij, R. (1986), Adaptation of secondary flow in nearly horizontal flow, J. of Hydr. Res., Vol. 24, no. 1.

[10] Latteux, B., (1987), Transport modelling of parparticulate matter-methodology of long-term simulation of bed evolutions, Lab. Nat. d'Hydraulic, Chatou, France, Rept. HE-42/87.25 (in French).

[11] Lin, Pin-nam, Huang, J. and Li, X. (1981), Unsteady transport of suspended load at small concentrations, J. of Hydr. Eng., Vol. 109, No. 1.

[12] O'Connor, B., Nicholson, J. and Rayner, R. (1990), Estuary as a function of tidal range, Proc. 22nd ICCE, Delft.

[13] Rijn, L.C. van (1984a), sediment transport, part I: Bed load Transport, J. of Hydr. Eng., Vol. 110, nr. 10.

[14] Rijn, L.C. van (1984b), sediment transport, part II: Suspended load Transport, J. of Hydr. Eng., Vol. 110, nr. 11.

[15] Steijn, R.C. (1991), Some considerations on tidal inlets, Report no. H840, DELFT HYDRAULICS

[16] Steijn, R.C., Louters, T., van der Spek, A.J.F and De Vriend, H.J., (1989), Numerical model hindcast of the ebb-tidal delta evolution in front of the Deltaworks, In: Falconer R.A. et al, "Hydraulic and Evironmental Modelling of Coastal, Estuarine and River Waters", Gower Technical, Aldershot.

[17] Stelling, G.S. and Leendertse, J.J. (1991), Modelling of convective processes in two- and three-dimensional models, Conf. on Estuarine and Coastal Modelling, to be held in Florida, Nov. 1991.

[18] Struiksma, N., Olesen, K.W., Flokstra, C. and Vriend, H.J. de (1985) Bed deformation in curved alluvial chan-

311

nels, J. of Hydr. Res., Vol. 23, No. 1.

[19]   Teison, C. (1991), Cohesive suspended sediment transport: feasiblity and limitations of numerical modelling, J. of Hydr. Res. Vol. 29, Nr. 6.

[20]   Tetzlaff, D.M. and Harbaugh, J.W., (1989), Simulating Clastic Sedimentation, Van Nostrand Reinhold, New York.

[21]   Vriend, H.J. de (1981), Steady flow in shallow channel bends, Doctoral thesis, Delft University of Technology.

[22]   Vriend, H.J. de (1985), Flow formulation in mathematical models for 2DH morphological changes, Report no. R1747-5, DELFT HYDRAULICS.

[23]   Vriend, H.J. de (1991), Mathematical modelling and large scale coastal behaviour, Part I: Physical processes, Part 2: Predictive models, J. of Hydr. Res. Vol. 29, Nr. 6.

[24]   Vriend, H.J. de, Louters, T., Berben, F. and Steijn, R.C., (1989), Hybrid prediction of a sandy shoal in a mesotidal estuary, In: Falconer R.A. et al, "Hydraulic and Evironmental Modelling of Coastal, Estuarine and River Waters", Gower Technical, Aldershot.

[25]   Vries, M. de (1959), Transients in bed-load transport in open channels (basic considerations), Report no. R3, DELFT HYDRAULICS.

[26]   Wang, Z.B. (1989), Mathematical modelling of morphological processes in estuaries, doctoral thesis, Delft University of Technology.

[27]   Wang, Z.B. (1991a), A morphodynamic model for a tidal inlet, Report H840, DELFT HYDRAULICS.

[28]   Wang, Z.B. (1991b), Morphodynamic modelling for a tidal inlet in the Wadden Sea, Report H840, DELFT HYDRAULICS.

[29]   Wang, Z.B., Vriend, H.J. de and Louters, T. (1991), A morphodynamic model for a tidal inlet, Proceedings of the second international conference on computer modelling in ocean engineering, Barcelona, 1991.

[30]   Winterwerp, J.C. (1989), Flow induced erosion of cohesive beds, A literature survey, DELFT HYDRAULICS, Cohesive sediments report 25.

26 A three-dimensional model of suspended sediment transport based on the advection–diffusion equation

J. N. Aldridge

Abstract

We describe here a numerical model of the transport of suspension sediment. The starting point is the advection–diffusion equation which we solve using a finite-volume formulation and operator splitting method in which the vertical and horizontal transport are considered separately. Some results from example calculations are shown.

1 Introduction

Most studies involving the modelling of suspended sediment transport using the advection—diffusion equation have solved a two—dimensional form in which one of the spatial directions has been averaged out. From the point of view of obtaining qualitative information, and for answering questions typically required for engineering applications, this is generally a perfectly adequate approach. From a scientific point of view it is of some interest to see if a fully three—dimensional model can give more insight into the physical processes involved, even if the computer resources required and the shear quantity of information provided may preclude the use of such models on a regular basis.

A number of three—dimensional models have been described in the literature e.g. [4], [8]. Of these, the former solves an advection–diffusion equation on an equally spaced vertical grid using operator splitting, the later uses a Lagrangian/Eulerian formulation on a sigma transformed grid. As in [4] we use a Eulerian operator splitting technique (although based on finite—volume rather than finite—difference considerations) and in common with [8] we employ a transformation to the vertical coordinate to accommodate the effects of an irregular bottom topography. We also employ a further transformation to compress the grid in the near–bed region to enable the high concentration gradients in the vicinity of the bed to be resolved.

An alternative to the advection-diffusion equation for modelling particle dispersion is to employ particle techniques, whereby a large number of particles are tracked as they are advected by the flow field. Turbulent diffusion is simulated by adding a random component to each particles velocity. Although this technique has many advantages for simulating dispersion from point sources, it seems likely that the number of particles required for a 3D calculation where the particle source covers a wide area (as would generally be the case for sediment pickup) would entail an inordinate number of particles. A further complication arises when diffusion with a non-constant eddy diffusivity is required, such as wall bounded flows where the eddy diffusivity increases linearly with distance from the boundary. Simply making the random step size dependent on the wall distance does not seem to work; see the discussion in [3, p. 549]. By contrast, the use of a linearly increasing or parabolic eddy diffusivity in the advection–diffusion equation for vertical transport of sediment is well accepted and gives acceptable agreement with experimental measurements [6], [9]. Additional processes, such as the effects of stratification caused by the suspended sediment, can also be incorporated readily in the advection–diffusion formulation; in this case by making the diffusion coefficient a function of the Richardson number [10].

2 Model description

Velocity, surface elevation, and bed stress fields are taken from an existing three—dimensional hydrodynamic model [2]. Alternatively, the sediment transport code can take the surface elevations and bed stresses only and calculate its own 3D velocity field. For consistency with the hydrodynamic model, the sediment transport equations are solved in spherical polar coordinates, although for most of the applications envisaged differences between a polar and a Cartesian grid will be small. In the approximate form that we implement the polar coordinate transformation the x step (longitudinal direction) becomes dependent on the latitude (the 'y' direction).

2.1 *Transport equations*

If c and $\boldsymbol{u}$ are the particle concentration and velocity, averaged over a given volume, it is straight–forward [1] to show they satisfy an advection equation

$$\frac{\partial c}{\partial t} + \nabla.(c\boldsymbol{u}) = 0, \tag{1}$$

The particle velocity is assumed to be

$$\boldsymbol{u} = \boldsymbol{v} + \boldsymbol{w}_0, \tag{2}$$

where $\boldsymbol{v}$ is the local fluid velocity and $\boldsymbol{w}_0 = (0, 0, -w_0)$ is the particle fall velocity. Relation (2) can be derived on a more or less rigorous basis on neglect of particle inertia [1].

Averaging (1) gives

$$\frac{\partial \bar{c}}{\partial t} + \nabla.(\bar{c}\bar{v} + < c'v' >) = 0. \tag{3}$$

where the correlation $< c'v' >$ is modelled with a gradient diffusion law

$$< c'v' >= -\boldsymbol{K}.\nabla \bar{c}, \tag{4}$$

314

with K, a second order tensor which is assumed to be diagonal

$$K = \begin{pmatrix} \kappa_h & 0 & 0 \\ 0 & \kappa_h & 0 \\ 0 & 0 & \kappa_v \end{pmatrix}.$$

In all subsequent developments we assume the vertical eddy diffusion coefficient $\kappa_h = 0$, so that the horizontal transport is purely advective. The vertical eddy diffusivity κ_v is assumed to be parabolic in form and given by $\kappa_v = v_* l(1 - l/H)$ where v_* is the friction velocity (calculated from the bed stress supplied by the hydrodynamic model), $l = \kappa z_w$, where κ is the von–Karman constant, z_w measures distance from the sea bed, and H is the total water depth.

Anticipating the separate treatment of horizontal and vertical processes, we write the modelled form of equation (3) as

$$\frac{\partial \bar{c}}{\partial t} + \nabla_h.(\bar{c}\bar{v}_h) + \frac{\partial}{\partial z}(\bar{w} - w_0)\bar{c} - \frac{\partial}{\partial z}(\kappa_v \frac{\partial \bar{c}}{\partial z}) = 0, \tag{5}$$

where ∇_h is the horizontal gradient operator, and $\bar{v}_h$, $\bar{w}$ are the horizontal and vertical components of velocity.

To deal with the irregular domain encountered in modelling representations of real bathymetry, a transformation of the vertical coordinate is used. This is a combination of the well-known 'sigma transformation' and a logarithmic compression in the near bed region to resolve the large concentration gradients appearing there. Let

$$z(\boldsymbol{x}, \zeta, t) = \sigma(\zeta, \boldsymbol{x})[\eta(\boldsymbol{x}, t) + h(\boldsymbol{x})] - h(\boldsymbol{x}) \tag{6}$$

where $\eta(\boldsymbol{x}, t)$ is the surface elevation above mean sea level (MSL), obtained from the hydrodynamic model, and $h(\boldsymbol{x}) > 0$ is the depth of the sea bed from MSL. Now compress σ by letting

$$\sigma(\zeta, \boldsymbol{x}) = a(\boldsymbol{x}) \left[\left(1 + \frac{1}{a(\boldsymbol{x})}\right)^\zeta - 1 \right]. \tag{7}$$

where the new independent variable is ζ and where $a(\boldsymbol{x})$ determines the severity of the compression and is a function of the water depth. In terms of ζ the equations can now be solved on a regular grid.

We have quite straight–forwardly

$$\frac{\partial f}{\partial z} = \frac{1}{z_\zeta} \frac{\partial f}{\partial \zeta}, \tag{8}$$

where the subscript denotes a derivative taken with respect to the subscript variable. The time and horizontal derivatives need to be taken holding ζ rather than z constant. Thus we have the following, where the vertical bar indicates that the associated variable is being held constant,

$$\begin{aligned} \frac{\partial f}{\partial t}\bigg|_z &= \frac{\partial f}{\partial t}\bigg|_\zeta - \frac{z_t}{z_\zeta} \frac{\partial f}{\partial \zeta} \\ &= \frac{\partial f}{\partial t}\bigg|_\zeta - \frac{1}{z_\zeta} \left[\frac{\partial}{\partial \zeta}(z_t f) - f \frac{\partial z_t}{\partial \zeta} \right], \end{aligned} \tag{9}$$

315

$$\nabla_h \cdot \boldsymbol{f}|_z = \nabla_h \cdot \boldsymbol{f}|_\zeta - \frac{1}{z_\zeta} \frac{\partial \boldsymbol{f}}{\partial \zeta} \cdot \nabla_h z$$

$$= \nabla_h \cdot \boldsymbol{f}|_\zeta - \frac{1}{z_\zeta} \left[\frac{\partial}{\partial \zeta} (\boldsymbol{f} \cdot \nabla_h z) - \boldsymbol{f} \cdot \frac{\partial}{\partial \zeta} (\nabla_h z) \right]$$

$$= \nabla_h \cdot \boldsymbol{f}|_\zeta - \frac{1}{z_\zeta} \left[\frac{\partial}{\partial \zeta} (\boldsymbol{f} \cdot \nabla_h z) - \boldsymbol{f} \cdot \nabla_h z_\zeta \right]. \tag{10}$$

Using (9) and (10) we can derive the conservation form of (5) in terms of the transformed coordinates as

$$\frac{\partial}{\partial t} (\bar{c} z_\zeta) + \nabla_h \cdot (\bar{c} \bar{\boldsymbol{v}}_h z_\zeta) + \frac{\partial}{\partial \zeta} (\bar{w}^* \bar{c}) - \frac{\partial}{\partial \zeta} \left(\kappa_v^* \frac{\partial \bar{c}}{\partial \zeta} \right) = 0, \tag{11}$$

where

$$\bar{w}^* = \bar{w} - w_0 - z_t - \bar{\boldsymbol{v}}_h \cdot \nabla_h z, \tag{12}$$

$$\kappa_v^* = \kappa_v / z_\zeta. \tag{13}$$

The vertical velocity component $\bar{w}$ is calculated diagnostically from the horizontal components via the continuity equation. Application of (10) to the continuity equation yields

$$\nabla_h \cdot (\bar{\boldsymbol{v}}_h z_\zeta) + \frac{\partial}{\partial \zeta} (\bar{w} - \bar{\boldsymbol{v}}_h \cdot \nabla_h z) = 0,$$

from which we obtain, since $\bar{w} = \bar{\boldsymbol{v}}_h \cdot \nabla_h z$ at $z(\boldsymbol{x}, 0, t) = -h(\boldsymbol{x})$,

$$\bar{w}(\boldsymbol{x}, \zeta) = \bar{\boldsymbol{v}}_h \cdot \nabla_h z - \int_0^\zeta \nabla_h \cdot (z_\zeta \bar{\boldsymbol{v}}_h) \, d\zeta. \tag{14}$$

2.2 *Numerical scheme*

The finite–volume method starts from a set of *control volumes*, with the values of the approximated function associated with the centre of the volume. The mesh is defined in terms of the position of the control volume *faces*. By contrast the starting point for a finite–difference scheme is the set of points that define the positions where the function is to be approximated and which then define the grid.

The finite–volume formulation for numerical solution of partial differential equations is particularly well suited to solving transport equations of the type required here, since it is based on approximating directly the integral balance relation in a control volume Ω

$$\frac{\partial}{\partial t} \int_\Omega c \, dV = -\int_{\partial \Omega} c \boldsymbol{u} \cdot \boldsymbol{n} \, ds, \tag{15}$$

from which the continuum equations were derived in the first place. In principle it is unnecessary to consider the continuum equations at all but to use the physical reasoning embodied in (15) directly even with the distorted mesh caused by the transformation from z to ζ. That is, the numerical equations are set up based on physical intuition about the flux of material entering the control volume. It is then essential to have the continuum version (11) to ensure that this has been done correctly and the resulting equations are consistent with the correct continuum equation in the limit of the finite—volume mesh becoming infinitely fine.

A more formal approach is to start with the continuum equations, integrate them over a control volume and use the divergence theorem to obtain the flux on the surface of the volume. We will not use this approach here.

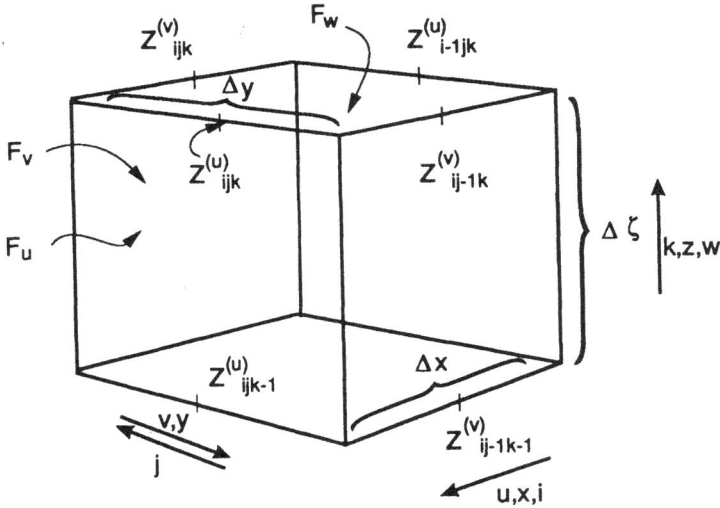

Figure 1: 3D control volume.

A typical control volume is shown in figure 1. Two arrays $z_{ijk}^{(u)}$ and $z_{ijk}^{(v)}$ are defined at the positions shown, enabling the quantities z_ζ, z_x, z_y required in (11) to be calculated at the centre of the control volume face, thus for example

$$(z_\zeta)_{F_u} = (z_{ijk}^{(u)} - z_{ijk-1}^{(u)})/\Delta\zeta,$$
$$(z_y)_{F_w} = (z_{ijk}^{(v)} - z_{ij+1k}^{(v)})/\Delta y.$$

The concentration $\bar{c}$ and velocities $\bar{u}, \bar{v}, \bar{w}$ are defined at the centre but will be required at the cell faces.

For clarity we consider the two-dimensional control volume in figure 2 in formulating the finite—volume equations. The vertical diffusive flux is given by

$$d = \kappa_v^* \frac{\partial \bar{c}}{\partial \zeta} \equiv \kappa_v^* \bar{c}_\zeta \tag{16}$$

and is required at the top and bottom cell faces. We neglect for the moment the time dependence of the mesh arising from changes in the sea surface elevation and approximate (15) based on the control volume in figure 2

$$\frac{c_{ik}^{n+1} - c_{ik}^n}{\Delta t} V = [(\bar{u}\bar{c}\Delta z)_{\mathrm{R}} - (\bar{u}\bar{c}\Delta z)_{\mathrm{L}}] - [(\bar{u}\bar{c}\delta z)_{\mathrm{T}} - (\bar{u}\bar{c}\delta z)_{\mathrm{B}}]$$
$$+ [(\bar{w}\bar{c}\Delta x)_{\mathrm{T}} - (\bar{w}\bar{c}\Delta x)_{\mathrm{B}}] + [d_{\mathrm{T}} - d_{\mathrm{B}}]\Delta x. \tag{17}$$

Here the volume V is given by $V = \frac{1}{2}[(\Delta z)_{\mathrm{R}} + (\Delta z)_{\mathrm{L}}]\Delta x$ and $\Delta z = z_\zeta \Delta\zeta$, $\delta x = z_x \Delta x$. Note that the 2nd term on the right hand side of (17) is due to the top and bottom faces not being parallel to the horizontal flow direction which in turn is due to the transformation applied to the vertical coordinate.

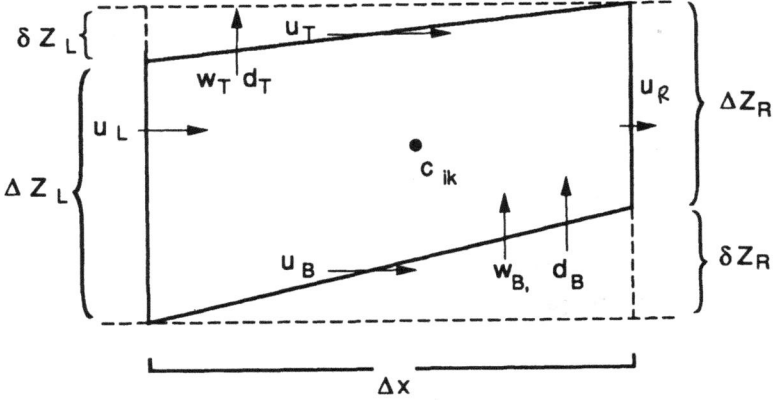

Figure 2: Simplified 2D control volume.

We write (17), with the above relations for V, Δz, δz, as

$$
\frac{z_\zeta c_{ik}^{n+1} - z_\zeta c_{ik}^n}{\Delta t} = \frac{(\bar{u}\bar{c}z_\zeta)_{\text{R}} - (\bar{u}\bar{c}z_\zeta)_{\text{L}}}{\Delta x} + \frac{(\bar{v}\bar{c}z_\zeta)_{\text{F}} - (\bar{v}\bar{c}z_\zeta)_{\text{Bk}}}{\Delta y}
$$
$$
+ \frac{[\bar{c}(\bar{w} - \bar{u}z_x - \bar{v}z_y)]_{\text{T}} - [\bar{c}(\bar{w} - \bar{u}z_x - \bar{v}z_y)]_{\text{B}}}{\Delta \zeta}
$$
$$
+ \frac{(\kappa_v^* \bar{c}_\zeta)_{\text{T}} - (\kappa_v^* \bar{c}_\zeta)_{\text{B}}}{\Delta \zeta} \tag{18}
$$

The contribution arising from the $\bar{v}$ component of velocity has been included here for completeness; the subscripts F and Bk refer to front and back control volume faces not shown in figure 2. It's form is entirely analogous to that for the $\bar{u}$ component .

Equation (18) is compatible (neglecting the time variation of z_ζ) with the continuum equations (11)—(13); this can be verified formally using Taylor expansions. In addition we now have a physical interpretation for the terms in (11) arising from the transformation from z to ζ. The occurrence of z_ζ in the horizontal gradient comes about because of the change in area of the left and right faces in figure 2, while the form of w^* is due to a contribution from the horizontal velocity through the top and bottom faces. Conversely, we have the continuum equation as a check that our physically derived control volume version is correct, since the two agree in the limit Δt, $\Delta x, \Delta y$, $\Delta \zeta \to 0$.

The effects of the mesh expanding and contracting with time due to the change in $\eta(\boldsymbol{x}, t)$ is more difficult to visualise, so we rely on the continuum forms (11), (12)

318

to account for this correctly, giving

$$\frac{(z_\zeta c)_{ik}^{n+1} - (z_\zeta c)_{ik}^n}{\Delta t} = \frac{(\bar{u}\bar{c}z_\zeta)_R - (\bar{u}\bar{c}z_\zeta)_L}{\Delta x} + \frac{(\bar{v}\bar{c}z_\zeta)_F - (\bar{v}\bar{c}z_\zeta)_{Bk}}{\Delta y}$$
$$+ \frac{[\bar{c}(\bar{w} + z_t - \bar{u}z_x - \bar{v}z_y)]_T - [\bar{c}(\bar{w} + z_t - \bar{u}z_x - \bar{v}z_y)]_B}{\Delta \zeta}$$
$$+ \frac{(\kappa_v^* \bar{c}_\zeta)_T - (\kappa_v^* \bar{c}_\zeta)_B}{\Delta \zeta} \qquad (19)$$

To derive the difference relations it is necessary to define how the quantities required at the control volume faces in (19) are to be calculated. It is here that we have the flexibility, as with finite—differences, of choosing different approximations to obtain numerical schemes of various orders of of accuracy, or to make the scheme implicit or explicit. All velocities are defined at the control volume centre and the values at the face are calculated as a simple average between the values in adjacent control volumes. The approximations for the concentration are discussed later.

The idea of operator—splitting is now introduced. In one form of the technique an equation such as

$$\frac{\partial u}{\partial t} + A + B = 0, \qquad (20)$$

where A, B are differential operators, is split as

$$\frac{\partial u}{\partial t} + A = 0$$
$$\frac{\partial u}{\partial t} + B = 0$$

and solved numerically via a two step process in which each equation is solved over the complete time step from t to $t + \Delta t$ thus

$$\hat{u} = u^n - \frac{\Delta t}{\Delta x} A^+(u^n)$$
$$u^{n+1} = \hat{u} - \frac{\Delta t}{\Delta y} B^+(\hat{u})$$

Here $A^+/\Delta x$, $B^+/\Delta y$ are the numerical representations of A and B in (20).

The flexibility in the choice of how to split the original equation is one of the attractions of the method. It can be done based on grouping all the terms associated with a particular direction together; in which case a multi-dimensional problem can be reduced to a series of 1-D equations for which powerful schemes may be available. Another alternative is to make the split dependent on the nature of the physical processes represented by particular terms.

In the case of our equation (5), the two approaches coincide if we split the equation into terms involving the horizontal direction and terms involving the vertical direction. The horizontal processes are purely advective, described by a 2D hyperbolic system of equations, while the physics of the vertical transport involves a competition between turbulent diffusion and particle fall velocity, with the resulting equations being 1D and parabolic. The transformation to ζ coordinates in the vertical blurs to some extent this neat partition, as the transformed horizontal terms generate a component of vertical transport (equation 10). Intoduction of the vertical velocity component however, cancels out the troublesome vertical gradient terms as can be seen if (14) is substituted into (12).

319

An added advantage of operator–splitting is the ease with which a much larger time step can be used in one of the sub–problems if it is describing a process that is varying on a slower time scale. Thus it is hoped in our case that the expensive advection calculations may be calculated with a significantly larger time step than that required by the vertical processes.

2.2.1 *Vertical scheme* An implicit scheme is used to avoid the $1/(\Delta z)^2$ diffusion limit on the time step. This is particularly important because the ζ transformation results in very small vertical grid spacing near the bed.

The value of $\bar{c}$ at the top and bottom faces is approximated assuming a linear variation so that $\bar{c}_\mathrm{T} = \frac{1}{2}(c_{ijk} + c_{ijk-1})$. The concentration gradient for the diffusion flux, equation (16), is approximated $(\bar{c}_\zeta)_\mathrm{T} = (c_{ijk} - c_{ijk-1})/\Delta\zeta$. The resulting scheme is second order in space and, for constant vertical velocities, is the same as that obtained using central finite—differences. For non–constant vertical velocities the two formulations differ slightly.

Expressing the concentration $\bar{c}$ as a weighted sum of the value at the current and next time step, a linear tri—diagonal system of equations for the concentration at the next time step is obtained of the form

$$- \theta f_k c^{n+1}_{ijk-1} + [(z_\zeta)^{n+1}_{ijk} + \theta g_k]c^{n+1}_{ijk} - \theta h_k c^{n+1}_{ijk+1} = r_k. \tag{21}$$

where

$$
\begin{aligned}
f_k &= \lambda w_\mathrm{B} + \mu(\kappa_v^*)_\mathrm{B}, \\
g_k &= \lambda(w_\mathrm{B} - w_\mathrm{T}) + \mu[(\kappa_v^*)_\mathrm{B} + (\kappa_v^*)_\mathrm{T}], \\
h_k &= -\lambda w_\mathrm{T} + \mu(\kappa_v^*)_\mathrm{T}, \\
r_k &= (1-\theta)f_k c^n_{ijk-1} + [(z_\zeta)^n_{ijk} - (1-\theta)g_k]c^n_{ijk} + (1-\theta)h_k c^n_{ijk+1}, \\
\lambda &= \Delta t/(2\Delta\zeta), \qquad \mu = \Delta t/(\Delta\zeta)^2.
\end{aligned}
$$

The scheme is unconditionally stable for values of the weighting factor $\theta \geq 1/2$, and is second order accurate in time for $\theta = 1/2$. However, there is a Peclet number (Pe) restriction

$$Pe = \frac{\Delta\zeta|w|}{\kappa_v^*} \leq 2 \tag{22}$$

which, if violated, yields unacceptable negative and oscillatory solutions. In practice this occurs when the turbulence levels drop to very low levels (e.g. at slack water) and particles settle out under the action of the advective fall velocity term. At present we force (22) to hold by increasing the diffusion coefficient artificially if necessary. Clearly this is not a satisfactory solution and indicates that further work is required to design an *implicit* scheme that is useable over a wide range of Peclet numbers. In connection with this we note that even for modest values of $Pe \approx 1/2$, the accuracy of the present scheme is degraded when comparison is made to an analytic solution.

2.2.2 *Horizontal scheme* The advection equation is solved using the scheme of Smolarkiewicz. Details can be found in [7] but in brief the scheme works as follows.

An initial upwind step is used. In terms of the control volume this corresponds to assuming a piecewise constant variation of $\bar{c}$ between control volumes (this is Roache's "2nd Upwind Differencing Method" [5, p. 73]). The artificial diffusion introduced by this crude approximation to the variation of $\bar{c}$ is removed by defining 'anti-diffusive' velocities, calculated from the analytically derived expression for the discretisation error of the upwind scheme. A second upwind step using these velocities reverses the effects of the numerical diffusion. The error introduced by this 2nd step can then be removed by a subsequent step, and so on. In practice, computer time limitations mean only one or two anti-diffusion steps can be applied, although it is shown in [7] that little benefit is gained from using more than about three anyway.

The method seems to offer many advantages over other advection schemes on offer. It is easy to program with relatively few problems at boundaries (the full anti-diffusive velocity field cannot be calculated at the boundaries and the 'cross terms' have to be omitted). It is relatively cheap to compute, provided the number of anti-diffusive iterations is kept small, and was found to be very accurate in tests involving advection of 2D Guassian profiles. Also very important is property of positive definiteness possessed by the scheme which guarantees non—negative concentrations, provided the usual Courant number restriction on the time step is observed.

3 Test problems

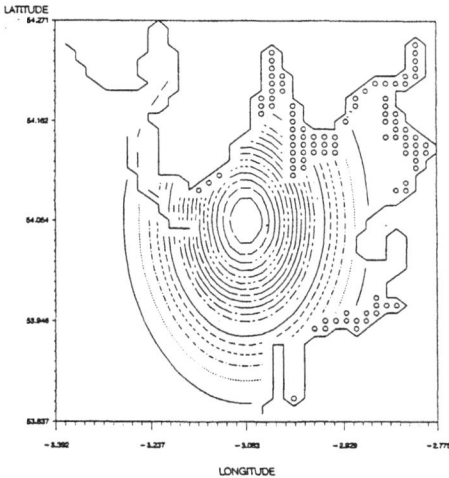

Figure 3: Initial Gaussian distribution.   Figure 4: Fixed advecting velocity field.

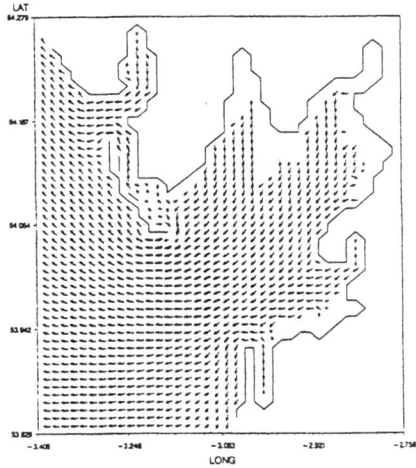

3.1 *Example I: Pure horizontal advection*

Figure 3 shows an initial Gaussian distribution together with a representation of a realistic coastline (Morecambe Bay). This distribution of initial concentration is used for example only and is not intended to be realistic distribution of suspended sediment in the region. This is advected with the fixed velocity field shown in figure 4 which

LATITUDE
64.27
48.06
41.86
35.65
29.44
-1.39   6.44   14.27   23.11   31.94
LONGITUDE

LATITUDE
64.27
48.06
41.86
35.65
28.44
-1.39   6.44   14.27   23.11   31.94
LONGITUDE

C   ——— 0.05   ---- 0.10   ·-·-· 0.15   ······ 0.20   --- 0.25
     ——— 0.30   ---- 0.35   ·-·-· 0.40   ······ 0.45   ·---· 0.50
     ——— 0.55   ---- 0.60   ·-·-· 0.65   ······ 0.70   ---- 0.75
     ······ 0.80 ······ 0.85 --- 0.90 --- 0.95 ---- 1.00

C   ——— 0.05   ---- 0.10   ·-·-· 0.15   ······ 0.20   --- 0.25
     ——— 0.30   ---- 0.35   ·-·-· 0.40   ······ 0.45   ·---· 0.50
     ——— 0.55   ---- 0.60   ·-·-· 0.65   ······ 0.70   ---- 0.75
     ······ 0.80 ······ 0.85 --- 0.90 --- 0.95 ---- 1.00

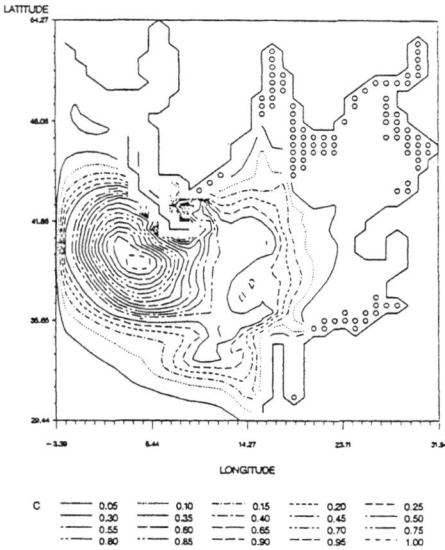

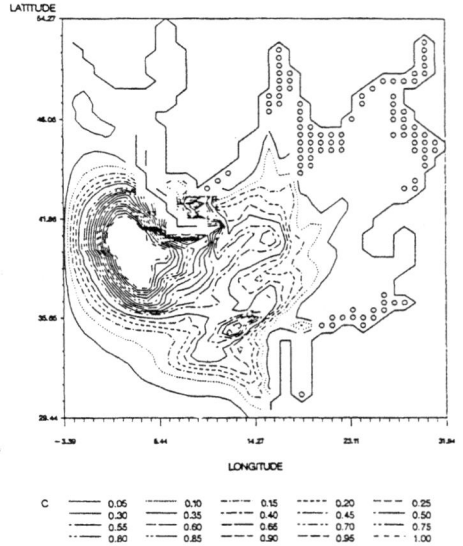

Figure 5: Result with pure upwind.   Figure 6: Result with anti-diffusive correction.

was derived from a tidal simulation of the region. No vertical terms are included in the equations and so this is purely a test of the horizontal advection scheme. Figure 5 shows the resulting distribution 6 hours later using no anti-diffusive correction i.e. a pure upwind solution. The effect of including a single correction at each step is shown in figure 6. As we would expect, the peak concentrations are much higher and concentration gradients much steeper in the solution shown in figure 6. Note because of the non—uniform velocity field peak concentrations can exceed the initial value of unity. The open circles indicate grid points which have dried as the tidal elevations change.

3.2   Example II: 2D advection–diffusion

Figure 7 shows a vertical slice through a slope at the top of which sediment is being picked up, diffused vertically, and advected from right to left. At the bottom of the slope it settles out. The boundary condition at the top is one of zero flux

$$w_0 + \kappa_v^* \frac{\partial \bar{c}}{\partial \zeta} = 0.$$

At the bed, the boundary condition at the top of the slope is a reference concentration related to the bed stress, while in regions where sediment is settling out we impose

$$\frac{\partial \bar{c}}{\partial \zeta} = 0.$$

322

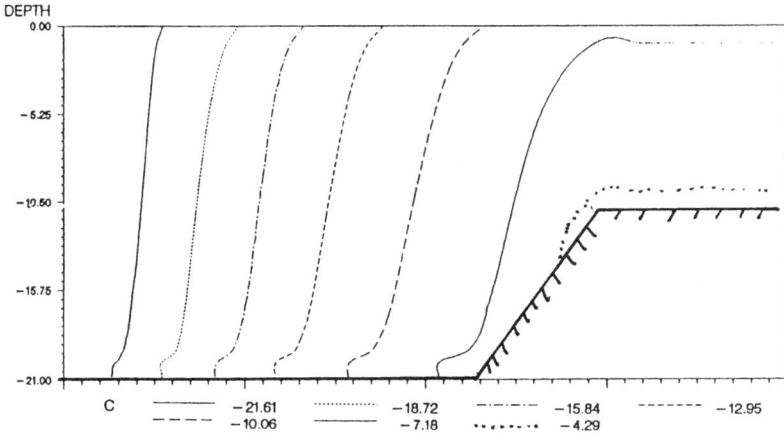

Figure 7: Advection–diffusion over a slope. Countors are $\log_{10}$ of concentration.

so that the flux at the bed arises purely from the fall velocity. In this particular case we have decided arbitrary the region where sediment is picked up and where it settles, although in practice this would be determined by the level of the bed stress. Note the build up of a layer of almost constant concentration near the bed in regions where sediment is settling out.

4 Summary

The numerical approach used in a full three-dimensional solution to the advection–diffusion equation governing the transport of suspended sediment has been described in some detail. Examples of the results obtained from running the numerical model on some test problems have been presented.

Acknowledgments

I would like to thank Mr. J. Wooltorton, who managed to resurrect the file containing this document from a PC disk after it was overwritten!

References

[1] Aldridge J.N. (1990) *A continuum mixture theory approach to sediment transport with application to oscillatory turbulent boundary layers*, Thesis presented at Polytechnic South-West, Plymouth, England.

[2] Davies A.M.,(1983), "Formulation of a linear three-dimensional hydrodynamic sea model using a Galerkin–Eigenfunction method", *Int. J.Num.Meth.in Fluids*,Vol **3**,pp 33—60.

[3] Holloway G.,(1989),"Subgridscale representation", in *Oceanic Circulation Models:Combining Data and Dynamics*, Ed A.L.T. Anderson & J. Willebrand, Kluwer Academic Publishers, Dordrecht, pp 513–587.

[4] Nicholson J.M.C. (1983) *Three–Dimensional Models of Particulate and and Cohesive Suspended Sediment Transport*, Thesis presented at University of Manchester, England.

[5] Roache P.J. (1972) *Computational Fluid Dynamics*, Hermosa Publishers, Albuquerque New Mexico.

[6] Rouse H. (1937) "Modern conceptions of the mechanics of turbulence, numerical model of particulate transport for coastal waters", *Trans.Am.Soc.Civ.Eng*,Vol **102**, pp 436—505.

[7] Smolarkiewicz P.K. (1984) "A fully multidimensional positive definite advection transport algorithm with small implicit diffusion", *J.Computational Physics*,Vol **54**, pp 325—362.

[8] Spaulding M.L., Pavish D. "A three-dimensional numerical model of particulate transport for coastal waters", *Continental Shelf Research*,Vol **3**(1), pp 55—67. Finite

[9] Smith J.D. (1977) "Modeling of sediment transport on continental shelfs", In *The Sea, Vol 6*, Ed Goldberg et al, Wiley-Interscience, New York, pp 579–601

[10] Taylor P.A., Dyer K. (1977) "Theoretical models of flow near the bed and their implications for sediment transport." In *The Sea, Vol 6*, Ed Goldberg et al, Wiley-Interscience, New York, pp 579–601

27 A comparison of *in situ* and laboratory methods to measure mudflat erodibility

C. L. Amos, H. A. Christian, J. Grant and D. M. Paterson

Abstract

Four methods for measurement of surface sediment erodibility (critical bed shear stress) are compared. Three were based on *in situ* technology that used successively a benthic flume (Sea Carousel), a shear pad (INSIST) and a vertical water jet (CSM). The fourth adopted a method of sampling natural sediment and transporting them to a laboratory flume where they were analyzed in a more conventional manner. The comparison was undertaken at three sites across a littoral mudflat in Minas Basin, Bay of Fundy, and comprised a part of a much larger multi-disciplinary study of mudflat stability (LISP). Sea Carousel detected critical shear stresses that varied between 0.5 and 2.5 Pa. No spatial trends in this critical stress were found, however, a systematic increase with time was detected on the central mudflat. INSIST gave values approximately 1 order of magnitude higher than Sea Carousel but reflected the increase in strength of the flat with time. The overprediction has been overcome by design changes to INSIST. CSM gave results that were a factor of two larger than those derived from Sea Carousel and also showed similar trends with time. The laboratory flume results showed consistently low critical shear stresses (*circa* 0.5 Pa) with no trends spatially or through time. Low values in the laboratory experiment reflect the onset of bedload transport of a non-cohesive population not evident on the mudflat during exposure. No erosion of the cohesive portion of the bed was observed. The CSM and Sea Carousel monitored the erosion and suspension of the cohesive portion of the bed only and so gave higher values of critical stress. Our combined results showed that mudflat stability is controlled by the production of diatomaceous polysaccharide mucilage and the presence of *Corophium volutator*. Furthermore, increases in mudflat critical shear stress during tidal exposure were quickly eliminated during subsequent inundation.

325

1 Introduction

The evaluation of sediment stability from littoral marine environments is an essential aspect of environmental impact assessment. It necessarily follows a evaluation of the hydrodynamic regime and is an essential precursor to a full biological or chemical evaluation. The constants and coefficients that define sediment behaviour are largely based upon studies carried out under laboratory conditions using either natural or artificial material. Recent evidence shows that laboratory measures of sediment stability are only representative of nature if they include complexities arising from chemical, biological, and hydrodynamical influences (Grant *et al.*, 1982; Amos and Mosher, 1985; Grant *et al.*, 1986; Grant and Gust, 1987; Paterson *et al.*, 1990). This is particularly evident in the study of fine-grained sediments. Consequently, we emphasise the need for *in situ* measurements to realistically determine bed stability.

Black (1989) has reviewed methods to measure *in situ* bed stability. Such methods include geotechnical tests (Pilcon shear vane tests, Daborn, 1991; *in situ* (INSIST) shear testing, Christian *et al.*, 1990, cone penetrometer testing, Christian, 1991), benthic flumes of a variety of types (see Amos *et al.*, in press), and other innovative hydrodynamically-based tests such as Microcosm (Buchholtz-Ten Brink *et al.*, 1989) and Cohesive Strength Meter (Paterson, 1989). The purpose of this study was to evaluate three of these *in situ* techniques by testing them against each other across a littoral mudflat in Minas Basin, Bay of Fundy. The three systems tested were: INSIST, CSM, and Sea Carousel (Amos *et al.*, in press). Results from these *in situ* tests were then compared against measures of erosion from a laboratory flume study of our natural sediment (Muschenheim *et al.*, 1986).

Three reference sites were selected for the comparison. They were situated on a transect of the

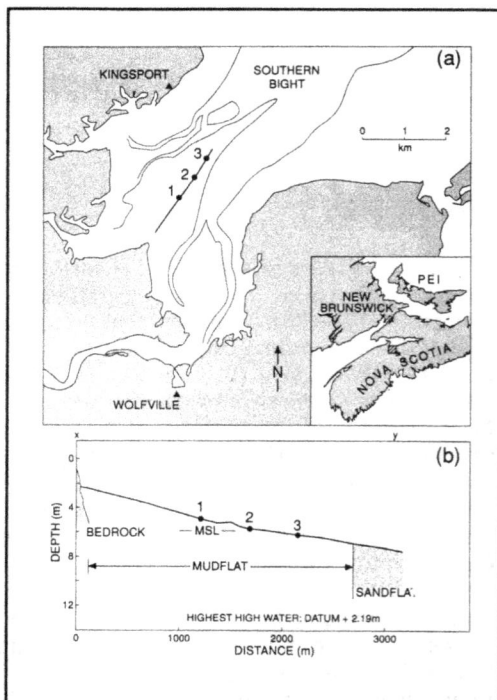

Figure 1 A location diagram of the three stations on the littoral mudflat in Minas Basin, Bay of Fundy.

littoral mudflat in Minas Basin, Bay of Fundy (Figure 1). The location of these sites were chosen on the basis of sediment bulk properties reported in Amos *et al.*, (1988). They found significant changes in bed vane shear strength and bulk density during the summer of 1984. The questions raised by that paper were: (1) does the vane shear strength adequately reflect sediment erodibility, and if not what *in situ* measures were made immediately after the mudflat exposure proceeding each Sea Carousel survey and again immediately before inundation. In this manner, varying effects of drying were evaluated. CSM was deployed only for the interval between 22th and 27th July, 1989 at sites adjacent to INSIST. Measurements were carried out 15 minutes after exposure, and at 75 and 135 minutes. Triplicate core samples for the laboratory flume analysis were collected and analyzed between 15th and 25th July, 1989. Samples were transported within hours the 100 km to the laboratory where they were immediately tested. The results of all methods are summarized in Table 1 (TAU - surface cohesive strength).

The Pilcon vane and cone penetrometer showed no measurable resistance within the top 0.10 m of the sediment column i.e. no shear strength. This was contrary to the other methods of measurement that in all cases detected a significant and varying sediment strength, but was not unexpected given the original purpose of these instruments.

4.1 Sea Carousel

Details of results obtained from this survey are presented in Amos et al., (in review). An example of a deployment at station 1 during 22nd July, 1989 is shown in Figure 2. Figure 2A shows the incremental increases in azimuthal flow over a 1-hour interval to a maximum of 0.9 m/s. Notice the increases in suspended sediment concentration (Fig. 2B) particularly at the times of flow increase. Also notice the gradient in sediment concentration

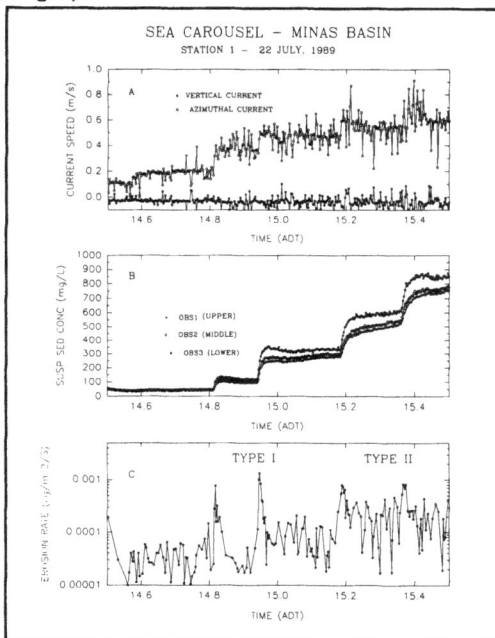

Figure 2 A time-series of Sea Carousel results at station 1, Minas Basin during 22 July, 1989.

within the flume being highest at the bed and least at the surface. Highest erosion rates occurred within 2 minutes of velocity increases and thereafter dropped rapidly with time at an exponentially decaying rate. The peak erosion rate was circa 0.001 kg/m²/s. The decay in erosion rate through time was termed "Type I" erosion and was evident in this example between 14.8 and 15.2 ADT. Bed erosion stopped usually within 2 - 5 minutes of the application of the applied stress. We called this type of erosion "benign" as the bed armoured itself, and thus the amount of material released to the water column was limited provided the bed stress does not increase further.

The peak in erosion rate at higher stresses was no greater in duration or magnitude than that at lower stresses, suggesting that it was the excess bed shear stress rather than the absolute that controlled erosion rate. This erosion rate, however, remained at a finite value between $1 - 2 \times 10^{-4}$ kg/m²/s. Notice the relatively high variability in erosion rate through time. This was interpreted to be methods do ?; and (2) are the inferred spatial and temporal changes in mudflat erodibility realistic, and if so, what controls them ? The first of these questions is addressed in this comparison, the second was addressed as part of an intensive multidisciplinary study: the Littoral Investigation of Sediment Properties (LISP). The major findings of the LISP study are reported in Daborn *et al.* (1991).

2 Methodologies

2.1 *Sea Carousel*

Sea Carousel is a benthic annular flume that is 2 m in diameter and has an active annulus 0.15 m wide and 0.30 m high. It is deployed during mudflat submersion and is monitored and controlled interactively from the sea surface. Erosion is inferred from changes in suspended sediment concentration (S) detected within the annulus using three Optical Backscatter Sensors (OBS's). The OBS's are calibrated by means of samples pumped from the flume at intervals throughout each experiment. Bed shear strength (τ_b) is equated with the fluid bed shear stress (τ_o) which is defined as critical (τ_c) at the onset of erosion: $\tau_o = \tau_b = \tau_c$. Successively increasing flows are induced by a rotating lid equipped with paddles. Azimuthal and vertical flows are detected with a Marsh-McBirney electromagnetic flow meter and lid rotation is detected by means of a rotating shaft end-coder. Data from all channels are monitored and logged at 1Hz using a Campbell Scientific CR10 data logger. A Sony V101 video camera monitors seabed response through a window in the side of the flume. The window sections the upper 50 mm of the seabed and lower 0.10 m of the flume water column. The video illustrates the nature of the bed and its failure and the size and number of eroded aggregates in transport. The boundary layer velocity structure is digitized from the trajectories of suspended particles. From this the thickness and nature of the viscous sub-layer are determined together with estimates of bed shear stress. Spatially-averaged depth of bed erosion (z) is determined from total eroded mass (SV, where V is the flume volume: 0.217 m^3) and independent measures of sediment bulk density (1800 kg/m^3). Depth is transformed to effective stress (σ') using the expression of Terzaghi and Peck (1967):

$$\sigma' = \gamma'z = 7.59z \text{ kPa/m} \tag{1}$$

where γ' is the buoyant unit weight of sediment ($\rho_s - \rho_w$)g, that is the buoyant density times gravity (g). By plotting τ_o against σ', a Mohr-Coulomb diagram is generated wherein the intercept is surface cohesive strength (or the critical shear stress for erosion) and the slope is the arctangent of the internal friction angle (Lambe and Whitman, 1969). This plot is directly comparable to plots generated using INSIST and the CSM. The details of the system and its calibration are given in Amos *et al.* (in press) and descriptions of its results are given in Amos *et al.* (in review). With this method, bed erodibility is measured over an area of 0.873 m^2.

2.2 *INSIST*

INSIST (*in situ* simple shear test) is a mechanical device used to measure the drained shear strength of surface sediment at very low consolidation stresses. It consists of a flat pad 0.3 x 0.3 m in size which sits on the natural sediment surface and is attached via a gallows to a set of adjustable weights. The weights apply a horizontal force to the pad. Sediment failure is defined at the point of motion of the pad over the sediment. Surface cohesive strength and the internal friction angle are derived in the same way as Sea Carousel. The procedure is repeated for differing loads (consolidation stresses) on the pad. The force at failure is plotted against consolidation stress to produce a Mohr-Coulomb diagram. Consolidation stress is equated with effective stress to define the increase in sediment shear strength with depth. Details of the methodology and initial results are given in Christian *et al.* (1990). Bed strength is defined over a bed area of 0.093 m², approximately 10% of the area of Sea Carousel. It also differs from Sea Carousel in being deployed subaerially. Independent samples for bulk density were collected adjacent to each experiment site to check that the bed remained saturated (as the water content of the bed decreased with time of exposure). Otherwise an apparent cohesion would develop due to the negative pore pressure induced by a falling water table.

2.3 *Cohesive Strength Meter*

The Cohesive Strength Meter (CSM) is a small portable devise that uses a vertical jet of water to erode surface sediment (Paterson, 1989). It too is deployed subaerially. It consists of a water-filled chamber 26 mm in diameter that is pushed into the sea bed. The jet of water comes from a downward-directed nozzle in the chamber. The force of the jet is controlled with a mercury manometer and is increased systematically through each experiment. The applied vertical force of this nozzle has been transformed to a horizontal bed shear stress in a series of experiments at Polytechnic Southwest (Christian and Paterson, unpublished data, 1991). Bed erosion is inferred from the suspended sediment concentration detected within the chamber by a transmissometer. Normalized attenuation is plotted against bed shear stress in the form of a pseudo- Mohr-Coulomb diagram. Again, the intercept is the surface cohesive strength. Bed strength is integrated over an area of only 0.00001 m². This is considerably less than the other two *in situ* methods.

2.4 *The Dalhousie Laboratory Flume*

The Dalhousie laboratory flume is a 3.25 m long, straight flow-through channel 0.35 m wide and 0.20 m high. Non-recirculating filtered seawater is used in the system. Flow is regulated by the head within an inflow tank and by adjustment of a set of baffles at the lower end of the flume. Natural sediment cores are emplaced in a 7.0 cm diameter circular recess of the flume floor approximately 2 m below the intake. Flow is increased in small increments until erosion of the natural sediment begins. Shear stress is measured with a hot-film anemometer near the core (Grant and Gust, 1987). The point of bed erosion is evaluated visually and is defined as the point at which general detachment of bed material occurs. The point of erosion is equated with the critical shear stress for incipient motion. Bed properties are integrated over an area of 0.00625 m².

The above experiments use differing techniques to derive surface sediment erodibility. Also the area of the bed affected varies by 5 orders of magnitude. INSIST and CSM are deployed subaerially, and so can evaluate effects of drying. They also offer the flexibility of providing small-scale spatial variations in erodibility not possible with Sea Carousel. Sea Carousel, on the other hand, is more representative of conditions during natural erosion. The laboratory flume represents techniques most commonly reported in the literature and is valuable in linking such measures with the newer field techniques. Finally, a cone penetrometer and Pilcon shear vane were deployed at the same times and locations as INSIST (Christian, 1991). The purpose of this was to compare standard geotechnical soil measures with those described above.

3 The Study Region

This study was undertaken on a littoral mudflat in the Southern Bight of Minas Basin, Bay of Fundy (Figure 1). The mudflat was 4500 m wide and dipped gradually seawards. The region was devoid of an inner salt marsh, creeks or reclamation schemes, and was protected from wave influences; it was, therefore, a relatively simple, accessible mudflat. The flat was inundated by a mixed semi-diurnal tide that averaged 11.5 m in range. The three reference sites (stations 1 - 3) occupied in this study were on the inner (-6.92 m[1]), central (-7.69 m) and outer (-8.53 m) mudflats respectively. The sediments on this flat varied between sandy silt to silty sand, it was therefore a siltflat. The sand content varied from 34% (station 1) to 55% (station 3): the clay content varied from 16% (station 1) to 10% (Station 3). The clays were non-reactive being composed largely of illite (19%), chlorite (14%) with small amounts of smectite and vermiculite (4%; Amos and Mosher, 1985). The water content at all sites varied between 25 and 40% and bulk density between 1800 and 1850 kg/m[3]. There was little variation in these properties with depth below the seabed (Daborn et al., 1991). Surface vane shear strengths reported in Amos et al. (1988) varied between 5.5 and 22.9 kPa. The highest values generally occurred at the centre of the mudflat (our station 2). Significant changes were observed both across the flat and throughout the period of detection. These changes were speculated to be related to relative exposure of the mudflats and the associated drying effects. This notwithstanding, shear vane measurements are considered unreliable due to the coarse texture of the bed (Christian et al., 1991).

Biological activity on the flats was high. Chlorophyll A ranged between 3 and 7 $\mu g/cm^2$ and carbohydrate ranged up to 4.5 mg/cm^2. The region was heavily colonized by the tubicolous amphipod *Corophium volutator* in densities up to 10,000/m^2. The amphipods fed on benthic algae and diatom mucilage and were in turn fed upon by large numbers of migrating Semipalmated Sandpipers. It is the selective feeding upon *Corophium volutator* by the Sandpipers that influenced most greatly mudflat stability through reduction in numbers of *Corophium volutator* that normally fed on benthic diatoms. This led to a consequent bloom in diatomaceous mucilage that increased measurably sediment cohesion (Daborn et al., in press) and thereby stabilized the mudflat.

4 Results

The field program was undertaken between 15th July and 2nd August, 1989. Sixteen deployments of Sea Carousel were undertaken from which reliable results were obtained on 14 occasions (4 at station 1; 8 at station 2; and 2 at station 3). INSIST was deployed daily adjacent to the Sea Carousel sites until 31st July, 1989. INSIST

---

[1] Elevations of stations are reported in metres below highest high water mark.

quantum releases of sediment from the bed in the form of aggregates (also seen in video records). This form of erosion ("Type II") was termed "chronic" as erosion appeared to continue as long as the applied bed stress was maintained. The time-series of surface cohesive strength is shown in Figure 4A. All stations plotted began with Type I erosion. Cohesive strength increased linearly with depth under these conditions and the surface intercept was clearly defined.

4.2 INSIST

Figure 3 illustrates results measured at station 1 using INSIST. Shear stress (the horizontal force per unit area required to move the pad) is plotted in each case against vertical consolidation stress (pad load per unit area) for measurements made immediately following site exposure (ebb) and immediately prior to the subsequent submergence (flood). In all cases the plots were highly linear and conformed with theory. The friction angle varied between 35.3 and 42° (FA, Table I). This falls within the range of expected values and was consistently lower than at the end of the exposure period. Indeed the flood tide friction angle was remarkably similar throughout, largely due, we feel, to the foraging effects of *Corophium volutator*

Cohesion immediately upon exposure varied significantly up to 36 Pa (Fig. 4B). Notice that this cohesion built significantly during the exposure period. The low cohesion on 22nd July corresponded to a period of high rainfall and wave activity. These changes in cohesive strength were due to the photosynthetic diatom blooms that took place during exposure of the flat (Daborn *et al*. 1991). These effects were generally reversed during the subsequent inundation. However a residual long-term change was detected that represents changes related to shorebird feeding (Daborn *et al*., in press).

4.3 Cohesive Strength Meter

Meaningful results from the CSM were obtained at station 2 for the period 22nd - 27th July, 1989. Regression of applied stress against normalized attenuance was highly linear and gave a consistent positive intercept (the critical shear stress). These intercepts are plotted in Figure 5A. The range in surface cohesive strength was 1.0 to 4.3 Pa. As with INSIST, the lowest measured values corresponded to a period of rainfall (22nd). Thereafter, a steady increase in bed strength was detected that corresponded closely with trends from Sea Carousel and INSIST. The range of critical stress values overlapped those from Sea Carousel though they were 50 - 100% higher.

4.4 Dalhousie Laboratory flume

The results of the laboratory flume experiments on the threshold for erosion were consistent for all stations over the duration of the study (*circa* 0.5 Pa; Fig. 5B). This threshold was observed as the initiation of bedload transport in the form of small (1-2 mm high) ripples. The core surface erosion rate (based on the sediment trap results) was ○ 1.8 x 10^{-5} kg/m^2/s. This is *circa* one order less than the erosion rate measured by Sea Carousel. The material observed in transport was predominantly sand sized and was composed of individual grains, aggregates of fines, and probably faecal pellets. As erosion proceeded, the upper portions of *Corophium volutator* tubes became exposed and occasionally eroded. No sediment was observed

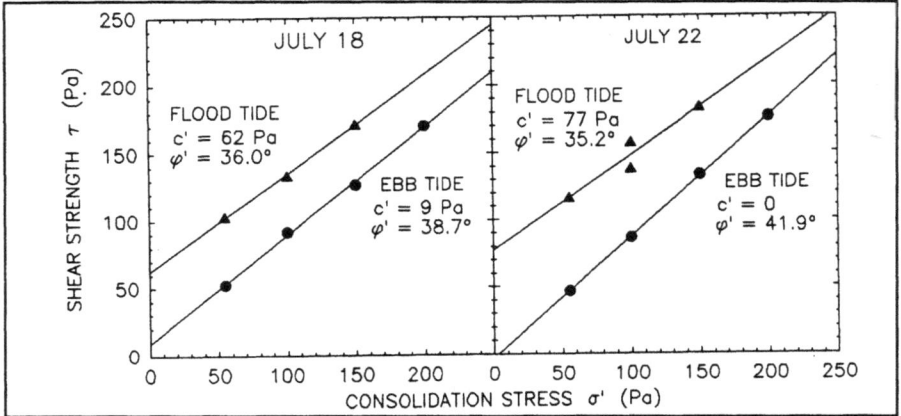

Figure 3 A Mohr-Coulomb plot of data from INSIST collected at station 2 on 18th and 22nd July, 1989 (c' - cohesion; φ' - friction angle)

to be in suspension.

5 Discussion

Type I and Type II erosion were both observed to occur in this study. Type II erosion usually superseded Type I at higher bed stresses. The comparisons made herein are, therefore, applicable to Type I erosion only. The nature of Type II erosion is more complex and probably related to the microfabric of the sediment (and associated planes of weakness) rather than to cohesion or bulk density of the sediment as a whole.

The time-series of results from the four methods are shown respectively in Figures 4 and 5 and are based on the data given in Table I. Sea Carousel (Fig. 4A) showed a range in critical surface shear stresses for erosion (cohesion or shear strength) that varied between 0.5 and 2.5 Pa. Whilst differences between the three sites were detected no systematic trends were evident. Station 2 (dots), however, showed evidence for a strengthening with time through the experiment. Though not as significant as that observed in 1984 (Amos *et al.* 1988) the trend occurred at the same time of year as the earlier survey and corresponded to increases in diatom abundance and decreases in numbers of *Corophium volutator* (Daborn *et al.* 1991). This trend was representative of realistic changes in bed stability brought about by biological controls. The same trend in bed stabilization was also apparent in the INSIST results (Fig. 4B) although the scatter appeared to be greater. Also, the magnitude of the surface cohesive strength was approximately 1 order of magnitude larger than those measured by Sea Carousel. The higher values were explained as due to sediment "bull-dozing" in front of the pad. Recent innovations of INSIST design have overcome this offset. Best results were obtained when INSIST was deployed immediately upon exposure of the mudflat. The significant increases in cohesion over the exposure appear to be controlled by biological processes. It could not be due to evaporation as the measured friction angle (and bulk density) decreased during this period. This was further verified by experimentation wherein a site near station 2 was treated with a biocide (formalin)

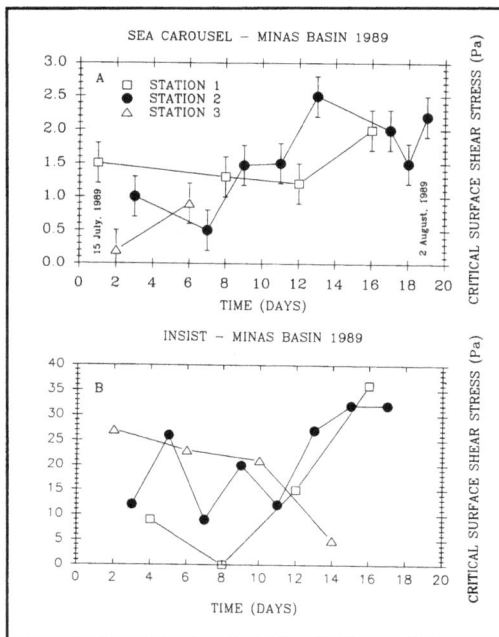

CRITICAL SURFACE SHEAR STRESS (Pa)

Figure 4 The time-series of critical shear stress for erosion for (A) the Sea Carousel and (B) INSIST. Notice the increase in bed strength between days 8 and 14 in both plots.

to selectively remove biota. This caused a reduction in friction angle. After several days, diatoms recolonized the affected area and underwent a bloom in the absence of *Corophium volutator* (Faas *et al.* in press). The highest recorded values of cohesion were recorded at the site by both Sea Carousel (2.9 Pa) and INSIST (45 Pa). The daily increases in cohesion point to dangers of subaerial measures of sediment stability as recorded strength values may be higher than realistic for fluid dynamical erosion.

The results of CSM and the laboratory flume are shown in Figure 5. The CSM measurements were limited to station 2 during the central portion of the experiment (Fig. 5A). It corresponded to the time when an increase in cohesion was observed by Sea Carousel and INSIST and so the results are considered valid. The cohesion values obtained by CSM were consistently 50-100% greater than those measured by the Sea Carousel. We note that the peak pressure of the CSM jet was used to define applied bed stress. These results suggest that perhaps the rsm or mean value of the jet pressure would produce closer results to the laboratory and Sea Carousel findings.

The laboratory flume data were collected for the early portion of the experiment when long-term changes in cohesion (determined by the other methods) was not taking place (Fig. 5B). Flume results on critical threshold stress correlated moderately well with the Sea Carousel data for stations 1 and 2. However, the lower laboratory values reflect onset motion of a non-cohesive population as bedload. In nature the importance of this population to total sediment transport is difficult to determine as it would not be detected by either CSM or Sea Carousel that monitor suspensions only. The flume results show a constant threshold for erosion for all stations and through time that is independent of biological activity or exposure time. This suggests the co-existence of a cohesive and non-cohesive bed composition, only the cohesive portion of which is changing with time.

McCave (pers. comm.) suggested that the direction of bed stress application is important as sediment strength is anisotropic, even in the horizontal plane. The direction of application of the bed stress varies with the methods tested. Sea Carousel is omnidirectional, CSM is downward and INSIST and the laboratory flume are unidirectional. Our results indicate that this effect is a second order one at most as the long-term variability is larger than differences between

flume are unidirectional. Our results indicate that this effect is a second order one at most as the long-term variability is larger than differences between hydrodynamical methods.

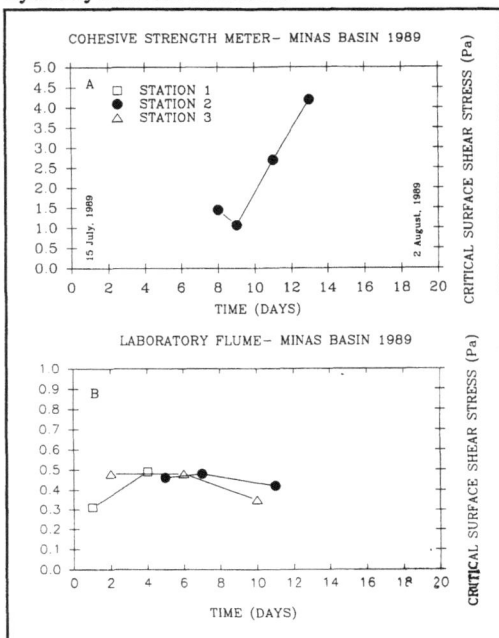

Figure 5 The time-series of critical shear stress for erosion for (A) the CSM and (B) the Dalhousie laboratory flume

6 Conclusions

The following are the main conclusions of this study: 1. Type I (asymptotically decaying) and Type II (constant) erosion patterns were detected on the Minas Basin mudflat. Type I was related to bulk surface properties of the sediment (cohesion, bulk density), whereas Type II erosion was related to weaknesses in the sediment brought about by microfabric. 2. Short-term build up of cohesion was detected during tidal exposure because of photosynthetic diatom blooms. This build up was reversed upon subsequent inundation. Sediment cohesive strength measurements made during exposure may thus overpredict sediment stability. 3. Long-term build up of cohesion (doubling of strength) was detected that appeared to be related to shorebird feeding habits. Bed strength is therefore not constant through time. 4. Sea Carousel, CSM and the Dalhousie laboratory flume gave similar measures of the surface cohesive strength (within a factor of 2). INSIST gave values that were an order of magnitude higher than other methods. This was due to "bull-dozing" of sediment and has since been overcome by redesign. 5. Sea Carousel, CSM and INSIST gave similar long-term trends and so are considered comparable. These trends are similar to trends observed by Amos *et al.* (1988), however, the strength build-up is related to biological activity and not to desiccation of the flats during exposure. 6. The shear vane results bore no relationship to measures of sediment stability made by other methods.

Acknowledgements

We thank G.R. Daborn, A. Robertson and M. Brylinski, Acadia University, R. W. Faas, Lafayette College, G. Drapeau and B.F. Long, INRS-Oceanologie, G. Perillo and C. Piccolo, IADO, and B. Tsinman, DSS, T.E.R. Jones Polytechnic Southwest and A Griswold, Dalhousie University.

References

Amos, C.L. and , Mosher, D.C. (1985), Erosion and deposition of fine-grained sediments from the Bay of Fundy. Sedimentology 32, pp. 815-832.

Amos, C.L., Van Wagoner and Daborn, G.R. (1988), The influence of subaerial exposure on the bulk properties of fine-grained intertidal sediment from Minas Basin, Bay of Fundy. Estuarine, Coastal and Shelf Science vol 27, pp. 1-13.

Amos, C.L., Grant, J., Daborn, G.R., and Black, K. (in press), Sea Carousel - a benthic annular flume, Estuarine, Coastal and Shelf Sciences.

Amos, C.L., Daborn, G.R., Christian, H.A., Atkinson, A. and Robertson, A. (in review), *In situ* erosion measurements of fine-grained sediments from the Bay of Fundy. Marine Geology.

Black, K. (1989), The in situ measurement of sediment erodibility: a review, Unpublished report to ETSU, Department of Energy, Harwell, U.K., 31 pp.

Buchholtz-Ten Brink, M.R., Gust, G. and Chavis, D. (1989), Calibration and performance of a stirred benthic chamber, Deep-Sea Research, pp 1083-1101.

Christian, H.A. and Daborn, G. R. (1990), Shear strength testing at the sediment-water interface. American Society of Limnologists and Oceanographers, College of William and Mary, Williamsburg, Virginia, June 10-14.

Christian, H.A., Gillespie, D. and Amos, C.L. (1990), Results of a new device to measure the in-situ shear strength of cohesive sediments, Minas Basin, Bay of Fundy. Thirteenth International Sedimentological Congress, Nottingham, U.K.

Christian, H.A., Piper, D.J.W. and Armstrong, R. (1991), Strength and consolidation properties of surficial sediments, Flemish Pass: effects of biological processes. Deep-Sea Research vol 38(6), pp. 663-676.

Christian, H.A. (1991), Geomechanics of the Starrs Point tidal flat in Daborn, G.R. (ed), LISP91, Littoral Investigation of Sediment Properties, Acadia Centre for Estuarine Research Publication No. 17, pp. 113-144.

Daborn, G.R. (ed), (1991), LISP91, Littoral Investigation of Sediment Properties, Acadia Centre for Estuarine Research Publication No. 17, 239 pp.

Daborn et al. (in press), An ecological 'cascade' effect: migratory birds affect stability of intertidal sediments. Limnology and Oceanography

Faas, R.W., Christian, H.A. and Daborn, G.R. (in press), Biological control of mass properties of surficial sediments: an example from Starr's Point tidal flat, Minas Basin, Bay of Fundy. in Nearshore and Estuarine Cohesive Sediment Dynamics. Publ. Springer-Verlag.

Grant, J., Bathmann, U.V. and Mills, E.L. (1986), The interaction between benthic diatom films and sediment transport. Estuarine, Coastal and Shelf Science vol 23, pp. 225-238.

Grant, G. and Gust, G. (1987), Prediction of coastal sediment stability from photopigment content of mats of purple sulphur bacteria. Nature vol 330(6145), pp. 244-246.

Grant, J., Daborn, G.R., Amos, C.L. and Griswold, A. (in review), The effects of bioturbation on sediment transport on an intertidal mudflat.

Grant, W.D., Boyer, L.F. and Sanford, L.P. (1982), The effects of bioturbation on the initiation of motion of intertidal sands. Journal of Marine Research vol 40(3), pp 659-677.

Gust, G. (1988), Skin friction probes for field applications. Journal Geophysical Research vol 93, pp. 14121-14132.

Lambe, T.W. and Whitman, R.V. (1969) Soil Mechanics. John Wiley & Sons, New

York, 553 pp.

Muschenheim, D.K., Grant, G. and Mills, E.L. (1986), Flumes for benthic ecologists: theory, construction and practice. Marine Ecology Progress Series vol 28, pp. 185-196.

Paterson, D.M. (1989), Short-term changes in the erodibility of intertidal cohesive sediments related to the migratory behaviour of epipelic diatoms. Limnology and Oceanography vol 34(1), pp. 223-234.

Paterson, D.M., Crawford, R.M. and Little, C. (1990), Subaerial exposure and changes in the stability of intertidal estuarine sediments. Estuarine, Coastal and Shelf Sciences vol 30, pp. 541-556.

Terzaghi, K. and Peck, R.B. (1967), Soil Mechanics in Engineering Practice. Publ. John Wiley & Sons, 729 pp.

Table I Results of the 3 *in situ* methods to measure seabed stability and the laboratory test: (1) Sea Carousel, (2) INSIST, (3) CSM, and (4) Dalhousie laboratory flume; (*) rain fell during exposure of flat.

DATE	STA	(1) TAU	(1) FA	(2) TAU	(2) FA	(3) TAU	(4) TAU
15 JULY	1	1.5	49	--	--		0.31
16 JULY	3	0.2	81	27	27		0.48
17 JULY	2	1.0	75	12	40		--
18 JULY	1	--	--	9.0	39		0.49
19 JULY	2	--	--	26	36		0.46
20 JULY	3	0.9	73	23	32		0.48
21 JULY	2	0.5	85	9.0	41		0.48
22 JULY	1*	1.3	69	0	42	2.0	--
23 JULY	2	1.5	--	20	37	1.5	--
24 JULY	3	--	--	21	37		0.35
25 JULY	2	1.5	68	12	42	1.0	0.42
26 JULY	1	1.2	76	15	42		--
27 JULY	2*	2.5	37	27	38	3.0	--
28 JULY	3	--	--	5.0	40		--
29 JULY	2	--	--	32	39		--
30 JULY	1	2.0	43	36	35		--
31 JULY	2	2.0	71	32	35		--
1 AUG	2	1.5	73	--	--	--	--
2 AUG	2	2.2	64	--	--	--	--

28 Application of the Watanabe cross-shore transport model to prototype-scale data

B. A. O'Connor, J. Nicholson, N. MacDonald and K. O'Shea

ABSTRACT

This paper describes the application of a modified version of the Ohnaka & Watanabe (1990) cross-shore transport model to a prototype-scale laboratory situation. The results of this exercise are promising, although it has yet to be established that the current version of the model is valid for other laboratory and field situations.

1 Introduction

The shape of a beach profile is governed by the spatial distribution of the cross-shore sediment transport. The latter, in turn, is produced by wave assymetry in combination with the flow pattern in the vertical plane normal to the coastline. This flow pattern consists of an off-shore-directed undertow, located in the lower part of the water column, and an onshore-directed bore, located in the upper part of the water column. Cross-shore transport is the dominant process in short-term variations in coastal morphology but its importance descreases with lengthening of the time-scale. It follows, therefore, that this type of transport must be included in any mathematical model used to predict short-term or medium-term bed level changes but that its effects can be neglected in the case of long-term

337

prediction.

Considerable advances in cross-shore transport modelling have been made in recent years. The resulting models fall into three categories, namely empirical schemes, one-dimensional schemes and two-dimensional schemes. In the empirical approach, the equilibrium profile associated with a given wave condition is known a priori. The existing profile is then assumed to approach its equilibrium shape at a rate which is a function of the difference between these two profiles. Examples of empirical models are those set up by Swart (1976) and Perlin & Dean (1985). In the one-dimensional approach, all processes taking place in the vertical plane are collapsed onto a line and are represented by analogues, which, by their nature, are highly empirical. Models of this type have been developed by Bailard (1982), Watanabe (1985) and Ohnaka & Watanabe (1990). Two-dimensional models, on the other hand, attempt to simulate all the relevant processes. Such schemes have been set up by various authors, including Nairn (1990), Roelvink (1991), Broker-Hedegaard et al. (1991), Southgate (1991) and Latteux & Peltier (1992).

One-dimensional models, because of their simplicity, are computationally more efficient than their two-dimensional counterparts. Hence, if the simpler schemes can be shown to be sufficiently accurate, then they can still be used to predict morphological changes, especially those of a medium-term nature. The aim of the work described in this paper, therefore, is to investigate the accuracy of the one-dimensional approach to cross-shore transport modelling. To that end, a modified version of the Ohnaka & Watanabe (1990) model was tested against prototype-scale laboratory data (Dette & Uliczka, 1986).

2. Model Description

The model developed by Ohnaka & Watanabe (1990) consists of three components, namely a wave sub-model, a current sub-model and a sediment sub-model. However, the situation considered in the present study only involved waves approaching a coastline normally, which obviated the need for a current sub-model.

The wave sub-model is based on the work of Watanabe & Dibajnia (1988). It comprises a set of two equations equivalent to a time-dependent version of the mild-slope equation with an additional term representing energy dissipation in the surf zone. The equations are as follows:

$$\frac{\partial q}{\partial t} + C^2 \frac{\partial \zeta}{\partial x} + f_D \, q = 0 \tag{1a}$$

$$\frac{\partial \zeta}{\partial t} + \frac{1}{n} \frac{\partial}{\partial x}(nq) = 0 \tag{1b}$$

where t is time (s), x is the horizontal co-ordinate (m), C is the phase velocity (m/s), n is the ratio of group velocity to phase velocity, f_D is the energy dissipation factor (s^{-1}), ζ is the instantaneous water surface displacement from the mean water level (m) and q is a representative depth-integrated flow rate (m^2/s). The latter variable is given by:

$$q = q' - \frac{C^2}{n\sigma^2} \, \nabla n \, . \, \frac{\partial \zeta}{\partial t} \tag{2}$$

where σ is the wave frequency (s^{-1}) and q' is the actual depth-integrated flow rate (m^2/s):

$$q' = \int_{-h}^{\zeta} u \, dz \tag{3}$$

where u is the orbital velocity (m/s) and h is the water depth (m).

The energy dissipation term f_D represents the rate at which energy is dissipated through breaking. As such, f_D is maintained at a value of zero outside the surf zone but after breaking is given by the following expression:

$$f_D = \alpha_D \, \tan \beta \left[\frac{g}{h} \left(\frac{q_M - q_R}{q_S - q_R} \right) \right]^{\frac{1}{2}} \tag{4}$$

where g is the acceleration due to gravity (m/s^2), $\tan\beta$ is the bottom slope at the break point, α_D is a dimensionless coefficient and q_M is the amplitude of the flow rate q (m^2/s). q_R is the amplitude of the flow rate of a recovered or stable wave given by Watanabe & Dibajnia (1988) as:

$$q_R = 0.4 \left(\frac{a}{h} \right)_B C \, h \tag{5}$$

where a is the wave amplitude (m) and subscript B indicates the break point. q_S is the amplitude of the flow rate of a wave inside the surf zone on a constant slope (m²/s) given by Watanabe and Dibajnia (1988) as:

$$q_S = 0.4 \ (0.57 + 5.3 \tan \beta) \ C \ h \tag{6}$$

When the wave energy has been dissipated sufficiently for the amplitude of the flow rate q_M to be less than the corresponding stable value q_R then the waves are considered to be fully reformed and f_D is set equal to zero.

The location of the break point is determined by the following criterion:

$$\gamma = u_c/C \geq 1 \tag{7}$$

where u_c is the wave crest velocity (m/s). Values for the ratio γ have been determined empirically by Watanabe et al. (1984).

In order to simulate wave setup and setdown, the mean surface elevation is adjusted according to a simple momentum balance:

$$\frac{\partial S}{\partial x} = - \ \rho \ gh \ \frac{\partial \eta}{\partial x} \tag{8}$$

where ρ is the fluid density (kg/m³), η is the mean surface elevation (m) and S is the radiation stress (N/m) given as:

$$S = \rho \ \frac{g}{2} \ a^2 \ (2n - \tfrac{1}{2}) \tag{9}$$

The solution method is as follows. Equations 1a and 1b are solved on a staggered grid using a centred finite difference scheme to determine q and ζ. Since period-averaged values, such as the wave and flow rate amplitudes, are required for determination of the dissipation and mean surface elevation expressions, the latter can only be calculated after the first wave period has elapsed. A running average technique is used to ensure that the wave and flow rate amplitudes are as representative as possible of the current conditions. Since the only mechanism for wave breaking is the energy dissipation term, the model must be operated from a cold start (i.e. zero surface displacements throughout the domain) to ensure that larger than allowable waves do not proceed past the break point during the first wave period.

At the shoreline, a boundary condition of q=0 is applied.

At the offshore boundary, an absorbing boundary condition based on the method of characteristics is imposed. The value of ζ for the present step at the offshore boundary is determined as:

$$\zeta_1{}^t = \zeta_2{}^{t-\Delta x/Co} - a_i \sin(\sigma t) - \sin\{k\Delta x - \sigma(t - \Delta x/C_o)\} \quad (10)$$

where a_i is the incident wave amplitude (m), k is the wave number (rad/m), Δx is the grid spacing (m) and C_o is the phase velocity at the offshore boundary (m/s). The value of $\zeta_2{}^{t-\Delta x/Co}$ is determined from a time history of surface displacements at grid point 2.

The sediment sub-model is based on the sediment transport rate due to wave action proposed by Ohnaka & Watanabe (1990):

$$Q = (A_w(\tau_B - \tau_C) + A_{wB} \tau_T) F_D \hat{u}_B/(\rho g) \quad (11)$$

where Q is the sediment transport rate (kg/m/s), A_w and A_{wB} are coefficients (kg/m³), τ_B is the maximum near-bed shear stress due to wave action (N/m²), τ_C is the threshold of movement shear stress (N/m²), τ_T is an additional near-bed shear stress produced by breaker turbulence (N/m²), F_D is a dimensionless directional function and $\hat{u}_B$ is the maximum near-bed orbital velocity (m/s). The near-bed shear stress, τ_B, is computed using a method developed by Tanaka and Shuto (1981), the threshold shear stress, τ_C, is given as a function of the median grain size and the critical Shields parameter by Watanabe et al. (1986) and the shear stress due to breaker turbulence, τ_T, is based on the energy dissipation rate so that:

$$\tau_T = \rho^{1/3}(n\ f_D\ E)^{2/3} \quad (12)$$

where E is the energy density (N/m) and f_D is defined by Equation 4. In addition, the direction function is given by (Watanabe and Dibajnia, 1988) as:

$$F_D = \tanh(K_D(\Pi - \Pi_C)/\Pi_C) \quad (13)$$

where K_D is a dimensionless coefficient, Π is a direction indicator based on the ratio of the relative intensity of the orbital motion to the Ursell number (Watanabe et al., 1986) and Π_C is the value of Π at the null point for sediment transport. Lastly, the maximum near-bed orbital velocity, $\hat{u}_B$, is computed using linear theory.

Having derived the sediment transport rate with the aid of Equation 11, bed level changes are computed from the sediment mass continuity equation:

$$\partial z / \partial t = -\partial Q_M / \partial x \qquad (14)$$

where z is the bed elevation above an arbitrary datum (m) and Q_M is the sediment transport rate modified to allow for bed slope effects. This variable is described by:

$$Q_M = Q - \epsilon |Q| \tan \beta \qquad (15)$$

where ϵ is a dimensionless coefficient. Eq. 15 is solved using a finite difference approach which contains a two-step modified Lax-Wendroff scheme. The latter feature was included in order to reduce truncation errors introduced by the lack of time-centring of the sediment transport gradient. At the landward boundary, Q_M is set equal to zero while sediment is permitted to flow through the seaward boundary.

An additional feature of the present version of the sediment sub-model is the incorporation of a breaker transition length. This allows for the fact that turbulence generated by the surface roller after breaking is only distributed to the base of the water column whenever the broken wave has advanced a certain distance towards the shoreline. This distance, or transition length, was assumed to be synonymous with the hydrodynamic transition zone, which was first reported by Svenden et al. (1978) and is associated with a rapid decrease in wave height without a corresponding increase in turbulent energy dissipation. An investigation of the transition zone was carried out by Nairn et al. (1990) and a re-analysis of the relevant data by O'Shea et al. (1991) indicated that:

$$L_T = L_B (0.56 \; \xi^{-1.47}) \tan \beta \qquad (16)$$

where L_T is the transition length (m), L_B is the wave length at breaking (m) and ξ is the Iribarren Number defined by:

$$\xi = [\tan \beta / (H/L)^{\frac{1}{2}}]_B \qquad (17)$$

In the above equation, H is the wave height (m). The effect of the transition length is included in the model by switching off the contribution of the breaker turbulence to the sediment transport rate (see Equation 11) in the section of the surf zone immediately landwards of the break point.

The arrangement of the constituent parts of the model is set out in the flow diagram contained in Figure 1. As this figure shows, the sediment sub-model is operated more frequently than the wave sub-model. During each run of the former, the transport rates are updated to reflect the change in the local bed slopes (see Equation 15).

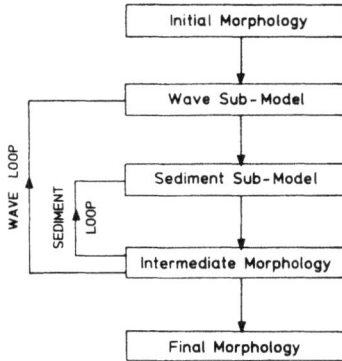

FIGURE 1   Model flow diagram

3 Model Tests

The model was tested against data measured in the proto-type-scale wave flume located at the University of Hannover (Dette & Uliczka, 1986).   Sand with a median grain size of $330\mu m$ was placed in the flume as shown in Figure 2 and this initial profile was then subjected to the action of monochromatic waves with a period of 6s and an incident amplitude of 0.75m.

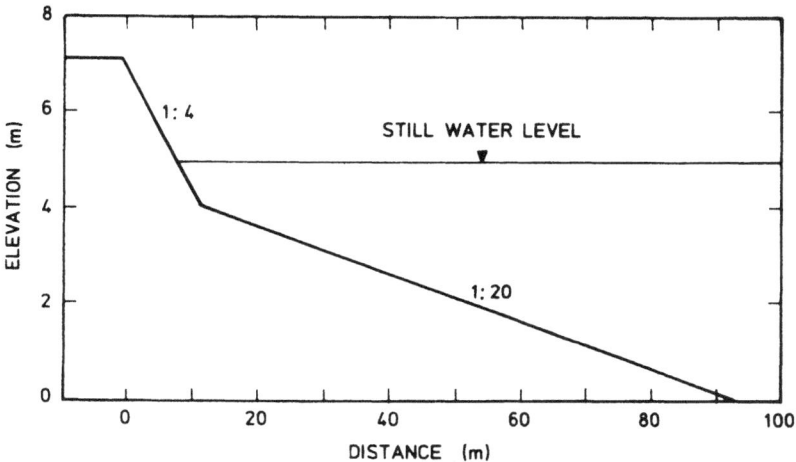

FIGURE 2   Layout of the Hannover flume

Testing of the model consisted of a comparison of the computed and measured beach profiles after 2730 cycles of wave action. However, in order to avoid operating the model with the highly artificial initial beach profile, that produced by 50 cycles of wave action was used as a starting condition. Following a sensitivity analysis, the best match between the computed and the measured profiles is as depicted in Figure 3.

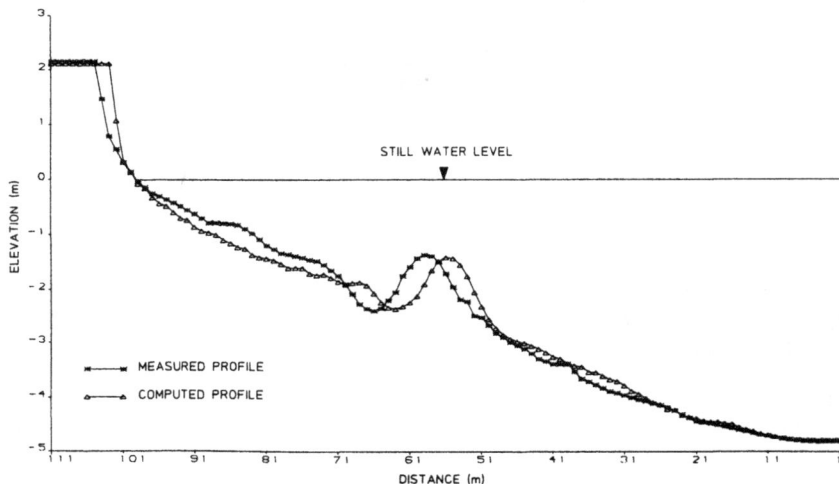

FIGURE 3 *Comparison of computed and measured beach profiles*

As Figure 3 indicates, the results of the comparison are satisfactory, relatively minor discrepancies in the predicted profile being the location of the break point bar too far offshore, a surfeit of erosion in the central part of the surf zone and a lack of erosion in the upper beach. Values used in this study for the various model coefficients described in Section 2, and the corresponding values recommended in the literature, are listed in Table 1.

TABLE 1   *Empirical model coefficients*

Coefficient	Value	
	Used	Recommended
α_D (Eq. 4)	5.0	2.5 (Ref. 7)
γ (Eq. 7)	0.42	≈ 0.4 (Ref. 17)
A_w (Eq. 11)	0.10	0.15 (Ref. 7)
A_{wB} (Eq. 11)	0.03	0.03 (Ref. 7)
K_D (Eq. 13)	0.5	≈ 1 (Ref. 16)
Π_c (Eq. 13)	0.16	0.16 (Ref. 18)
ϵ (Eq. 15)	3.0	≈ 1 (Ref. 7)

4 Discussion

The solution produced by a simple model, such as the one in question, may be so sensitive to small changes in the empirical model coefficients as to render the scheme of no practical use. This potential problem, therefore, was investigated by carrying out a sensitivity analysis. Typical results obtained from this exercise are contained in Figure 4, which shows the variation in the predicted beach profile with the coefficient A_{wB} (see Equation 11).

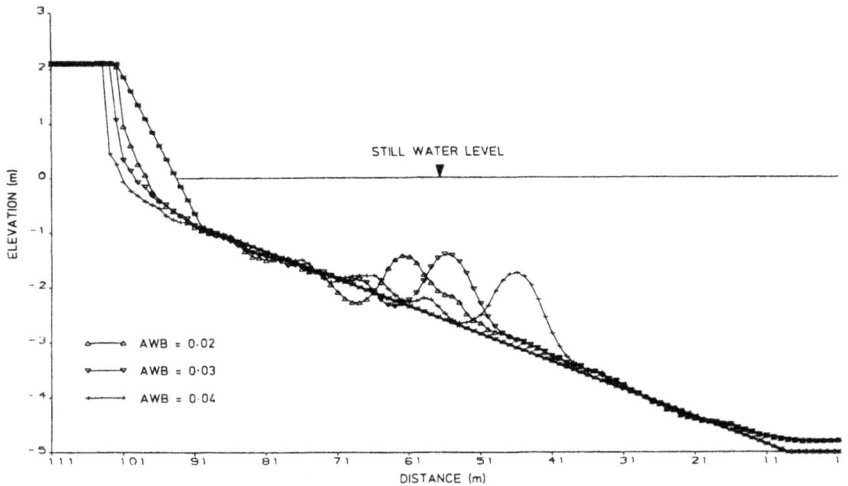

FIGURE 4 *Sensitivity of the model output to changes in the coefficient A_{wB}*

The contents of Figure 4, together with the results of similar sensitivity analyses involving the other empirical coefficients, suggested that the proposed scheme was not over-sensitive to the values adopted for these coefficients.

Another potential problem associated with simple models stems from the fact that the values chosen for the empirical coefficients may only be valid for a limited range of hydrodynamic and sedimentological conditions. This topic, however, can only be investigated by applying the scheme to other laboratory or field situations.

The proposed model generates sediment transport rates which exhibit large spatial gradients; for this reason, it was necessary to employ smoothing techniques. Various techniques were considered, the variables involved being the transport rates themselves, the bed level changes and the bed levels. In the event, a variant of bed level smoothing was adopted in which a limiting value was imposed on the change of slope at the point of interest. This technique was chosen, not only because it is effective, but also because it conserves sediment mass. The optimum limiting value for the change in slope was found to be 0.1 radians.

5 Conclusions

An existing simple one-dimensional model of cross-shore transport, that of Ohnaka & Watanabe (1990), has been modified by the inclusion of a transition length to allow for the delay in the distribution of breaker turbulence throughout the water column after breaking. The results obtained by testing this model against a prototype-scale laboratory situation are encouraging but it remains to ensure that the model has universal applicability by carrying out similar tests on other laboratory or field situations.

6 Acknowledgement

This work was undertaken as part of the MAST G6 Coastal Morphodynamics research programme. It was funded by the Commission of the European Communities, Directorate General for Science, Research and Development, under MAST contract no. 0035.

REFERENCES

[1] Bailard, J.A. (1982), Modeling On-Offshore Sediment Transport in the Surf Zone, *Proc. Eighteenth Coastal Engineering Conf.*, ASCE, pp 1419-1438.

[2] Broker-Hedegaard, I., Deigaard, R. and Fredsoe, J. (1991), Onshore-Offshore Sediment Transport and Morphological Modelling of Coastal Profiles, *Proc. "Coastal Sediments '91"*, Seattle, pp 643-657.

[3] Dette, H.H. and Uliczka, K. (1986), Velocity and Sediment Concentration Fields across Surf Zones, *Proc. Twentieth Coastal Engineering Conf.*, ASCE, pp 1062-1076.

[4] Latteux, B. and Peltier, E. (1992), A Two-Dimensional Finite Element System of Sediment Transport and Morphological Evolution, *Proc. Int. Conf. on Computer Modelling of Seas and Coastal Regions*, Southampton.

[5] Nairn, R.B. (1990), Prediction of Cross-Shore Sediment Transport and Beach Profile Evolution, *Ph.D. Thesis*, Imperial College, Univ. of London.

[6] Nairn, R.B., Roelvink, J.A. and Southgate, H.N. (1990), Transition Zone Width and Implications for Modelling Surface Hydrodynamics, *Proc. Twenty-Second Coastal Engineering Conf.*, ASCE, pp 68-81.

[7] Ohnaka, S. and Watanabe, A. (1990), Modeling of Wave-Current Interaction and Beach Change, *Proc. Twenty-Second Coastal Engineering Conf.*, ASCE, pp 2443-2456.

[8] O'Shea, K., Nicholson, J. and O'Connor, B.A. (1991), The Transition Zone Length in Cross-Shore Sediment Transport Modelling, *Report MCE/1*, Dept. of Civil Engineering, Univ. of Liverpool.

[9] Perlin, M. and Dean, R.G. (1985), 3D Model of Bathymetric Response to Structures, *Journal of Waterway, Port, Coastal and Ocean Engineering*, ASCE, Vol. 111, No. 2, pp 153-170.

[10] Roelvink, J.A. (1991), Modelling of Cross-Shore Flow and Morphology, *Proc. "Coastal Sediments '91"*, Seattle, pp 603-617.

[11] Southgate, H.N. (1991), Beach Profile Modelling: Flume Data Comparisons and Sensitivity Tests, *Proc. "Coastal Sediments '91"*, Seattle, pp 1829-1841.

[12] Svendsen, I.A., Madsen, P.A. and Hansen, J.B. (1978), Wave Characteristics in the Surfzone. *Proc. Sixteenth Coastal Engineering Conf.*, ASCE, pp 520-539.

[13] Swart, D.H. (1976), Predictive Equations regarding Coastal Transports, *Proc. Fifteenth Coastal Engineering Conf.*, ASCE, pp 1113-1132.

[14] Tanaka, H. and Shuto, N. (1981), Friction Coefficient for a Wave-Current Coexistent System, *Coastal Engineering in Japan*, Vol. 24, pp 105-128.

[15] Watanabe, A. (1985), Three-Dimensional Predictive Model of Beach Evolution around a Structure, *Proc. Water Wave Research Conf.*, Hannover, pp 121-141.

[16] Watanabe, A. and Dibajnia, M. (1988), A Numerical Model of Wave Deformation in Surf Zone, *Proc. Twenty-First Coastal Engineering Conf.*, ASCE, pp 578-587.

[17] Watanabe, A., Hara, T. and Horikawa, K. (1984), Study on Breaking Condition for Compound Wave Trains, *Coastal Engineering in Japan*, Vol. 27, pp 71-82.

[18] Watanabe, A., Maruyama, K., Shimizu, T. and Sakakiyama, T. (1986), Numerical Prediction Model of Three-Dimensional Beach Deformation around a Structure, *Coastal Engineering in Japan*, Vol. 29, pp 179-194.

29 Inception of sediment on a level or sloping bed due to waves

J. C. Morfett

ABSTRACT

This paper considers the problem of evaluating the threshold of movement of sediment on beaches under incoming waves. It is suggested that a threshold criterion should be based upon energy dissipation, in preference to shear stress. A model is developed in which the rate of energy dissipation due to waves is related to the conditions for the initiation of particle movement for horizontal and sloping beds..

Introduction

Most sediment transport formulae are based on an "excess value" of some parameter of the flow. A typical example is the excess shear stress $(\tau - \tau_{cr})$ at the bed. The subscript "cr" refers to the threshold condition, below which no transport occurs.

There have been a number of studies of the threshold of motion under waves based on the assumption that shear stress is the cause of sediment transport [eg. 3, 4, 5, 6, 7]. However, in a coastal environment sediment transport may be produced by currents, by unbroken waves and currents or by breaking waves with or without currents. For the first two cases the use of shear stress is reasonable but wave breaking generates intense turbulence, which must affect the threshold of movement. For example, on a typical gravel beach, with a slope of about 10%, plunging breakers form. The resultant force at the beach surface at the point of impact is due mainly to the rate of change of momentum rather than shear force, though further up the beach (in the uprush) and in the backwash the sediment will be subjected to shear.

In this environment there are significant advantages in using energy dissipation, Dw as a basic parameter of the coastal flow field, rather than shear stress. Energy dissipation occurs whether waves are breaking or unbroken and it is possible to base a transport parameter on energy dissipation rates [8, 9].

In principle such a parameter should be equally applicable outside or inside the surf zone. A corresponding quantity, which is usually known as the 'dissipation velocity', may be defined as $u_+ = (D_w/\rho)^{0.33}$, where ρ = fluid density. In turn this may be used to produce a version of the Reynolds Number ($Re_+ = \rho u_+ D/\mu$, where D is the sediment size, μ = the viscosity). This is analogous to shear velocity (u_*).

Another reason for a continuing interest in the threshold is that the amount of research that has been undertaken into the threshold for coarse sediments is limited [2, 7, 13]. This paper is therefore concerned with the definition of a suitable threshold criterion for transport on sand or gravel beaches under attack from incoming waves.

A simple model of wave action on a granular bed

Model of an unbroken wave

Seaward of the break point, the passage of a wave produces an oscillatory motion at the sea-bed, so the energy is dissipated by bed friction alone. The peak dissipation due to friction, D_{fw}, may be estimated from the product of shear stress and the maximum near-bed velocity, U_w of the wave. The shear stress is estimated from an appropriate friction factor, C_{fw}. Hence the frictional dissipation is:

$$D_{fw} = \rho \, C_{fw} \, U_w^3 \tag{1}$$

The method for calculating the friction factor is given by O'Connor and Yoo [11]:

$$C_{fw} = \exp [5.213 \, (a/K_s)^{-0.194} - 6.67] \quad \text{(for } a/K_s > 2\text{)}$$
or $\hspace{9cm}$ (2)
$$C_{fw} = 0.12 \text{ (for } a/K_s \leq 2\text{)}$$

a is the maximum length of orbital excursion at the bed and K_s is the effective (Nikuradse) bed roughness height.

The velocity U_w is calculated using linear wave theory. This leaves the problem of determining K_s. A simple approach due to van Rijn [14] is to take K_s as $3D_{90}$ ($\approx 6D_{50}$), though it must be pointed out that this relationship was derived for uni-directional flows.

Model of a breaking wave

When waves break energy losses occur due to a combination of bed shear and the dissipation of turbulent kinetic energy. The dissipation due to bed shear may be calculated by equation (1) above. This leaves the problem of estimating the effect of wave breaking. The idea of using bore theory to model breakers originated with LeMehaute and was subsequently developed by Battjes and Janssen [1]. The model used here is an adaptation by Morfett [10] of the concept of Battjes and Janssen which is designed to provide a measure of the maximum dissipation rather than the average dissipation over a wave length.

Taking a bore for which the upstream depth is Y_1, and the downstream depth is Y_2 (see figure 1), the dissipation rate (per unit width across the front of the bore) is

$$D_B = 0.25 \, \rho \, g^{1.5} \, (Y_2 - Y_1)^3 \, ((Y_1 + Y_2)/2Y_1Y_2)^{0.5} \tag{3}$$

where g is gravitational acceleration.

350

Making the assumptions that wave height $H = (Y_2 - Y_1)$, water depth, $h = Y_1$ and that $H = \gamma h$, (3) may be written as

$$D_b' = 0.177 \, \rho \, g^{1.5} \, H^{2.5} \, (\gamma(\gamma + 2)/(\gamma + 1))^{0.5} \tag{4}$$

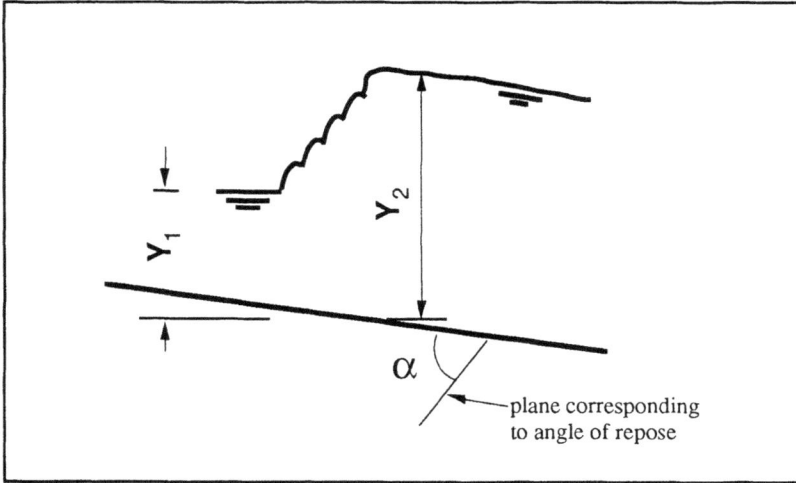

Figure 1. *Definition sketch*

If all the dissipation due to breaking is confined within the 'roller' length L_B, then the rate of dissipation per unit area is $D_b'/L_B = D_{bw}$. An expression for L_B due to Sarma and Newnham [12] is $L_B = 6.71 \, h \, (F - 1)$, where F is the Froude Number for the flow upstream of the bore. Assuming the standard relationship

$$1 + 8F^2 = ((2Y_2 + Y_1)/Y_1)^2 \tag{5}$$

it can easily be shown that

$$F = (((2\gamma + 3)^2 - 1) /8)^{0.5} \tag{6}$$

Hence if γ is known D_{bw} can be evaluated.

A number of formulae for γ are available, the simplest being $\gamma = 0.78$, however, the equation due to Morfett [8] has the advantage of being applicable to wave breaking over level and sloping beds.

The model is only conceptual, since the distribution of dissipation is assumed to be uniform over the roller length, which is not physically correct. From the foregoing the total dissipation for the incoming breaker is $D_{bw} + D_{fw} = D_w$.

The simplest approach to modelling the downslope (backwash) flow on a beach is to assume that the maximum velocity may be obtained from linear wave theory, and hence to evaluate the dissipation rate from equation (1). Clearly this is only a rough and ready approximation, as the actual flow in the backwash would be a function of the incident wave (height etc.), the sediment size, porosity and beach slope.

The immersed weight of a particle in liquid is

$$W = f(D^3, (\rho_s - \rho), g) \tag{7}$$

where ρ_s is the density of sediment. The weight of the grain may be resolved into components perpendicular to the beach slope (= W cos S) and parallel to the slope (W sin S). Also parallel to the slope is a component of the force due to friction. Thus (assuming a constant friction angle) the resultant force $R = f(D^3, (\rho_s - \rho), g, S)$.

The usual definition of the beach slope is as the tangent of the angle (or angle in radians) between the plane of the beach and the horizontal. However, this means that a horizontal bed has a slope of zero. This raises a problem in incorporating results for the threshold of movement on a horizontal bed into a functional relationship derived from dimensional considerations, since a product involving zero is a trivial result. Furthermore, the threshold dissipation presumably reduces to zero if the slope of the beach is at the angle of repose for the sediment. For this analysis the definition has therefore been modified such that the angle of slope, α, is the included angle between the plane corresponding to the angle of repose and the plane of the beach surface as shown on figure 1. (The angle of repose is assumed to be 0.52 radians (30 degrees) for gravel and for coarse sand).

At the threshold a small proportion of the particles at the bed (p) will just be levered over neighbouring particles (through height δy) during time δt. For unit area the number of particles involved is proportional to p/D^2. The time interval is presumably related to the time scale of turbulent eddies, so $\delta t = f(g\, D\, T/h)$, where T is the periodic time and the height δy is assumed to be related to D. The rate of energy input required per particle is $f(R\delta y/\delta t)$, so the energy input required per unit area is

$$Dp = f(D^3 (\rho_s - \rho)\, g\, (p/D^2)\, \alpha\, (g\, D\, T/h)) \tag{8}$$

This is equated to the "output" (dissipation rate) from the waves, D_w.

Data used for calibration of model

Because of the limited knowledge of flow and sediment transport processes, empirical techniques are used to develop and calibrate the threshold equation. The data used in the analysis have been derived from two sets of experiments. The first (designated HR) was derived from some water tunnel experiments undertaken by Hydraulics Research Ltd [2]. The bed was horizontal. The second set (designated BP) was derived from experiments in the wave flume facilities at Brighton Polytechnic. Waves were produced by wave paddles, at the upstream end of the flume, providing a monochromatic wave field. Experiments were undertaken for the following:

Horizontal bed of sediment under non-breaking waves.

Horizontal bed of sediment under breaking waves. To produce wave breaking a horizontal platform was positioned in the flume with the sediment bed laid out over the platform.

Sloping bed of sediment under breaking waves.

In developing the threshold criterion the following ranges have been covered: Still water depth: $0.1 < h < 2m$, Sediment particle size: $0.0002 < D < 0.06m$, Wave height: $0.01 < H < 1.7m$, Wave period: $1 < T < 8s$, Bed slope (relative to the angle of repose): $0.8 < \alpha < 1.31$ radians.

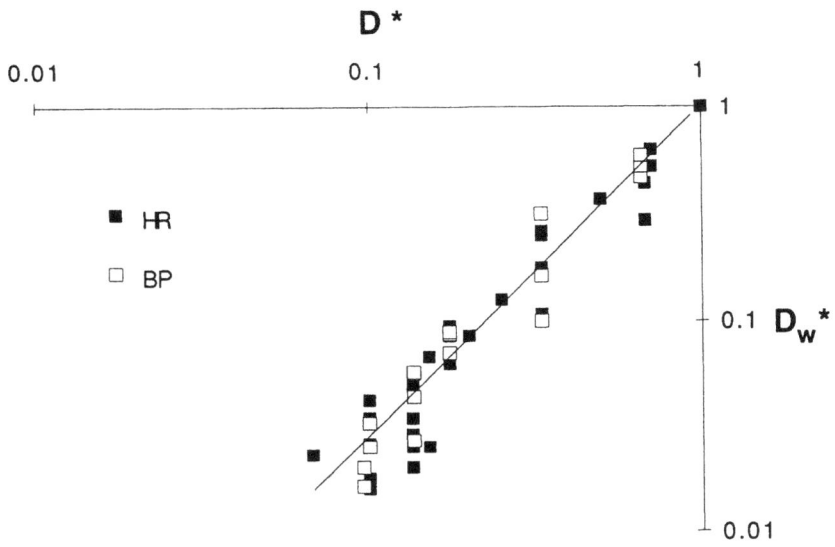

(a) Sediment size and threshold energy dissipation

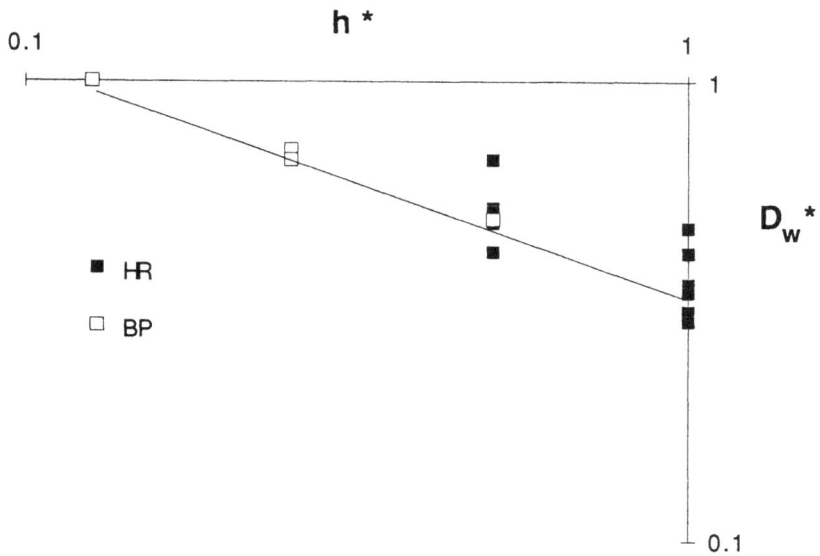

(b) Water depth and threshold dissipation

Figure 2. *Illustrative summary of regression analysis*

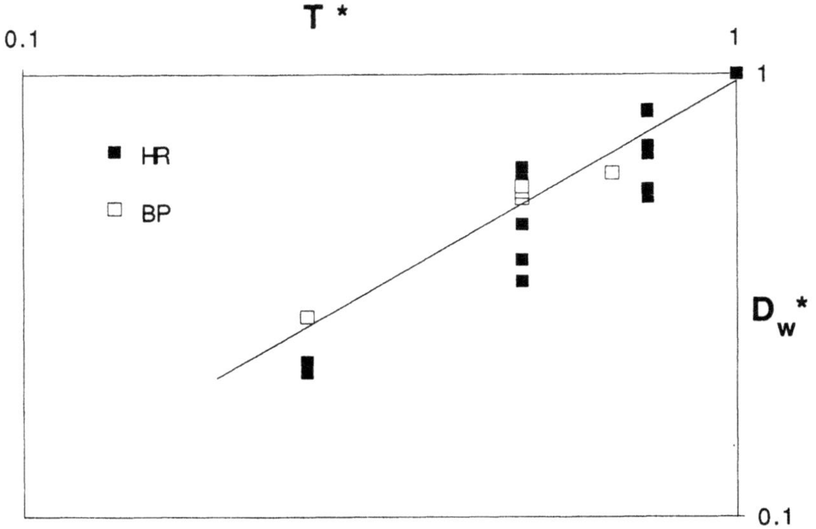

(c) Wave period and threshold dissipation

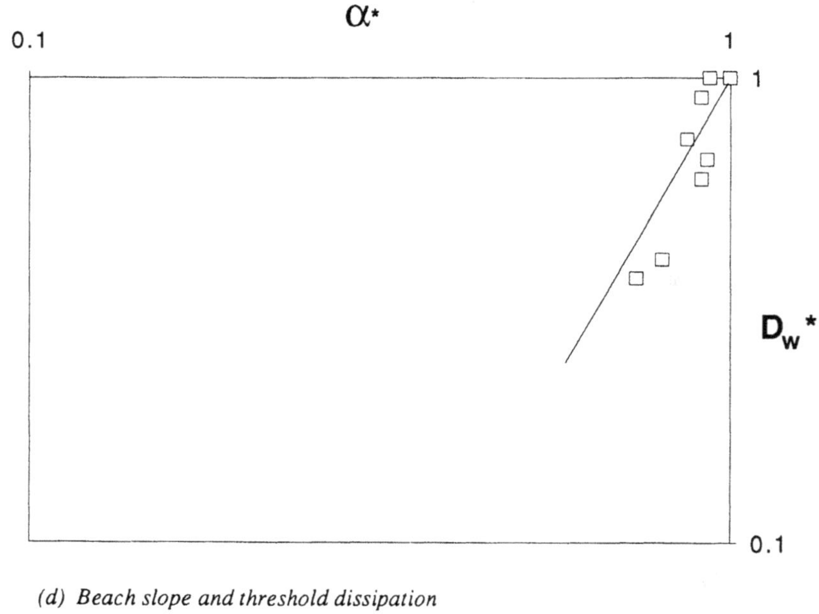

(d) Beach slope and threshold dissipation

Figure 2 (continued) *Illustrative summary of regression analysis*

Model calibration

It is convenient to reduce the data to a dimensionless form for analysis. Non-dimensional relationships are produced by using ratios, eg. dimensionless depth, h^*, is the ratio h/h_{REF}, where h_{REF} is some reference value.

For each particle size, slope, water depth, wave height and period the threshold conditions are evaluated. Hence a series of sub-sets of the data can be identified to isolate the influence of each variable in turn. The results of the regression analysis are illustrated on figure 2.

To determine the relationship between threshold dissipation and particle size, the ratio $D^*(=D/D_{REF})$ is plotted against $D_w^*(=D_w/D_{wREF})$. The reference value (of D_w^* or D^*) is the maximum from the sub-set. Hence it is found (see figure 2a) that D_w varies with $D^{1.62}$. A similar procedure is adopted to determine the relationship between D_w^* and water depth (h^* defined above), as shown on figure 2b from which the threshold varies with $h^{-1/2}$. For a given particle size, slope and water depth, the effect of wave period is shown on figure 2c ($T^*=T/T_{REF}$). The effect of beach slope (represented by the dimensionless ratio $\alpha^* = \alpha/\alpha_{REF}$) on the threshold is given on figure 2d. Hence the threshold dissipation varies as T and α^3

Based on the foregoing the threshold function is in the form:

$$F_T = K\left(\left(\frac{D_w}{(\rho_s - \rho)gD^{1.12}}\right)\left(\frac{h^{1/2}}{gD^{1/2}T}\right)\left(\frac{1}{\alpha^3}\right)\right) \qquad (9)$$

where K is a number which may not be a constant. Equation 9 is not quite dimensionless, as there is a residual $D^{0.12}$.

The threshold diagram

The threshold diagram (figure 3) has been laid out in a form which is analogous to the Shields diagram. F_T is plotted against Re_+, and it is evident that most of the data lie within a well defined band.

Discussion

The flow at the seabed due to the passage of a wave is unsteady, hence D_w will vary with time. Clearly sediment threshold will first be attained where D_w is a maximum, and it can be argued that a threshold criterion should be based on peak values rather than on values which are assumed to be uniform over L_B. However, the supposition has to be made that there is a consistent relationship between the values thus calculated and the peak values.

No allowance has been made for the presence of the longshore currents which are generated by wave breaking. Where these are known to be present the value of the bed shear stress will be modified (O'Connor and Yoo [11]).

The range of values of Re_+ covered here is 4 to 30000. The mean value of F_T from figure 3 is 0.0116, for Re_+ greater than 100. The data points exhibit a trend to increase at the lower values of Re_+, rather like the line on Shields diagram but the slope is less pronounced. It is usually assumed that fully turbulent boundary layers occur for Re_* greater than 400, which very roughly corresponds to $Re_+ = 600$. (Though it should be emphasised that the relationship between Re_* and Re_+ is not a simple one which could be represented by a ratio)

According to Komar and Miller [3], for grain sizes less than 0.5mm, the threshold of motion

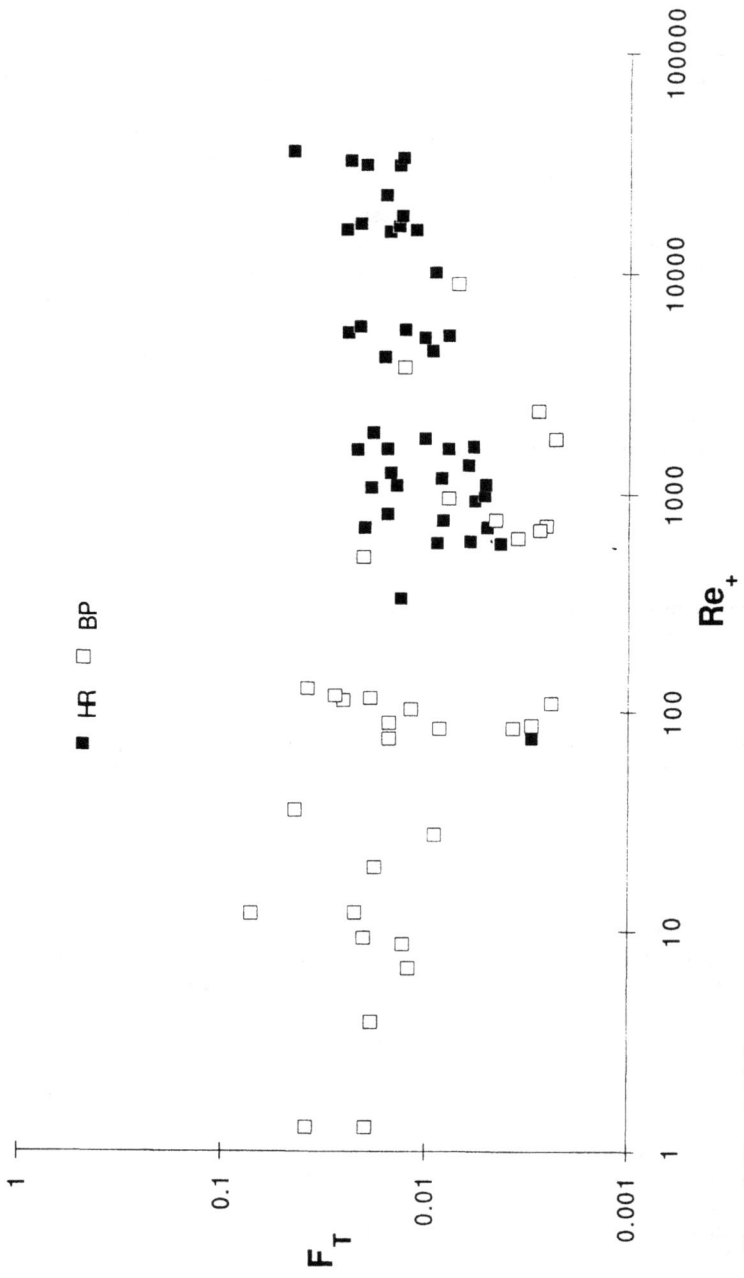

Figure 3 Threshold diagram

is attained before the transition from a laminar to a turbulent boundary layer. From the threshold experiments undertaken at Brighton this corresponds to $Re_+ < 10$.

One parameter which is not explicitly considered is the angle of attack (ie. the angle between the wave crest and a beach contour line. If the angle of attack is zero, then the uprush travels up the slope orthogonally to the contour lines. If the angle of attack has some finite value, than the effective slope (ie. the slope up which the uprush travels) is correspondingly reduced (in the limit for an angle of attack of 90 degrees the wave crest would be orthogonal to the contour, and the slope relative to the wave front would be zero). It is usually assumed that in the uprush sediment transport is related to angle of attack, but in the backwash it is orthogonal to the beach contours. Hence it is possible to postulate that the threshold as defined above would be valid for waves at an angle to the beach contour, but with effective beach slope (in the direction of the uprush) substituted for α. However, it was not possible to take measurements to confirm this in the wave flume.

It is not possible to show that the method of evaluating K_s is satisfactory. There are obvious difficulties in defining and measuring K_s on beaches. If berms and bars are included then K_s would exceed the value of $3D_{90}$ adopted here. For the present it has to be assumed that the influence of K_s is accounted for by C_{fw} and D in the threshold relationship (equation (9)).

Attempting to determine the threshold on a sloping bed under waves necessarily implies an ability to discriminate between initiation of upslope sediment motion and initiation of downslope movement. In practice this can be extremely difficult. For example, with the steeper slopes the tendency is for downslope movement to occur first. Hence if the incident wave conditions are sufficient for threshold of motion for upslope movement, considerable downslope transport will be occurring.

The difficulties in obtaining reliable readings are clearly greater for a sloping bed than a flat bed, so it is not surprising to note the extent of the scatter on most of the graphs

The combination of bore theory and linear wave theory to produce a model of a breaking wave is attractive because it means that the model is relatively easy to use. Nevertheless, it is only a conceptual approach to the problem of simulating the complex flow field produced when waves break over a sloping bed. From observations of the breaking waves in the laboratory experiments and in the field, the method of calculating the roller length, L_B, may tend to produce an overestimate. The shortcomings of the model only serve to emphasise the need for further advances in our understanding of the physics of the processes involved in the breaking of waves.

The threshold formula clearly cannot be applied to the zone which is above the mean still water level, but within the uprush

Conclusions

The model derived above, based on the use of bore theory, permits a unified treatment of threshold conditions under breaking and non-breaking waves. The scatter on figure 3 is not unreasonable in view of the difficulties for an observer in identifying the point at which sediment motion is commencing under the waves.

References

1. Battjes, J A, and Janssen, J P F M,. (1978). *Energy loss and set-up due to breaking of random waves*. ASCE, Proc 16th Conf. Coastal Engineering, pp569 - 587.

2. Hydraulics Research Ltd (No date). *Movement of shingle subject to wave action.* Hydraulics Research Ltd. Summary 115.

3. Komar, P D, and Miller, M C, (1973). "The threshold of sediment movement under oscillatory waves". *Journal of Sedimentary Petrology*, Vol 43, pp1101 - 1110.

4. Komar, P D, and Miller, M C, (1974). *Sediment threshold under oscillatory waves.* ASCE, Proc 14th Conf. Coastal Engineering, pp765 - 775.

5. Komar, P D, and Miller, M C, (1975). "On the comparison between the threshold of motion under waves and currents with a discussion of the practical evaluation of the threshold". *Journal of Sedimentary Petrology,* Vol 45, pp362 - 367.

6. Madsen, O S, and Grant, W D, (1975). "The threshold of sediment movement under oscillatory waves: a discussion". *Journal of Sedimentary Petrology,* Vol 45, pp360 - 361.

7. Manohar, M, (1955). *Mechanics of bottom sediment movement due to wave action.* US Army Corps Engineers, Beach Erosion Board, Technical Memorandum No 75.

8. Morfett, J C, (1989). "The development and calibration of an alongshore shingle transport formula. *Journal of Hydraulic Research ,* Vol 27, No 5, pp717 - 730.

9. Morfett, J C, (1990). "A 'virtual power' function for estimating the alongshore transport of sediment by waves". *Coastal Engineering,* Vol 14, No 3, pp439 - 456.

10. Morfett, J C, (1992). "Threshold of motion of sediment under waves in shallow water" (to be published).

11. O'Connor, B A, and Yoo, D, (1988). "Mean bed friction of combined wave/current flow". *Coastal Engineering,* Vol 12, No 1, pp1 - 21.

12. Sarma, K V A, and Newnham, D A, (1973). "Surface pofiles of hydraulic jump for Froude number less than four". *Water Power,* April 1973, pp139 - 142.

13. van Hijum, E and Pilarczyk, K W, (1982). *Equilibrium profiles and longshore transport of coarse material under regular and irregular waves.* Delft Hydraulics Laboratory Publication NO 274.

14. van Rijn, L C, (1984), "Sediment transport, part III, bed forms and roughness". ASCE, *Journal of Hydraulic Engineering,* Vol 110, No 12, pp1733 - 1754.

30 Beach profile changes

F. M. Abdel-Aal

ABSTRACT

This paper investigates the changes that occur in beach profiles. The onshore-offshore sediment transport and beach change models are also discussed. The Nile Delta beach profiles are presented.

1. INTRODUCTION

Beaches change in response to the interaction of waves. Since the type of waves that break on the beach are not the same from day to day, the beach profile does not remain the same. In general, beaches are in a state of dynamic equilibrium. Their deformation due to change in wave climate is rather seasonal causing erosion or accretion. Prediction of beach changes is essential in order to take protective measures.

Beach profiles in nature are continuously evolving under the varying action of waves, currents, tides, and sediment transport. When these effects are maintained constant, the profile will stabilize into a so-called "equilibrium beach profile". This could be defined as the condition at which the supply of sediment at a given reach is equal to the loss of

sediment from that reach. The equilibrium slope of
the beach is a useful design parameter, since this
slope along with the berm elevation. determines the
minimum beach width.

2. BEACH PROFILE

The slope of the beach depends on the grain size of
the sediment and on the incident wave steepness.
However, beach slopes are difficult to define since
they are seldom planar and more than one slope can be
present across the beach face. Figure (1) shows a
series of results for beach slope against sediment
size and wave steepness summarized by Carter, Liu,
and Mei (4). Generally, the larger the sediment grain
size, the steeper the beach slope. Beaches with
gently sloping foreshores and inshore zones usually
have preponderance of finer sediment sizes (22).
Figure (2) gives the effect of increased wave energy
to cause a milder slope for a given sediment size
(23).
    Using data from twenty-seven beaches, King (9)
obtained the empirical formula,

Beach slope = 407.71 + 4.2 Φ (D) - 0.17 Log E   (1)

Where (E) is the wave energy, and (D) the grain size
in phi units.

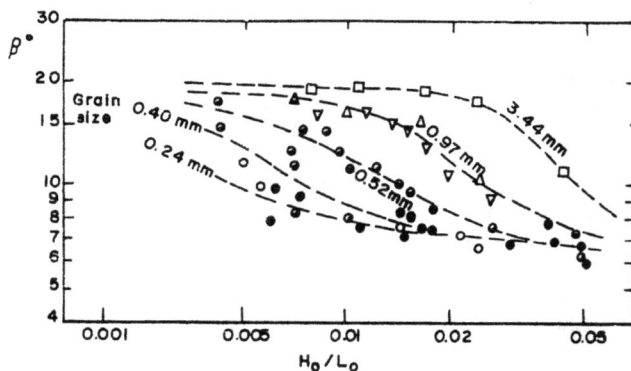

Fig. (1) Beach Slope vs. Wave Steepness [4].

The majority of the sediment that is moved on a beach is moved by the forces generated by the waves impinging on the beach. The sediment moves primarily in two directions, perpendicular to the face of the beach by onshore-offshore transport; and parallel to the face of the beach by longshore transport. The former motion occurs temporarily and is normally restored, whileas the latter motion has a much more continuous character. Sediment movement resulting in erosion or accretion of beach section is actually an interaction of these two types of motion.

There are two broad classifications which are commonly used to describe most beach profiles, the summer profile and the winter profile. According to Johnson (15), summer profiles are created by gentel waves (Ho/Lo < 0.025), where (Ho) and (Lo) are the wave height and length in deep water. Winter profiles are created by steep waves (Ho / Lo > 0.025).

One factor which has an effect on the stability of a beach is the level of the water table. Grant (12) stated that there was a definite relationship between the height of the water table under the beach and the stability of the beach. He found that a high water table accelerated erosion and a low water table retarded erosion and aided accretion on the foreshore.

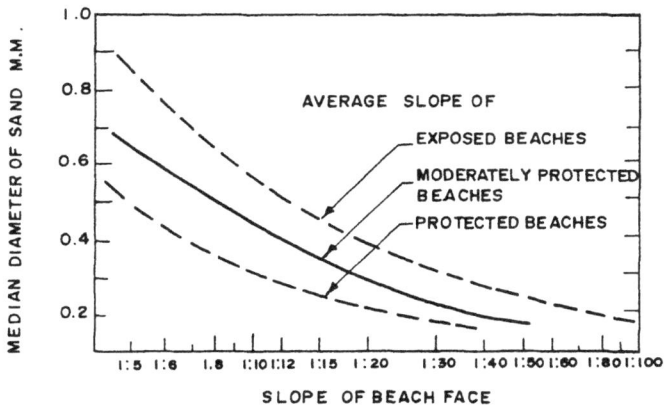

Fig(2) Size_Slope Relationship of Beach [23]

3. ONSHORE-OFFSHORE TRANSPORT

In order to maintain an equilibrium beach profile, considerable onshore-offshore sediment transport is necessary under conditions of varying incident wave steepness (Ho/Lo). Here, Ho and Lo are the wave height and wave length in deep water. In general, high steep waves move material offshore, and low steep waves of long period move material onshore (22).

It was found from laboratory experiments that, the transition from "bar profiles" associated with high steepness ratio to "swell profiles" which has no bar but a berm on the beach, takes place at (Ho/Lo)≈(0.024-0.028). In the field a similar tendency was noted but with a much lower (Ho/Lo) values (3).

There is a tendency for the foreshore to become steeper as grain size increases; and to become flatter as mean wave height increases. For flat slopes with bed sediment being smaller than the boundary layer thickness, net sediment transport is mainly onshore because of weak reflection. If the reflection coefficient is large as on steep slopes or with low amplitude waves, sediment movement may occur offshore. Manohar (19) indicated that offshore movement eventually stops in deep water when mass transport becomes negligible at the bottom.

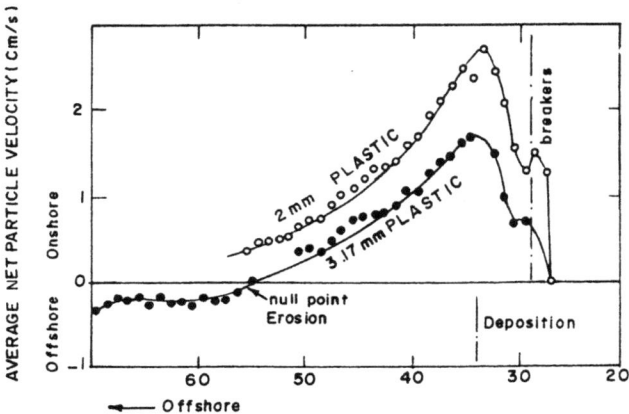

Fig.(3) Movement of Sediment Sizes [14].

Laboratory experiments have shown that seaward of the breaker line there is a neutral line, or null point, where though there is an oscillation of the sediment particles, there is an equilibrium between the force of gravity and the wave-induced forces-seawards of the null point, there is an offshore movement and land ward of it movement towards the shore. The position of the null point varies with the wave steepness, for increased steepness the null point moves shoreward.

Ippen and Eagleson (14) showed that inside the null point the shoreward directed grain velocities increased to a maximum just seaward of the break line, Fig (3). This would lead to an accumulation of sediment just shoreward of this maximum where the velocities diminish, and leads to the formation of a break point bar. The null point would be an area of net erosion with material moving both seaward and landward from it. Ippen and Eagleson's measurements were for a single beach slope of (1:15) but they found the neutral line occurred where,

$$(\frac{H}{d})^2 \; (\frac{L}{H}) \; (\frac{C}{v_f}) \; = \; 11.6 \tag{2}$$

where (H, L, and C) are the local wave height, wave length, and wave celerity; (d) is the water depth; and (v_f) is the sediment settling velocity. This result implies that each grain size has a different null point, the one for fine grains being further seaward than that for coarser grains, which could lead to a sorting of the different size fractions.

Hattori and Kawamata (13) derived a relationship indicating the important parameters controlling onshore- offshore sediment transport as,

$$\frac{(H_0/L_0) \; \tan \beta}{v_f \; / \; gT} \begin{array}{l} < \quad \text{(onshore transport)} \\ = c \quad \text{(neutral)} \\ > \quad \text{(offshore transport)} \end{array} \tag{3}$$

in which (β) is the beach slop, (v_f) is the fall velocitysediment grains determined for the median diameter, (g) is the acceleration of gravity, (T) is the wave period, and (c) is a constant to be determined from laboratory and field data.

They classified beach profiles into accretive and erosive profiles. From experimental data, the two profiles were found to be separated by the line of c= 0.5. This c- value also represents the equilibrium condition of profile change of two-dimensional beaches. The obtained criterion is,

$$\begin{array}{ll} & \text{(onshore transport, accretive profile)} \\ c = 0.5 & \text{(neutral, equilibrium profile)} \qquad (4) \\ & \text{(offshore transport, erosive profile)} \end{array}$$

4. EQUILIBRIUM BEACH PROFILE

Beach profiles are continuously changing under the action of waves, currents, tides, and sediment transport. If these variables were maintained constant, the profile would stabilize into a so-called "equilibrium beach profile." . Although beaches in nature may never achieve equilibrium, the concept of an equilibrium beach profile is generally useful in problems and interpretations involving beaches (8).

Dean (8) analyzed many beach profiles from nature and laboratory and found that the profiles could be described by,

$$h(x)=A.\chi^{m} \qquad (5)$$

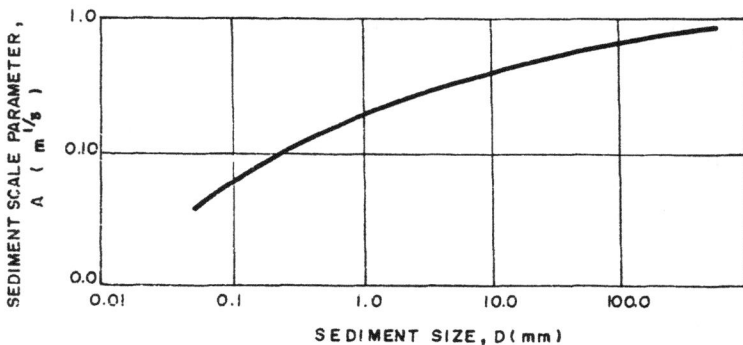

Fig.(4) Beach Profile Parameter ,A , [8] .

in which (χ) is the distance offshore to a water depth (h), (A) is a so-called "Scale Parameter" which depends primarily on sediment size, and m is a shape parameter and equals $^2/3$.

Figure (4) presents A vs. D which is recommended if no information is available describing the particular wave height characteristics. It is clear from this figure that beaches composed of larger diameter sediments are steeper, whereas finer grained beaches are characterized by smaller A values and thus are milder in slope. The second representation of (A) is in Fig (5) where the effect of the sediment size is represented by the fall velocity (v_f), and the effect of wave is given by the breaking wave height (H_b), and wave period (T). It is obvious from this figure that steep slopes are associated with large fall velocities, small wave heights, and long wave periods. Whileas, mild slopes are associated with small fall velocities, large wave heights, and short wave periods.

Eagleson, Glenne, and Dracup (10) attempted to calculate equilibrium beach profile under certain specified wave conditions starting from the null point in the offshore region to the nearshore. Their computed curve seems to fit the experimental profile up to a certain point, then the discrepancy between the predicted curve and the experimental curve becomes fairly large.

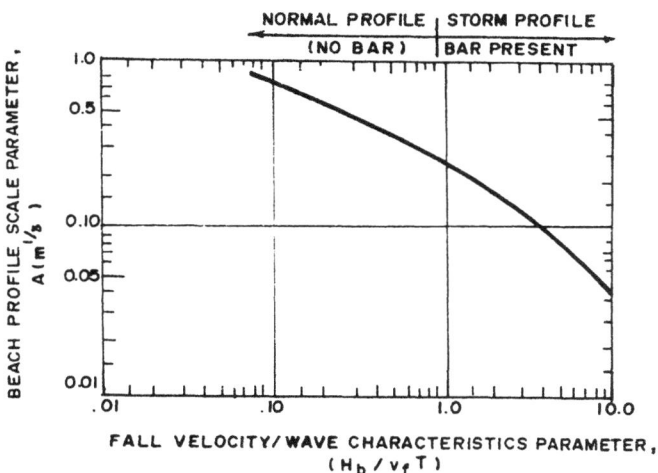

Fig. (5) Beach Profile Scale Parameter , A ,[8]

365

Dalrymple and thompson (6) proposed the use of a dimensionless fall velocity of the sediment for determining equilibrium foreshore slope angle (θ), Fig. (6). The parameter was introduced by Nayak (20) on dimensional grounds, and by Dean (7) based on physical argument. Dean found that the wave steepness (H_o/L_o) separating bar profiles from normal profiles of the beach varied with the fall velocity of the sediment (v_f) divided by the deep water wave celerity ($C_o = L_o/T$), where (T) is the wave period,

$$\frac{H_o}{L_o} = 0.85 \frac{v_f}{C_o} \tag{6}$$

or $\quad \dfrac{H_o}{v_f \cdot T} = 0.85 \tag{7}$

The fall velocity (v_f) was calculated using the median grain size (D_{50}). Kohler and Galvin (16) using different data recommended that,

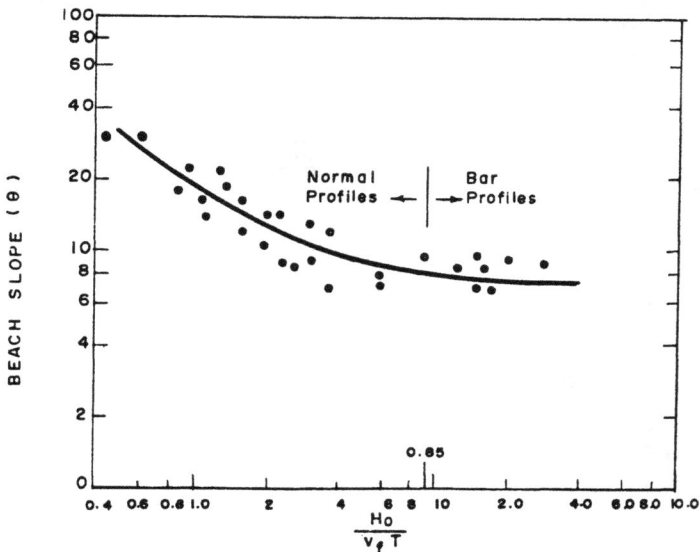

Fig. (6) Beach Slope Versus $\dfrac{H_o}{v_f T}$ [6].

366

$$\frac{H_0}{v_f \cdot T} = 0.7 \qquad\qquad\qquad (8)$$

The CERC Shore   Protection Mannual (22) recommends,

$$\frac{H_0}{v_f \cdot T} = (1.0\text{-}2.0) \qquad\qquad\qquad (9)$$

5. BEACH CHANGE MODELS

Beach changes are determined by the interaction of sea waves and the sediment forming the beach. Prediction of the beach change is essential in order to take protective measures. This can be achieved by the use of physical, analytical, and numerical models.

Bakker, Bretler and Ross (2) extended the theory of Pelnard Considere by representing the beach profile by two contours. In this case there are two governing equations which incorporate the effect of the onshore or offshore motion between the two contours due to nonequilibrium beach slope.

Fleming and Hunt (11), using both bed-load and suspended load mechanism, formulated a three-dimensional model for a specific beach segment. Changes in the beach profile were caused by longshore sediment motion.

Sunamura (21) proposed a model for two-dimensional beach changes based upon laboratory onshore-offshore transport measurements.

Dally (5) developed a numerical model for beach profile evolution. This model is based on suspended sediment transport mechanism. The sediment concentration profile is assumed to be exponential and the water column is divided into two layers in which the time averaged sediment transport velocity is developed separately.

Kyungduck (17) showed that the equilibrium beach profile resulting from an initially plan beach could be schematized by two different cases. The beach advancement as well as recession is given by,

$$\frac{R}{W_c} = \frac{1}{2} \left(\frac{6m_e}{5m_i} - 1 \right) \qquad\qquad\qquad (10)$$

in which (R) is the beach advancement with negative

(R) indicating recessing, (Wc) is the offshore distance to the depth of closure (h_c), (m_i) is the initial beach slope and,

$$m_e = \frac{h_c}{W_c} \tag{11}$$

For erosional beach,

$$\frac{m_i}{m_e} \geq \frac{6}{5} \qquad \text{(For } R \leq 0) \tag{12}$$

For accretional beach,

$$\frac{m_i}{m_e} \leq \frac{6}{5} \qquad \text{(For } R \geq 0) \tag{13}$$

6. NILE DELTA BEACH PROFILES

They have various categories of gradient namely a steep one (1:5 to 1:75) in the breaker zone, followed seaward by a flat one (1:150 to 1:350) and still flatter one offshore. Overnourished profiles are nearly horizontal (18).

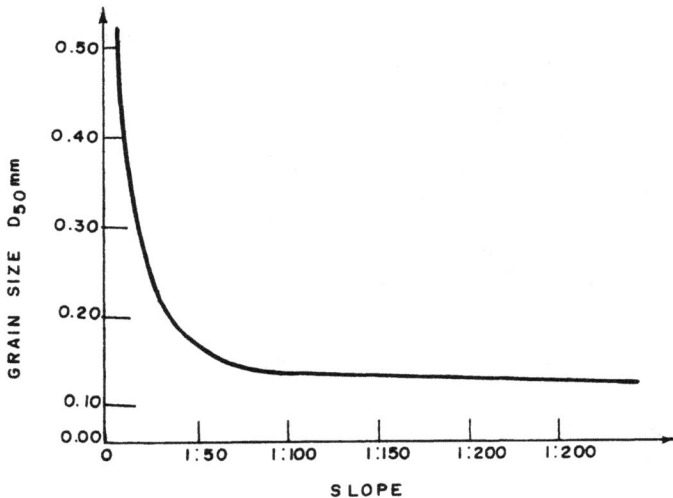

Fig.(7) Sediment Size vs. Slope [18]

The Delta coast is undernourished except for a short reach (3-4 km) in the Abuquir Bay, and a longer reach (4) km at a distance of 35 km) east of Burullus outlet. The profiles tend to show that with no Nile Sediment, even the present nourished ones eventually attain the status of undernourished (1).

In general, the coarse material tends to accumulate in the zone of maximum wave energy, namely the breaker zone. Seaward direction where the wave energy reduces, the sediment size gets progressively finer. Fig. (7) guies the relationship between grain size and slope (18).

7. CONCLUSIONS

This paper investigates beach profile changes.
The Nile Delta beach profiles are discussed. The following conclusions are reached :
1. The larger the sediment grain size, the steeper the beach slope.
2. The larger the wave steepness, the flatter the beach slope.
3. A high water table under the beach accelerates erosion, and a low water table retardes erosion and aides accretion on the foreshore.
4. The foreshore slope angle is related to the nondimensional parameter $(H_o/V_f.T)$ which accounts for the major wave and sediment properties.
5. The Nile Delta beach profiles are undernourished. This is due to the lack of Nile sediment after the construction of the High Aswan Dam.

8. REFERENCES

1.Abdel-Aal, F.M.," Erosion of the Nile Delta Coast," Paper Presented at the 1984 International Symposium on Urban Hydrology, Hydraulics, and Sediment Control, University of Kentucky, Lexington, Kentucky, July 23-26, 1984.
2.Bakker, W.T.,Klein Breteler, E.H.J., and Roos, A.,"The Dynamics of Coast with a Groyne System," Proc.1 2 th C.E. Conf., Washington, D.C., Vol. , pp. 1001-1020 , 1970.
3.Bruun, P.,"Port Engineering," Gulf Publishing Company, Houston, Texas, 1976.
4.Carter,T.G.,Liu, P.L-F,and Mei, c.c.,"Mass Transport by Waves and Offshore Bed Forms,"J.Waterways, Harbors , and Coastal Engineering,ASCE,Vol. 99,No. WW2, May, 1973.

5.Dally, W.,"A Numerical Model for Beach Profile Evolution,"Master's Thesis, University of Delaware, 1980.
6.Dalrymple,R. A.and Thompson,W.W.,"Study of Equilibrium Beach Profiles," Coastal Engineering,Chapter 75,1976.
7.Dean,R.G.,"Beach Erosion: Causes, Proc., and Remedial Measures, "Reprinted from CERC Critical Reviews in Enviromental Control, Vol. 6, Issue 3, Sept. 1976.
8.Dean, R. G., "Shoreline Erosion Due to Extreme Storms and Sea Level Rise,"Report No. UF/COEL-83/007, Coastal and Oceanographic Engineering Department, University ofFlorida, Gainesville, FL., 1983.
9.Dyer,K. R., "Coastal and Estuarine Sediment Dynamics,"Wiley-Interscience Publication, John Wiley & Sons,Chichester, 1986.
10.Eagleson,P.S.,Glenne,B.and Dracup,J.A.,"Equilibrium Characteristics of Sand Beaches,"Proc.ASCE,Vol.89, No. HYI, PP. 35-57, Jan., 1963.
11.Fleming, C. A. and J.N. Hunt, "Application of a Sediment Transport Model,"In Proc. 15th Coastal Engng. Conf.Honolulu, pp. 1184-1202, 1976.
12. Grant,u. s.,"Influence of the Water Table on Beach Aggradation and Degradation," Journal of Marine Reseach,VOL. 7, PP. 655-660, 1984.
13.Hattori,M.,and Kawamata,R.,"Onshore-Offshore Tran. and Beach Profile Change,"In Proc.17thCoastal Engng. Conf. Sydney, pp. 1175-1195, 1980.
14.Ippen,A.T.,and Eagleson, P.S.,"A Study of Sorting by Waves Shoaling on Plane Beach,"U.S. Army Corps of Engineers,Beach Erosion Board Tech Memo. No.63,83 p.1955.
15.Johnson,J.,"Scale Effects in Hyd. Models Involving Wave Motion," Transactions of AGU,Vol. 30,No.4, pp. 517-527, 1949.
16.Kohler, R.R., and Galvin, C.J., "Berm-Bar Criterion,"Unpublished Memorandom for the Record,U.S. Army CERC,Washington, Aug., 1973.
17.Kyungduck,S.,R.Dalrymple,"Expression for Shoreline Advancement of Initially Plan Beach," Jour. Waterway, Port,Coastal and Ocean Engng,Vol.114,No.6,Nov.,1988.

18.Manohar,M.,"Beach Profile,Proc.of Seminar on Nile Delta Sedimentology,"Unesco Proj., Alex.,Egypt,1976.

19.Manohar, M., "Undulated Bottom Profiles and OnShore-Offshore Transport," Coastal Engng, 1978.
20.Nayak,J.V.,"Equilibrium Profile of Model Beaches, "Ph. D. Thesis,Univ. of Calif.,at Berkeley,1970.

21.Sunamura,T.,"A Lab.Study of Offshore Transport of Sediment and a Model for Eroding Beaches,' Proc.17th Coastal Engng. Cnof. Sydney, pp. 1051, 1980.
22.U.S. Army Coastal Engineering Research Center (CERC)Shore Protection Manual,Washington,D.C.,1977.
23.Wiegel,R. L.,"Oceanographical Engng," Prentice Hall Englewood Cliffs, N.J., p. 359, 1964.

31 The concept of a single representative wave for use in numerical models of long term sediment transport predictions

T. J. Chesher and G. V. Miles

Abstract

A method is proposed for the schematisation of wave fields into one single, representative field for use in long-term sand transport predictions. The concept comprises determining the optimum wave (in terms of sediment transport) from each direction and combining them, taking into account their relative frequency.

Combination of the waves into one wave field requires the determination of weighting factors. Sensitivity tests based on a coastal study area indicated that weighting factors based on the longer, "classical" approach gave qualitatively the same result, although quantitatively the total sedimentation prediction was underestimated.

Errors as a result of the schematisation are considered in relation to the overall accuracy of the sediment transport system, which includes both process filtering (e.g. the use of a 2DH model) and other input filtering (e.g. single grain size).

Introduction

In order to make long term sand transport predictions in the coastal zone it is often important to consider the effects of both currents and waves. By the very nature of a typical wave climate it is clear that some degree of schematisation is required so that the various combinations of tides and currents is reduced to an acceptable level.

Typically the schematisation of waves takes the form of a choice of a particular wave height (e.g. significant wave height) or a particular exceedence (eg. 1 in 1 year event). Having determined the input wave conditions the wave, tidal current, and sediment transport models are run for each wave direction and long term predictions are made by considering the relative frequency of the particular waves from each direction.

A method is proposed whereby this schematisation process is carried one stage further and, having chosen a particular wave type, waves from all directions are combined into one single representative wave field (SRW), again taking into account the relative occurrence of each wave direction.

The concept represents a means of substantially reducing the computational expense incurred in making long term predictions by averaging over the directional space of a particular wave type.

Philosophy of the method

The SRW method is presented as a schematisation based on an order-of-operators concept; rather than predicting sand transport patterns for a particular wave type by combining the individual results for each wave direction, the individual wave fields are first combined into one wave field by averaging over the directions. This is then used to represent the whole wave field in the flow and sand transport models.

Figure 1 shows in schematic form how these two predictive approaches compare. With the "classical" approach the final sediment transport patterns are combined, with weighting factors (WC) determined by their occurrence in order to generate an overall prediction. With the SRW approach new weighting factors (WS) must be applied to generate a single wave field, which is then employed in the flow and sand transport models.

CLASSICAL APPROACH

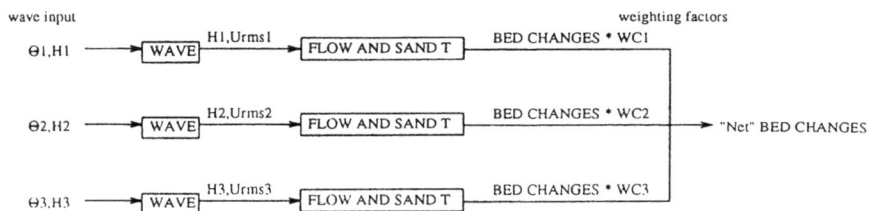

SINGLE REPRESENTATIVE WAVE METHOD

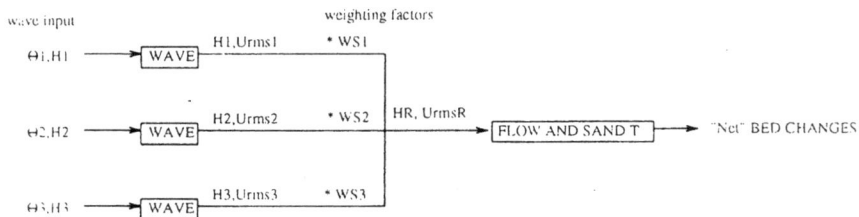

Figure 1 The concept of the single representative wave

Clearly, the method would lead to considerable saving in time and computational expense since the flow and sand transport models would need to be run once only. The efficacy of this method in providing a valid means of schematising the complete wave climate into this more acceptable and manageable field depends on how the waves affect the sediment transport patterns.

Waves affect sediment transport in two main ways ; firstly by affecting the tidal currents via wave radiation stresses, mass transport and enhanced friction at the bed, and secondly by stirring up the bed and increasing the mass of suspended sediment which is subsequently advected away by the flow.

The stirring effect of even relatively small waves can give rise to vast increases in sediment transport [5] and it is this mechanism which is exploited in the SRW concept since the stirring is independent of direction.

Model details

The numerical modelling of sediment transport under combined waves and currents can take a variety of levels of complexity, eg fully coupled flows considering the effects of both the waves on the currents and the currents on the waves followed by a sediment transport relation including wave stirring, wave asymmetry and gravity (slope) effects. Conversely a simplified approach including waves as a stirring parameter in the sand transport relation may be adopted.

The choice of method will be determined by the relative importance of the waves and the currents, as well as by other external factors such as financial and time management considerations.

This schematisation study is based on a semi-coupled approach whereby 2DH models are coupled to include the effects of waves on currents as above, followed by sediment transport based on these combined flows with the inclusion of wave stirring. A short description of each module follows.

Wave module

The wave fields are determined by solution of the parabolic approximation to the mild slope equation

$$\nabla(C \ C_g \ \nabla\phi) + \frac{C_g}{C}\omega^2\phi = 0 \tag{1}$$

where ∇ is the horizontal gradient operator
C is the wave celerity (= ω/k)
C_g is the group velocity $d\omega/dk$
$\phi = \phi(x)$ a complex velocity potential at the mean free surface,
ω is the angular frequency

After further approximation the mild slope equation is solved on a finite difference grid to yield wave heights and periods over the model area.

Using these fields wave radiation stresses and orbital velocities are derived based on the method of Longuet-Higgins and Stewart [4] and Soulsby and Smallman [6] respectively.

Tidal current module

The appropriate equations for studying water movements in tidal areas are the shallow water equations. These are obtained by vertically integrating the equation of motion governing mass and momentum and making the following simplifying assumptions:
(i) the flow is incompressible;
(ii) it is well mixed (no variations in density);
(iii) vertical accelerations are negligible (the hydrostatic pressure assumption);

(iv)   the effective lateral stresses associated mainly with shearing in the horizontal, and to a small extent with the averaging of sub-grid scale turbulence, may be approximated by a constant eddy viscosity;

(v)   bed stress can be modelled using a quadratic friction law.

The equations then take the following form:

Conservation of mass:

$$\frac{\partial z}{\partial t} + \frac{\partial}{\partial x}(ud) + \frac{\partial}{\partial y}(vd) = 0 \tag{2}$$

Conservation of momentum:

$$\frac{\partial u}{\partial t} + u\frac{\partial u}{\partial x} + v\frac{\partial u}{\partial y} =$$

$$-g\frac{\partial z}{\partial x} - f\frac{u}{d}(u^2 + v^2)^{\frac{1}{2}} + D[\frac{\partial^2 u}{\partial x^2} + \frac{\partial^2 u}{\partial y^2}] + \Omega v + \frac{\tau_{sx}}{\rho d} + \frac{\tau_{wx}}{\rho d} \tag{3}$$

$$\frac{\partial v}{\partial t} + u\frac{\partial v}{\partial x} + v\frac{\partial v}{\partial y} =$$

$$-g\frac{\partial z}{\partial y} - f\frac{v}{d}(u^2 + v^2)^{\frac{1}{2}} + D[\frac{\partial^2 v}{\partial x^2} + \frac{\partial^2 v}{\partial y^2}] - \Omega u + \frac{\tau_{sy}}{\rho d} + \frac{\tau_{wy}}{\rho d} \tag{4}$$

where   t   = time (s)
         z   = elevation above datum (m)
         u,v   = depth averaged component of velocity in x,y direction (m/s)
         d   = total depth
         f   = friction coefficient
         D   = horizontal eddy viscosity coefficient (m²/s)
         Ω = Coriolis parameter (s⁻¹)
         τ_{sx}, τ_{sy} = x and y components of surface windstress. (N/m²)
         τ_{wx}, τ_{wy} = x and y components of the wave radiation stress. (N/m²)

For conditions with wave activity the calm conditions friction factor is modified to

$$f_w = f(1+\alpha(\frac{W}{U})^n) \tag{5}$$

based on the method of Yoo and O'Connor [8] where

$$(8f)^{-\frac{1}{2}} = 2\log_{10}(14.8d/k_s) \tag{6}$$

and   W   = wave orbital velocity (m/s)
       k_s = roughness length (m)
       U   = current speed (m/s) $=(u^2 + v^2)^{0.5}$
       α and n are set to 0.72 and 1 respectively.

Sediment transport module

Although sand transport in estuaries is really an unsteady, 3D problem it can be dealt with using a 2D, depth-averaged model provided special provision is made to account for the vertical profile effects of the sediment concentration. Under these circumstances the depth-averaged, suspended solids concentration c(x, y, t) satisfies the conservation of mass equation

$$\frac{\partial}{\partial t}(dc) + \alpha[\ \frac{\partial}{\partial x}(duc) + \frac{\partial}{\partial y}(dvc)\ ]$$

$$= \frac{\partial}{\partial s}(\ dD_s\frac{\partial c}{\partial s}\ ) + \frac{\partial}{\partial n}(\ dD_n\frac{\partial c}{\partial n}\ ) + \beta w_s(c_s - c) \tag{7}$$

where D_s = longitudinal (shear flow) dispersion coefficient (m^2/s)
D_n = lateral (turbulent) diffusivity (m^2/s)
(s,n) = natural coordinates (parallel and normal to mean flow (m))
w_s = settling velocity (m/s).

The parameters α and β are introduced to account for vertical profile effects lost during the depth-averaging.

$$\alpha = \int_0^d u(z)c(z)dz/(ucd) \tag{8}$$

represents the factor required to recover the true transport from the product of depth-averaged quantities.

β represents the factor to recover the correct rate of exchange of sediment at the bed. In its simplest form $\beta = c_{bed}/c$ but a better relation can be derived from the underlying 3D equation

$$\frac{\partial c}{\partial t} + u\frac{\partial c}{\partial x} + v\frac{\partial c}{\partial y} + (w - w_s)\frac{\partial c}{\partial z}$$

$$= \frac{\partial}{\partial x}(D_x\frac{\partial c}{\partial x}) + \frac{\partial}{\partial y}(D_y\frac{\partial c}{\partial y}) + \frac{\partial}{\partial z}(D_z\frac{\partial c}{\partial z}) \tag{9}$$

with the usual meaning for the variables. By considering uniform horizontal flow conditions this reduces to

$$\frac{\partial c}{\partial t} = \frac{\partial}{\partial z}(D_z\frac{\partial c}{\partial z}) + w_s\frac{\partial c}{\partial z} \tag{10}$$

An analytical solution by Laplace Transform is possible for a depth independent eddy viscosity (D_z) which can be integrated to yield an expression for β in terms of error functions for each velocity, depth and sand size [2].

c_s is the depth-averaged concentration when the flow is saturated with sediment. Deposition or erosion takes place if the instantaneous model concentration exceeds or falls short of the saturated load at each point.

The saturation concentration, c_s is calculated by appealing to an empirically based sand transport law after van Rijn [7] with enhancement due to wave stirring after Grass [1].

The van Rijn formula for currents only is

$$\text{Bedload: } q_b = 0.005\rho_s\ Ud[\frac{U - U_{cr}}{[(s-1)gd_{50}]^{0.5}}]^{2.4}(\frac{d_{50}}{d})^{1.2} \tag{11}$$

$$\text{Suspended load: } q_s = 0.012\rho_s\ Ud[\ \frac{U - Ucr}{[(s-1)gd_{50}]^{0.5}}]^{2.4}\ D_*^{-0.6} \tag{12}$$

where q_b, q_s represent mass of sediment/unit width/unit time
d_{50} = median grain diameter (m)
ρ_s = sediment density (kg/m^3)
s = sediment specific gravity
g = gravitational acceleration (m/s^2)
U_{cr} = threshold speed (m/s)

375

$$D_* = d_{50} \left[\frac{(s-1)g}{v^2}\right]^{1/3} \tag{13}$$

v = kinematic viscosity of water (m²/s)

These formulae for currents only are written in total load flux form as

$$q_t = Au\,(U-U_{cr})^{n-1} \text{ (kg/m/s)} \tag{14}$$

where   A is a constant for a given grain size
n = 3.4

By appealing to the method of Grass, the wave enhanced sediment transport formula, including a threshold velocity is defined as

$$q_{t+} = Au[(U^2 + BW^2)^{0.5} - U_{cr}]^{n-1} \text{ (kg/m/s)} \tag{15}$$

where   B = 0.08/C_D
W = wave orbital velocity (m/s)
C_D = drag coefficient

Methodology

Having defined the SRW concept it remains to determine the method of choosing the particular wave type and estimating the new weighting factors, WS. As expected, the manner in which the choice of wave is made would have a direct consequence on the determination of the weighting factors. The method described in this study is clarified by referring to the arbitrary annual wave height exceedence data in Table 1.

Table 1 Annual wave height exceedence diagram for wave model point X

H1	to H2	Wave angle (degrees N)											
		-30 / 0	0 / 30	30 / 60	60 / 90	90 / 120	120 / 150	150 / 180	180 / 210	210 / 240	240 / 270	270 / 300	300 / 330
0.00	0.20	3044	2888	1088	1344	1149	705	654	1566	2347	530	1444	3345
0.20	0.40	2144	1599	2200	1666	2072	983	1018	2221	2734	678	2200	3065
0.40	0.60	1611	1010	1665	2389	1557	756	779	1999	4000	0	1546	1965
0.60	0.80	268	250	575	1524	1949	860	1208	2446	2934	0	547	560
0.80	1.00	47	49	103	544	887	444	437	929	1577	0	213	139
1.00	1.20	9	10	81	612	410	514	686	851	1367	0	444	35
1.20	1.40	2	2	4	266	411	240	588	585	999	0	134	15
1.40	1.60	0	0	5	198	357	303	468	287	567	0	15	6
1.60	1.80	0	0	9	85	221	289	416	178	202	0	6	2
1.80	2.00	0	0	0	16	59	440	402	199	117	0	0	0
2.00	2.20	0	0	0	55	109	227	328	21	156	0	0	0
2.20	2.40	0	0	0	11	115	167	317	30	11	0	0	0
2.40	2.60	0	0	0	23	40	148	361	5	20	0	0	0
2.60	2.80	0	0	0	10	109	205	234	5	1	0	0	0
2.80	3.00	0	0	0	0	121	157	232	2	6	0	0	0
3.00	3.20	0	0	0	0	43	132	124	1	2	0	0	0
3.20	3.40	0	0	0	0	47	87	193	2	1	0	0	0
3.40	3.60	0	0	0	0	35	57	88	0	0	0	0	0
3.60	3.80	0	0	0	0	23	35	134	0	0	0	0	0
3.80	4.00	0	0	0	0	14	29	53	0	0	0	0	0
4.00	4.20	0	0	0	0	26	22	33	0	0	0	0	0
4.20	4.40	0	0	0	0	14	18	19	0	0	0	0	0
4.40	4.60	0	0	0	0	8	29	8	0	0	0	0	0
4.60	4.80	0	0	0	0	0	7	10	0	0	0	0	0
4.80	5.00	0	0	0	0	0	3	6	0	0	0	0	0
5.00	5.20	0	0	0	0	0	3	5	0	0	0	0	0
parts per thousand for each direction		73	59	58	89	100	70	90	115	174	12	67	93

Table 1 Annual wave height exceedence diagram for wave point X

The first stage of the schematisation comprises selection of a particular wave height from each direction. The wave model is then run for each chosen wave height from each direction.

Since the schematisation comprises averaging over the wave direction field each individual wave direction must be weighted according to its relative frequency. The most simple schematisation comprises combining all waves from a particular direction into one representative wave height, H_R for that direction. This is calculated by appealing to the sediment transport law and weighting according to the wave height, H and occurrence, f in the following way :

$$H_\theta = [\Sigma(f.H^{2.4})/\Sigma f]^{1/2.4} \tag{16}$$

Further filtering of the wave input data could be carried out by neglecting those waves that are below a certain threshold and as such are considered to play only a minor role in the overall system.

Having generated the input wave fields for the tidal current and sand transport models in the classical approach each wave directional case H_θ would be run and the sediment transport patterns combined using as weighting factors the relative frequencies shown at the base of the table.

In the SRW approach the waves are firstly combined into one wave field H_R by averaging over the directional space with suitable weighting factors. The method of choice of weighting factors is not altogether clear since the whole sediment transport system is highly non-linear with respect to the wave input.

Results are presented in this paper for a sensitivity test on the sediment transport patterns for a particular study area between the classical approach and the SRW approach using the same weighting factors (i.e. assuming a linear response) based on the occurrence and the wave height for each direction.

Model Application

The schematisation concept was applied to an open coastal study area at Poole Bay in S. England where the sedimentation regime is strongly influenced by the effects of both currents and waves. This was confirmed by initial sedimentation sensitivity tests with and without waves where the sediment transport was minimal under currents alone but increased by an order of magnitude after inclusion of average-sized waves [3]. That there is considerable sedimentation at the site is confirmed by the need for capital dredging in order to keep the approach channel clear.

The wave data in Table 1 is based on data from a waverider buoy at this site. It can be seen that the waves are largest from the south and the east. Using this information it was assumed that waves below a wave height of 1m would play only a minor part in the overall sedimentation prediction and hence only three direction bands were considered, viz. 090-120, 120-150 and 150-180 degrees (N).

The weighting factors WC1, WC2, WC3 were calculated from the relative frequencies (f) from table 1 (ie assuming only these waves were present).

As indicated above the first guess at the schematised method weighting factors WS1, WS2, WS3 was made by setting them equal to WC1, WC2, WC3 respectively.

Net sedimentation patterns based on a spring tide using the classical approach and the SRW approach (Fig 1) were calculated and the results for each test are presented in Figures 2 and 3 respectively. Differences between the two approaches are presented in Figure 4.

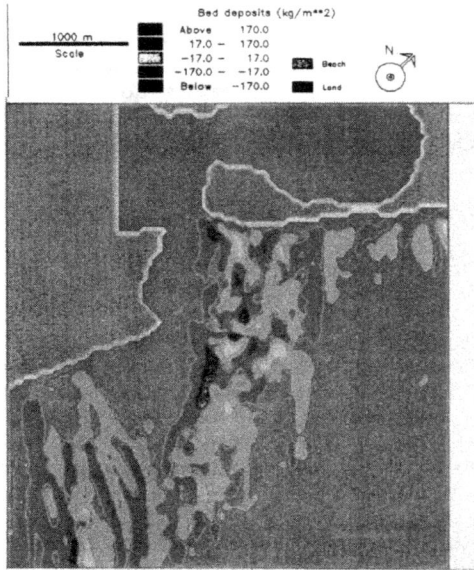

Figure 2 Net sedimentation based on the classical approach

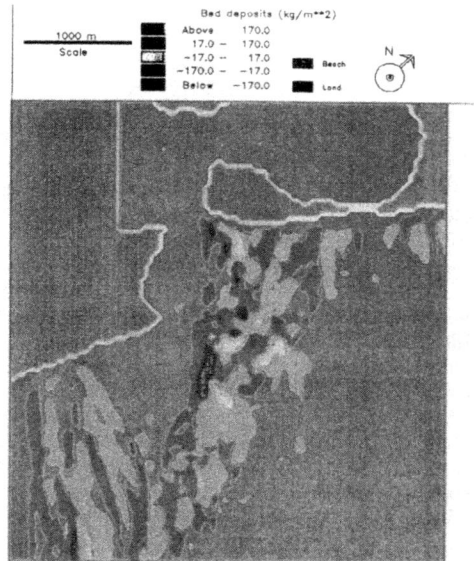

Figure 3 Net sedimentation based on the single representative wave approach

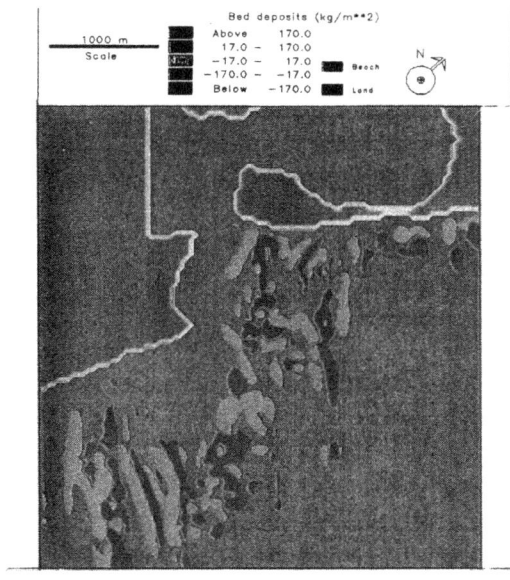

Figure 4 Net sedimentation differences (SRW - classical)

Discussion

Clearly there are differences (errors) as a result of the schematisation. However these errors must be considered in relation to the many other deficiencies involved even in this relatively elaborate system. Also, these errors should be considered against the expense of the classical method, the flow model being by far the most expensive module to run (in terms of time and expenditure).

The first check of the SRW method against the classical method is purely qualitative whereby one approach predicts erosion while the other predicts deposition (or vice versa). In this test 524 model cells out of a total of 14490 cells conflicted, which is less than 4%.

A closer quantitative assessment identified the cell with the worst error of 145 kg/m^2 deposition versus 44 kg/m^2 erosion. This is equivalent to an error of about 10cm.

Globally, a comparison of $\sum|\text{deposits(kg/m}^2)|$ indicated rather more (20%) total bed changes with the classical method than with the SRW method, which suggests that the weighting factors should be adjusted.

Conclusions

A schematisation method has been devised which appears to give rise to comparable predictions of sedimentation with the more detailed, full analysis, based on semi-coupled waves+currents at a particular test site.

The method comprises the generation of a single representative wave field by schematising results from repeated applications of the wave module (for a particular wave from a particular direction), prior to inclusion of the wave field into flow and sediment transport modules. It should be stressed that more than one wave propagation exercise is considered so that, for example, a long breakwater would not be mis-represented unduly since the effect of the breakwater on the waves from all relevant directions would be included in the definition of the SRW.

The validation of the method is by no means rigorous, with a sensitivity test based on identical weighting factors for the classical method and the SRW method (i.e. assuming a linear response). Further tests, or a more detailed analysis of the method should be carried out to yield a more satisfactory set of weighting factors.

The test considered only waves from three dominant directions. A more open site would require consideration of waves from more directions. However, in these cases the successful derivation of a satisfactory SRW field would result in a considerable time and cost saving.

It is appreciated that the method will break down in the case of large waves with small currents, since the wave-driven currents from two or more directions could not be represented by one single wave from one single direction.

In the sensitivity test a representative wave from each direction was derived based on the stirring term in the sediment transport algorithm. Storm waves present an additional problem. In many coastal areas the sediment regime is dominated by the effects of storm conditions. For the reasons given above it is not expected that the method would give a satisfactory prediction under these conditions. However, in the cases where storm waves occur from a dominant direction the SRW schematisation for typical waves (using the method described in this paper) could be supplemented with individual "storm" cases. More importantly, in these storm cases the prediction of sediment transport remains a more elusive problem.

Acknowledgements

The author gratefully acknowledges financial support for research underpinning this work from the UK Department of the Environment under Research Contract PECD 7/6/161, and from the Commission of the European Communities Directorate General for Science, Research and Development under Contract No. MAST-0035-C as part of the G6M Coastal Morphodynamics research programme.

References

[1]   Grass AJ (1981). *Sediment Transport by Waves and Currents*. SERC London Centre for Marine Technology Report FL29, University College, London.
[2]   HR Wallingford (1988). *A numerical sand transport model with time dependent bed exchange*. Report No. SR 148.
[3]   HR Wallingford (1991). *Poole Bay Phase II - Hydraulic Study*. Report No. EX 2228
[4]   Longuet-Higgins MS and Stewart RW (1964). "Radiation stress in water waves; a physical discussion with applications". *Deep-Sea Research*, **XI.**
[5]   Owen MW and Thorn MFC (1978). "Effects of Waves on Sand Transport by Currents." *Proc 16th Coast Eng Conf Hamburg.*
[6]   Soulsby RL and Smallman JV (1986). *A Direct Method of Calculating Bottom Orbital Velocity under Waves*. HR Wallingford Report No. SR 76.
[7]   van Rijn LC (1984). "Sediment Transport, Part III.   Bed Forms and Alluvial Roughness." *J. Hydr Eng*, **110**, 12.
[8]   Yoo D and O'Connor BA (1987). "Bed Friction Model of Wave-current Interacted Flow." *ASCE Spec Conf on Coastal Hydrod, Delaware, USA.*

32 3-D mathematical modelling of current and pollutant transport in Izmir Bay

E. Özhan and L. Hapoğlu

ABSTRACT

A 3-D model for wind induced currents and pollutant transport is discussed. Preliminary applications of the model to Izmir Bay are presented.

1   Introduction

Izmir Bay is a major bay located on the Turkish coast of the Aegean Sea. Its length is about 60 km. The inner bay where the metropolitan city of Izmir (the third largest Turkish city) is located, is densely populated. A moderate level of industrial facilities is also present. Furthermore, the Izmir Port, a major port which is especially important for export goods, is located in the inner part of the Bay.

The northeastern sector of the inner bay is notably shallow. A major salt mining facility exists in this area. Furthermore, the neighboring marshes are among the important Aegean bird habitats.

Izmir Bay has traditionally been an important area for local fisheries. However, excessive water pollution in the inner bay which has amplified in the last two decades, has almost completely destroyed the resource value of the inner bay for commercial fishing.

Various initiatives have taken place during the recent years for water quality management. On one side, an extensive sewerage project financed by a World Bank loan, has been under construction. This project aims to collect all municipal sewage for subsequent treatment and disposal [1]. However, there has been an ongoing debate , concerning the disposal alternatives. This has mainly concentrated on the location of disposal point in the bay on one side, and on the methods and levels of pretreatment on the other.

A second effort towards water quality management has been through the project named " Coastal Zone Management at the Izmir Bay ". This is a priority action project (PAP) being carried out in the scope of Mediterranean Action Plan (MAP) activities. This PAP project is basically research oriented. A number of studies (mostly involving field measurements) have been carried out. However, the level of information obtained up to date, concerning the hydrodynamics and waste assimilation capacity of Izmir Bay is far from being sufficient for rational water quality management decisions.

Tides of Izmir Bay are very weak (the tidal range is around 15-20 cm). Therefore, wind induced currents are of primary importance. This type of currents, showing significant variation in the vertical direction, require 3-D hydrodynamic treatment.

The modeling study presented in this paper was conducted for producing a mathematical tool which would be useful for future water quality management efforts at Izmir Bay. The use of 3-D models for similar practical purposes is being common [2],[3],[4].

2   Mathematical Model and Solution Procedures

For computing water level elevations and current velocities in space, the depth integrated continuity equation is solved together with the Navier-Stokes equations. Resulting velocity components are used in the three dimensional convective diffusion equation to determine the transport of pollutant.

The flow in a 3-D domain with hydrostatic pressure distribution along the vertical, due to smallness of the vertical accelerations (i.e. nearly horizontal flow), is described by the following set of equations:

$$L_x(u,v) = \frac{\partial u}{\partial t} + u\frac{\partial u}{\partial x} + v\frac{\partial u}{\partial y} - f_c v + g\frac{\partial \xi}{\partial x} - \left[2\frac{\partial}{\partial x}\left(\varepsilon_x \frac{\partial u}{\partial x}\right)\right.$$

$$\left. + \frac{\partial}{\partial y}\left(\varepsilon_y \left(\frac{\partial u}{\partial y} + \frac{\partial v}{\partial x}\right)\right) + \frac{\partial}{\partial z}\left(\varepsilon_z \frac{\partial u}{\partial z}\right)\right) = 0 \qquad (1)$$

$$L_y(u,v) = \frac{\partial v}{\partial t} + u\frac{\partial v}{\partial x} + v\frac{\partial v}{\partial y} + f_c u + g\frac{\partial \xi}{\partial y} - \left[2 \frac{\partial}{\partial y}\left(\varepsilon_y \frac{\partial v}{\partial y}\right)\right.$$

$$\left. + \frac{\partial}{\partial x}\left(\varepsilon_x\left(\frac{\partial v}{\partial x} + \frac{\partial u}{\partial y}\right)\right) + \frac{\partial}{\partial z}\left(\varepsilon_z \frac{\partial v}{\partial z}\right)\right) = 0 \tag{2}$$

$$\frac{\partial \xi}{\partial t} + \frac{\partial (H\ U)}{\partial x} + \frac{\partial (H\ V)}{\partial y} = 0 \tag{3}$$

where,

$$U = \frac{1}{H} \int_{-h}^{\xi} u\ dz \ ; \quad V = \frac{1}{H} \int_{-h}^{\xi} v\ dz \ ; \quad f_c = 2\ Q\ Sin\ \phi \tag{4}$$

The lateral shear terms are included in Equations (1) and (2) in addition to the vertical shear.

3-D convective diffusion equation which describe the transport of a non-conservative pollutant is:

$$\frac{\partial C}{\partial t} + u\frac{\partial C}{\partial x} + v\frac{\partial C}{\partial y} = D_x \frac{\partial^2 C}{\partial x^2} + D_y \frac{\partial^2 C}{\partial y^2} + D_z \frac{\partial^2 C}{\partial z^2} - kC \tag{5}$$

where; ξ = water surface elevation above still water level, x,y = horizontal cartesian coordinates, z = vertical coordinate, t = time, H = total depth, U,V = depthaveraged current velocity components in x,y directions respectively, u,v = velocity components in x, y directions respectively, $\varepsilon_x, \varepsilon_y, \varepsilon_z$ = eddy viscosities in x,y,z directions respectively, f_c = Coriolis coefficient, g = gravitational acceleration, C = pollutant concentration, D_x, D_y, D_z = turbulent diffusion coefficients in x,y,z directions respectively, Q = earths angular velocity, ϕ = latitude of observation point and k is decay rate coefficient.

Flows induced by wind and tidal forces are assumed to be approximately horizontal flows, and thus vertical velocity component is neglected. Turbulence is taken as homogeneous. Usual boundary conditions for flow velocity components and pollutant concentrations are used at the free surface, the sea bed, the open sea and the land boundaries.

The fractional step finite difference model producing small sized solution matrices is found to be an appropriate model [5]. Therefore, a solution procedure which is similar to that described by Koutitas and O'Connor (1980) is used.

Equations (1) and (2) are written in the form of :

$$\frac{\partial u}{\partial t} = L_1(u) + \alpha_1 \tag{6}$$

$$\frac{\partial v}{\partial t} = L_2(v) + \alpha_2 \tag{7}$$

in which L_1, L_2, α_1, α_2 are defined as ;

$$L_1 = -u\frac{\partial}{\partial x} - v\frac{\partial}{\partial y} + \frac{\partial}{\partial z}\left(\varepsilon_z\frac{\partial}{\partial z}\right) + \frac{\partial}{\partial y}\left[\varepsilon_y\frac{\partial}{\partial y}\right] + 2\frac{\partial}{\partial x}\left[\varepsilon_x\frac{\partial}{\partial x}\right] \tag{8}$$

$$L_2 = -u\frac{\partial}{\partial x} - v\frac{\partial}{\partial y} + \frac{\partial}{\partial z}\left(\varepsilon_z\frac{\partial}{\partial z}\right) + \frac{\partial}{\partial x}\left[\varepsilon_x\frac{\partial}{\partial x}\right] + 2\frac{\partial}{\partial y}\left[\varepsilon_y\frac{\partial}{\partial y}\right] \tag{9}$$

$$\alpha_1 = f_c v - g\frac{\partial \xi}{\partial x} + \frac{\partial}{\partial y}\left[\varepsilon_y\frac{\partial v}{\partial x}\right] \tag{10}$$

$$\alpha_2 = -f_c u - g\frac{\partial \xi}{\partial y} + \frac{\partial}{\partial x}\left[\varepsilon_x\frac{\partial u}{\partial y}\right] \tag{11}$$

Following the procedure of Koutitas and O'Connor [6], the operators L_1 and L_2 are split into two components L_1^1, L_2^2 and L_2^1, L_2^2 which are assumed to be independent of u and v over a small time interval Dt. Then it can be shown that [7];

$$\frac{\partial u^n}{\partial t} = L_1^2(u^n) + \alpha_2^{n+1/2} \quad ; \quad \frac{\partial u^{n+1}}{\partial t} = L_1^1(u^{n+1}) \tag{12}$$

$$\frac{\partial v^n}{\partial t} = L_2^1(v^n) + \alpha_2^{n+1/2} \quad ; \quad \frac{\partial v^{n+1}}{\partial t} = L_2^2(v^{n+1}) \tag{13}$$

in which n : an integer, showing that a variable is to be evaluated at time n Dt. The parts of operators L_1^1, L_1^2 and L_2^1, L_2^2 are chosen as;

$$L_1^1 = -u\frac{\partial}{\partial x} - v\frac{\partial}{\partial y} + \frac{\partial}{\partial y}\left[\varepsilon_y\frac{\partial}{\partial y}\right] + 2\frac{\partial}{\partial x}\left[\varepsilon_x\frac{\partial}{\partial x}\right] \tag{14}$$

$$L_2^2 = L_1^2 = \frac{\partial}{\partial z}\left(\varepsilon_z\frac{\partial}{\partial z}\right) \tag{15}$$

$$L_2^1 = -u\frac{\partial}{\partial x} - v\frac{\partial}{\partial y} + \frac{\partial}{\partial x}\left[\varepsilon_x\frac{\partial}{\partial x}\right] + 2\frac{\partial}{\partial y}\left[\varepsilon_y\frac{\partial}{\partial y}\right] \tag{16}$$

The computed velocity components are used in the three-dimensional convective diffusion equation for solving the pollutant concentrations. For solving this equation, an implicit difference approximation is used to overcome the restriction of using a small time increment in order to have stable computation.

It should be noted that if turbulent diffusion is too small, then, the static instability may occur due to numerical errors during computations [8].

3 Application of the Model to Izmir Bay

3.1 Wind Induced Currents and Set Up

Izmir Bay has a length of 60 km. and a width of 39 km. In this work, the coastal area is schematized by a square grid as shown in Figure (1). The water density is taken uniform and equal to 1025 kg/m^3. For the preliminary computation presented in this paper, constant eddy viscosities are used as 10 m^2/s and 0.1 m^2/s respectively in horizontal and vertical directions. Time step is used as 10 s. Flow velocities are computed at 6 elevations along the water depth.

The water mass is subjected to the free surface shear induced by a uniform and steady wind of 10 m/s blowing from NNW. Steady-state circulation pattern is established approximately 6 hours after the beginning of the storm. These steady-state flow patterns at the sea surface and at the bottom (one sixth of the water depth above the sea bed) are sketched in Figure (2).

Steady state depth averaged flow velocities and water level changes are presented in Figures (3-4). Figures (5) shows the velocity profiles at two nodes (nodes (37,15), (3,24) shown in Figure 1). Similar results are given in Figures (6-7) also for a wind speed of 10 m/s from SW.

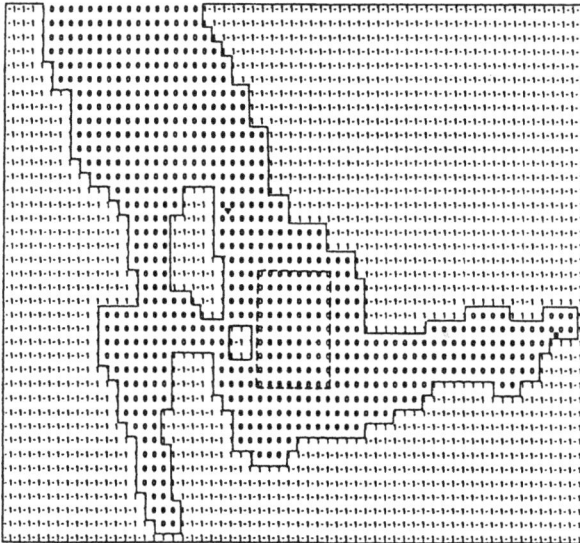

FIGURE 1   *Shematization of Izmir Bay where,   ▼:node (37,15), ■:node (3,24), 1 : land-water boundary, 0:sea region, -1:land region, dashed grid : grid for pollutant transport*

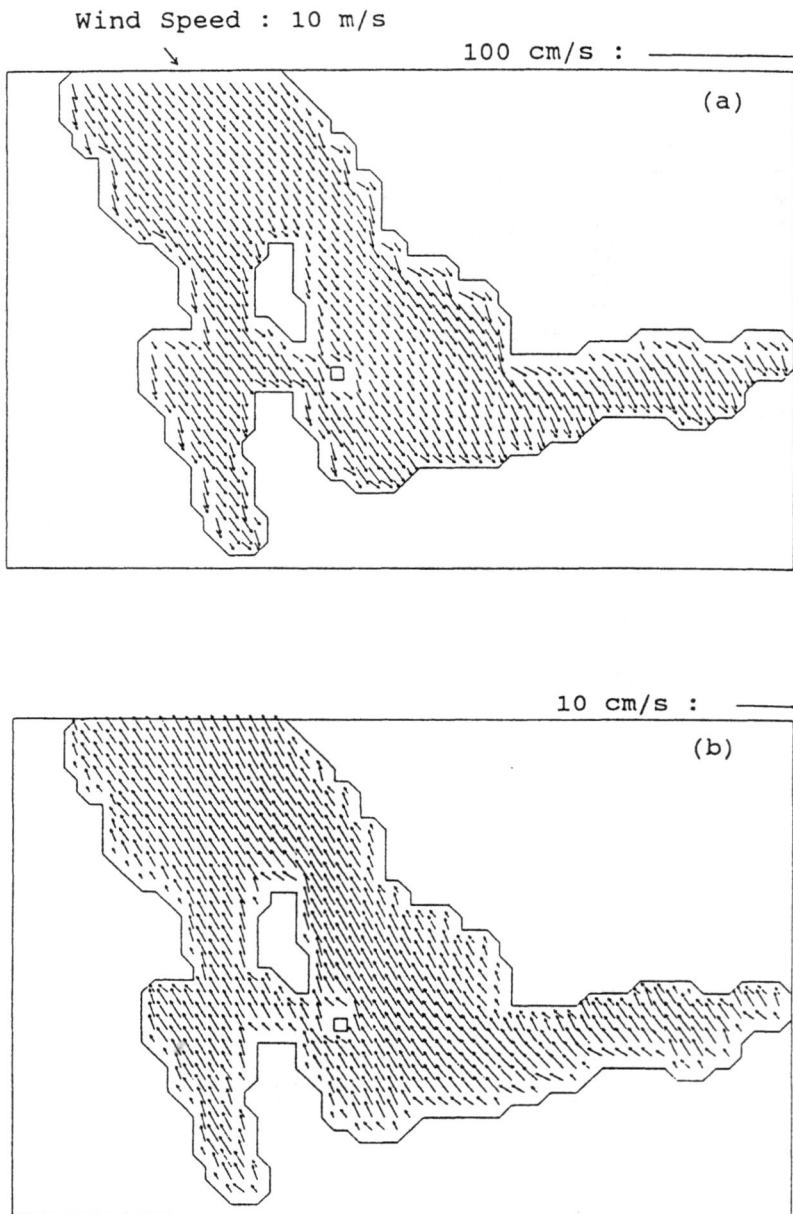

FIGURE 2   *Steady state current pattern in Izmir Bay (NNW wind speed : 10 m/s), (a) at sea surface,(b) at bottom layer*

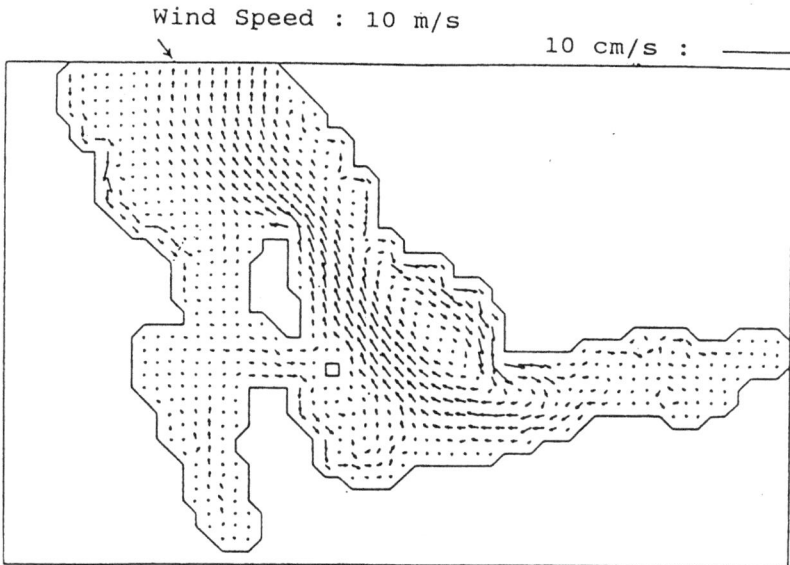

Wind Speed : 10 m/s

10 cm/s : ⎯⎯⎯

FIGURE 3   *Steady state depth averaged current pattern in Izmir Bay (NNW wind speed : 10 m/s)*

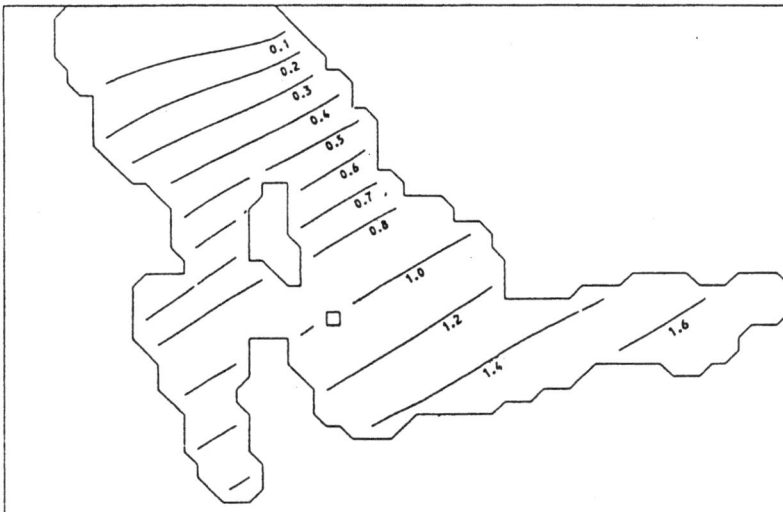

FIGURE 4   *Steady state water fluctuations (m) in Izmir Bay (NNW wind speed : 10 m/s)*

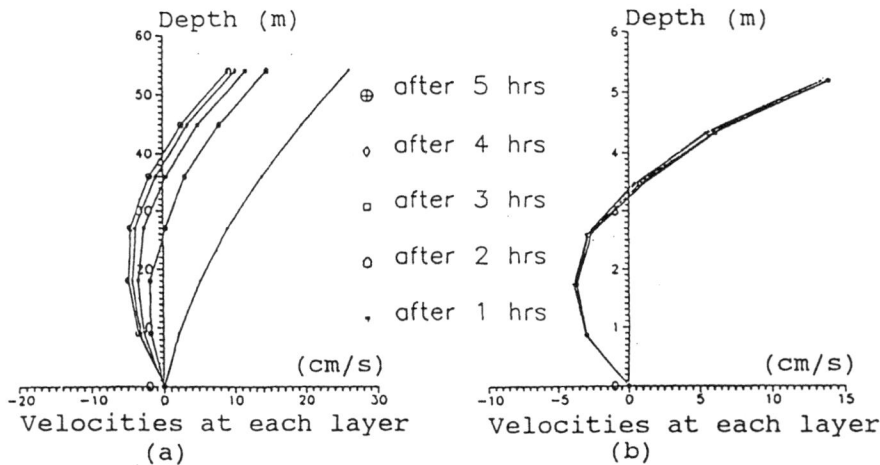

FIGURE 5 *Vertical velocity profiles,(a) at node (37,15), (b) at node (3,24) (NNW wind speed : 10 m/s)*

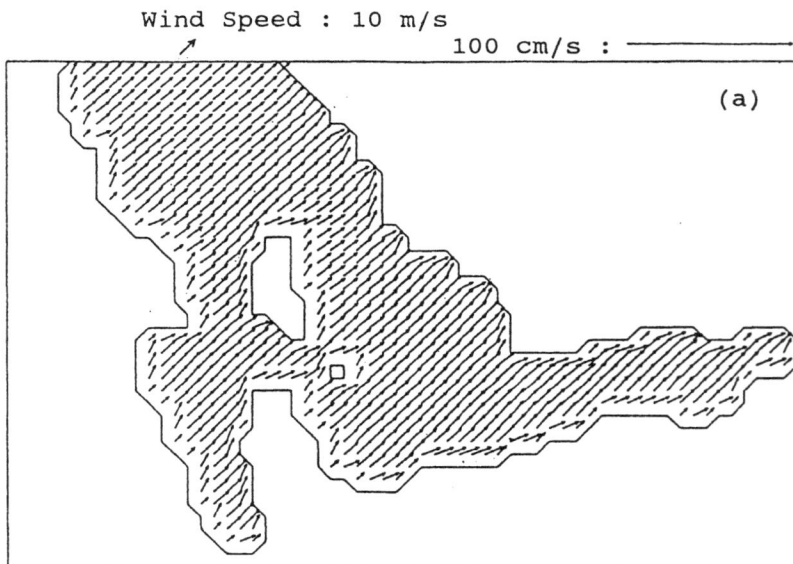

FIGURE 6 *Steady state current pattern in Izmir Bay (SW wind speed :10 m/s), (a) at sea surface*

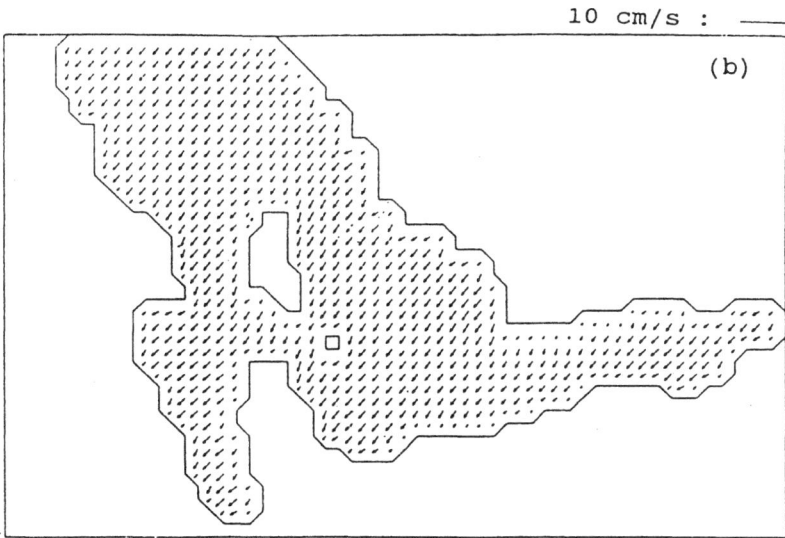

10 cm/s : ——

(b)

FIGURE 6   *(cont'd)   (b) at the bottom layer*

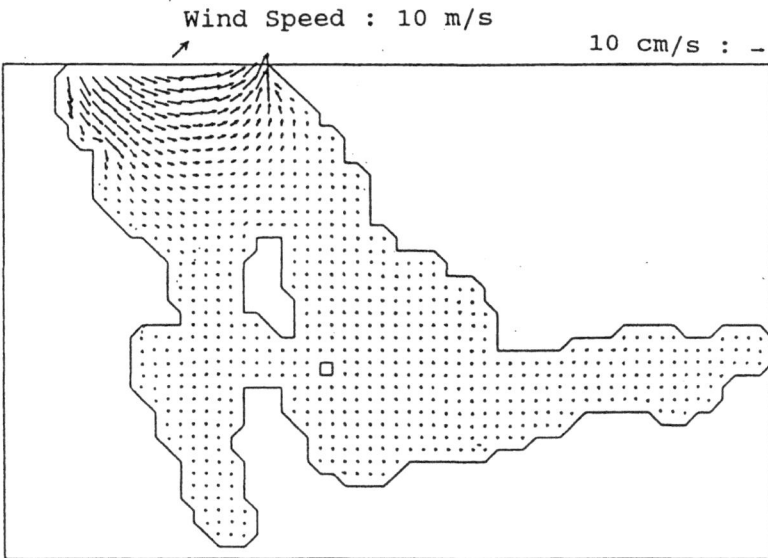

Wind Speed : 10 m/s

10 cm/s : —

FIGURE 7   *Steady state depth averaged current pattern in Izmir Bay (SW wind speed : 10 m/s)*

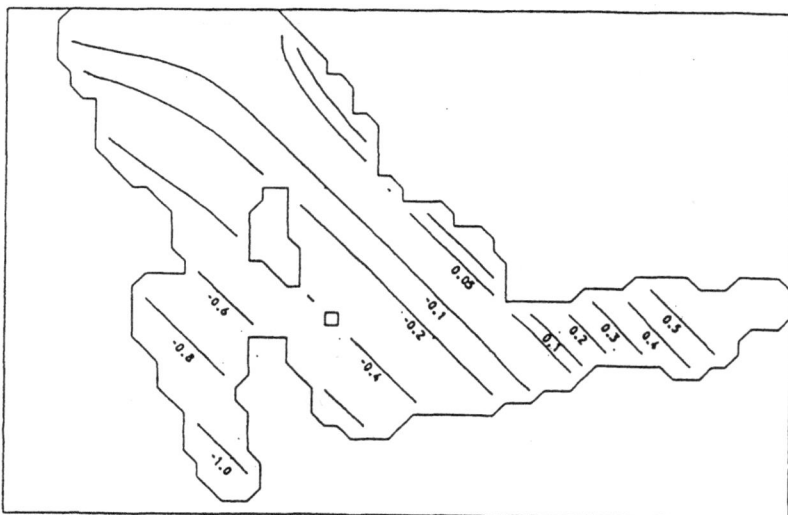

FIGURE 8   *Steady state water fluctuations (m) in Izmir
Bay (SW wind speed : 10 m/s)*

Figure (4), and Figure (8) show that water level
elevations increase along the wind direction almost
uniformly. A significant set up of about 1.6 m. is
observed at the end of the bay in the computations
with the northwesterly wind. The flow structures in
the surface and bottom layers are typical of wind
induced currents. A horizontal gyre is not observed
in any of the layers. However, depth averaged velocities (two dimensional treatment) given in Figure (3)
and Figure (7) show a gyre at the middle bay.

3.2 Pollutant Transport

For simulating the transport of raw sewage from an
ocean outfall discharging into the middle bay, the
coliform count is used as the tracer. The initial
concentration is taken as 10^6 bac/ml. The discharge
is assumed to take place steadily into the surface
layer (i.e.the buoyant jet behavior is not simulated).
Sewage discharge is assumed to start at the same time
as the storm.
    Since the value of diffusion coefficient needs to be
chosen in relation to the grid length, a much smaller
grid length is necessary for computing the pollutant
transport. Therefore, a new square grid with a side
of 100 m is located in the area shown in Figure (1).

Horizontal and vertical diffusion coefficients are used respectively as 25 m^2/s and 0.1 m^2/s. The decay rate constant of the bacteria is taken as 0.003194 s^{-1}. Equilibrium state is reached after 5 hours.

Distribution of pollutant concentrations at the surface (from inner contour to outer one in bac/ml: 900000, 300000, 100000, 50000, 25000, 10000, 5000, 2500, 1000, 500, 100, 10) and bottom (from inner contour to outer one in bac/ml: 100000, 50000, 25000, 10000, 5000, 2500, 1000, 500, 100, 10) layers are given at two successive times (1hr and 3 hrs) before the equilibrium is reached, in Figure (9). The equilibrium state is sketched in Figure (10) which shows the areas having bacteria concentrations in excess of 10 bac/ml. It is seen that at the surface layer advection is predominant in the direction of current velocity and dispersion is small compared to the advection. At the bottom layer, concentration decreases about 10 times.

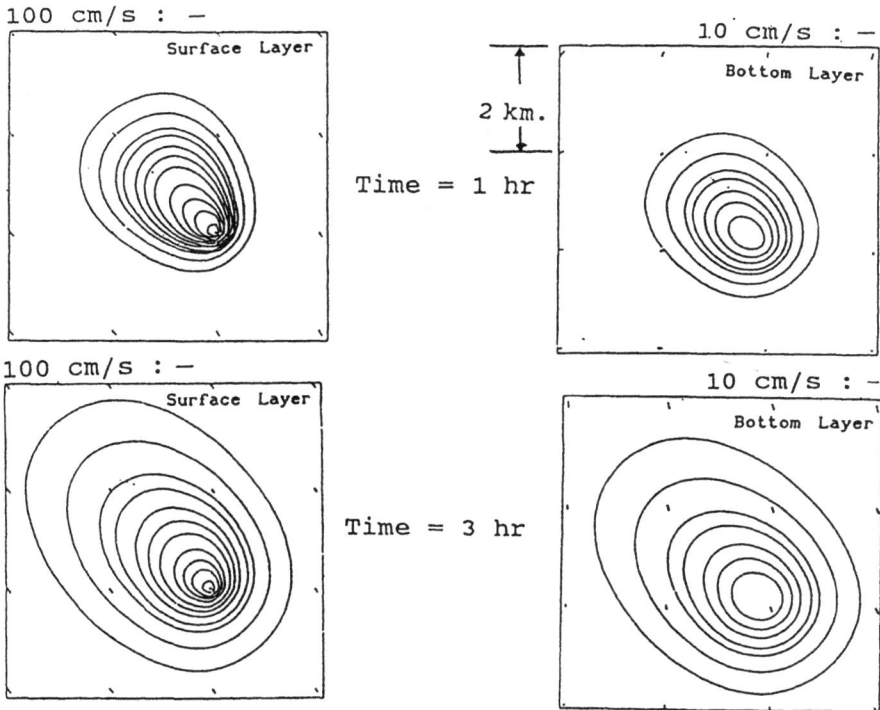

Figure 9 Progress of concentration contours (NNW wind speed : 20 m/s)

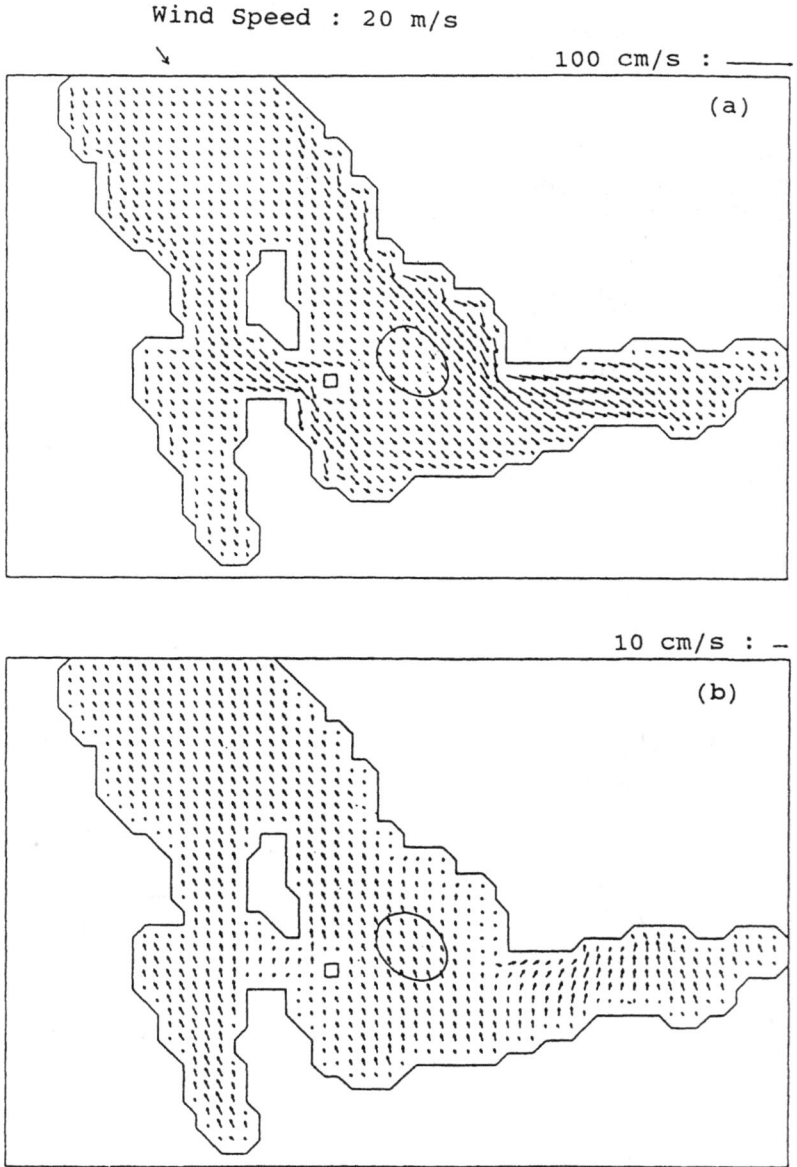

FIGURE 10   *10 bac/ml pollutant concentration contour at the steady state after five hours in Izmir Bay (NNW wind speed : 20 m/s),(a) at the surface layer, (b) at the bottom layer*

4 Concluding Remarks

The 3-D models of circulation and pollutant transport prepared for application to Izmir Bay are powerful management tools. Preliminary calculations with the models provide the essential features of wind induced currents, together with advection and diffusion of a pollutant plume.

The hydrodynamic model will be verified by using the results of current measurements which are in progress.

The model presented in this paper may also be applied for flushing of small water basins, such as harbors, by wind due to shearing effect at the entrances since lateral shears are included.

REFERENCES

[4] Borella,A. et.al.,(1992),"An Analysis of the Mechanisms of Water Exchange between the Inner Gulf and the Open Sea",*Marina Technology*, Ed. W.R. Blain, Computational Mechanics Publications-Thomas Telford, pp.297-312.

[8] Chapra,S.C.,Canale,P.R.,(1989), *Numerical Methods for Engineering*, Mc Graw-Hill Book Company,New York, pp.669-773.

[2] Falconer,R.A.,(1991),"Review of Modeling Flow nd Pollutant Transport Processes in Hydraulic Basins", *Proceedings of the First International Conference on Water Pollution: Modeling Measuring and Prediction* , Southampton, UK, Computational Mechanics Publications, pp.3-23.

[6] Koutitas,C.G.,O'Connor,B.,(1980),"Modeling of Three-Dimensional Wind Induced Flows", *Journal of the Hydraulic Division*, pp.1843-1865.

[5] Mashroutechi,F.,(1991)," Mathematical Modeling of Three-Dimensional Coastal Circulation and Application to Marmaris Bay",M.S. Thesis,Middle East Technical University.

[3] Rajar,R.,Cetina,M.,(1991),"Modeling Wind Induced Circulation and Dispersion in the Northern Adriatic",*XXIV. Congress of International Association for Hydraulic Research* .

[1] Uslu,O.,Sengul F.,(1987),"Description of Izmir Great Sewerage Project",*Environment and People (Cevre ve Insan)*,no.4,pp. 18 - 23

[7] Verboom,G.,(1975),"The Advective-Dispersion Equation for an Isotropic Medium Solved by the Fractional Step Method", *Mathematical Models for Environmental Problems*, pp.229-312, London, England.

Part 4
WAVE KINEMATICS AND MODELLING

33 The impulsive pressure due to wave impact on a sea wall

M. J. Cooker

ABSTRACT

Pressure impulse theory is explained and used to model the impulsive pressures and total force on a vertical seawall. The theory can also estimate the fluid velocity after impact from simple data about the incident wave immediately before impact. The vertical rise of spray is modelled and an estimate is gained for the cross-sectional area of the fluid in the rising plume.

1. Introduction

When an ocean wave breaks against a coastal structure very large forces of short duration are exerted. The liquid pressure at a vertical wall struck by an overturning wave can be ten times greater than the hydrostatic pressure due to the head of water in the wave, and the pressure can rise and fall in a time of less than 1 millisecond. These high pressures are associated with high pressure gradients, which in turn are related to large fluid accelerations, which are manifested in the vertical plumes of spray thrown upward by wave impacts during storms.

This work follows on from Cooker and Peregrine (1991,1992a,b) and describes a mathematical model of the pressure field and the abrupt changes in the flow field which occur during wave impact on a vertical wall. We only give a summary of the theory here and pass on to show a comparison with experiment and to discuss vertically projected spray in the context of the overtopping of vertical seawalls.

2. Pressure Impulse Theory

This theory rests on the idea of pressure impulse $P(\underset{\sim}{x})$ which is the time-integral of the pressure, p, at a point, $\underset{\sim}{x}$, in the fluid from the start of impact at time $t = t_b$ to the end of impact at time $t = t_a$ (throughout the subscripts b and a refer to "before" and "after" impact, respectively). That is

$$P(\underset{\sim}{x}) = \int_{t_b}^{t_a} p(\underset{\sim}{x},t) \, dt \ . \tag{1}$$

Bagnold (1939) used this concept in his experimental study of wave impacts on vertical walls. He found that the peak pressure p_{pk} measured at the wall varies greatly among apparently identical waves, but that P (at a given point) is more repeatable.

We assume that the pressure rises and falls in time in a triangular spike of width Δt (= t_a-t_b) and height p_{pk} , so that (1) gives

$$p_{pk}(\underset{\sim}{x}) = 2P(\underset{\sim}{x})/\Delta t. \tag{2}$$

398

Pressure impulse is discussed by Batchelor (1973, p471) in the context of the initiation of flow in a fluid region initially at rest by the impact of a moving body. Here we extend the theory to model the sudden change of flow when moving liquid strikes a fixed wall. Suppose U_o is a typical fluid speed for the flow before the impact of the wave, and that the wave has a length scale (eg. wave height) L_o, and suppose the short time of impact is Δt. Then provided $U_o \Delta t / L_o$ is much less than one the pressure impulse approximately satisfies Laplace's equation:

$$\frac{\partial^2 P}{\partial x^2} + \frac{\partial^2 P}{\partial y^2} + \frac{\partial^2 P}{\partial z^2} = 0 \ . \tag{3}$$

Also
$$\underset{\sim}{u}_a = \underset{\sim}{u}_b - \frac{1}{\rho} \left(\frac{\partial P}{\partial x} \ , \ \frac{\partial P}{\partial y} \ , \ \frac{\partial P}{\partial z} \right) \tag{4}$$

where ρ is the fluid density and $\underset{\sim}{u}_b$, $\underset{\sim}{u}_a$ are the fluid velocities before and after impact, respectively.

Since Δt is relatively short it is appropriate to solve (3) in a fixed domain. Waves which approach the shallows in front of a seawall, from deep water, tend to be much higher than the local water depth; they also have long backs, and the most severe impacts come from waves at normal incidence to the wall. We idealize the wave to a semi-infinite rectangle (see figure 1) inside which we solve (3), subject to boundary conditions discussed in the next section.

3. The Boundary-Value Problem for Wave Impact

We take atmospheric as a zero reference pressure, so that $P = 0$ on the free surface ($y = 0$ in fig. 1). Now consider $\underset{\sim}{n}$,the unit normal at a point on the solid boundary of the domain ($x > 0$,

399

$-H \leq y \leq 0$). Resolving the terms in equation (4) along $\underset{\sim}{n}$ gives

$$\frac{\partial P}{\partial n} = \rho(u_{bn} - u_{an}) , \tag{5}$$

where u_{bn} and u_{an} are the normal components of $\underset{\sim}{u}_b$ and $\underset{\sim}{u}_a$. If the fluid stays in contact with the boundary during impact then $u_{bn} = u_{an} = 0$, and on such a boundary (5) shows that $\partial P/\partial n = 0$. More interestingly at a point where the fluid strikes the wall with the given speed u_{bn} , we may take $u_{an} = 0$, and (5) states

$$\partial P/\partial n = \rho u_{bn}. \tag{6}$$

(These impact conditions appear to be unaffected by the

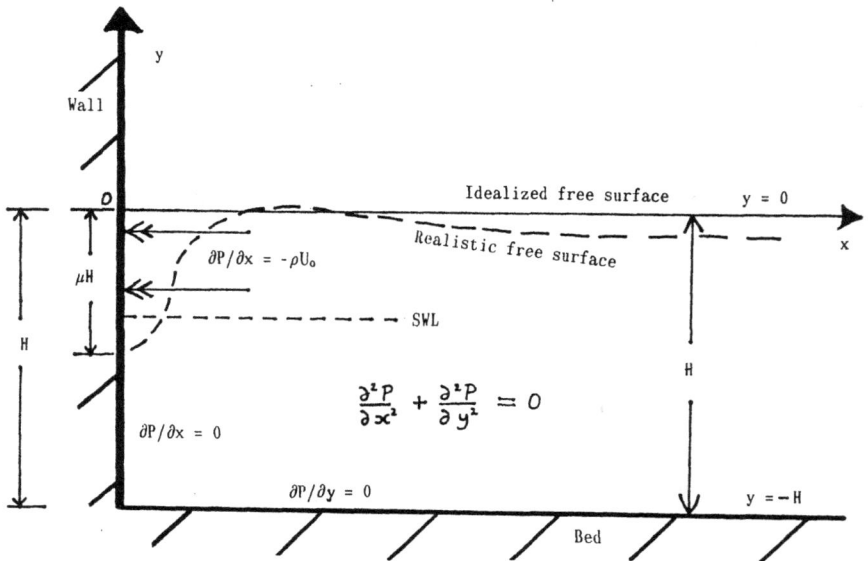

FIGURE 1 *Boundary value problem for pressure impulse*, P. *The fluid domain at impact is idealised to* $x \geq 0$, $-H \leq y \leq 0$

presence or absence of any air cushion trapped by the wave against the wall, although such details may affect the size of ·Δt in equation (2).) Far from the wall P is vanishingly small. In figure 1 the boundary conditions are shown. Note that we model the wave impact as a zone on the wall which stretches down the wall from $y = 0$ to $y = -\mu H$ (where μ is a fraction between 0 and 1 which must be chosen). In this zone u_{bn} is a function of y. Below the impact zone $\partial P/\partial x = 0$.

A solution of (3) which satisfies the conditions at the bed, on the free surface, and in the far field, and which retains a dependence on the z-coordinate parallel to the vertical wall is

$$P = \sum_{j=1}^{\infty} \sum_{i=1}^{\infty} \sin b_i y \ (A_{ij} \sin c_j z + B_{ij} \cos c_j z) \ \exp(-d_{ij}x)$$

(7)

where $d_{ij}^2 = b_i^2 + c_j^2$, $b_i = (i-\frac{1}{2})\pi/H$ and the c_j's are given by the geometry bounding the flow in the z-direction. The Fourier coefficients A_{ij} and B_{ij} are determined by the known values of $\partial P/\partial x$ on the verticl wall. For simplicity we treat a problem in the x-y plane in which the wave is at normal incidence to the wall. Equation (7) simplifies to

$$P(x,y) = \sum_{i=1}^{\infty} A_i \sin b_i y \ \exp(-b_i x) \ ,$$

(8)

where

$$A_i = \frac{-2\rho}{H b_i} \int_{-\mu H}^{0} u_{bn}(y)\sin(b_i y) \ dy \ .$$

(9)

For the sake of simplicity alone we take $u_{bn}(y) = -U_0$ a constant, in which case (8) becomes

$$P(x,y) = \frac{-2\rho U_0}{H} \sum_{i=1}^{\infty} \frac{(1 - \cos \mu H b_i)}{b_i^2} \sin b_i y \ \exp(-b_i x)$$

(10)

where $b_i = (i-\frac{1}{2})\pi/H$.

Provided we can estimate Δt, equations (2) and (10) give a

401

theoretical estimate of the peak pressure distribution throughout the fluid domain. We compare our estimate of p_{pk} with the
experimental results of Partenscky and Tounsi (1989). We choose
H=2.45m, U_0= 4.9m/s, ρ=1035kg/m^3, μ=0.24, and Δt = 0.015s
(given). Figure (2) shows p_{pk} at the wall. The theory and
measurements both show the characteristic maximum, high up on

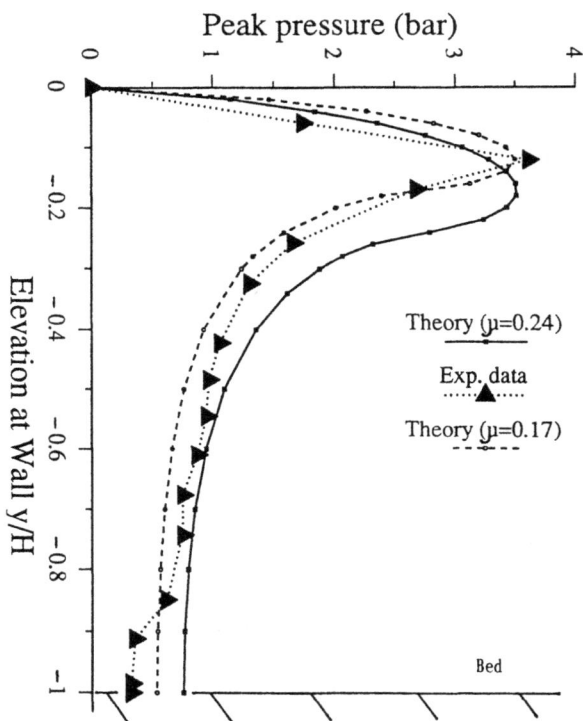

FIGURE 2 *Comparison between the peak pressure data reported by*
[7] *and pressure impulse theory.*

the wall (see Nagai, 1973), but we have under-predicted the
maximum value because we chose too large a value of Δt. The
theory for μ=0.17 is also shown.

Here we do not attempt to model Δt because its value is bound
up in an unknown way with the *details* of wave impact. A minimum
possible value of Δt is $\mu H/c$ where c is the fluid sound
speed. As Bagnold rightly remarked c =1500m/s yields an

unreasonably small Δt. But an air bubble content of only a few percent can reduce c to only 100m/s, and bubbles may strongly modify Δt in real waves.

4. The Momentum Length

The total impulse, I, on the wall is obtained by integrating P up the wall:

$$I = \rho U_0 H \left[\frac{2H}{\pi^3} \sum_{i=1}^{\infty} \frac{\{1 - \cos \mu(i-\frac{1}{2})\pi\}}{(i-\frac{1}{2})^3} \right] . \qquad (11')$$

The peak force F on the wall can be estimated from $F = 2I/\Delta t$. The term [ ] in (11) has dimensions of length, and we call it the momentum length, L_m. Horizontal momentum is lost from all parts of the wave, but most is lost from a region next to the wall which has width L_m. The momentum length (for fixed U_0) is a function of μ and takes a maximum value of 0.57H when $\mu = 1$. (See fig. 6 of Cooker and Peregrine, 1991.) The surprisingly small width of fluid which actively imparts momentum to the wall explains why the impulse is insensitive to the shape of the wave far from the wall, and this is borne out by the study of other geometries by Cooker et al (1992a).

5. The Flow After Impact

We now examine the flow after impact in order to model the amount of fluid thrown vertically upward by the wave. Equations (4) and (10) give the flow after impact $\underset{\sim}{u}_a = (u_a , v_a)$ in terms of the incident flow $\underset{\sim}{u}_b = (u_b , v_b)$. For simplicity we suppose that the top of the wave moves only in the horizontal before impact, so that $\underset{\sim}{u}_b = (-U_0, 0)$. $\underset{\sim}{u}_a$ is most interesting to evaluate at the free surface where the vertical velocity component v_a is the following function of x (at y=0)

$$v_a(x) = 2U_0 \; \frac{1}{\pi} \sum_{i=1}^{\infty} \frac{[1 - \cos \mu(i-\tfrac{1}{2})\pi \;] \; \exp(-x(i-\tfrac{1}{2})\pi/H)}{(i-\tfrac{1}{2})}$$

which can be summed analytically to give

$$v_a(x) = \frac{-U_0}{2\pi} \; \ln \left| \frac{r^2(1 + r^2s^2)(1-s^2)}{(r^2 - s^2)(1+s^2)} \right| \;, \tag{12}$$

where $r = \tanh(\pi x/4H)$ and $s = \tan(\pi\mu/4)$. According to (12) v_a is logarithmically singular as we approach the wall (at x=0) along the free surface: this is a consequence of the over-simple impact boundary condition that u_{bn} is a *constant*, $-U_0$. Nevertheless we do expect very large vertical velocities near the wall after impact, and we can still use (12) to find the vertical rise of the free surface a time T after impact, assuming that the fluid rises in free fall under gravitational acceleration g. See figure 3. Consider those particles which rise above the top of a wall which has height D above y=0. Now v_a decreases with increasing x so there is a particular particle at x=X which ascends under gravity, with *initial* speed $v_a(X)$, to the height D in a time T, at which time its speed has decreased to *zero*. For x>X, $v_a(x)$ is too small for the particle to reach the top of the wall, and for x such that $0 \le x \le X$ the fluid ascends higher than D. The position X is found from the relation $v_a(X) = \sqrt{(2Dg)}$ and we also have $T = \sqrt{(2D/g)}$. At time T after impact the cross-sectional area, A, of fluid above the elevation y = D in figure 3 is

$$A = \int_0^X v_a(x) \; T \; dx. \tag{13}$$

Equation (12) shows that

$$X = \frac{4H}{\pi} \operatorname{arctanh} \left\{ \left[\frac{E - 1 + \sqrt{((1-E)^2 + 4s^4 E)}}{2s^2} \right]^{1/2} \right\}$$

where $\quad E = \exp \left\{ \dfrac{-2\pi U_0}{\sqrt{2Dg} - \dfrac{U_0}{2\pi} \ln \left| \dfrac{1+s^2}{1-s^2} \right|} \right\} \quad$ and $\quad s = \tan(\pi\mu/4)$.

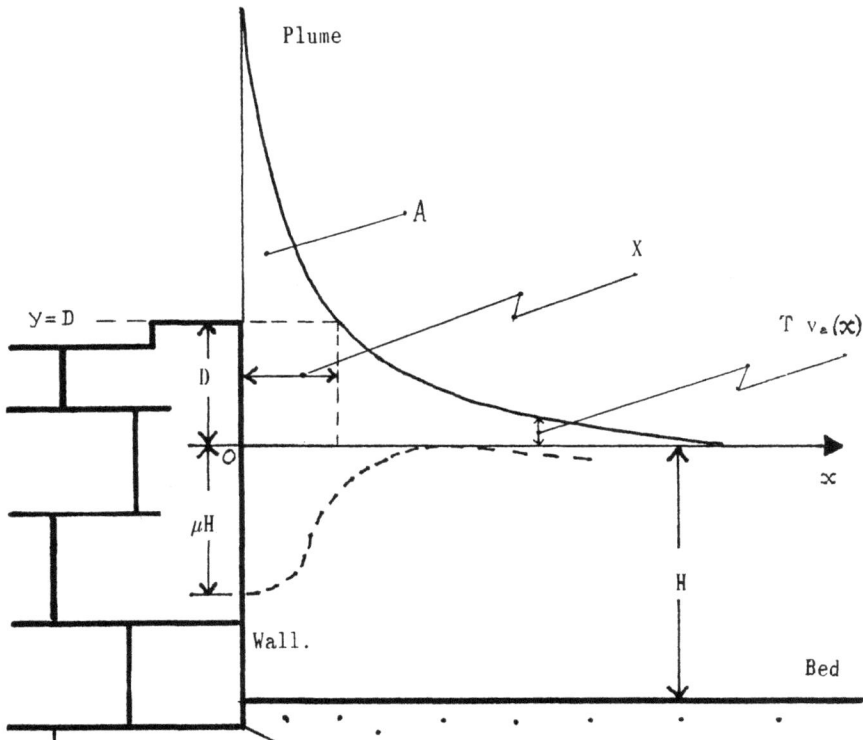

FIGURE 3   *The vertical rise and fall of fluid at the wall. For x>X the surface cannot ascend to a height more than* D.

Hence from (13) we have

$$A = \frac{\sqrt{(2D)}\; 2HU_0}{\pi^2 \sqrt{g}} \sum_{i=1}^{\infty} \frac{[1 - \cos \mu(i - \frac{1}{2})\pi]}{(i - \frac{1}{2})^2} [1 - \exp\{(\frac{1}{2} - i)\pi X/H\}]$$

$$(14)$$

We can estimate an upper limit to the area A. It is difficult to model horizontal speeds for breaking waves but experiments suggest that approximately the wave face impact speed is $U_0 = \sqrt{(gH)}$ and we note that the summation in (14) is always less than the value obtained by setting $X = \infty$ and $\mu = 1$. Under these conditions the summation is $\pi^2/2$ and we can state that in general $A \leq H^{3/2}\sqrt{(2D)}$. Note that after time T has elapsed the particles closer to the wall than those at $x=X$ will begin to fall; so A may be a rough estimate of the cross-section of water which is available to be blown over the seawall by an onshore wind.

6. Conclusions

Bagnold (1939) showed that the pressure impulse P measured at the wall is a more repeatable quantity among a sequence of similar wave impacts than the peak pressure. We have extended the idea of pressure impulse to a quantity which varies throughout the domain of fluid during wave impact. The spacial gradients of P/ρ give the sudden change of fluid velocity during impact. Examination of the solution in a special case enables us to find the impulse distribution on a vertical wall and the total impulse. This shows that only the fluid close to the wall loses significant horizontal momentum to the wall; so that only the shape of the wave close to the wall need be accurately described for this model. From $P(\underset{\sim}{x})$ we can estimate a range of possible values of the peak pressure distribution through relation (2). The value of Δt for any single impact may depend on the fine details of the process by which the horizontal motion of the wave face is brought to rest at the wall. This is a complex issue which may have to be addressed through a probabilistic model of Δt. Finally the flow speed distribution after impact against the wall can be used to roughly estimate the cross-section of fluid which ascends to a height greater than the elevation of the wall.

ACKNOWLEDGEMENTS: Valuable discussions with Prof. D.H. Peregrine and financial support from U.K. S.E.R.C. grants GR/F 28298 and GR/G 21032 are gratefully acknowledged.

REFERENCES

[1] Bagnold, R.A. (1939) "Interim report on wave pressure research" *J. Institute of Civil Engineers* 12 , pp201-226.

[2] Batchelor, G.K. (1973) *An Introduction to Fluid Dynamics*, Cambridge University Press, 615pp.

[3] Cooker, M.J. and Peregrine, D.H. (1991) "A model for breaking wave impact pressures" Proc. 22nd Intl. Conf. on Coastal Engineering, Delft 1990, pp1473-1486.

[4] Cooker, M.J. & Peregrine, D.H.(1992a) "A mathematical model for liquid impact problems" submitted to *J. Fluid Mech.* 31pp.

[5] Cooker M.J., and Peregrine, D.H. (1992b) "Wave impact pressure and its effect upon bodies lying on the bed", to appear in *Coastal Engineering*, 23pp.

[6] Nagai, S. (1973) "Wave forces on structures" in *Advances in Hydroscience*, ed. Ven Te Chow, 9, pp253-324.

[7] Partenscky, H.W. and Tounsi, K. (1989) "Theoretical analysis of shock pressures from waves breaking at vertical structures" XXIII Congress I.A.H.R., Ottawa 1989, 6pp.

34 Wave-induced oscillations in Visakhapatnam fishing harbour

M. L. Narasimham and D. Apparao

ABSTRACT

The problem of harbour resonance has been studied by several investigators in the past, who considered both regular and arbitrary shaped harbours. Curves predicting the harbour oscillations for different values of wave numbers have been obtained by them. A study has been presently taken up by considering the basic equations proposed by Lee (1971) with a view to develop a mathematical model for evaluation of wave induced oscillations in Visakhapatnam harbour located in the east coast of India. In this study the waves are considered to be approaching the harbour entrance in a normal direction. The geometry of the harbour is the basic input to the model along with the wave number corresponding to the approaching wave. The computer program developed has been checked for its performance by validating the same using Lee's results. Later the model has been used to obtain the harbour oscillations in the outer harbour and fishing harbour as well. The results for the latter location viz., the fishing harbour are discussed in this paper.

Introduction

The phenomenon of harbour resonance causes significant damage to moored ships and adjacent structures, especially if the resonant periods of the harbour are close to those of the mooring and other systems. For an effective design of a harbour to be free from oscillations or to avert the resonance problem in a proposed harbour, the coastal engineer must first be able to predict the resonance corresponding to the expected wave periods (or wave numbers).

The problem of oscillations in a circular harbour of small entrance was first considered by Mcnown in 1952. In this study a standing wave was assumed to be located near the entrance. This method was later extended to rectangular harbours by Kravechenko and Mcnown in 1953. Biesel and LeMehaute (1955 and 1956) and LeMehaute (1960 and 1961) studied the oscillation pattern in rectangular harbours with various types of entrances connected to an infinitely long but narrow channel. The problem of rectangular harbours connected directly to the open sea has been investigated by Miles and Munk in 1961. In their study the effect of wave radiation from harbour mouth to open sea has been included there by limiting the maximum wave amplitude inside the harbour. Ippen and Goda in 1963 also studied the problem of rectangular harbour connected to the open sea, considering the waves to be radiating from the entrance in to the open sea. Leendreste in 1967 developed a numerical scheme based on finite differences for propagation of long period waves in an arbitrary shaped basin given the water surface elevations at the open boundary.

Lee in 1971 (2) developed a theory for wave induced oscillations in harbours of arbitrary geometry with constant depth by applying Weber's solution to Helmholtz equation in regions both inside and outside the harbour. The final solution was obtained by matching the results near the entrance. In the present instance Lee's approach has been considered for evaluating the wave induced oscillations in the Visakhapatnam fishing harbour.

410

Basic Equations And Methodology

For a wave propagating over constant depth in a harbour (Fig.1a), within the limitations of small amplitude wave theory, the velocity potential satisfying Laplace equation is given as

$$\phi(x,y,z,t) = \frac{1}{i\sigma} A_i g \frac{\cosh k(h+z)}{\cosh kh} f(x,y)\exp(-i\sigma t) \quad \ldots 1$$

in which the wave function $f(x,y)$ represents the variation of velocity potential ϕ in x-y direction, which in turn has to satisfy the Helmholtz equation:

$$\frac{\partial^2 f}{\partial x^2} + \frac{\partial^2 f}{\partial y^2} + k^2 f = o \quad \ldots 2$$

It has to further satisfy the following boundary conditions:

(i) No flow passes through any solid boundary, such as coast line of the harbour (i.e. f/n=0)

(ii) Waves radiated from the harbour entrance decay to zero at an infinite distance from harbour - Radiation condition (Fig.1b)

To obtain a solutaion to equation 2, the domain of interest has been divided in to two regions: infinite ocean (region -I) and the region bounded by the limits of the harbour (region -II) as shown in figure 1b. The function f which is to be determined for solving the harbour oscillation problem is designated as f_1 and f_2 in regions I and II respectively. These functions and their normal derivatives are equal along the entrance to the harbour. The solution to equation 2 in region II has been obtained (2) by considering the Hankel function of the 1st kind to be the fundamental solution of the two-dimensional Helmholtz equation. (Equation A2 of Appendix). A simlar expression has been obtained for f_1 in region-I.

411

The harbour boundary is divided in to sufficiently large number of segments and the value of f or ∂f /∂n for each segment is considered constant and equal to the value at the mid point of each segment. Thus the governing equations are written in matrix form both on the boundary and near the entrance (equation A3).

The vector C of the normal derivatives of f_2 (∂f_2/∂n) given in equation A5 is evaluated by expressing the function f_1 in region I as a function of the same normal derivatives of C and matching them at the entrance.

The function f_1 is expressed as

$$f_1 (x,y) = f_i (x,y) + f_r(x,y) + f_3 (x,y)$$

in which f_i= incident wave functin, f_r=reflected wave function considered to occur as if the entrance were closed and f_3=a correction to f_r due to wave radiation in to the open sea. For the case of a periodic incident wave with a wave ray perpendicular to the coast line, the function $f_i(x,y)$ can be choosen as $\frac{1}{2} e^{-1ky}$. For this case with f_3 neglected the value of f_1 is given by equation A10.

Finally the values of $f_2(x)$ at any point x inside the harbour is given by the following equation which is same as equation A11.

$$f_2 (x) = -\frac{1}{4} i\left\{ \sum_{j=1}^{N} f_2 (X_j) \left[-kH_1^{(1)}(kr)\frac{\partial \Upsilon}{\partial n} \right]\Delta s_j \right.$$

$$\left. -\sum_{j=1}^{p} H_0^{(1)} (kr) C_j\Delta s_j \right\} \qquad \ldots 3$$

where X_j is at the midpoint of the jth boundary sigment, $r=|X-X_j|$, p=no. of segments along the entrance and C= ∂f_2/∂n along the harbour entrance.

The responce factor, R of the harbour to the incident waves is given by the ratio of wave amplitude at any·point (x,y) inside the harbour to the sum of the incident and reflected wave amplitudes at the coast line (with the harbour entrance closed). Thus the amplification factor R is given by

$$R = |f_2(x,y)| \qquad \text{------- (4)}$$

Discussion of Results

The model developed in the present study has been validated by obtaining results using Lee's laboratory data (2). Figure 2 given here shows the variation of the amplification factor R with dimension less wave mumber parameter, k.l (l being an arbitrary length) for the case of a circular harbour with 10° opening. From this figure it can be clearly seen that the results obtained during the present investigation are coinciding well with the experimental and theoretical results of Lee. A similar agreement was observed for circular harbour with 60° opening and a rectangular harbour fully open to sea (1).

After establishing the validity of the present model, results for Visakhapatnam outer harbour (figure 3) have been obtained. Figure 4 shows the wave induced oscillations at the point 'A' located near the off-shore tanker terminal. In the range of $0 < kl < 5$ a number of peaks have been observed. The wave periods characteristic of the site are between 7.5 see and 8.5 sec and inorder to closely study the the response pattern in this range, more number of points were taken in this range of kl ($1.5 < kl < 1.9$). The results obtained are shown in the inset of figure 4, which indicate 5 peaks in this range. This reveals that the results are to be obtained with more number of points in the range of kl values characteristic to the location.

The study was later extended to the fishing harbour which more or less satisfies the condition of constant depth. The harbour boundary has been divided into 47 segments including 3 segments near the entrance. Initially the results were obtained in a wider range of kl values as carried out for the outer harbour case. Subsequently the amplification factors in a close range of wave periods (7.5 sec $<$ T $<$ 8.5 sec) were obtained. These results are shown in figure 5 and inset of figure 5

413

respectively, which indicate three predominant peaks occuring at the values of kl equal to 3.975, 4.20 and 4.75. The corresponding time periods are 6.55 sec, 6.26 sec and 5.68 sec. The magnitudes of these peaks are seen to be equal to 3.85, 6.0 and 7.9 respectively. From the inset of this figure it can be clearly seen that there are no additional peaks in the range 7.5 sec $\angle$ T $\angle$ 8.5 sec, which establishes that the fishing harbour is free from severe wave induced oscillations in this range of wave periods.

The present study has been carried out assuming a normal incidence and in order to ascertain the possible oscillations, the study has to be carried out further by considering waves approaching the entrance in various possible directions.

Conclusions

From the present investigation the following conclusions are made.

(i)   The model that has been developed for computing the wave induced oscillations in Visakhapatnam harbour has been found to be satisfactorily performing based on the comparisons made using Lee's results.

(ii)   The results obtained at the off shore tanker terminal site show many predominant peaks in the range of wave periods studied. The necessity to limt the rage of kl and to take more number of points in the interested range has been established.

(iii)   The results for the fishing harbour indicate no predominant peaks in the range of wave periods characteristic to the location. This aspect has to be further established by carrying out studies with different wave rays entering the harbour.

REFERENCES

1. Appa Rao, D. (1990) "Some Studies on Wave Induced Oscillation in Harbours of Arbitrary Shape"; M.E. Thesis, Andhra University, Visakhapatnam, India.
2. Lee, J.J. (1971) "Wave Induced Oscillations in Harbours of Arbitrary Shape"; Jl. Fluid Mechanics, 45 Part 2, PP 375-394.

APPENDIX

HELMHOLTZ EQUATION :
$$\frac{\partial^2 f}{\partial x^2} + \frac{\partial^2 f}{\partial y^2} + k^2 f = 0$$
$$\ldots \text{A1}$$

SOLUTION OF HELMHOLTZ EQUATION:

$$f_2(x_i) = -\frac{1}{2} i \int_S \left\{ f_2(x_o) \frac{\partial}{\partial n} \left[H_o^{(1)}(k \mid x_i - x_o \mid) \right] \right.$$

$$\left. - H_o^{(1)}(k \mid x_i - x_o \mid) \frac{\partial}{\partial n} \left[f_2(x_o) \right] \right\} ds(x_o) \qquad \ldots \text{A2}$$

(Along the Harbour boundary)

MATRIX FORM OF EQUATION A2:

$$X = -\frac{1}{2} i (G_n X - G P) = \sum_{j=1}^{P} M_{ij} C_j \qquad \ldots \text{A3}$$

Whare $\quad X = f_2(x_i), \quad i = 1, 2 \ldots N \qquad \ldots \text{A4}$

$$P = \frac{\partial}{\partial n} f_2(x_j) = C_j, \quad j = 1, 2 \ldots N \qquad \ldots \text{A5}$$

$$(G_n)_{ij} = -k H_1^{(1)}(k r_{ij}) \left[-\frac{x_i - x_j}{r_{ij}} \left(\frac{\partial y}{\partial s} \right)_j \right.$$

$$\left. + \frac{y_i - y_j}{r_{ij}} \left(\frac{\partial x}{\partial s} \right)_j \right] \Delta s_j$$

$$(i, j = 1, 2 \ldots N ; i \neq j) \qquad \ldots \text{A6}$$

$$(G_n)_{ii} = -\frac{i}{\pi} \left(\frac{\partial x}{\partial s} \frac{\partial^2 y}{\partial s^2} - \frac{\partial^2 x}{\partial s^2} \frac{\partial y}{\partial s} \right)_i \Delta s_i$$
$$(i = 1, 2 \ldots N) \qquad \ldots \text{A7}$$

$$(G)_{ij} = H_o^{(1)}(kr_{ij}) \Delta s_j,$$

$$(i, j = 1, 2 \ldots N; \ i \neq j) \qquad \ldots A8$$

$$(G)_{ii} = 1 + i \frac{2}{\pi} [\log(\frac{k \Delta s}{4} i) - 0.42278] \Delta s_i$$

$$(i = 1, 2 \ldots N) \qquad \ldots A9$$

Δs = Segment length, $H_o^{(1)}$ and $H_1^{(1)}$ are Hankel functions of zeroth and first order respectively.

$$f_1(x_i) = 1 + (-\frac{1}{2} i) \sum_{j=1}^{p} H_{ij} C_j \quad (i \neq 1, 2 \ldots p) \qquad \ldots A10$$

(Along the Harbour entrance)

Whare $H_{ij} = H_o^{(1)}(kr_{ij}) \Delta s_j$ for $i, j = 1, 2 \ldots p$; $i \neq j$ and

$$H_{ii} = [1 + i(2/\pi)(\log(\frac{1}{4} k \Delta s_i) - 0.42278)] \Delta s_i$$

for $i = 1, 2 \ldots p$

FINAL SOLUTION FOR POINTS INSIDE HARBOUR

$$f_2(x) = -\frac{1}{4} i \left\{ f_2(X_j) [-kH_1^{(1)}(kr) \frac{\partial r}{\partial n}] \Delta s_j - \sum_{j=1}^{p} H_o^{(1)}(kr) C_j \Delta s_j \right\} \qquad \ldots A11$$

in which $r = |x - x_j|$, xj being the mid point of the jth boundary segment.

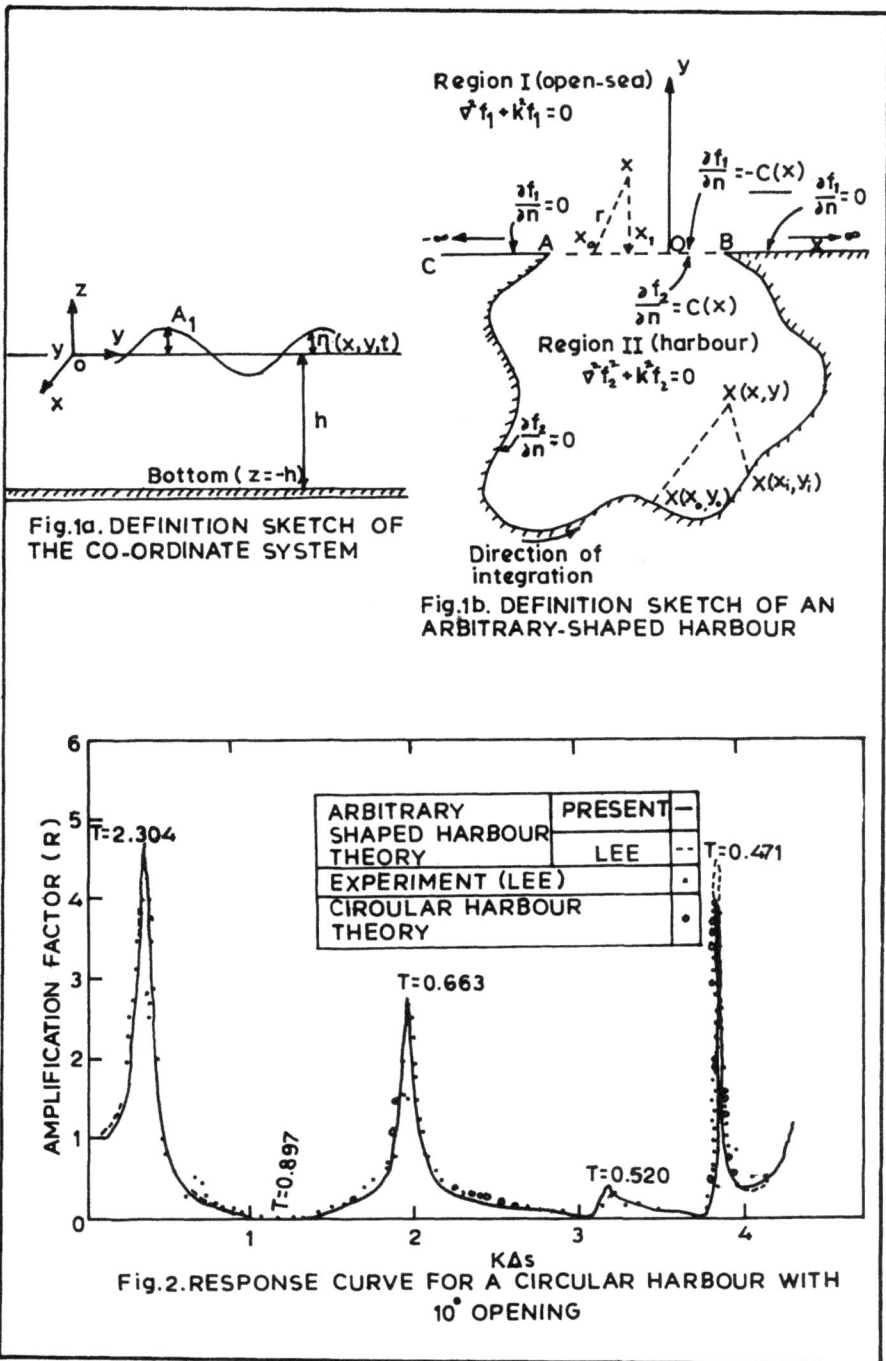

Fig.1a. DEFINITION SKETCH OF THE CO-ORDINATE SYSTEM

Fig.1b. DEFINITION SKETCH OF AN ARBITRARY-SHAPED HARBOUR

Fig.2. RESPONSE CURVE FOR A CIRCULAR HARBOUR WITH 10° OPENING

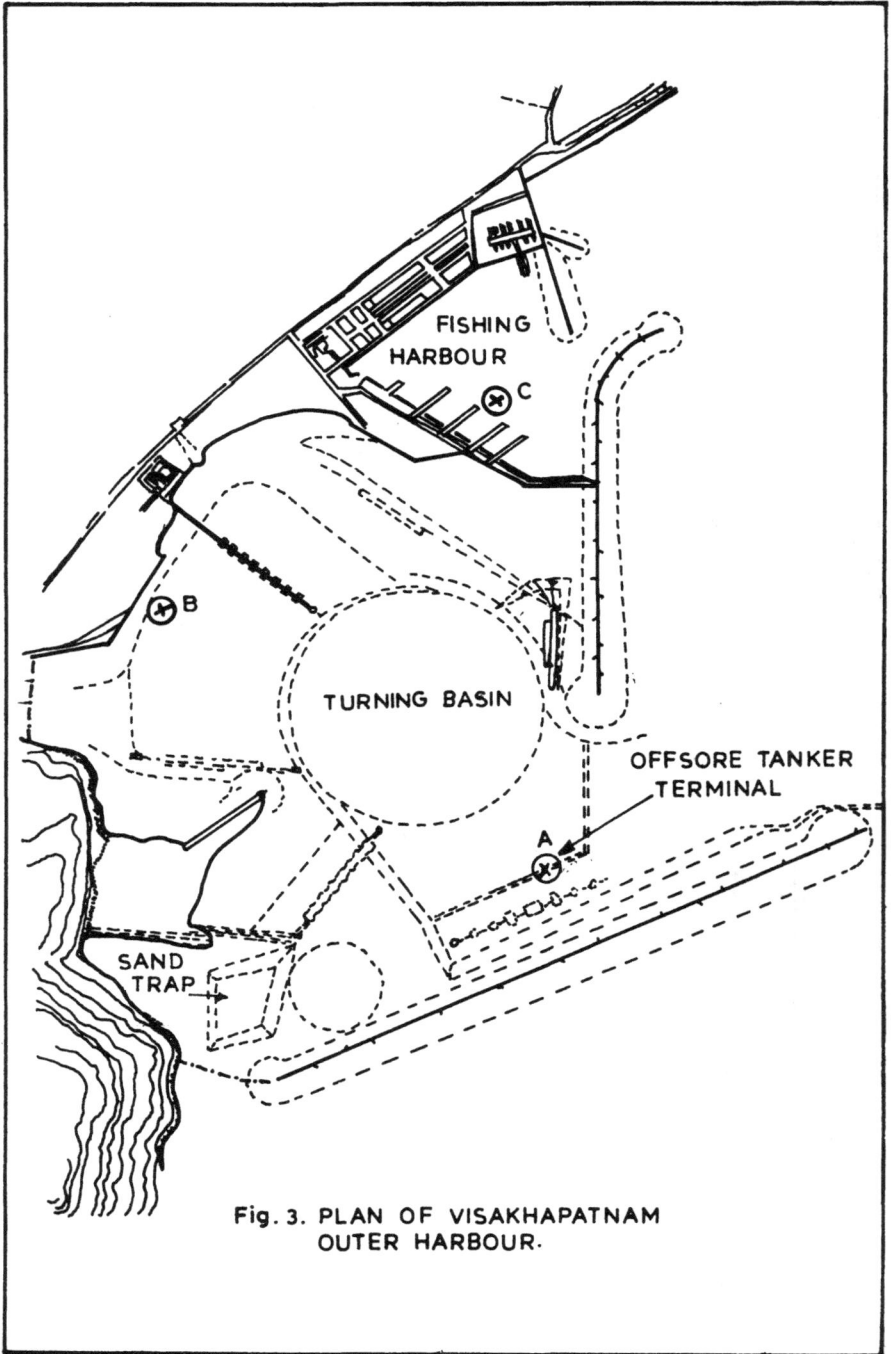

Fig. 3. PLAN OF VISAKHAPATNAM OUTER HARBOUR.

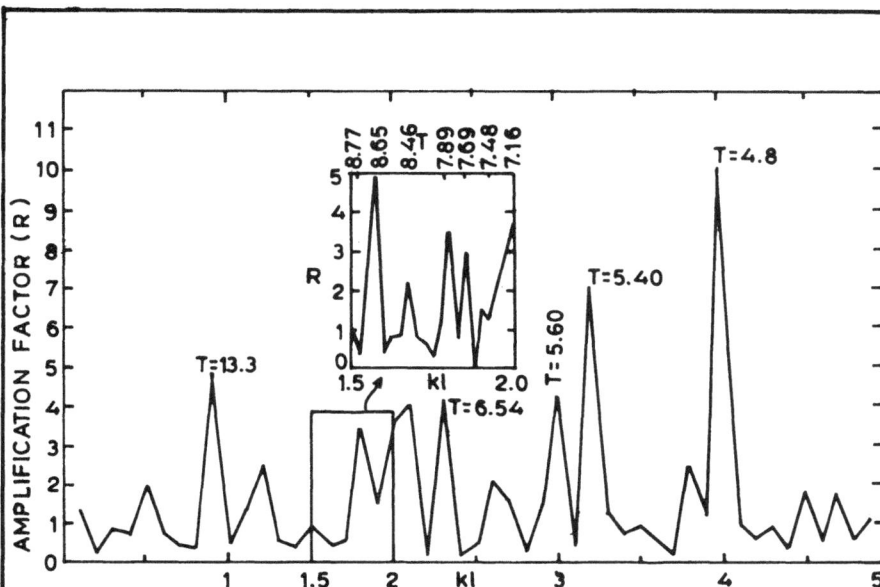

Fig.4. RESPONSE CURVE OF OFFSHORE OIL TANKER TERMINAL
(POINT-A)

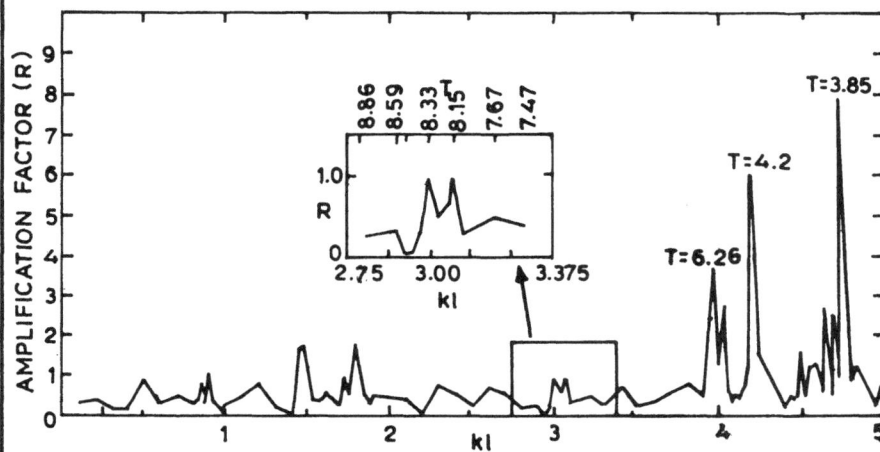

Fig.5. RESPONSE CURVE OF FISHING HARBOUR
(POINT-C)

35 A two-layer numerical model of wave motion over soft mud

J. C. I. Piedra Cueva

ABSTRACT

Motion within a soft muddy bottom under small amplitude wave action is briefly examined using an analytical approach to gain insight into the mechanism by which coastal mud responds to water waves. The bed is modelled as a viscoelastic body using the Voigt model. The model's results on interface shear stress and wave damping show acceptable agreement when compared to laboratory data. This model is an extension of the one developed by MacPherson(3), since it considers the continuity of the shear stress through the water-mud interface, taking into account a thin water boundary layer upon the interface. It can be concluded that: a) the wave damping coefficient can be one or two orders higher than the one corresponding to the rigid bed theory, and it depends on the mud's elasticity; b) the interface shear stress can be higher or lower than the one calculated when the bed is considered rigid; and it depends on mud properties, wave frequency and mud depth.

1. Introduction

High turbidity levels and strong wave attenuation at muddy coasts are two noteworthy phenomena which stem from the interaction between surface water waves and soft, movable beds (2).

It is necessary to consider a non-rigid bed to obtain a more realistic picture of the water waves-mud interaction, and thus a better approach to estimate wave damping and shear stress at the water-mud interface.

This paper describes the analysis of the coupled interaction between a non-rigid bed which responds in both an elastic and a viscous manner, and the water waves which advance on its surface. This paper is an extension of MacPherson's paper (3), since it considers the continuity of the shear stress through the water-bed interface, introducing a thin water wave boundary layer. The flow in the boundary layer and in the mud layer are considered as laminar. The paper focuses on the interface shear stress and on the surface wave amplitude damping.

The motion within a soft mud bed is analyzed in an analytical approach to acquire a better understanding of the mud response to water waves.

2. The two layer model

Assuming that the disturbances vary sinusoidally in time such that the displacements and velocities vary as $\exp(-i\sigma t)$, the linear equation for small disturbances of an incompressible viscoelastic medium is obtained (3):

$$\rho \frac{\partial u}{\partial t} = -\nabla P + (\mu + iG/\sigma)\, \nabla^2 u - \rho g \qquad (1)$$

which only differs from the linear Navier Stokes equation for a viscous fluid in the parameter
$\nu e = \mu + iG/\sigma$, called complex viscosity; the real part being the viscosity and the imaginary part being a measure of the elasticity.

Cartesian coordinates (x-z) are introduced in a manner that the origin is at the undisturbed interface between the two layers and "z" is positive upwards. Subindexes "a" and "b" refer to water and mud. The system is forced by a small amplitude surface wave of frequency σ which propagates in a water layer of depth h1 and above a bed of depth h2. The pressures and velocities are expressed in the following complex form:

$$f(x,z,t) = \hat{f}(z) \exp[i(kx - \sigma t)] \tag{2}$$

where $\hat{f}(z)$ usually is complex. Only the real part of this expression should be taken up. The elevation of the surface and interface wave are defined as:

$$\eta = a \exp[i(kx - \sigma t)] \ , \ \zeta = b \exp[i(kx - \sigma t)] \tag{3}$$

In general a is real and b and k are complex numbers. The surface wave can be restated as:

$$\eta = a \exp(-k_j x) \cdot \exp[i(k_r x - \sigma t)] \tag{4}$$

where the real part of k is the wave number and the complex part is the damping coefficient of wave height.

The linear Navier Stokes equations for both layers allow to split the velocity field (u,w) into a potential part (U_p, W_p) and a rotational part (U,W) (3). Thus, the resulting equations are:

2.1 Water

For the potential flow outside of the boundary layer:

$$\nabla^2 \varphi_a = 0 \qquad U_p = -\frac{\partial \varphi_a}{\partial x} \qquad W_p = -\frac{\partial \varphi_a}{\partial z} \tag{5}$$

$$P_a(z) = \rho_a \frac{\partial \varphi_a}{\partial t} - \rho_a g \, z$$

For the rotational flow inside of the boundary layer:

$$\frac{\partial U}{\partial t} = \nu_a \frac{\partial^2 U}{\partial z^2} \tag{6}$$

The vertical velocity w inside of the boundary layer is obtained from the mass conservation equation:

$$w(z) = \int_{-0}^{z} - \frac{\partial u}{\partial z} \, dz + cte \tag{7}$$

2.2 Mud

The velocity field (u,w) is stated using a velocity potential and a stream function in the following way:

$$u_b = -\frac{\partial \varphi_b}{\partial x} - \frac{\partial \psi_b}{\partial z} \qquad\qquad w_b = -\frac{\partial \varphi_b}{\partial x} + \frac{\partial \psi_b}{\partial z} \qquad (8)$$

and the resulting equations are:

$$\nabla^2 \varphi_b = 0 \qquad P_b(z) = \rho_b \frac{\partial \varphi_b}{\partial t} - \rho_b g z \qquad (9)$$

$$\frac{\partial \psi_b}{\partial t} = \nu_e \nabla^2 \psi_b \qquad (10)$$

2.3 Boundary conditions

$$z = h1 \qquad \dot{\eta} = -\frac{\partial \varphi_a}{\partial z} \qquad \phi - g\eta = 0 \qquad (11)$$

$$z = 0 \qquad \begin{array}{ccc} \zeta = -\dfrac{\partial \varphi_a}{\partial z} & u_a = u_b & w_a = w_b \\[2mm] \tau_a = \tau_b & \sigma_{(zz)a} = \sigma_{(zz)b} & \end{array} \qquad (12)$$

$$z = -h2 \qquad u_b = 0 \qquad w_b = 0 \qquad (13)$$

3. Solution

System 5,6,9 and 10 with the boundary conditions 11-13 can be solved straightforward. The resulting velocities are:

$$\hat{U}_\infty = \hat{U}_p(0) = -\sigma a \sinh kh1 + \frac{gka}{\sigma} \cosh kh1$$

$$\hat{u}_a(z) = \hat{U}_\infty + B e^{-qz} \qquad (14)$$

$$\hat{w}_a(z) = -i\sigma b - ik\hat{U}_\infty z + \frac{B_a ki}{q}(e^{-qz} - 1)$$

$$\hat{u}_b - D_b \; [\, l \, \cosh k\,(z+h2) - l \, \cosh l\,(z+h2)\,]\; +$$
$$+ \; C_b \; [\, k \, \sinh k\,(z+h2) - l \, \sinh l\,(z+h2)\,]$$

$$\hat{w}_b - iD_b \; [\,-l \, \sinh k\,(z+h2) + k \, \sinh l\,(z+h2)\,]\; +$$
$$+ \; iC_b \; [\,-k \, \cosh k\,(z+h2) + k \, \cosh l\,(z+h2)\,]$$

(15)

and the shear stress in the mud layer is:

$$\hat{\tau}_b(z) - \rho_b \nu_e \; \Big(\frac{\partial \hat{u}_b}{\partial z} + \frac{\partial \hat{w}_b}{\partial x}\Big)$$

$$\hat{\tau}_b - \rho_b \nu_e \; k \; \{kC_b \, [\, 2\cosh k\,(z+h2) - (m^2 + 1) \, \cosh l\,(z+h2)$$

$$+ \quad kD_b \, [\, 2m \, sinkk\,(z+h2) - (m^2 + 1) \, \sinh l\,(z+h2)\,]$$

(16)

$$D_b - \frac{X1 \, Z3 \, R - X3 \, (A1 + Z1R)}{X1 \, (A2 + Z2R) - X2 \, (A1 + Z1R)}$$

$$C_b - \frac{X3 - DX2}{X1} \qquad B_a - -\frac{k}{R} \; (C_b \, A1 + D_b \, A2)$$

(17)

where $m=l/k$,

$$l^2 - (k^2 - \frac{i\sigma}{\nu_e}) \qquad R - \frac{\rho_a \, \nu_a \, q}{\rho_b \, \nu_e \, k} \qquad q - (1-i) \; (\frac{\sigma}{2\nu_a})^{0.5}$$

and:

$$A1 - 2 \, \cosh kh2 - (m^2+1) \, \cosh lh2$$
$$A2 - 2m \, \sinh kh2 - (m^2+1) \, \sinh lh2$$

(18)

$$Z1 - \sinh kh2 - m \, \sinh lh2 \;, \quad Z2 - m(\cosh kh2 - \cosh lh2)$$
$$Z3 - -\frac{\sigma a}{k} \, \sinh kh1 + \frac{ga}{\sigma} \, \cosh kh1$$

(19)

$$X1 - -\cosh kh2 + \cosh lh2 \;, \quad X2 - -m \, \sinh kh2 + \sinh lh2$$
$$X3 - -\frac{\sigma a}{k} \, \cosh kh1 + \frac{ga}{\sigma} \, \sinh kh1$$

(20)

The complex wave number k is obtained by applying
the condition of continuity of normal stress through

the interface, which can be stated as:

$$- \rho_a \frac{\partial \varphi_a}{\partial t} + \rho_a g \zeta + 2 \rho_a v_a \frac{\partial w_a}{\partial z} -$$
$$= - \rho_b \frac{\partial \varphi_b}{\partial t} + \rho_b g \zeta + 2 \rho_b v_e \frac{\partial w_b}{\partial z} \qquad (21)$$

This is the dispersion equation from which it is possible to obtain the wave number and the damping coefficient in terms of the physical quantities: σ, h1, h2, ρ_a, ρ_b and v_e.

4.   Results and conclusions

4.1 Velocity profiles

Figure 1 shows the velocity profiles in the two layers, at different times and for an elasticity modulus G=100 N/m². The typical overshoots in the water boundary layer just above the interface can be observed. The maximum velocity in the mud layer decreases when the elasticity goes up. The velocities in the water layer are $\pi/2$ out of phase with the mud velocities for this case. The results of velocity profiles agree reasonably well with the data published by Maa & Mehta(1990).

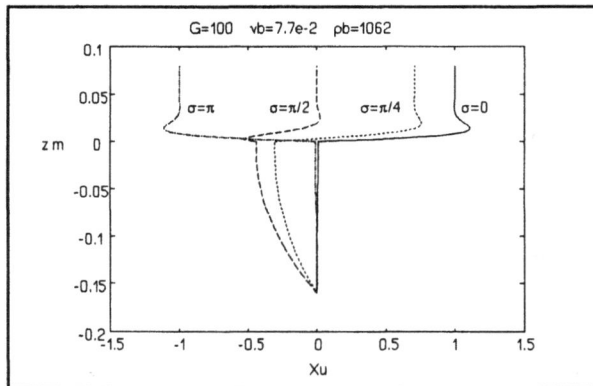

FIGURE 1   Velocity profiles (G=100 N/m², Xu=û/û∞)

4.2 Wave damping

Figure 2.a shows the variation of the dimensionless wave amplitude damping coefficient ($Xkj=kj.h1$) against the dimensionless wave frequency ($X\sigma=h1.\sqrt{(\sigma/\nu_a)}$), for three different mud viscosities and zero elasticity G. The value of kj was obtained solving Eq. (18) in dimensionless form by using the Power Hybrid Method. The curve corresponding to $\nu_b=0.015$ agrees with Sakakiyama and Bijker's results (4). The maximum damping is obtained with the highest mud viscosity.

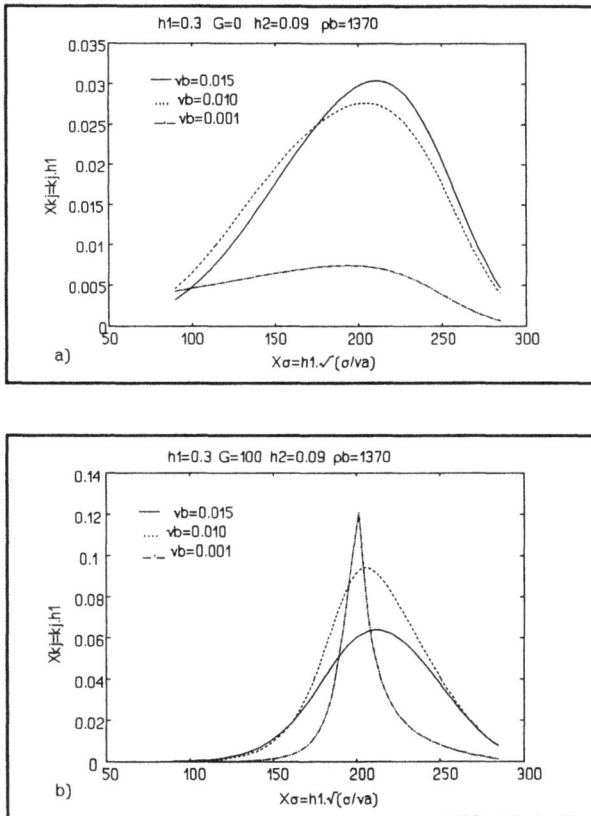

FIGURE 2   *Damping coefficient against frequency, a) G=0,   b) G=100*

Figure 2.b shows the same as before but for G=100 N/m2. The wave damping is higher in this case than in the former one. The wave damping coefficient increases when the viscosity goes down when elasticity G >0; the opposite occurring when G=0. This is due to the fact that the elasticity increases the phase lag (69° for G=0 and 91° for G=100), and thus the velocity gradients in the mud layer. Depending on the physical parameter values, kj can reach values greater than 1, and so the system's resonance occurs due to the elasticity of this viscoelastic model.

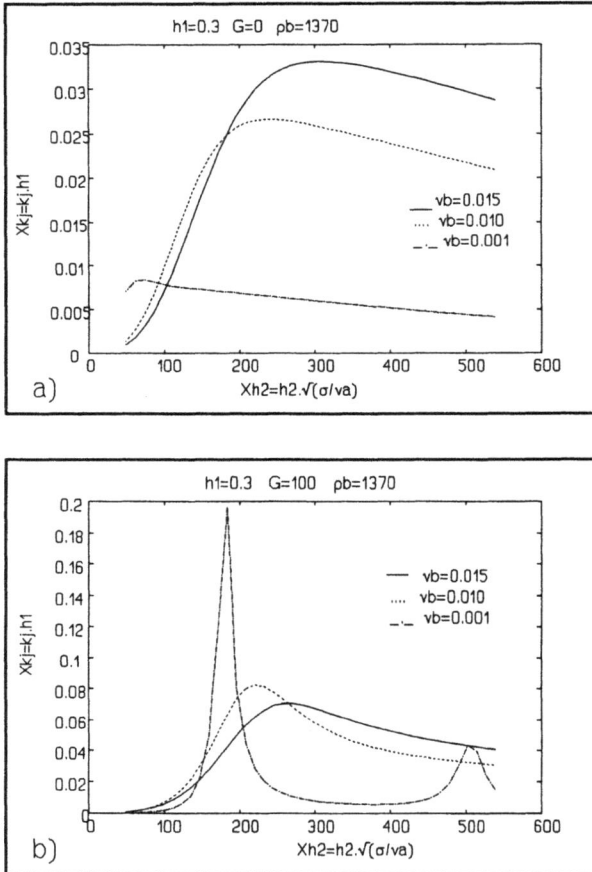

FIGURE 3 *Wave damping coefficient against mud depth*

Figure 3 shows the dimensionless wave damping against the dimensionless mud depth ($Xh2=h2.\sqrt{(\sigma/\nu_a)}$), for two elasticity values: a) $G=0$ and b) $G=100$.

We can see again that without elasticity, damping increases when the viscosity does so, and with elasticity the opposite occurs, for values of $Xh2$ near the main peak. For $Xh2$ near the small peak, the same performance as without elasticity is obtained. Figure 3.b shows only the presence of two peaks, but searching in the range of $Xh2$ up to 2500, another one is found at $Xh2=900$. The presence of various peaks could be related to the resonance phenomenon.

4.3 Shear stress

The interface shear stress - Eq. (16) - is dimensionless with the shear stress as given by the rigid bed theory (ie. assuming a rigid bed located at the interface):

$$\tau_r - \frac{\rho_a \, (\sigma \nu_a)^{0.5} \, a\sigma}{\sinh(k_o h1)} \tag{22}$$

Figure 4 shows the dimensionless interface shear stress against the dimensionless wave frequency, for three values of dimensionless elasticity:

$$M - G/\rho_b \nu_b \sigma \tag{23}$$

For zero elasticity, i.e $M=0$, the shear stress decreases when the frequency increases, reaching an approximately constant value. For a low $X\sigma$, the dimensionless Tci approaches 1, so the bed responds as a rigid body. The curve corresponding to $M=5$, shows a different trend. Tci increases as $X\sigma$ increases; the opposite to the former case, reaching a value of 1.3. Finally, for $M=50$, the shape of the curve changes again, in this case being almost independent of the frequency.

Figure 5 shows the shear stress against the dimensionless elasticity M for three values of mud viscosity. For high viscosities (the bed responds as rigid), Tci remain near to 1. For lower viscosities, Tci is lower than 1 for M near zero, reaching a peak and decreasing towards 1, for large M. The Tci peak increases when the mud viscosity decreases.

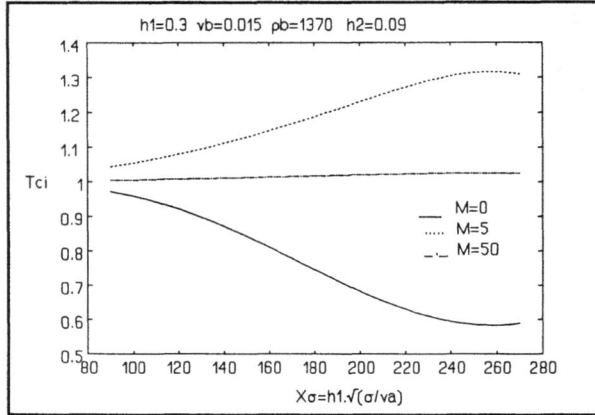

FIGURE 4 *Dimensionless interface shear stress against dimensionless frequency*

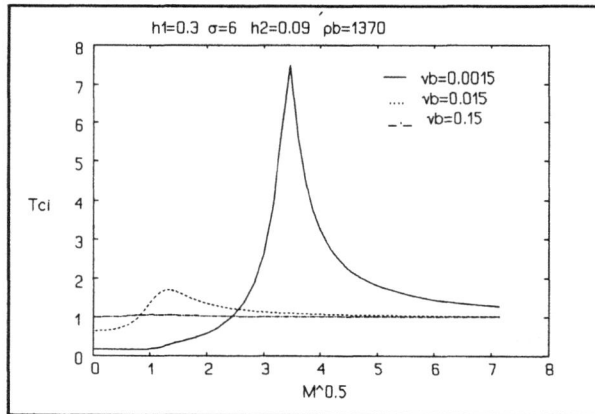

FIGURE 5 *Dimensionless shear stress against dimensionless elasticity M*

Now the effect of mud thickness on the shear stress can be seen. Figure 6 a) and b) show the variation of shear stress with dimensionless mud depth $Xh2$, with and without elasticity. The figure a) where $M=0$, Tci decreases towards values near zero -depending on the

430

wave frequency- when Xh2 increases. For Xh2 near zero, the interface shear stress is near 1. Figure b) where M=1, has a different trend. For low Xh2, the shear stress increases when Xh2 increases, reaching a peak and then decreasing for high Xh2, until values lower than 1. Higher peaks and minimum values are obtained in this case with lower frequency.

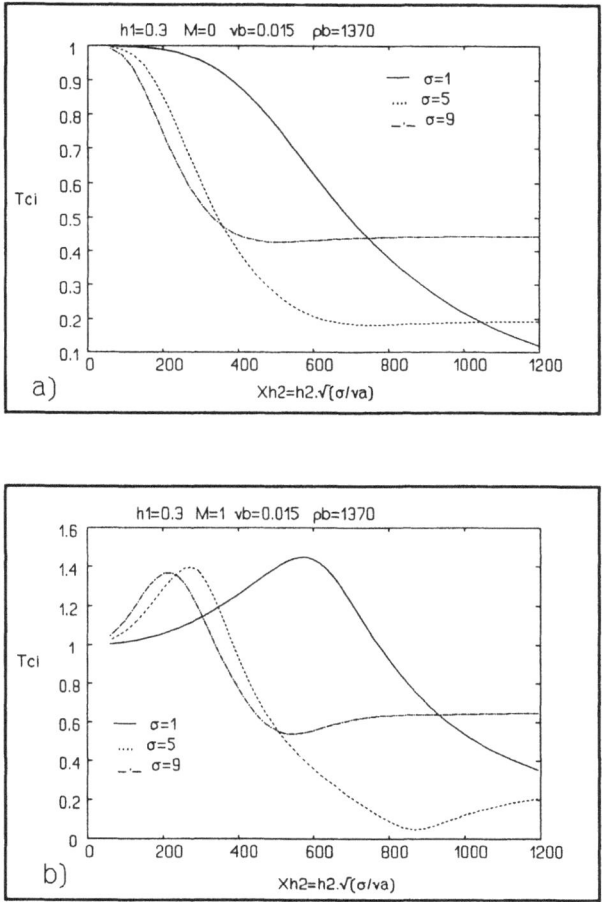

FIGURE 6   *Dimensionless   shear   stress   against dimensionless mud thickness Xh2*

A model has been developed to determine the response of a muddy bottom by wave action. An analytical solution was obtained considering a thin boundary layer over the water-mud interface. The Voigt model was selected for modelling mud rheology. The calculated wave damping amplitude shows the possible occurrence of resonance which could explain high damping rates of the surface wave. This highlights the importance of studying mud rheology and its thixotropic property to obtain more realistic results as concern the response of muddy bottoms when forced by water waves. The calculated interface shear stress shows a strong dependence on mud properties and mud thickness. The dimensionless shear stress reaches values higher or lower than 1, depending on mud depth, elasticity, viscosity and wave frequency.

This result is very important when considering the mud erosion rate, for the shear stress can vary significantly from the one in which the bed is a rigid body.

REFERENCES

[1] Dalrymple,R. and Liu,P., (1978), Waves Over Soft Mud: A two Layer Fluid Model, *Journal of Physical Oceanography*, **Vol. 8**, pp 1121-1131.
[2] Maa,P-Y. and Mehta,A.J., (1990), Soft Mud Response to Water Waves, *Journal of Waterway, Port, Coastal and Ocean Engr.*, **Vol. 116**, No 5, pp 634-649.
[3] MacPherson,H., (1980), The Attenuation of Water Waves over a Non-rigid Bed. *Journal of Fluid Mechanics*, **Vol. 97**, part 4, pp 721-742.
[4] Sakakiyama,T. and Bijker,E., (1985), Mass Transport Velocity in Mud Layer due to Progressive Waves, *Journal of Waterway, Port, Coastal and Ocean Engr.*, **Vol. 115**, No 5, pp 614-633.

36 ARMADA: An efficient spectral wave model

M. Al-Mashouk, D. E. Reeve, B. Li and C. A. Fleming

ABSTRACT

A newly developed monochromatic wave model is utilised to formulate an efficient spectral model for wave propagation over large regions. The spectral implementation involves discretising the directional spectrum into components, transforming each component inshore and then assembling these by linear superposition to form the resultant wave field. Results from an application to a real bathymetry indicate that monochromatic waves could represent a poor approximation to irregular waves when modelling combined refraction/diffraction effects.

INTRODUCTION

In the past two decades there has been a considerable amount of research directed towards developing and refining numerical techniques for the solution of wave propagation problems. This effort has been primarily driven by the increasing importance of wave modelling in coastal engineering projects. The significant technical advances achieved in computer hardware architecture, in terms of both processing speed and memory storage, have also greatly contributed to this research impetus.

The need for efficient wave models arises as in most coastal engineering studies, wave data is available in the form of hindcast deep water wave climates. These would have to be transferred inshore to obtain appropriate conditions for the design of coastal structures. The output from wave models can also form the basis for the driving conditions for any sedimentation or pollutant dispersion models that may need to be implemented within a particular project. Models that lead to better predictions of the nearshore wave climate are therefore valuable tools that can greatly aid the design process.

This paper presents initially a review of wave models and assesses their suitability to practical application. It proceeds to describe the theoretical structure of the spectral implementation of ARMADA, a multigrid-type wave model. Results are then presented from an application which demonstrates the need for spectral, rather than monochromatic, models.

REVIEW OF WAVE MODELS

One of the first practical tools developed for general wave refraction analysis is the ray tracing model which calculates the paths of wave rays as they propagate over a digital model of the seabed. This could be used in either forward tracking mode, where the path of the wave is followed as it propagates towards the shore from deep water, or in backward tracking mode where waves are tracked from any inshore point of specific interest back to deep water. The former gives a synoptic picture of wave conditions over a wide area and can be particularly advantageous in identifying the location of caustics, ie where wave rays cross, as these can lead to significant increases in wave height. The latter approach however is more useful for site - specific design purposes. Ray tracing models are quite versatile and economical but they lack the theoretical description of bathymetric diffraction effects. It should be noted however that Southgate (1985) developed a model that accounts for diffraction due to surface-piercing structures.

Grid wave models provide a significant advantage over ray models in that the wave field is derived over an area rather than at a single point. This provides a spatial representation of the wave heights which may then be incorporated into other models. Such models are usually based on the solution of different forms of the mild-slope equation. This equation was first derived by Berkhoff (1976) and includes the combined effects of wave refraction, shoaling, diffraction and reflection. In its original form, the mild-slope equation is an elliptic partial differential equation, the solution of which requires prohibitive amounts of computer resources. This is due to the minimum resolution criterion associated with the numerical scheme which requires no fewer than eight grid points per wavelength for reasonable accuracy. This renders most engineering problems relating to large coastal areas unpractical for solution by this method. Copeland (1985) derived a transient, or hyperbolic, formulation of the mild-slope equation which possesses certain computational advantages over the elliptic model, but which does not ease the grid resolution restriction. To overcome this problem, Radder (1979) proposed a parabolic formulation based on a description of wave behaviour in terms of amplitude and phase. The parabolic approximation adopted implied that a resolution of two nodes per wavelength was sufficient for most practical problems. It however imposed a rather severe restriction on the variation of the direction of wave propagation; this was limited to approximately $\pm$ 15° from the grid axis. The mathematical transformation derived retained the ability to solve for the combined effects of refraction, shoaling and forward diffraction, but reflection could not be modelled. Ebersole (1985) and Panchang et al (1988) both put forward similar models, RCPWAVE and the Error Vector Propagation (EVP) model, which may be considered 'economical', but which suffered similar constraints as Radder's parabolic model. More recently, Li and Anastasiou (1992) successfully applied the multigrid method, which was originally developed by Brandt (1977), to the solution of the elliptic mild-slope equation. In implementing the multigrid technique, approximate solutions of the governing equation are obtained for a number of increasingly finer grid meshes. A sufficiently accurate solution for the finest mesh is obtained based on the residual errors of the approximate solutions for the coarser meshes. The scheme was extensively tested on real and idealised bathymetries and was found not to impose any of the constraints associated with the models due to Radder and Ebersole while being just as efficient computationally. Although wave reflections are still not represented, it would nevertheless seem that among the monochromatic models, the multigrid model, ARMADA, is the most appropriate for general application.

In reality, however, ocean waves are not monochromatic. They may exhibit a wide range of frequencies and directions in accordance with established spectral definitions. It is usual practice in coastal engineering to adopt the representative parameters of these spectra, such as significant wave height, peak energy period and the predominant energy direction for use in the monochromatic wave propagation models. These key parameters have been thought to provide an adequate description of the offshore wave climate and therefore, the transformation inshore of this single representative wave train has been the most widely used method for assessing the nearshore wave climate. This approach, however, has been found to be deficient in some circumstances where bathymetric features lead to strong wave convergence.

Vincent and Briggs (1989) conducted experiments on wave propagation over an elliptic shoal in a tank which was equipped with a directional spectral wave generator. Monochromatic waves as well as spectra with both narrow and broad frequency and directional spreads were generated in this controlled laboratory experiment. The wave height patterns associated with the monochromatic waves showed a strong amplification region behind the mound whereas the directional spectral waves were significantly less perturbed by this bathymetric feature. The main conclusions from this study were that monochromatic waves tend to overestimate the maximum amplification of irregular waves by 50 to over 100%. Furthermore, the degree of directional spread in the irregular wave case was found to be a more significant parameter than the spread of energy in frequency space. It is apparent therefore that there exists a need for the development of numerical spectral wave models. The ray tracing models described above have been used in conjunction with spectral transfer programs to obtain more realistic estimates of inshore wave heights. However, as mentioned previously, these suffer from certain drawbacks which necessitates the use of grid models. Panchang et al (1990) developed one such model based on the linear superposition of monochromatic calculations as discretised from the appropriate spectra. It is interesting to note that in this work Radder's parabolic method was chosen as the basis for the monochromatic predictions in spite of its requirement for paraxiality. The data of Vincent and Briggs was used for validating this model and it was found that the irregular sea state was satisfactorily simulated.

The multigrid model was adopted by Halcrow (1992) to formulate a similar spectral model. Again, it was tested extensively against the experimental data of Vincent and Briggs and the comparison was very favourable.

FORMULATION OF SPECTRAL MODEL

The spectral version of ARMADA was developed by the coastal modelling team at Halcrow to meet part of the increasing technical challenges set by projects. The theoretical structure of the monochromatic version is described in some detail by Li and Anastasiou and hence our scope here shall be confined to outlining only the spectral implementation. For this purpose, the monochromatic model may therefore be treated as a 'black box' which requires as input the parameters of the deep water incident wave (period, direction, height) and which produces as output a grid of wave heights. The spectral modification essentially involves discretising frequency and directional spectra into components and then passing each combination of components in turn to the 'black box'. The set of wave height grids obtained in this manner are then assembled by a process of linear superposition to form the resultant wave field. The model requires the user to specify the required directional spectrum, $S(f,\theta)$; this is normally regarded as the product of a frequency spectrum, $F(f)$, and a directional spreading function, $G(\theta,f)$. The latter is often prescribed as a function of direction only and its integral over $-\pi < \theta < \pi$, normalised to unity.

$$S(f, \theta) = F(f)\, G(\theta,f) \tag{1}$$

The model provides the user with a choice of frequency spectra and spreading functions. For the purposes of this work, however, only one combination has been used. This comprised the TMA frequency spectrum (Bouws et al, 1985):

$$F(f) = \alpha\, g^2\, (2\pi)^{-4} f^{-5} \exp\left(-1.25\left(\frac{f_m}{f}\right)^4 + \ln\gamma \exp\left[\frac{-(f-f_m)^2}{2\sigma^2 f_m^2}\right]\right) \phi\,(f,h) \tag{2}$$

In which the spectral energy density $F(f)$ for frequency f depends on the parameters α = Phillips constant, f_m = peak frequency, γ = peak enhancement factor, σ = shape parameter and $\phi(f,h)$ is a factor that incorporates the effect of the depth, h.

and the Borgman spreading function (Borgman, 1984):

$$G(\theta) = \frac{1}{2\pi} + \frac{1}{\pi} \sum_{j=1}^{20} \exp\left[\frac{-(j\sigma_m)^2}{2}\right] \cos j(\theta-\theta_m) \tag{3}$$

Where θ_m = mean wave direction and σ_m = spreading parameter.

This particular directional spectrum was chosen because it was extensively tested when validating the model against the experimental data of Vincent and Briggs.

The process of calculating the resultant wave heights was adopted from the method of Goda (1985) where the combined diffraction/refraction/shoaling coefficient of random waves is given by:

$$K_{random} = \left[\frac{1}{m_o} \int_o^\infty \int_{-\pi}^\pi S(f,\theta) \, K^2(f,\theta) \, d\theta \, df \right]^{\frac{1}{2}}$$ (4)

$K(f,\theta)$ being the coefficient of the component waves with frequency, f, and direction, θ, and m_o, the integral of the directional spectrum,

$$m_o = \int_o^\infty \int_{-\pi}^\pi S(f,\theta) \, d\theta \, df$$ (5)

Assuming a Rayleigh wave height distribution, we may relate m_o to the significant wave height by:

$$Hs = 4\sqrt{m_o}$$ (6)

Choosing a finite number of frequencies and directions specifies a discretisation of the directional spectrum and the above integrals may then be replaced by double summations, in the usual manner. The incident wave height for the component waves is taken as unity for calculating $K(f,\theta)$.

Non-linear effects such as wave-wave interaction are not included, but wave breaking due to depth limitation is represented by ensuring that Hs < 0.78h throughout the model domain.

APPLICATION TO MARSAXLOKK BAY

Marsaxlokk Bay lies at the southernmost tip of the island of Malta and is exposed to a substantial fetch of the Eastern Mediterranean Sea. The bathymetry shown in Figure 1 has been digitised from the Admiralty Chart and is characterised by two features: firstly, at the offshore end, the water depths reach values of 100m, which is generally quite deep, and secondly, there exists a reef near the bay entrance where the depths are significantly reduced. The size of the bay is approximately 3km x 2km at its widest points.

A wave ray tracing model was set up to obtain an initial assessment of the wave conditions. Figure 2 shows an example of the results for forward tracking for a wave of period 8 seconds. It may be seen that a number of wave rays converge and cross in an area seawards of the entrance to the bay. This indicates a region of much increased wave height due to the convergence of wave energy, and came about because of the reef that exists in this area. Figure 3 shows an example of a back-tracked ray plot from a single point within the bay.

ARMADA was then applied to examine the wave field in more detail. A 4.0km x 5.6km grid was set up with a uniform resolution of 100m in both the offshore and longshore directions. This implied that the total number of grid nodes was 2337 (41 x 57). Results from three runs which illustrate the trends of spectral wave propagation are presented in this work. Initially, a monochromatic run was undertaken with the following parameters: wave height = 3.0m, wave period = 8.0s and wave angle = 0° from the normal to the offshore boundary. Figure 4 shows the computed wave field for this case where it is noticed that there is a significant increase of wave height outside the entrance to the bay in accordance with the wave ray plot of Figure 2. The wave height magnification factor, attains a maximum value of $\approx$ 2.0 at the centre of this amplification area.

The next two runs undertaken were with spectral waves, utilising the TMA - Borgman directional spectrum. A set of five frequency components was chosen with a peak energy period of 8.0s. Phillips' constant, α, was taken to be 0.002 and the peak enhancement factor, γ, was 20.0. These values yielded a significant deep water wave height of 3.0m when integrating the frequency spectrum. Five directional components were also chosen with the mean wave direction at 0° from the normal to the offshore boundary. The limits of the wave angles were set to $\pm$ 45°. Experience gained in the use of the model for this case has shown that there is no significant advantage in utilising more than the 5 x 5 components adopted for the spectral discretisation.

Two values for the spreading parameter were used corresponding to a narrow spreading case, σ_m = 10°, and a wide spreading case σ_m = 30°. Figure 5 shows the wave field resulting from the former which is seen to be similar to the monochromatic result. However, the focusing of wave heights at the entrance to the bay is not quite as severe since the magnification factor is approximately 1.6. There also appears to be an area of increased wave height well inside the bay which did not arise with the monochromatic waves. This is in fact due to the oblique waves which are incident from the SE quadrant and accordingly would only appear with the monochromatic model for these specific angles.

The result of the wide spreading case, shown in Figure 6, demonstrates a significant departure from the monochromatic trends in that the wave focusing at the entrance is very much diminished. The wave height magnification factor is reduced to approximately 1.3. The area of focusing inside the bay is also seen to be of a greater extent and intensity than its equivalent for the narrow spreading case. This is due to the greater energy attributed to the oblique waves with the wider spreading function.

In all cases, it is also noticed that there are amplifications of wave height at headland features, which are consistent with the usual refraction theory.

CONCLUSION

This work has presented an alternative approach for the modelling of water wave propagation based on discretising, transforming and reassembling the directional spectrum. The usual approach of representing the offshore wave climate by a single wave and transforming this inshore may be useful for preliminary purposes, but it could be deficient in certain circumstances. In situations where wave amplification is an important consideration, monochromatic models will tend to overestimate the wave heights which might lead to overdesign. It is important therefore to put forward robust spectral models, such as the one described in this paper, in order to arrive at more realistic assessments of nearshore wave climates.

ACKNOWLEDGEMENTS

The development of the spectral version of ARMADA was undertaken by Halcrow's Coastal Engineering Department as part of their ongoing research and development efforts. The monochromatic version was developed by the Civil Engineering Department of Imperial College, London.

REFERENCES

Berkhoff, JCW (1976), "Mathematical models for simple harmonic linear water waves. Wave refraction and diffraction", Publication 163, Delft Hydraulics Lab., Delft, Netherlands.

Brandt, A (1977), "Multi-level adaptive solutions to boundary value problems," Math. Comp., 31, pp 333-390.

Copeland, GJM (1985), "Refraction-diffraction model for linear water waves", J.Wtrway., Port, Coast. and Oc. Engrg., ASCE, 111 (6), pp 939-953.

Goda, Y (1985), Random seas and design of maritime structures, Univ. of Tokyo Press, Japan.

Halcrow (1992) ARMADA: Multigrid wave model: User manual and validation report.

Li, B and Anastasiou, K (1992), "Efficient elliptic solvers for the mild-slope equation using the multigrid technique", Coast. Engrg., to be published.

Panchang, VG, Cushman-Roisin, B and Pearce, BR (1988), "Combined refraction - diffraction of short waves for large domains", Coast. Engrg., 12, pp 133-156.

Panchang, VG, Wei, G, Pearce, BR and Briggs, MJ (1990), "Numerical simulation of irregular wave propagation over shoal", Coast. Engrg., 116, pp 324-340.

Radder, AC (1979), "On the parabolic equation method for water wave propagation: I", J.Fluid Mech., 95, pp. 159-176.

Southgate, H (1985), "A harbour ray model of wave refraction - diffraction", J.Wtrway., Port, Coast. and Oc. Engrg., ASCE, 111(1), pp 29-44.

Vincent, CL and Briggs, MJ (1989), " Refraction - diffraction of irregular waves over a mound", J.Wtrway., Port, Coast. and Oc. Engrg., 115(2), pp 269-284.

439

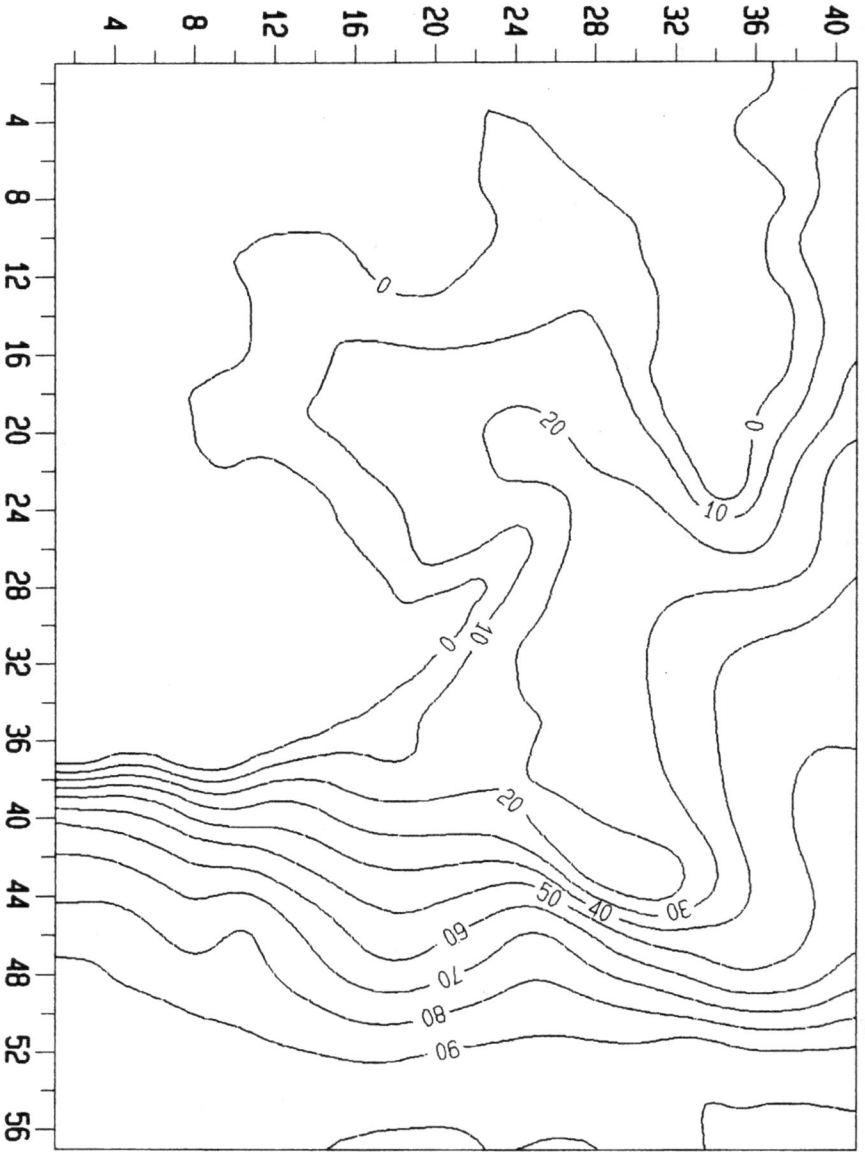

Figure 1: *Sea bed contours around Marsaxlokk Bay*

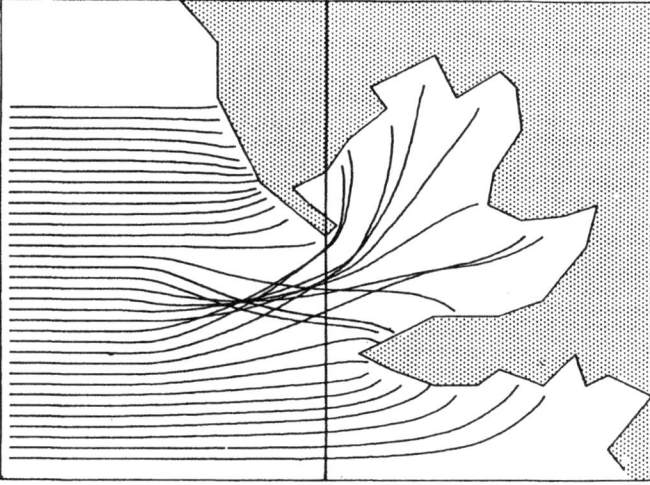

Figure 2: Forward tracking wave ray plot

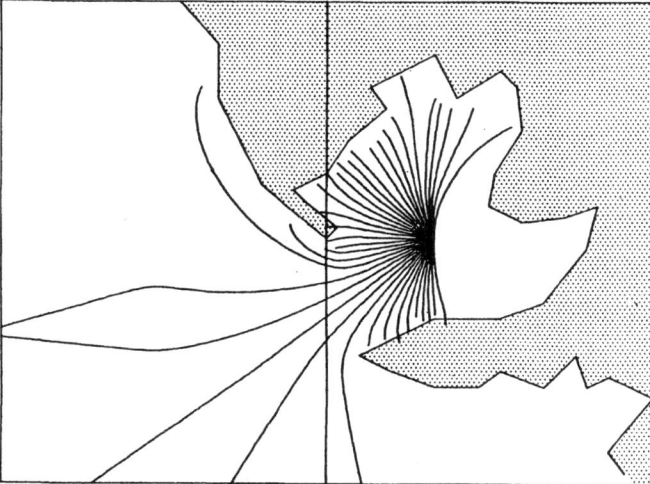

Figure 3: Backward tracking wave ray plot

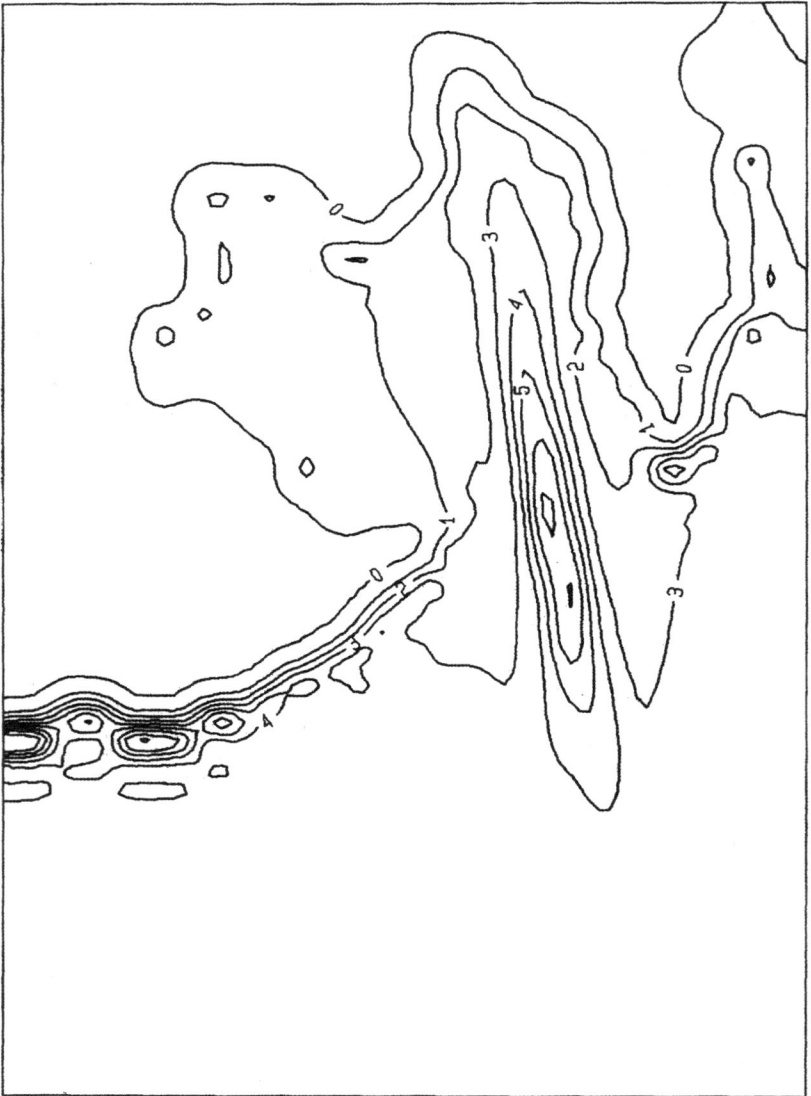

Figure 4: *Wave height contours for monochromatic waves*

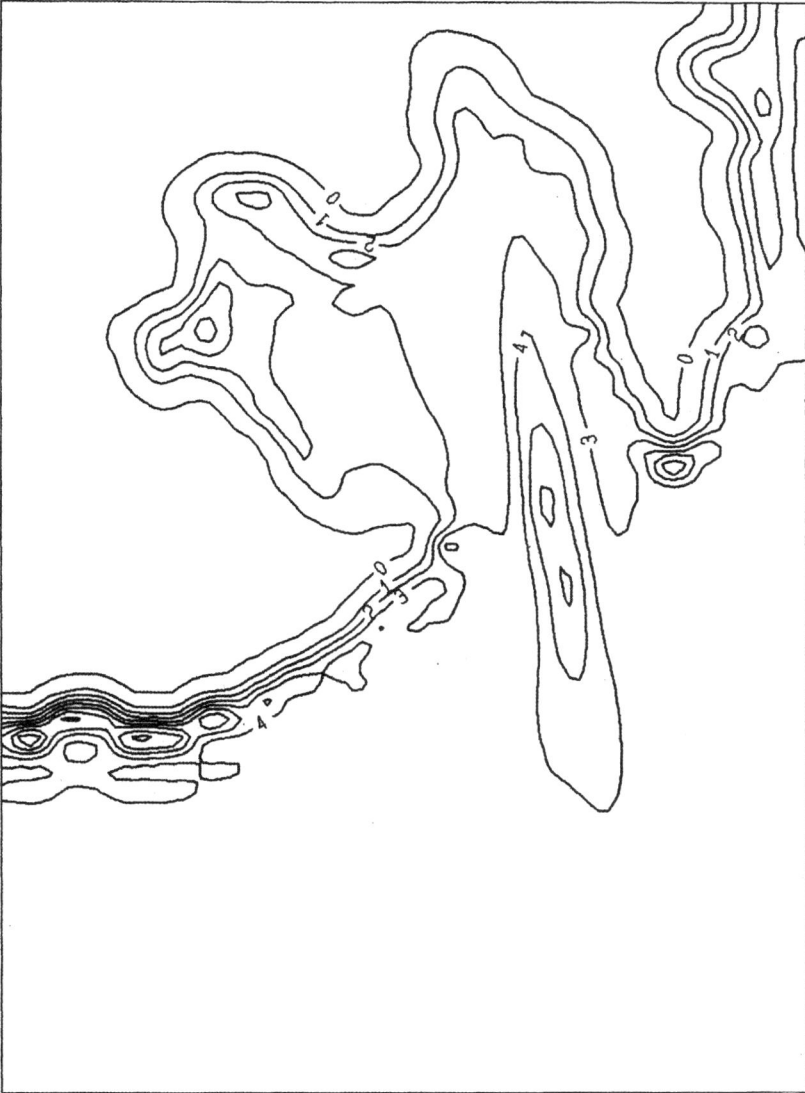

Figure 5: *Wave height contours for narrow-spreading waves*

Figure 6: *Wave height contours for broad-spreading waves*

37 Effects of submerged structures on coastal currents and geometry

T. Koike, K. Bando, M. Tanaka and K. Kudo

ABSTRACT

The nearshore currents and topography change caused by submerged breakwaters are investigated by the up-to-date numerical model. This model consists of the wave analysis, analysis of wave-driven current and topography change evaluation. Before applying it to the designated case, a preliminary study is made using the available laboratory and field data to validate the model and to determine some constants that are necessary to define the sediment transport rate. The numerical results predict strong current circulations and little topography change around the breakwaters at the present particular site.

1. Introduction

Beach erosion is a serious problem as environmental condition around the coastal area deteriorates. Recent development in this region and realization of beach itself as precious natural resource bring more attention to the beach and coastal condition. Especially, when a new structure is planned in the region, it should be examined that it would not cause negative impact on the ambient environment such as beach erosion.

Japan Marine and Science Technology Center plans to install prototype floating but submerged plates off the Japan Sea coast, which intend to attract the incoming waves behind them (Kudo et al., 1988). The concentrated wave may be used for the wave energy power station. Before construction, influence of the structures on the local topography was evaluated by the numerical model. This wave attracting structures, which are floating crescent-shaped boards, are modeled by submerged breakwaters in this analysis. By this assumption, the numerical analysis can be two-dimensional on the horizontal plane. We chose a two-

445

dimensional analysis rather than three, considering computational burden and the accuracy of the whole numerical system.

The configuration change in the nearshore region may be estimated either by laboratory tests, by field observations or by analytical approaches. The assessment by laboratory and field tests, which usually take long time and cost expensive, aims at the particular site, and its results may not be applied to other conditions. The laboratory tests also have the problem of the scale effect. On the other hand, sediment transport is such a complicated process that its hydrodynamic mechanism is not understood well enough for us to have a general analytical methodology. The main difficulty in the analytical methods is accurate estimation of the nearshore currents and sediment transport. This task becomes even more difficult in the surf zone where wave–current coexistent field is dynamic and complicated because of wave breaking and induced strong turbulent motion. In addition, it is in this surf zone that the nearshore current and sediment movement is dominant.

Although all approaches have the drawbacks discussed above, the analytical method was chosen in this study because it is practical if appropriately used and because it could be improved by critical applications and further theoretical research. In the usual analytical approach, the wave field and wave–driven nearshore currents are first evaluated, and estimate of sediment transport by waves and currents follows. Solution of the wave field by certain numerical methods is practically accepted. Some problems are in the estimation of the nearshore currents, while the hydrodynamic mechanism of the sediment transport is not well known. The radiation stress is the sole driving force for the currents, and an appropriate estimate of the radiation stress is a key in the prediction. Although the radiation stress is analytically obtained under undeformed regular waves, its estimate in disturbed wave field is not straight–forward. The horizontal diffusion, which is less important than contribution by the radiation stress, has the practical problem in assigning proper values to the diffusion coefficients. For evaluation of sediment transport rate, some crude assumptions must be made because of many unknown factors, and all available models are semi–empirical formulae. Further, most of them treat one–dimensional movement; on–offshore or alongshore, or represent the transport rate under the simple condition. To evaluate horizontal (i.e., two-dimensional) topography change in disturbed wave field, direct application of these formulae is not appropriate, and some compromise and modification must be made. Since the last problem in evaluation of sediment transport rate is much more serious than others, the topography change prediction is even less accurate than wave analysis and current prediction. See Horikawa (1988) for detailed discussions.

Watanabe et al.(1986a and 1986b) developed a system of numerical models to predict the topography change, which consists of analyses of wave field, near–shore circulation and sediment movement. They applied it to the coastal region around a detached breakwater, and the numerical result compared well with the experimental data. In the present study, this model was carefully applied to the construction site to investigate the effect of floating plates (modeled by submerged breakwaters) on the ambient topography.

2. Submerged breakwater

Because the solution by the numerical model to be used depends on some input data that are difficult to uniquely determine, it is more than useful to study the effect of submerged breakwaters in advance. Here, a number of studies on the effect of submerged breakwaters is reviewed.

A submerged breakwater, which induces wave breaking on it, is used to obtain calm area and to stabilize beach condition by preventing sand from being washed away. Because this

446

type of breakwater is constructed under water, it will not damage natural scenery. As application of submerged breakwater increases because of these advantages, assessment of how the submerged breakwater is functioning, and how it affects ambient circulations and bathymetry was tried both in laboratory and in the field (For example, Shoji et al. (1991) and Okada et al. (1991)). These efforts confirmed its effectiveness, but understanding of coastal currents and sediment transport is not satisfactory due to the complicated process of wave breaking and highly non-linear motion in the surf zone. Sasaki et al. (1990) obtained the current and bathymetry change around submerged breakwaters both by experiment and by the similar numerical approach to that by Watanabe et al (1986a and 1986b). Since the clearance between the top of the breakwater and sea surface is very small in their case (about 3% of water depth), the computed currents were not accurate on the breakwater. However, when the measured velocity was used at the top of the breakwater, agreement improved. Their observed flow pattern was characterized by strong onshore currents on the breakwater and rip currents at the opening. The topography change seemed to be strongly governed by this circulation system, leaving accretion behind the breakwater and erosion elsewhere.

Irie et al. (1985) proposed a transport model of the suspended sediment, and Iwatani et al. (1989) applied it to the data collected around submerged breakwaters in the field. The computed bathymetry exhibited similar pattern to that by Sasaki et al. but predicted little change around the submerged breakwater. On the other hand, the field data observed by Shoji et al (1991). showed very complicated temporal change, being eroded behind the breakwater after an entire winter period but deposited after one storm event. Okada et al. (1991) also reported a repeat of erosion and accretion behind the submerged breakwater.

Kohno et al.(1987) studied the effect of relative clearance above the submerged breakwater, wave steepness and opening size between breakwaters in laboratory. They reported that erosion took place in the region where the breakwater was absent, and that sediment transport was sensitive to the opening length and wave steepness.

3. Description of numerical model

3.1 General flow

The present numerical model consists of three analytical parts: wave height distribution, wave driven currents and topography change. Sediment transport rate is evaluated from the shear stress on the bottom and mean currents, which are the outputs of the first two analyses. Figure 1 illustrates the analytical flow of the model. This model is two-dimensional with variation in the horizontal direction. The analytical outline of each stage is described below.

3.2 Wave analysis

Berkhoff (1972) presented the basic equation for the periodic wave motion in mildly variable water depth. The equation, which is obtained by integrating the momentum equation over water depth, is

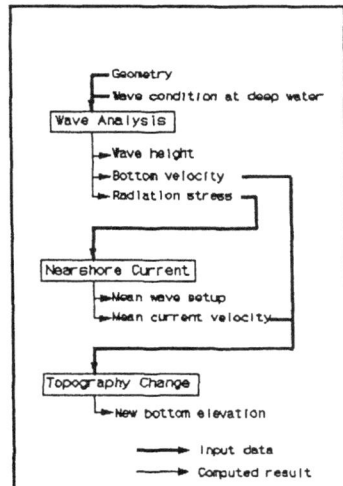

FIGURE 1   *Analytical flow of topography change estimation*

$$\nabla \cdot (CC_g \nabla \varphi) + \sigma^2 \frac{C_g}{C} \cdot \varphi = 0 \tag{1}$$

where φ, C, C_g and σ are velocity potential, phase velocity, group velocity and angular frequency, respectively. Because only two dimensions in the horizontal plane are considered, $\nabla = (\frac{\partial}{\partial x}, \frac{\partial}{\partial y})$. Since Eqn.(1) requires large matrix inversion, the unsteady version of Eqn.(1) was chosen for this study, which was derived by Watanabe et al. (1986a). The equations can easily include the effect of wave breaking, which is a very important factor to the coastal circulation.

$$\frac{\partial Q}{\partial t} + \frac{C^2}{n} \nabla (n\eta) + f_D Q = 0 \tag{2}$$

$$\frac{\partial \eta}{\partial t} + \nabla \cdot Q = 0 \tag{3}$$

where Q, n, η, and f_D are depth-integrated velocity vector, ratio of group velocity to phase velocity, surface elevation and decaying factor due to wave breaking, respectively. These equations are solved by the finite difference method until a steady state solution is reached. Components of the radiation stress are computed as a part of the solution, which is the only driving force for the coastal currents in this model.

3.3 Nearshore currents

When assuming a uniform flow in the vertical direction and averaging the equations in the wave-current coexistent field over a wave period, the mean current vector $U = (U, V)$ and wave setup $\bar{\eta}$ are determined by

$$\frac{\partial \bar{\eta}}{\partial t} + \frac{\partial \{U(h + \bar{\eta})\}}{\partial x} + \frac{\partial \{V(h + \bar{\eta})\}}{\partial y} = 0 \tag{4}$$

$$\frac{\partial U}{\partial t} + U\frac{\partial U}{\partial x} + V\frac{\partial U}{\partial y} + F_x - M_x + R_x + g\frac{\partial \bar{\eta}}{\partial x} = 0 \tag{5}$$

$$\frac{\partial V}{\partial t} + U\frac{\partial V}{\partial x} + V\frac{\partial V}{\partial y} + F_y - M_y + R_y + g\frac{\partial \bar{\eta}}{\partial y} = 0 \tag{6}$$

where the vectors $(F_x, F_y), (M_x, M_y)$ and (R_x, R_y) represent bottom friction, horizontal diffusion and radiation stress. These equations are also solved by the finite difference method. To expedite the computation, a steady state solution with no current condition is first solved for wave setup and is used as an initial condition for the original equations (4)-(6).

In the present computations, coarser grid was used than that in the wave analysis to decrease the computational burden, considering larger horizontal characteristic length of the nearshore currents. To achieve this transformation, the radiation stress computed by the wave analysis was averaged over the new larger grid size. This procedure is unnecessary if computational time is not a restraint.

3.4 Topography change

The main assumptions employed are that (1) sediment transport by bed load is dominant, (2) sediment transport rate is given by the local quantities and (3) sediment transport by currents and by waves is estimated independently, and total amount is linear sum of these two contributions.

The temporal change of sea bottom elevation, z_b, is evaluated from

$$\frac{\partial z_b}{\partial t} = -\frac{\partial}{\partial x}(q_x - \epsilon_s |q_x| \frac{\partial z_b}{\partial x}) - \frac{\partial}{\partial y}(q_y - \epsilon_s |q_y| \frac{\partial z_b}{\partial y}) \tag{7}$$

where $q = (q_x, q_y)$ is the sediment transport rate vector, and ϵ_s is a positive constant. The terms multiplied by ϵ_s are introduced empirically to explain the effect of local slope of sea bed. According to the assumptions, the transport rate is

$$q = q_c + q_w \tag{8}$$

where q_c and q_w are transport rate components by waves and currents, respectively. These transport rates are given by the conventional power law with some empirical coefficients to be determined by users based on the site condition.

$$q_c = A_c \frac{(u_*^2 - u_{*c}^2)U}{g} \tag{9}$$

$$q_w = F_D A_w \frac{(u_*^2 - u_{*c}^2)u_b}{g} \tag{10}$$

where u_*^2 and u_{*c}^2 are friction velocity and critical friction velocity for the onset of sediment movement in the wave–current coexistent field. A_c and A_w are semi–empirical constants, while F_D determines the direction of sediment movement and is defined as

$$F_D = \tanh(\kappa_d \frac{\Pi_c - \Pi}{\Pi_c}) \tag{11}$$

in which Π is an indicator to the transport direction, and Π_c and κ_d are another constants. This direction function, F_D, takes a positive value when the transport is in the wave propagating direction. Since this topography change model is semi–empirical, the value of some coefficients are not universal and can not be determined from the wave and geometry condition. Certain site data on sediment transport is required to determine the appropriate values for these parameters and to obtain a reasonable solution.

4. Verification of numerical model

This model attempts to simulate the process of sediment transport that is rather complicated and is not rigorously understood, yet. It also has some parameters, which usually cannot be specified uniquely in advance. Then, it becomes important to evaluate applicability and accuracy of this model before applying to actual cases. Two laboratory experiments on sediment transport around a detached breakwater (Watanabe et al., 1986b) and around submerged breakwaters (Kohno et al., 1987) were used to verify the model.

4.1 Detached breakwater case

A breakwater was placed parallel to the shoreline at depth of 9 cm on a 1/20 slope bottom, and regular waves were applied normal to the shoreline. Watanabe et al. (1986b) determined the parameters necessary for this computational model by simulating the natural beach without any structure and by comparing these preliminary results with collected data. They were found to be

$$A_c = 0.5 \quad A_w = 0.15 \quad \Pi_c = 0.16 \quad \kappa_d = 2 \quad e_s = 2 \tag{12}$$

Computed wave height distribution, coastal circulation due to the breakwater and resultant bottom topography are shown in Figure 2, while the contour of the measured depth is in Figure 3. The numerical results show a clear circulation and consequent sediment accretion behind the breakwater, simulating the measured data well.

4.2 Submerged breakwaters case

The layout of four submerged breakwaters on a constant slope is shown in Figure 4. Regular wave incident normal to the breakwaters was also used in this case. The water depth was measured along the lines parallel to wave propagation in the onshore region of the breakwaters, and the total amount of deposit and erosion on each line was reported.

The alongshore distribution of deposit has a peak at the line about a quarter opening length from the tip of a breakwater, while that of erosion occurs at the edge of the other breakwater. Although this observation is slightly different from the expectation of the maximum accretion being behind the middle of the breakwater and erosion being at the center of the opening, we must consider inhomogeneous conditions in laboratory tests in spite of symmetric geometry. This asymmetric variation is also seen in the numerical results. This is partly because the computation of the currents is not very stable and the averaging process of the radiation stress was not symmetric. Figure 5 compares the net amount of deposit sand. Both data differ at each measuring line, but the alongshore variation pattern and the maximum accretion and erosion agree fairly well. Note that the constants for this example are different from the previous case, and that the numerical results depend on these values. The constants used are

$$A_c = 0.05 \quad A_w = 0.015 \quad \Pi_c = 0.16 \quad \kappa_d = 2 \quad e_s = 2 \tag{13}$$

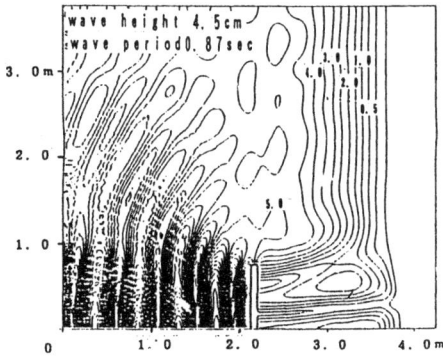

(a) Wave height distribution   (b) Nearshore currents

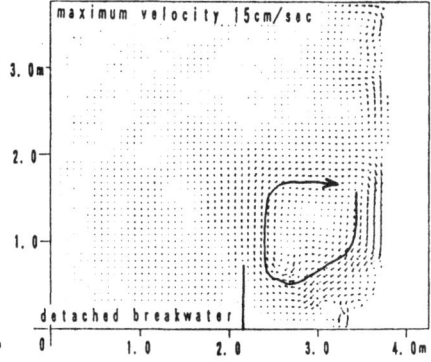

FIGURE 2   Numerical results for detached breakwater case

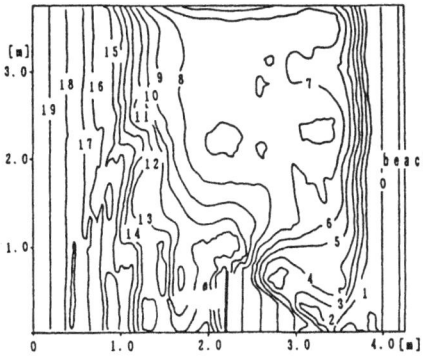

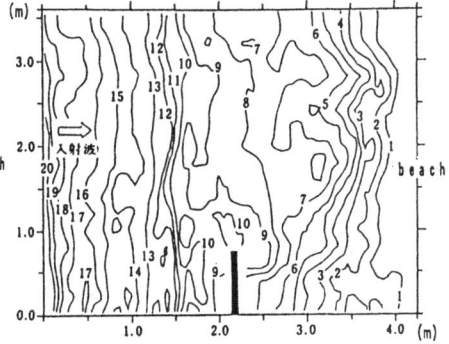

(c) Bathymetry

FIGURE 2   Numerical results
for detached breakwater case

FIGURE 3   Measured bathymetry
around detached breakwater

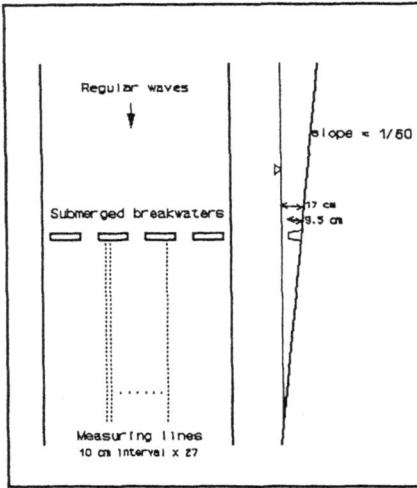

FIGURE 4 *Experimental layout of submerged breakwaters*

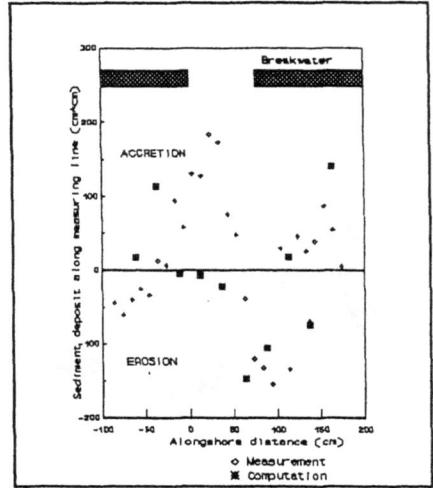

FIGURE 5 *Alongshore variation of sediment accretion and erosion*

Both examples shown here confirm that the present model practically predict the topography change only if the parameters in the model are properly specified.

5. Topography change in bay due to submerged breakwaters

5.1 Natural condition

The site of an interest is surrounded by a small beach, a tombolo and a head land as shown in Figure 6. The bottom slope in the bay is approximately 1/50 and isobathytherms run almost parallel to the shoreline. The submerged plates, which are to be modeled by the submerged breakwaters in this study, will be located at depth of 10m. Since the enough wave data at the site was not available, the wave condition was determined from the extreme condition observed at the nearby harbor.

5.2 Numerical condition

The analytical area, 1.5 km in the on-offshore direction and 0.8 km in the alongshore direction, is shown in Figure 7. Considering the computer capacity and the accuracy of the numerical model, the side boundaries are assumed to be symmetric and the line of symmetry goes through the center of one of the submerged breakwaters. Incident waves are imposed at the offshore boundary. The wave energy absorption layers were placed around the tombolo, the island and shoreline boundary, because reflecting waves are supposed to be small due to wave breaking and wave runup.

The other parameters including the coefficients in the equations and appropriate duration of the wave incident were determined by considering the field data observed at the site.

5.3 Recent temporal change of the bay configuration

To understand the physical condition in the bay, the aerial photos over 40 years and time history record of the sea bottom elevation at two locations were investigated. The series of pictures showed little noticeable variation in the shoreline location, suggesting weak sediment movement in the bay. The sea bottom elevation was measured at depth of 6m and 11m for a month of winter. Variation of approximately one meter was observed at both locations, and it implies some on–offshore sediment transport during winter.

In order to check if the numerical setting is reasonable and to identify the appropriate values for the coefficients included in the model, a series of computations were made for the natural case without any structure. The results computed by changing the values of the coefficients were compared with a time history record of the sea bottom elevation available at two points in the bay. It concluded that the numerical condition was appropriate and that the parameters for this case should be

$$A_c = 0.1 \quad A_w = 0.2 \quad \Pi_c = 1.0 \quad \kappa_d = 1 \quad e_s = 10 \tag{14}$$

The goal of this study is to predict the topography change over a year. The duration of the selected wave application to the site during one calendar year is another important parameter to be determined. The statistics on the wave condition in the nearby harbor were used in this study. According to the data, the waves as large as the selected wave come only during winter at this site, and elapsed time during which the selected wave was observed is accumulated. This estimate brought up a period of three days.

5.4 Topography change due to submerged breakwaters

The computed wave height distribution is shown in Figure 8. It is observed from this figure that the waves break at the submerged breakwater and the wave height behind the breakwater is smaller than it is without the breakwater. The on–offshore gradient of the radiation stress, S_{xx}, is large in front of the breakwater, and it is surmised that the force pushing water forward is strong there. On the other hand, S_{yy} is symmetric with respect to the line going through the opening between the breakwaters. This distribution implies the two alongshore current systems flowing away in the opposite directions originating from the center of the breakwater.

Figure 9 illustrates the computed current pattern. There are strong onshore currents on the breakwater and they turn to the alongshore direction beyond the breakwater and finally flow back through the openings. This strong circulation agrees well with that expected by the radiation stress distribution. There is another weak circulation in the surf zone, which is also observed in the natural case. The maximum speed of the circulation around the breakwater is approximately five times larger than that inside the breakwater.

The initial bathymetry and changed bathymetry after a three–day period (corresponds to one calendar year) are shown in Figure 10, while the cross–sectional profiles along the on–offshore line through the breakwater center before and after wave application are illustrated in Figure 11. Some erosion is observed in front of the breakwater and little change is recognized behind the breakwater probably because the breakwaters calm down the wave action there. Repeat of erosion and accretion (Figure 10) whose wavelength corresponds to that of the partial standing waves in front of the breakwater may be caused by these standing waves.

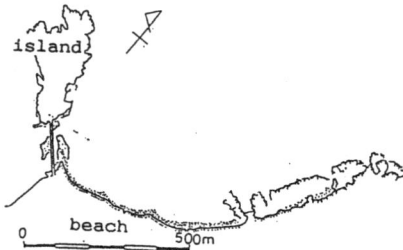

FIGURE 6 Map of the coastal region
where submerged plates will be placed

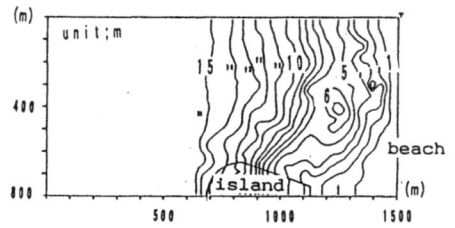

FIGURE 7 Analytical region
and input geometry

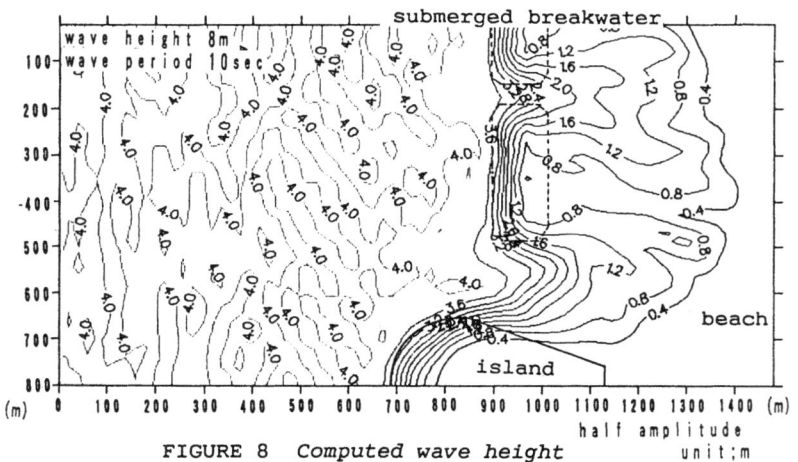

FIGURE 8 Computed wave height
distribution around breakwater

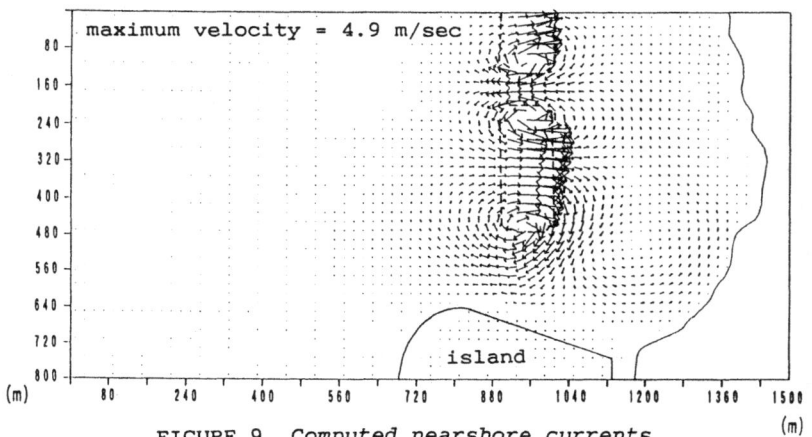

FIGURE 9 Computed nearshore currents
around breakwater

454

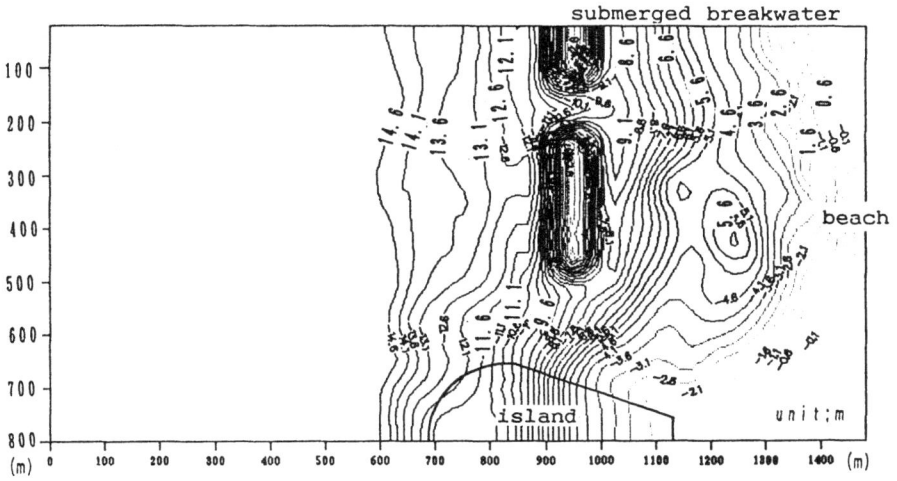

FIGURE 10 *Computed bathymetry after three-day wave application*

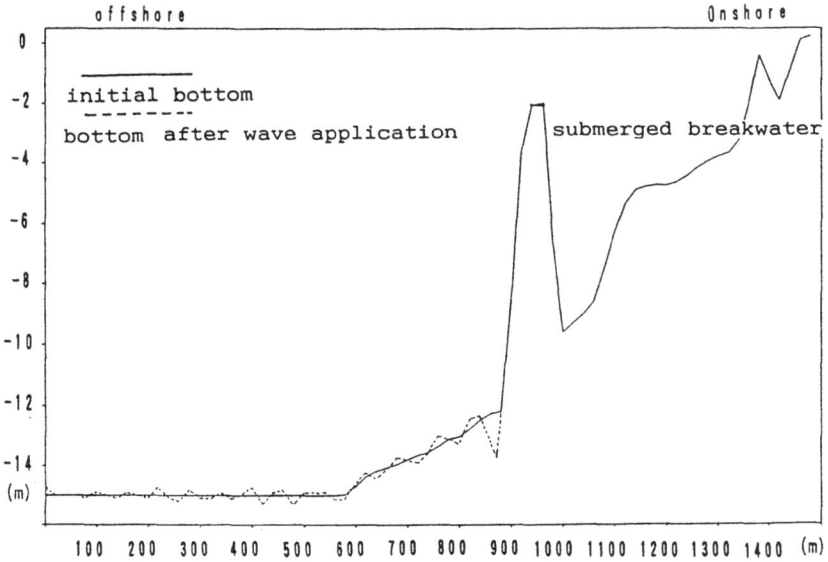

FIGURE 11 *Computed cross-sectional topography change*

6. Conclusions

The preliminary study, in which the predictions were examined by the measured data, suggests that this numerical model is capable of providing the practical estimate of topography change. It is also true that the model must be carefully applied because of some significant simplifications on the sediment transport model, and that the physical interpretation of the numerical results must always be made. In addition to establishing a more accurate evaluation method of sediment transport rate, the improvement in the wave analysis and nearshore current estimation in the random wave condition will upgrade the model.

The computed nearshore circulation pattern is very similar to those observed in laboratory. The computed flow pattern is anticipated by the distribution of the radiation stress, which is essentially determined by the wave breaking. This agreement supports that the wave breaking and its location, which are strongly influenced by the submerged breakwaters, are key factors to the nearshore circulation. From the prediction of the topography change at the planed coastal area, the installation of the floating plates is found not to disturb the ambient coastal condition. This is partly because this area is a relatively quiet area in terms of sediment movement as indicated by the available field data collected last 40 years.

REFERENCES

(1)Berkhoff, J.C.W., (1972), Computations of Combined Refraction–Diffraction, *Proc. 13th Coastal Eng. Conf.*, ASCE, pp. 471–490.

(2)Horikawa, K. (ed.), (1988), *Nearshore Dynamics and Coastal Processes – Theory, Measurement and Predictive Models*, Univ. of Tokyo Press.

(3)Irie, I. Kuriyama, Y. and Tagawa, M., (1985), Combination of Physical and Mathematical Models to Predict Change of Sea Bottom Elevation, *Proc. 32nd Japanese Conf. on Coastal Eng.*, pp. 345–349. (in Japanese)

(4)Iwatani, F., Miyamoto, T., Matsushita, M., Yoshinaga, S., Kawamata, R. and Adachi, Y., (1989), Prediction of Waves, Currents and Topographical Change around Submerged Offshore Breakwater, *Coastal Eng. in Japan*, JSCE, Vol. 32, pp.

(5)Kohno, F., Horikawa, T., Takano S. and Miyawaki, T., (1987), Experimental Study on Sediment Movement around Submerged Breakwaters, *Proc. 34th Japanese Conf. on Coastal Eng.*, JSCE, pp. 431–435.

(6)Kudo, K., Tsuzuku, T., Imai, K. and Akiyama, Y. (1988), Wave Focusing by a Submerged Plate, *Proc. OCEANS '88*, pp. 1061–1066.

(7)Okada, Y. and Kohno, F., (1991), Field Investigation of the Effect of the Submerged Breakwaters in the Aoshima coast, *Proc. 38th Japanese Conf. on Coastal Eng.*, JSCE, pp. 321–325. (in Japanese)

(8)Sasaki, M., Shuto, K. and Takeshita, A., (1990), Numerical Method to Compute Nearshore Currents and Topography Change around a Submerged Breakwater, *Proc. 37th Japanese Conf. on Coastal Eng.*, JSCE, pp. 404–408. (in Japanese)

(9)Shoji, N., Nakayama, H., Takiguchi, Y., Takahashi, T., Kuroki, K. and Sakai, T., (1991), Effect of Submerged Breakwater on Waves, Currents and Topography Change in the West Niigata Coast, *Proc. 38th Japanese Conf. on Coastal Eng.*, JSCE, pp. 429–433. (in Japanese)

(10)Watanabe, A. and Maruyama, K., (1986a), Numerical Modeling of Nearshore Wave Field under Combined Refraction, Diffraction and Breaking, *Coastal Eng. in Japan*, JSCE, Vol. 29, pp. 19–39.

(11)Watanabe, A., Maruyama, K., Shimizu, T. and Sakakiyama, T., (1986b), Numerical Prediction Model of Three–Dimensional Beach Deformation around a Structure, *Coastal Eng. in Japan*, JSCE, Vol. 29, pp. 179–194.

38 Effect of the submerged obstacle height on the wave energy distribution behind the detached breakwater

M. I. Balah and H. K. Mostafa

ABSTRACT

Due to the widely constructing of the detached breakwaters for the shore protection purposes, the necessity for improvement their efficiency highly increases. This can be done by constructing a supplementary structure , like an obstacle. Through this research, an attempt has been done to improve the distributions of the wave energy behind the detached breakwater. Submerged straight and broken obstacles have been experimentally investigated in different heights facing the detached breakwater gap at the shore ward direction. The drawn results showed the effect of the submerged obstacle relative height on the wave energy distribution at the shore ward direction behind the detached breakwater parts and at the gap in between. The critical value for the submerged obstacle relative height affecting the energy distribution has been determined.

NOMENCLATURE

D   Still water depth.
h   Height of submerged obstacle.

h/D	Relative height of obstacle
H_i	Incident wave height at the gap centre for the detached breakwater arrangement without obstacles.
H	wave height at any particular point.
H/H_i	Relative wave height at any particular point.
E_i	Wave energy at the gap centre for the detached breakwater arrangement without obstacles.
E	Wave energy at any particular point.
E/E_i	Relative wave energy at any particular point.
$\Delta E = E_i - E$	Wave energy losses at any particular point.
$\Delta E/E_i$	Relative wave energy losses at any particular point.

1 Introduction

In the last twenty years, the detached breakwater geometry and alignment and its effect on the wave properties have been studied. The effect of construction an obstacle at the detached breakwater gap higher than the still water level on the passing wave energy has been experimentally investigated in research (1) for the same detached breakwater model under investigation. The effect of the detached breakwater alignment on the shore line has been investigated in many researches (2,3,4). Also the effect of the breakwater parts spacing has been studied (2). Experimental and theoretical investigations on wave interaction with the submerged structures have been carried out by various investigators (5,6,7,8,9). The incoming energy through the detached breakwater gap subdivided mainly into three parts, reflected, transmitted over obstacle, and diffracted parts. The submerged obstacle usually are less costly to construct and maintenance than the conventional type of the obstacles extending above the highest expected water level . The submerged structures also are effective to control beach erosion. Through this research, submerged , straight and broken obstacles have been investigated experimentally with different height values to investigate their effect on wave energy distribution in the zone between the detached breakwater and shore line.

2   Experimental work

Experiments were conducted inside the main rectangular harbour basin of Suez Canal Research Centre At El Esmaylia with irregular wave generator system, wave sensors and computer facilities which explained in details in paper (1). The experimental investigation has been conducted using fixed detached breakwater dimensions, breakwater parts aligned at the same line with length equal to the inbetween gap width and equal to 75m. The breakwater was constructed at water depth of (-5.00) and 120m seaward from the shore line with crest level of (+1.50) and side slopes of 2:1. Fig. 1 shows the main arrangements of the investigated system. Obstacles have been tested with top level below still water level in two main geometries, stright with length of 26.25m and broken with total length of 26.25m also (8.75m horizontal part and two arms inclined at 45°, 8.75m each). Both the straight and broken obstacles, have been investigated in Seven different heights ranging from 1.05m to 4.20m in an increment of 0.525m. Fig. 2 shows the details of the different obstacles . The model scale has been chosen to be 1:35. The wave properties for the different runes have been recorded using a system of ten wave sensors aligned as shown in Fig. 1 and connected with a completely equiped control room.

3   Experimental results

The generated wave properties for the experimental runs were 1.0m height and 5.0 sec. period. The wave properties at location (1) were measured for the detached breakwater arrangement without obstacles. For the different Fourteen obstacles(straight and broken) the wave properties were recorded at different wave gauges. The wave spectrum corresponding to the different tested obstacles, straight and broken with different heights together with the similar spectrum of detached breakwater without obstacles have been recorded. For the space limit, Fig. 3 shows some wave spectrum selected for the location behind the obstacle with different geometries. The relations between the relative wave energy (E/E_i) against the relative obstacle height (h/D) are drawn for the different measuring locations in Figs. 4 and 5. Fig 4 shows the relation

459

for the straight obstacle configurations. While Fig. 5 shows the same relation for the broken obstacle configurations. The drawn results for both configurations, straight and broken showed that the relative wave energy decreases with increasing the relative obstacle height for the zone of gauges 5,6,8,10 and 9. The decreasing rate for locations 5,6,8 and 10 showed nearly the same value. While the decreasing rate for location 9, behind the obstacle, showed nearly zero decreasing rate up to a relative obstacle height of 0.45. Beyond the relative height value of 0.45, the relative energy highly decreases as the obstacle relative height increases. The relative energy showed very small decreasing rate by increasing the obstacle relative height for locations 4 and 7, behind the detached breakwater parts and at the same line of obstacle alignment.

At the gap entrance close to the detached breakwaters head, the relative energy showed the same value for the different obstacles of relative energy increases as the relative height value increases. The relative energy at location (1) showed nearly no variation with the increasing of the relative obstacle height up to a value of 0.67. Beyond this value the relative wave energy slightly increases as the obstacle relative height increases. The relation between the relative wave energy losses and the obstacle relative height are drawn in Fig. 6 for the straight obstacle configurations and in Fig. 7 for broken ones. The drawn results showed off course similar behaviour as the drawn relative energy losses relations.

4   Analysis of experimental results

In general the obstacles attenuate the incoming energy through the detached breakwater gap. This can be related to subdivision of the incoming wave energy mainly into there parts; reflected , overtopping and diffracted parts. As the obstacle relative height increases, its damping effect increases significantly. Since the wave energy is concentrated in upper portion of the fluid, the reflection of the wave and the interaction between the wave and the obstacle increases resulting in low transmission energy. For this reason, the high decreasing rate for the relative wave energy behind the obstacle for its relative height ≥ 0.45 can be refered to the effect of the obstacle in reducing the wave overtopping over it due to increasing its relative height. While for obstacles of relative height < 0.45, the overtopping wave does not affected by the obstacle height. The reason for the marginal relative wave energy decreasing rate as obstacle height increases at locations 4 and 7 can be refered to the highly protection effect of the detached breakwater water parts which overcomed on the obstacle effect. The increasing rate of the relative wave energy at the gap centre for the obstacle of relative height > 0.67 can be refered to the effect of the obstacle in reflecting of a percentage from the incident wave energy   The reflection percentage increases as the relative obstacle height increases. The same analysis can be explained for the relative energy losses relations.

5   Conclusions

From the experimental results and analysis, the outhors can be conclude the following :-
- The effectiveness of submerged obstacle in wave energy redistribution dependent on its relative height.
- Obstacle with relative height values larger than a certain critical value are more effective than those with relative height value smaller than critical value.
- The critical value of h/D was found to be equals to 0.45 and it is independent on obstacle geometry straight or broken.
- The wave reflection effect due to the obstacles appears for relative height higher than 0.67.

REFERENCES

[1]Balah,M.I.A. and Mostafa,H.K.,(1991), "Effect of constructing an obstacle at detached breakwater gap on the passing wave energy", *Frist International Conference on Engineering Research, Development and Application*, ERDA 91,26-28 Nov.1991,Civil Engineering proceeding, Faculty of Engineering, Suez Canal University.
[2]Harris,M.M. and Herbich,J.B.,(1986),Effect of breakwater spacing on sand entrapment,*journal of Hydraulic Research*,**VOL. 24,** No. 5.

[3]Rosen,D.S.and Vajida,M.,(1982),"Sedimentological influences of detached breakwaters" , Proc. *Conference on Coastal Engineering* ,ASCE.

[4]Shinohara, K. and tusbaki, T.,(1966), "Model study in the change of shoreline of sandy breach by the offshore breakwater ",*Proc. Conference on Coastal Engineering*, ASCE.

[5]Abdul Khader, M.H.and Rai,S.P.,(1980),"A study of

submerged breakwaters", *Journal of Hydraulic Research*, IAHR , **VoL 18**, No.2. , PP.113-121.

[6]Dattatri,J. , Raman, H. and Shankar, J. ,(1978) , "performance characteristics of submerged breakwaters",*Proc.* 16th *Coastal Engineering Conference* ,ASEC, Hamburg.

[7]Dick, T.M. and Brenber, A.,(1968)," Solid and permeable submerged breakwaters ",*Proc.* 11th *Conference of Coastal Engineering* ,ASCE,London.

[8]Garrison, C.J. and Rao,V.S.,(1971)," Interaction of waves with submerged objects", *Journal of Waterways, Harbours and Coastal Engineering Division* ,ASCE, **VoL 97.**

[9]Ouellet,Y. and Eubanks, P.,(1976),Overtopping of rubble-mound breakwaters by irregular waves", *Proc. of* 15th *Conference on Coastal Engineering* ,pp 2756-2776.

[10]U.S army, Corps of Engineers,(1984),Shore Protection Manual,Coastal Engineering Research Center ,**Vols. 1&2.**

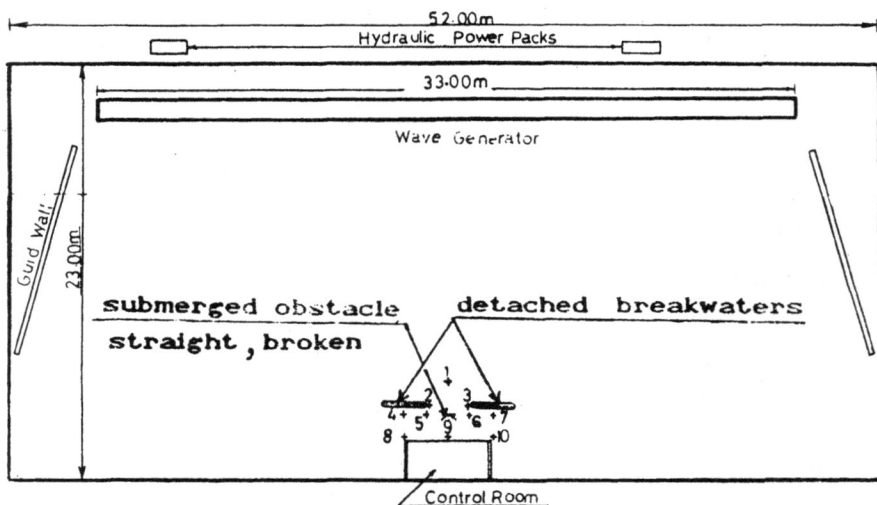

Fig 1 *Details of experimental basin mode. arrangement.*

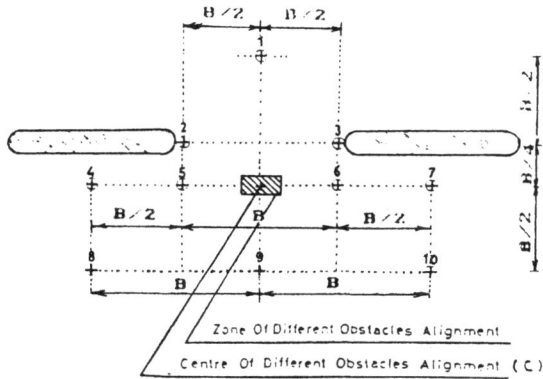

= Alignment of wave gauges

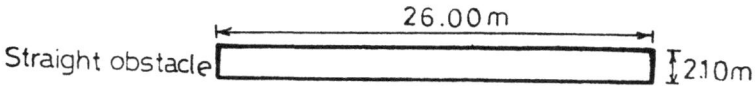

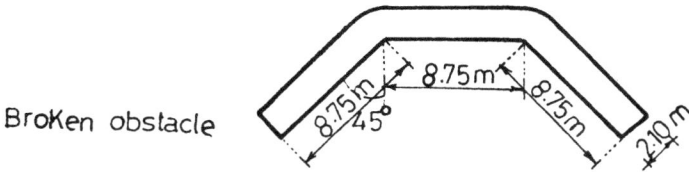

Straight obstacle 26.00 m, 2.10 m

Broken obstacle 8.75 m, 8.75 m, 8.75 m, 45°, 2.10 m

= Plan of stright and broken obstacles

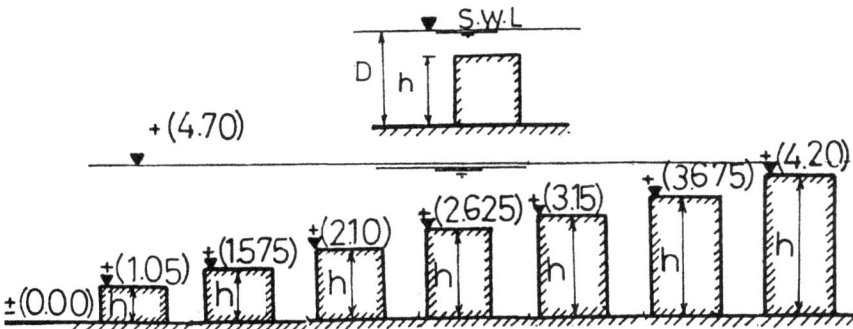

= Cross section of different obstacles

Fig 2 *Details of the tested obstacles*

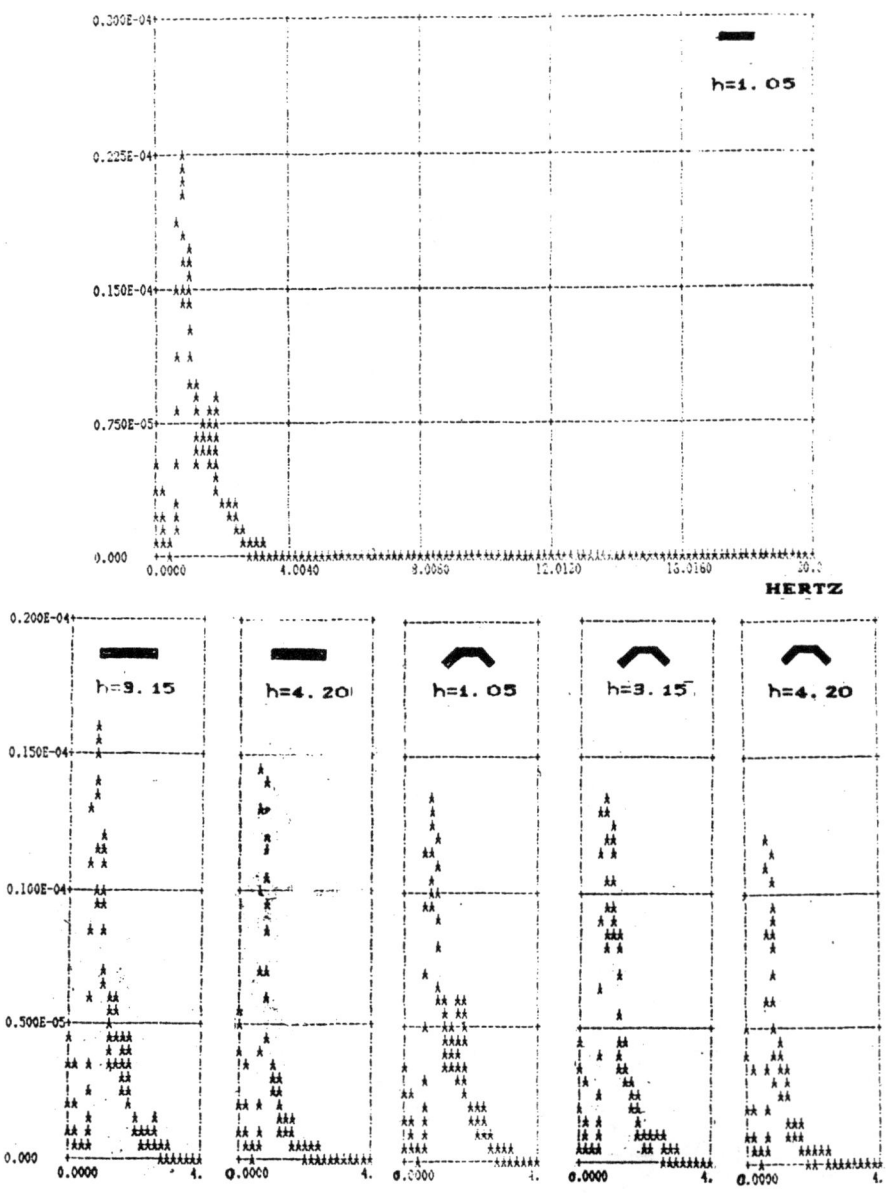

Fig 3 Wave spectrum for wave gauge number 9 for different obstacle arrangements.

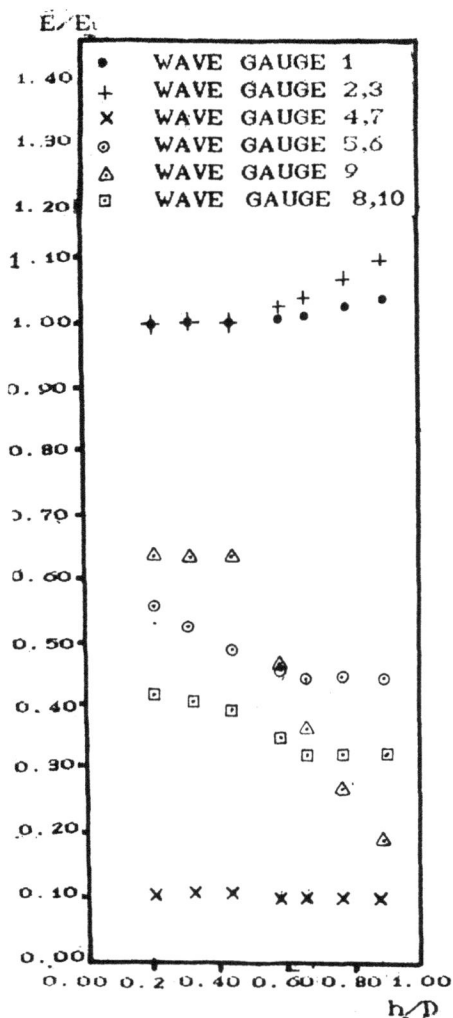

Fig. 4 Relation between the relative wave energy E/Eᵢ and the relative height of obstacle h/D for straight obstacle arrangements.

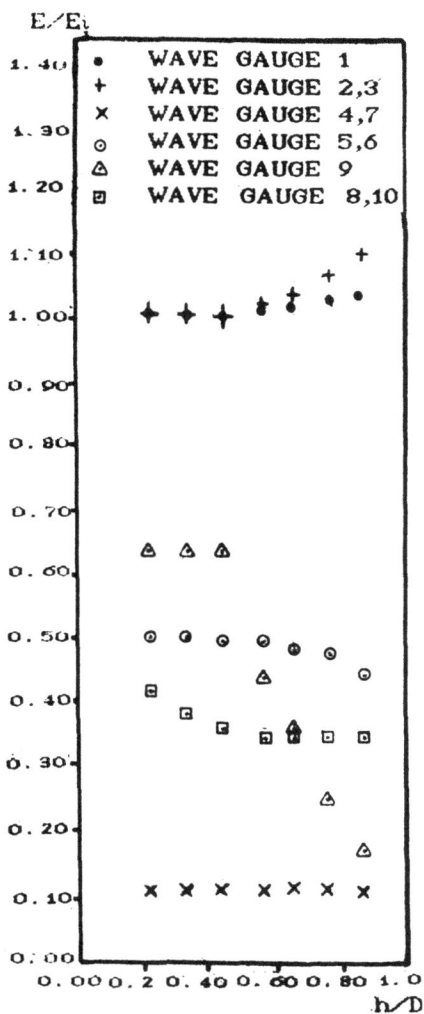

Fig. 5 Relation between the relative wave energy E/Eᵢ and the relative height of obstacle h/D for broken obstacle arrangements.

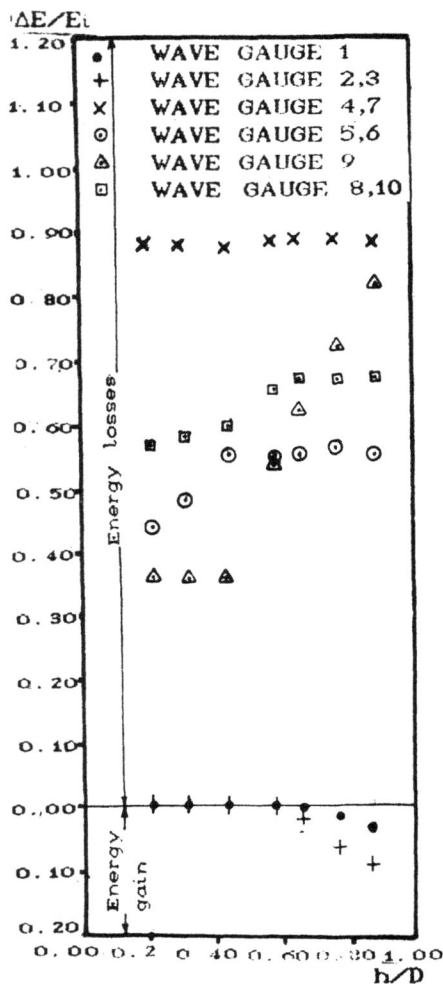

Fig 6   *Relation between the relative wave energy losses* ΔE/E₁   *and the relative height of obstacle* h/D *for straight obstacle arrangements.*

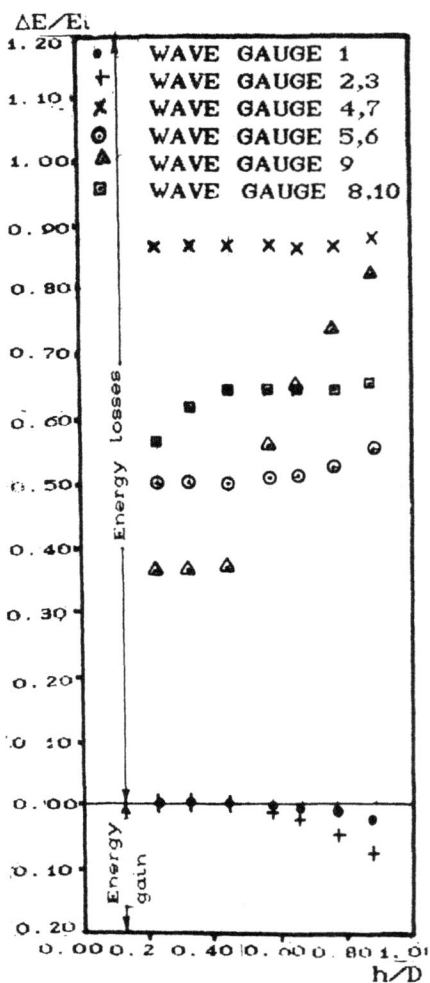

Fig. 7   *Relation between the the relative wave energy losses* ΔE/E₁ *and the relative height of obstacle* h/D *for broken obstacle arrangements*

466

Part 5
COMPUTATIONAL METHODS

39 Solving the shallow water equations using a non-orthogonal curvilinear coordinate system

R. W. Barber

ABSTRACT

This paper describes a semi–implicit finite–difference scheme for the solution of the depth–averaged shallow water equations on a non–orthogonal boundary–fitted coordinate system. The numerical model provides a method of dealing with the curved or irregular flow boundaries typically encountered in reservoirs and estuaries without having to implement the usual approximation of following the curvature of the flow perimeter by a 'staircase' of grid points. It is envisaged that the boundary–fitted technique could provide water resource engineers with a powerful alternative to Cartesian finite–difference methods.

1 Introduction

Most hydrodynamic models of reservoirs, tidal harbours and estuaries rely on finite–difference solutions of the depth–averaged shallow water equations expressed in a Cartesian reference frame [1,4,9,10]. In practice, a rectangular mesh with fixed grid spacing is placed over the domain of interest, usually resulting in non–alignment of the physical boundaries with the edges of the computational flow domain. This lack of alignment may give rise to major inaccuracies in the flow solution, particularly spurious vorticity generation at the sharp corners of the mesh. In addition, features such as islands and dredged shipping channels often require local grid refinement in order to increase the accuracy of the flow predictions in such areas.

Boundary–fitted curvilinear coordinate systems provide an approach which combines the best aspects of finite–difference discretisation with the grid flexibility usually attributed to finite–element procedures. The essential feature of a boundary–fitted coordinate system is that grid lines coincide with the flow perimeter, no matter how irregular the shape of the region.

The concept of boundary–fitted coordinate systems originated in the U.S. aerospace industry in the early 1970s as a response to NASA's requirements to predict high velocity flow patterns around irregularly–shaped space vehicles. Since then, boundary–fitted systems have been utilised for the solution of a wide variety of fluid dynamic problems in aeronautical and mechanical engineering. Surprisingly, even though numerical grid generation techniques have been thoroughly documented, little published research is currently available on the solution of the shallow water equations using curvilinear finite–difference schemes. The earliest studies on the transformed non–linear shallow water equations were conducted by Johnson [7,8] in the United States. Other curvilinear models include Haüser et al.'s [5,6] and Raghunath et al.'s [12] numerical scheme for the *linearised* shallow water equations. In addition, Wijbenga [18,19] and Willemse et al. [20], of Delft Hydraulics in The Netherlands, have developed an alternative numerical approach requiring an orthogonal curvilinear mesh.

2 Grid generation

At the onset of any numerical study using boundary–fitted coordinates, it is necessary to decide whether the grid system should be orthogonal or non–orthogonal. The former method is attractive since the governing hydrodynamic equations in the curvilinear coordinate reference frame are very similar to the standard Cartesian hydrodynamic expressions. Consequently existing solution algorithms developed for the Cartesian equations can be modified with little additional effort. However, the orthogonality constraint during mesh generation limits the distribution of coordinate lines around complex topographical features such as headlands and bays. On the other hand, non–orthogonal grid systems allow more flexibility in the internal grid point distribution but have the drawback that the governing equations of motion are considerably more complex than their orthogonal counterparts. In the present study, the ability to stretch and distort the internal grid nodes without placing any restriction on the disposition of the boundary points was judged to be sufficient incentive to tolerate the disadvantages of using a non–orthogonal mesh.

Following Thompson et al. [16], the non–orthogonal boundary–fitted coordinate meshes were generated by solving an elliptic system of the form:

$$\xi_{xx} + \xi_{yy} = P(\xi,\eta)$$
$$\eta_{xx} + \eta_{yy} = Q(\xi,\eta)$$
$$\left.\vphantom{\begin{array}{c}1\\1\\1\end{array}}\right\} \quad (1)$$

where the subscripts denote the usual shorthand notation for partial differentiation and (ξ,η) are the boundary–fitted curvilinear coordinates.

The weighting functions, P and Q, are the so-called 'attraction operators' or 'control functions' which cause the coordinate lines to concentrate as desired; they are useful to counteract grid skewness and excessive cell–size variation in regions of large boundary curvature. After interchanging the dependent and independent variables, equation (1) may be written as a quasi–linear elliptic system:

$$
\left.
\begin{aligned}
\alpha x_{\xi\xi} - 2\beta x_{\xi\eta} + \gamma x_{\eta\eta} + J^2(Px_\xi + Qx_\eta) = 0 \\
\alpha y_{\xi\xi} - 2\beta y_{\xi\eta} + \gamma y_{\eta\eta} + J^2(Py_\xi + Qy_\eta) = 0
\end{aligned}
\right\}
\quad (2)
$$

where $\quad \alpha = x_\eta{}^2 + y_\eta{}^2 \; ; \quad \beta = x_\xi x_\eta + y_\xi y_\eta \; ; \quad \gamma = x_\xi{}^2 + y_\xi{}^2$

and $\quad$ J = Jacobian of the transformation $= x_\xi y_\eta - x_\eta y_\xi$.

Equation (2) is solved iteratively using successive over–relaxation to obtain the physical (x,y) coordinates of the non–orthogonal grid in terms of the boundary–fitted coordinates (ξ,η).

3 Cartesian hydrodynamic equations

The depth–averaged continuity equation for a Cartesian reference frame is

$$
\frac{\partial \zeta}{\partial t} + \frac{\partial (UD)}{\partial x} + \frac{\partial (VD)}{\partial y} = 0
\quad (3)
$$

where x and y are the horizontal coordinates, t is the time, ζ is the surface elevation above an arbitrary datum, D (which equals h + ζ) is the local water depth where h is the distance between the bed and the datum, and U and V are depth–averaged velocity components.

The non–conservative depth–integrated momentum equations in the x– and y–directions are formulated as

$$
\frac{\partial U}{\partial t} + B\left[U\frac{\partial U}{\partial x} + V\frac{\partial U}{\partial y}\right] + g\frac{\partial \zeta}{\partial x} + \frac{\tau_{bx}}{\rho D} - \frac{1}{\rho D}\left[\frac{\partial (DTxx)}{\partial x} + \frac{\partial (DTxy)}{\partial y}\right] = 0
$$

and $\qquad\qquad\qquad\qquad\qquad\qquad\qquad\qquad\qquad\qquad\qquad\qquad\qquad (4)$

$$
\frac{\partial V}{\partial t} + B\left[U\frac{\partial V}{\partial x} + V\frac{\partial V}{\partial y}\right] + g\frac{\partial \zeta}{\partial y} + \frac{\tau_{by}}{\rho D} - \frac{1}{\rho D}\left[\frac{\partial (DTxy)}{\partial x} + \frac{\partial (DTyy)}{\partial y}\right] = 0
$$

$$\qquad\qquad\qquad\qquad\qquad\qquad\qquad\qquad\qquad\qquad\qquad\qquad\qquad (5)$$

where ρ is the fluid density, g is the acceleration due to gravity, B is the correction factor for the non–uniform vertical velocity profile, τ_{bx} and τ_{by} are bed friction stress components, and Txx, Txy and Tyy are the effective stress components (as defined by Kuipers and Vreugdenhil [9]).

Implementation of the Boussinesq eddy viscosity concept allows the effective stresses to be written as

$$Txx = 2\rho\nu_t\frac{\partial U}{\partial x}$$

$$Txy = \rho\nu_t\left[\frac{\partial U}{\partial y} + \frac{\partial V}{\partial x}\right] \qquad (6)$$

$$Tyy = 2\rho\nu_t\frac{\partial V}{\partial y}$$

where ν_t is the eddy viscosity coefficient.

4 Transformed governing equations

Before transforming the shallow water equations, it is necessary to decide whether Cartesian or contravariant velocity components are to be utilised as dependent variables in the transformed domain. From the point of view of accuracy, it is better to use contravariant components since these are always normal to the boundaries of the individual grid cells and can therefore better represent the mass and momentum fluxes across the cell faces. However, the disadvantage in choosing the contravariant method is the complexity of the transformed equations; Sheng and Hirsh [13] expanded the tensor invariant form of the shallow water equations and found that there were approximately sixty terms in each contravariant momentum expression. In view of this, Cartesian velocity components were adopted for the purposes of the present study.

The transformation of the Cartesian shallow water equations is accomplished using the non-conservative derivative relationships presented by Thompson et al. [16]:

$$f_x = \frac{\partial f}{\partial x} = \frac{\partial(f,y)}{\partial(\xi,\eta)} \div \frac{\partial(x,y)}{\partial(\xi,\eta)} = \frac{1}{J}(y_\eta f_\xi - f_\eta y_\xi)$$
and
$$f_y = \frac{\partial f}{\partial y} = \frac{\partial(x,f)}{\partial(\xi,\eta)} \div \frac{\partial(x,y)}{\partial(\xi,\eta)} = \frac{1}{J}(x_\xi f_\eta - f_\xi x_\eta) \qquad (7)$$

where f denotes a differentiable function of x and y.

Substituting the above transformation expressions into all partial derivatives involving x or y in the Cartesian equations of motion leads to the depth-averaged continuity equation being recast as

$$\frac{\partial\zeta}{\partial t} + \frac{1}{J}\left[y_\eta\frac{\partial(UD)}{\partial\xi} - y_\xi\frac{\partial(UD)}{\partial\eta} + x_\xi\frac{\partial(VD)}{\partial\eta} - x_\eta\frac{\partial(VD)}{\partial\xi}\right] = 0 \qquad (8)$$

whilst the non-conservative momentum relationships are rewritten as:

x—momentum

$$\frac{\partial U}{\partial t} + \frac{B}{J}\left[y_\eta U\frac{\partial U}{\partial \xi} - y_\xi U\frac{\partial U}{\partial \eta} + x_\xi V\frac{\partial U}{\partial \eta} - x_\eta V\frac{\partial U}{\partial \xi}\right] + \frac{g}{J}\left[y_\eta\frac{\partial \zeta}{\partial \xi} - y_\xi\frac{\partial \zeta}{\partial \eta}\right]$$

$$- \frac{1}{\rho DJ}\left[y_\eta\frac{\partial (DTxx)}{\partial \xi} - y_\xi\frac{\partial (DTxx)}{\partial \eta} + x_\xi\frac{\partial (DTxy)}{\partial \eta} - x_\eta\frac{\partial (DTxy)}{\partial \xi}\right]$$

$$+ \frac{\tau bx}{\rho D} = 0 \tag{9}$$

and

y—momentum

$$\frac{\partial V}{\partial t} + \frac{B}{J}\left[y_\eta U\frac{\partial V}{\partial \xi} - y_\xi U\frac{\partial V}{\partial \eta} + x_\xi V\frac{\partial V}{\partial \eta} - x_\eta V\frac{\partial V}{\partial \xi}\right] + \frac{g}{J}\left[x_\xi\frac{\partial \zeta}{\partial \eta} - x_\eta\frac{\partial \zeta}{\partial \xi}\right]$$

$$- \frac{1}{\rho DJ}\left[y_\eta\frac{\partial (DTxy)}{\partial \xi} - y_\xi\frac{\partial (DTxy)}{\partial \eta} + x_\xi\frac{\partial (DTyy)}{\partial \eta} - x_\eta\frac{\partial (DTyy)}{\partial \xi}\right]$$

$$+ \frac{\tau by}{\rho D} = 0 \tag{10}$$

The effective stress equations are also transformed into the (ξ,η) coordinate system:

$$\left.\begin{array}{l} Txx = 2\rho\nu_t\frac{1}{J}\left[y_\eta\frac{\partial U}{\partial \xi} - y_\xi\frac{\partial U}{\partial \eta}\right] \\[4mm] Txy = \rho\nu_t\frac{1}{J}\left[x_\xi\frac{\partial U}{\partial \eta} - x_\eta\frac{\partial U}{\partial \xi} + y_\eta\frac{\partial V}{\partial \xi} - y_\xi\frac{\partial V}{\partial \eta}\right] \\[4mm] Tyy = 2\rho\nu_t\frac{1}{J}\left[x_\xi\frac{\partial V}{\partial \eta} - x_\eta\frac{\partial V}{\partial \xi}\right] \end{array}\right\} \tag{11}$$

5 Numerical discretisation

The transformed governing equations (8, 9, 10 and 11) were discretised on a staggered (ξ,η) grid and solved using a semi—implicit alternating direction finite—difference scheme. In contrast to the earlier work of Johnson [7,8] and Häuser et al. [5,6] where the U— and V—velocity components were placed at the same location, the present scheme employs a fully staggered layout for the velocity variables as illustrated in Figure 1.

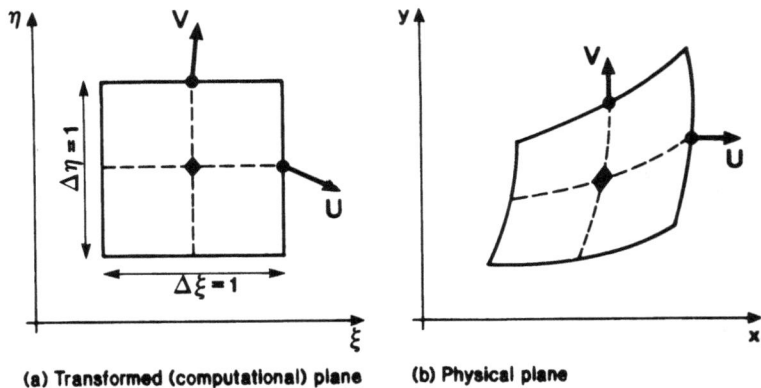

(a) Transformed (computational) plane   (b) Physical plane

FIGURE 1   *Cell definition*

Although this arrangement requires additional spatial averaging, the advantage of the present scheme is that it reverts to a standard Cartesian discretisation whenever a uniform orthogonal grid aligned with the coordinate axes is employed.

5.1 Control of non-linear instability

The inclusion of the non-linear advective acceleration terms in the x- and y-momentum equations produces destabilising effects in the numerical scheme. These instabilities are most troublesome when modelling hydraulic regimes with strong recirculating regions and low values of eddy viscosity. The present numerical model employs a second-order upwind differencing technique based upon proposals by Stelling [14], Stelling and Willemse [15] and Willemse et al. [20]. The treatment of the advective acceleration terms eliminates grid scale numerical instabilities whilst minimising artificial viscosity.

All spatial derivatives in the transformed momentum expressions are approximated by central differences except for the *'cross-derivative'* advective acceleration terms; for the x-momentum equation (9) these are $y_\xi.U.\partial U/\partial\eta$ and $x_\xi.V.\partial U/\partial\eta$ whilst for the y-momentum equation (10), the cross-derivative accelerations are $y_\eta.U.\partial V/\partial\xi$ and $x_\eta.V.\partial V/\partial\xi$. These four terms are evaluated either as weighted central differences or as second-order upwind differences, depending upon the particular stage of the computational scheme in a similar manner to Willemse et al. [20]. For example, in the first half time step of the A.D.I. scheme, the cross-derivative terms in the x-momentum equation are both approximated as weighted central differences i.e.,

$$y_\xi U\frac{\partial U}{\partial\eta} = y_\xi U\left[\ \frac{1}{3}\ D_2\eta U\ +\ \frac{2}{3}\ D_1\eta U\ \right] \tag{12}$$

474

where $D_{1\eta}U$ = 'standard' central difference approximation to $\partial U/\partial\eta$
(evaluated over 2 grid increments),

and $D_{2\eta}U$ = central difference approximation to $\partial U/\partial\eta$
(evaluated over 4 grid increments).

Thus,

$$y_\xi U\frac{\partial U}{\partial\eta} = y_\xi U_{i,j}\left\{\frac{1}{3}\left[\frac{U_{i,j+2} - U_{i,j-2}}{4}\right] + \frac{2}{3}\left[\frac{U_{i,j+1} - U_{i,j-1}}{2}\right]\right\} \qquad (13)$$

and similarly,

$$x_\xi V\frac{\partial U}{\partial\eta} = x_\xi \overline{V}_{i,j}\left\{\frac{1}{3}\left[\frac{U_{i,j+2} - U_{i,j-2}}{4}\right] + \frac{2}{3}\left[\frac{U_{i,j+1} - U_{i,j-1}}{2}\right]\right\} \qquad (14)$$

where the overbar, $^-$, indicates that the hydrodynamic variable is obtained by four–point averaging.

During the second half time step, the x–momentum cross–derivative terms are differenced using a combination of weighted central and quadratic upwind expressions. Since the boundary–fitted coordinate meshes are generated with the y– and η–coordinate axes in the same approximate direction, the cross–derivative term $x_\xi.V.\partial U/\partial\eta$ can use the sign of the V–velocity component to determine the direction of the upwind finite–difference expression for $\partial U/\partial\eta$. On the other hand, the derivative $y_\xi.U.\partial U/\partial\eta$ is calculated using a weighted central difference formula because the U–velocity direction is approximately normal to the η–direction. Consequently, the cross–derivative terms are discretised during the second half time step as:

$$x_\xi V\frac{\partial U}{\partial\eta} = \begin{cases} x_\xi\overline{V}_{i,j}\left[\dfrac{3}{2}\,U_{i,j} - 2\,U_{i,j-1} + \dfrac{1}{2}\,U_{i,j-2}\right] & \text{if } \overline{V}_{i,j} > 0 \\[4mm] x_\xi\overline{V}_{i,j}\left[-\dfrac{3}{2}\,U_{i,j} + 2\,U_{i,j+1} - \dfrac{1}{2}\,U_{i,j+2}\right] & \text{if } \overline{V}_{i,j} < 0 \end{cases} \qquad (15)$$

whilst $y_\xi.U.\partial U/\partial\eta$ is discretised as shown before in equation (13).

The same principle is employed for the y–momentum cross–derivative components; the first half time step utilises upwind differences for the term $y_\eta.U.\partial V/\partial\xi$ and weighted central differences for $x_\eta.V.\partial V/\partial\xi$ whilst the second half time step employs weighted central differences for both these terms. Further details concerning the numerical scheme, including the boundary conditions can be found in [2].

6 Results

6.1 Jet–forced flow in a circular reservoir

The prediction of jet–forced flow in a flat–bottomed circular reservoir provides an excellent opportunity to test the discretisation of the non–linear advective acceleration terms in the transformed shallow water equations;

the sudden expansion of flow geometry at the inlet causes flow separation and leads to the formation of two zones of recirculation. The rate of divergence of the high velocity throughflow jet is dependent upon the magnitude of the eddy viscosity coefficient as well as the level of artificial viscosity inherent in the numerical scheme.

Figure 2 illustrates the 93 x 93 node boundary–fitted coordinate system employed in the present study. The parallel–sided inlet and outlet channels each subtend an angle of $\pi/16$ radians and their centrelines are separated by $7\pi/8$ radians. This is representative of the geometry used by Mills [11] for the numerical investigation of low Reynolds number flow inside a circle. The basin was assumed to have a diameter of 1.5 m and a depth of 0.1 m. No–slip boundary conditions were applied to the perimeter walls of the flow domain, and bed friction was neglected for compatibility with an alternative computer simulation using the Navier–Stokes equations. A constant eddy viscosity of 2.92×10^{-4} $m^2 s^{-1}$ was utilised across the entire flow domain.

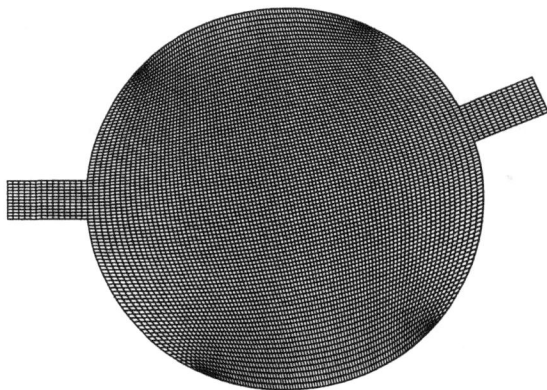

FIGURE 2   *Boundary–fitted coordinate system for a circular reservoir*

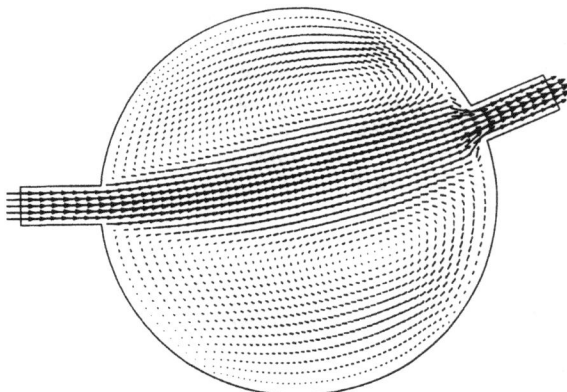

FIGURE 3   *Velocity vectors at steady–state*

Figure 3 illustrates the computed velocity vectors at steady–state for an inlet velocity of 0.1 ms^{-1}. The flow pattern predicted by the boundary–fitted shallow water equation solver shows good agreement with results from the standard orthogonal stream function/vorticity–transport (ψ,ω) solution technique presented by Borthwick and Barber [3]. This is illustrated in Figure 4 which compares non–dimensionalised velocities across the axis of symmetry of the reservoir. The solid curved line depicts the velocity distribution computed using the stream function/vorticity transport discretisation, whereas the crosses indicate the profile predicted by the boundary–fitted shallow water equation scheme. The velocity and coordinate variables, U' and y' are defined with respect to a second Cartesian coordinate system inclined at $\theta=+\pi/16$ radians to the primary (x,y) system; this enables the velocity profiling to be carried out along the axis of symmetry of the reservoir. Apart from slight discrepancies near the centreline of the jet, Figure 4 shows good agreement between the two numerical approaches, especially in the shear layers either side of the main throughflow. In fact, the agreement is remarkable when it is considered that the boundary–fitted mesh uses approximately one–sixth the number of grid points of the (ψ,ω) formulation.

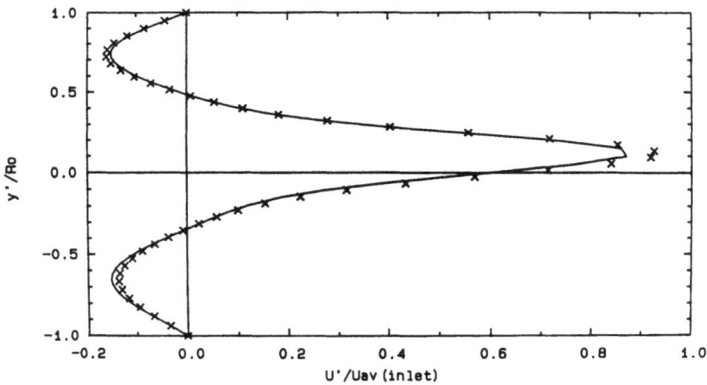

FIGURE 4 *Transverse velocity profile across reservoir*

6.2 *Irregular river geometry*

This section demonstrates the versatility of the curvilinear coordinate technique by examining the flow in an irregularly–shaped watercourse. Figure 5 illustrates the 121 x 31 node coordinate system used in the investigation of a river flood plane. The flow depth ranged from 0.5 m at the perimeter to approximately 3.5 m near the centre of the water course.

The velocity across the narrow inflow jet was set to 0.5 ms^{-1} and the Chezy roughness coefficient was assumed to be 45 m$^{\frac{1}{2}}$s^{-1}. In a similar manner to the circular reservoir flow study, described previously, the magnitude of the eddy viscosity coefficient is the critical parameter in determining the circulation pattern. As well as controlling the size and strength of secondary flow phenomena, the value of eddy viscosity affects the transfer of momentum between the main channel and the shallow flow areas near the banks, and therefore influences the transverse velocity profile across the river. From studies of rivers in flood (for example, Vreugdenhil and Wijbenga [17]), the eddy viscosity coefficient in the present example could be expected to range between 0.5 and 0.1 m^2s^{-1}. Figure 6 illustrates the steady–state circulation pattern in the vicinity of the inflow for a constant eddy viscosity coefficient of 0.25 m^2s^{-1} across the entire flow domain. It can be appreciated that the ability to model flows in irregular geometries without implementing the usual 'staircase' approximation for the boundary curve justifies the increased complexity of the computational technique.

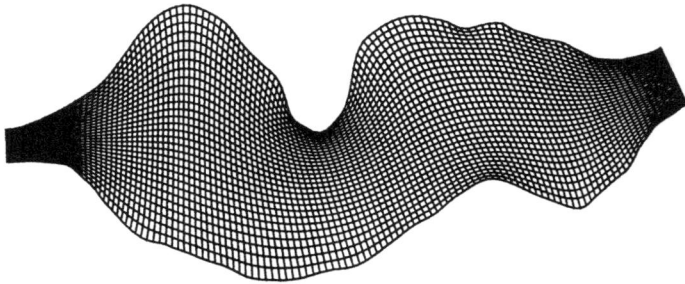

FIGURE 5 *Boundary–fitted coordinate system for an irregularly–shaped river*

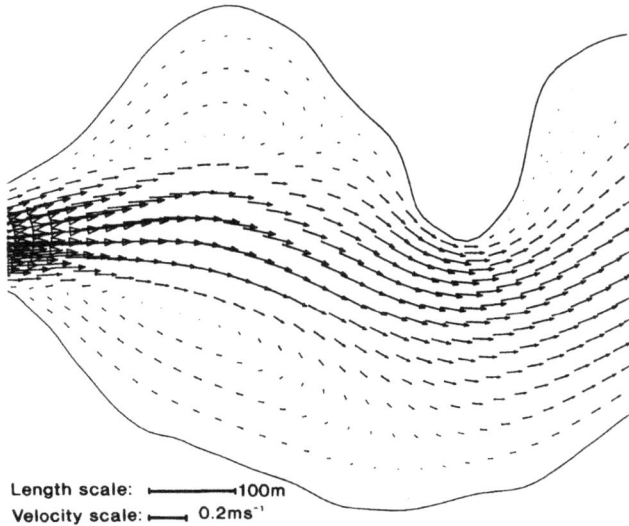

Length scale: ┣━━━━━┫100m
Velocity scale: ┣━━┫ 0.2ms⁻¹

FIGURE 6 *Velocity vectors at steady–state*

478

7 Conclusions and recommendations

The results indicate that the boundary-fitted shallow water equation methodology described in this paper has considerable promise for simulating the hydrodynamics in awkwardly shaped flow domains. It is therefore recommended that the curvilinear model should be refined to include a depth-averaged advection-diffusion equation for water quality and pollutant transport studies and a k-ϵ model to predict the spatial distribution of turbulence.

The next stage of the development process will be to use the numerical scheme to predict the tidal flow in an estuary. Instead of using a conventional flooding and drying technique to model the changes in wetted plan-form area of the estuary, a moving boundary-fitted coordinate system will be utilised which constantly adapts to the changing shape of the flow domain over each tidal cycle.

REFERENCES

[1]   Abbott, M.B., Damsgaard, A. and Rodenhuis, G.S. (1973), "System 21, 'Jupiter' (A design system for two-dimensional nearly-horizontal flows)", *J. of Hydraulic Research*, 11(1), pp. 1-28.

[2]   Barber, R.W. (1990), "Numerical modelling of jet-forced circulation in reservoirs using boundary-fitted coordinate systems", Ph.D. Thesis, University of Salford, England.

[3]   Borthwick, A.G.L. and Barber, R.W. (1990), "Prediction of low Reynolds number jet-forced flow inside a circle using the Navier-Stokes equations", *Int. J. Eng. Fluid Mech.*, 3(4), pp. 323-343.

[4]   Falconer, R.A. (1980), "Numerical modelling of tidal circulation in harbours", *J. Waterway, Port, Coastal and Ocean Div.*, Proc. ASCE, **106**, WW1, pp. 31-48.

[5]   Häuser, J., Paap, H.G., Eppel, D. and Mueller, A. (1985), "Solution of shallow water equations for complex flow domains via boundary-fitted coordinates", *Int. J. Num. Meth. in Fluids*, **5**, pp. 727-744.

[6]   Häuser, J., Paap, H.G., Eppel, D. and Sengupta, S. (1986), "Boundary conformed coordinate systems for selected two-dimensional fluid flow problems. Part 2: Application of the BFG method", *Int. J. Num. Meth. in Fluids*, **6**, pp. 529-539.

[7]   Johnson, B.H. (1980), "VAHM – a vertically averaged hydrodynamic model using boundary-fitted coordinates", Misc. Paper HL-80-3, U.S. Army Engineer Waterways Experiment Station Hydraulics Laboratory, Vicksburg, Mississippi, U.S.A.

[8]   Johnson, B.H. (1982), "Numerical modelling of estuarine hydrodynamics on a boundary-fitted coordinate system", in *Numerical Grid Generation*, J.F. Thompson (editor), Proc. of a Symp. on the Numerical Generation of Curvilinear Coordinate Systems and their Use in the Numerical Solution of Partial Differential Equations, held at Nashville, Tennessee, U.S.A., pp. 409-436.

[9] Kuipers, J. and Vreugdenhil, C.B. (1973), "Calculations of two–dimensional horizontal flow", Research Report S163, Part 1, Delft Hydraulics Laboratory, Delft, The Netherlands.

[10] Leendertse, J.J. (1970), "A water–quality simulation model for well–mixed estuaries and coastal seas: Vol. 1, Principles of computation", The Rand Corporation, RM–6230–RC.

[11] Mills, R.D. (1977), "Computing internal viscous flow problems for the circle by integral methods", *J. Fluid Mech.*, 79(3), pp. 609–624.

[12] Raghunath, R., Sengupta, S. and Häuser, J. (1987), "A study of the motion in rotating containers using a boundary–fitted coordinate system", *Int. J. Num. Meth. in Fluids*, 7, pp. 453–464.

[13] Sheng, Y.P. and Hirsh, J.E. (1984), "Numerical solution of shallow water equations in boundary fitted grids", Technical Memo. 84–15, Aeronautical Research Associates of Princeton Inc., Princeton, New Jersey.

[14] Stelling, G.S. (1983), "On the construction of computational methods for shallow water flow problems", Ph.D. Thesis, Delft University of Technology, Delft, The Netherlands.

[15] Stelling, G.S. and Willemse, J.B.T.M. (1984), "Remarks about a computational method for shallow water equations that works in practice", *Colloquium Topics in Applied Numerical Analysis*, J.G. Verwer (Editor), C.W.I. Syllabus No. 5, Centrum voor Wiskunde en Informatica, Amsterdam, The Netherlands, pp. 337–362.

[16] Thompson, J.F., Thames, F.C. and Mastin, C.W. (1974), "Automatic numerical generation of body–fitted curvilinear coordinate system for field containing any number of arbitrary two–dimensional bodies", *J. Comp. Physics*, 15, pp. 299–319.

[17] Vreugdenhil, C.B. and Wijbenga, J.H.A. (1982), "Computation of flow patterns in rivers", *J. Hyd. Div.*, Proc. ASCE, 108, HY11, pp. 1296–1310.

[18] Wijbenga, J.H.A. (1985), "Determination of flow patterns in rivers with curvilinear coordinates", Proc. 21st IAHR Cong., Melbourne, Australia, Aug. 1985. [reprinted as Publication No. 352, Delft Hydraulics Laboratory, The Netherlands, Oct. 1985.]

[19] Wijbenga, J.H.A. (1985), "Steady depth–averaged flow calculations on curvilinear grids", 2nd Int. Conf. on the Hydraulics of Floods and Flood Control, Cambridge, England, Sept. 1985, pp. 373–387.

[20] Willemse, J.B.T.M., Stelling, G.S. and Verboom, G.K. (1985), "Solving the shallow water equations with an orthogonal coordinate transformation", Presented at the Int. Symp. on Computational Fluid Dynamics, Tokyo, Sept. 1985. [reprinted as Delft Hydraulics Communication No. 356, Delft Hydraulics Laboratory, The Netherlands, Jan. 1986.]

40 A general mathematical model of tidal current in natural river

H. S. Jin, Y. M. Zheng and J. Ye

ABSTRACT

Based on the boundary–fitted orthogonal coordinate system and the grid "block" technique, a finite difference mathematical model for depth–averaged unsteady current in tidal river, in which the governing equations are solved by the time–splitting method with 3 steps, is developed. The model is calibrated and verified by some field data in Hwuangpu River, China. It is shown that the results computed by the present model agree fairly well with the field measurements.

1. Introduction

During the past decades, water quality modelling has been widely applied to engineering of water pollution control and water quality management[1]. However, a water quality model must be based on the correct description on hydrodynamics features, i.e., it is necessary to develop a good hydrodynamics model.

In numerical simulating of fluid flow, finite difference method is extensively used because it is simple and effective. However, the planar nature of river is generally complicated, its boundaries are irregular, it may also include some bends, shallows, sandbars and so on, and the length–width ratio is very great. Furthermore, the bank boundaries will vary gradually with fluctuations of tide, i.e., they are moving boundaries. Hence the discretization of the

481

computational domain and the application of numerical method can be very te-
dious and difficult for the development of a general computer code. Recent
years, great efforts for removing these problems have been made[2–5]. In this
paper, a numerically generated boundary–fitted orthogonal curvilinear
coordinate system and a grid "block" technique are used to settle the above difficulties. As a result, the numerical solving can be achieved in a fixed domain
which is a rectangle. Based upon the above techniques, a general
mathematical model for depth–averaged unsteady current in tidal river is presented.

2. Boundary–fitted orthogonal coordinate system and grid "block" technique

2.1 Boundary–fitted orthogonal coordinate system

Boundary–fitted coordinate system is one which coordinate lines completely coincide with the body boundaries. Generally, it is a nonorthogonal curvilinear
coordinate system. Nonorthogonal coordinate system can lead to great numerical errors in the computation. So we usually hope to achieve an orthogonal
curvilinear coordinate system to reduce the complexity and transformation errors involved in the solution of partial differential equations.
Many methods can be used to generate the boundary–fitted coordinate
system. One of the methods usually used is developed by Thompson who used
two Poisson equations to conduct the transformation between the physical
coordinate(x, y) and the boundary–fitted coordinate(ξ, η) system[4]. However
the orthogonal boundary–fitted coordinate system is not easily generated by the
Thompson method. For the generation of orthogonal coordinate system,
many scientists have presented a lot of methods[5–7]. Here, following two
elliptic equations are numerically solved to achieve a boundary–fitted
orthogonal curvilinear coordinate system.

$$\alpha^2 x_{\xi\xi} + x_{\eta\eta} + \alpha\alpha_\xi x_\xi - \frac{\alpha_\eta}{\alpha} x_\eta = 0 \tag{1}$$

$$\alpha^2 y_{\xi\xi} + y_{\eta\eta} + \alpha\alpha_\xi y_\xi - \frac{\alpha_\eta}{\alpha} y_\eta = 0 \tag{2}$$

in which, $\alpha = \dfrac{g_{22}}{g_{11}}$, $g_{11} = \sqrt{x_\xi^2 + y_\xi^2}$, $g_{22} = \sqrt{x_\eta^2 + y_\eta^2}$.

In transformed plane, taking ξ and η as monotonically increasing parameters,
the finite difference and the line by line iteration methods are employed to solve
the equations (1) and (2) with relevant Dirichelet boundary conditions, i.e.,
correspondent relations at the boundary grid points between the physical and
the transformed plane are given. Furthermore, following conditions are
adopted to fulfil the orthogonality more better at the boundaries,

$$x_\eta = -\alpha y_\zeta, \quad y_\eta = \alpha x_\zeta \tag{3}$$

If all nodes distributions on boundaries are kept unvaried, convergence rate will be low and orthogonality precision will be poor. In order to speed up the convergence rate and improve the orthogonality precision, the grid point distributions on a pair of boundaries (the upstream and the downstream boundaries of the domain) are throughout specified and the grid point distributions on the other pair of boundaries (the left and the right bank boundaries) are allowed to vary alternately along the relevant boundary for correcting the orthogonality. The algorithm is as below. (1) The node distribution on one boundary (e.g., the right bank) is specified and the distribution on the other boundary (the left bank) is allowed to alter along the boundary (the left bank), if the specified orthogonality precision is reached, the node distribution on the later boundary (the left bank) is fixed. (2) The node distribution on the former boundary (the right bank) is altered according to the orthogonality condition until the specified precision is satisfied. (3) If the specified ultimate convergence is reached, stop the numerical computation, otherwise go back step(1) and repeat the computation until convergence. Generally, only one cycle is needed for generating the nearly orthogonal coordinate system.
For the solution of 141×13 grids of the domain on fig.1(a) which is about the length 23.8km by the average width 400m (the maximum about 700m and the minimum about 300m), only about 120 iterations are required and the included—angles of coordinate lines are $85.7° \sim 94.3°$, i.e., a nearly orthogonal boundary—fitted curvilinear coordinate system is achieved.

2.2 Grid "block"

By the above coordinate transformation, the numerical computation can be carried out in a rectangle domain which is shown on fig.1(b), the representation of boundary conditions is more easy and simple, and the numerical methods well applied at present for paritial differential convection—diffusion equation can be used conveniently[8]. However, in the problems involved with moving boundary, it is very expensive that every new domain is transformed in order to adapt the variation of the boundaries and obtain a relevant boundary—fitted orthogonal coordinate system. In the present model, a grid "block" technique is introduced so that not only the computation can be conducted in a rectangle domain but also the computational cost is low.
Grid "block" means that no water will flow outfrom or into the grid and just as the grid is blocked. In the model, if river bottom at the water surface level node is above the water surface, i.e., the river bottom elevation is greater than the water surface level, the grid of which the node is at center is considered as a emergent one from water surface and the river bottom roughness(n) at the node will be given a very great number ($n = 10^{10}$, instead n equals about 0.015 at the normal state). Thus, because the resistance in the grid is enormous, the velocities at the grid periphery computed by the momentum equations will

(a) The physical plane

(b) The transformed plane

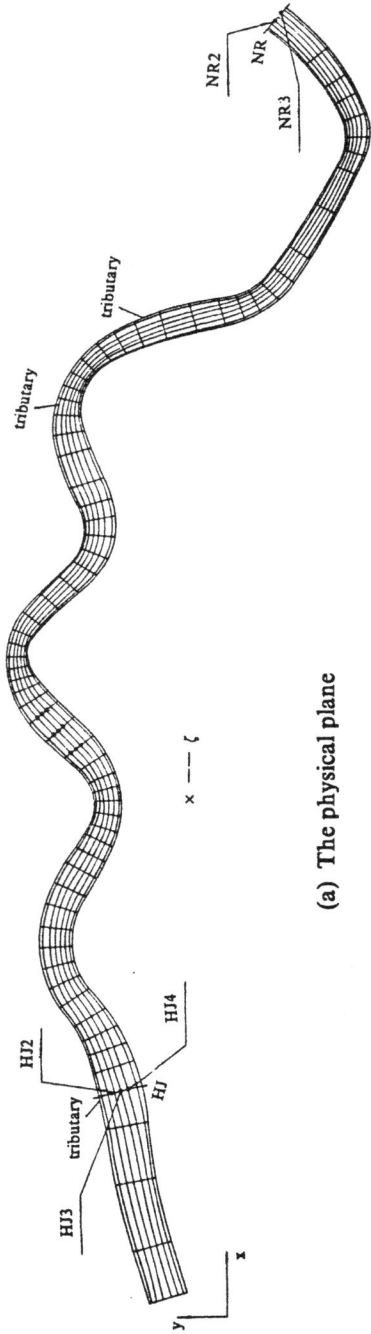

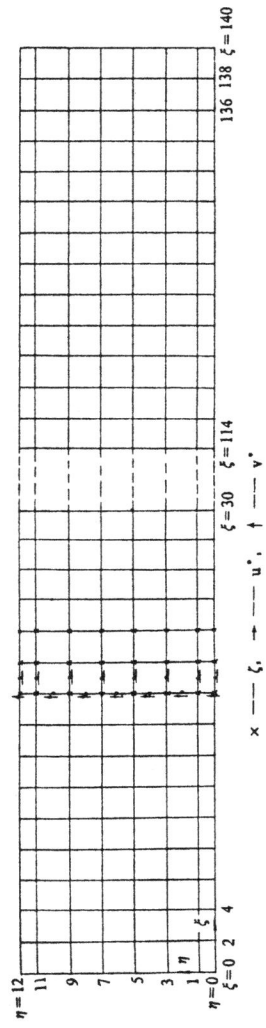

FIGURE 1 *Computational domain and grids*

approximately equal zero, i.e., the grid is blocked. Comparing the river bottom elevation at a node with the water surface level at the deepest channel of cross section, we can judge whether the bottom is emergent from water surface or not.

Using the grid " block" technique, the grid at which the river bottom is emergent from the water surface can be included in the computational domain, provided altering the resistance at the node which is at center of the grid and the numerical solving can be carried out in an unvaried domain. Although the computational load will be increased, the benefit obtained from the technique is distinct and the computational cost is still cheaper than that by some other methods.

3. Mathematical model

The boundary–fitted orthogonal curvilinear coordinate system and the grid "block" technique allow us to develop a general computer code for depth–averaged tidal current in natural water environments.

3.1 Governing equation

The governing equations consist of a depth–averaged continuity equation and two depth–averaged momentum equations. In the orthogonal curvilinear coordinated system, the equations can be expressed as follows:

$$\frac{\partial \zeta}{\partial t} + \frac{1}{J}[\frac{\partial(u^{\bullet} h g_{22})}{\partial \xi} + \frac{\partial(v^{\bullet} h g_{11})}{\partial \eta}] = 0 \tag{4}$$

$$\frac{\partial u^{\bullet}}{\partial t} + \frac{u^{\bullet}}{g_{11}}\frac{\partial u^{\bullet}}{\partial \xi} + \frac{v^{\bullet}}{g_{22}}\frac{\partial u^{\bullet}}{\partial \eta} + \frac{u^{\bullet} v^{\bullet}}{J}\frac{\partial g_{11}}{\partial \eta} - \frac{v^{\bullet 2}}{J}\frac{\partial g_{22}}{\partial \xi} =$$

$$-\frac{g}{g_{11}}\frac{\partial \zeta}{\partial \xi} - \frac{g u^{\bullet} \sqrt{u^{\bullet 2} + v^{\bullet 2}}}{C^2 h} + \frac{\varepsilon_t}{h}(\frac{1}{g_{11}}\frac{\partial A}{\partial \xi} - \frac{1}{g_{22}}\frac{\partial B}{\partial \eta}) + f v^{\bullet} \tag{5}$$

$$\frac{\partial v^{\bullet}}{\partial t} + \frac{u^{\bullet}}{g_{11}}\frac{\partial v^{\bullet}}{\partial \xi} + \frac{v^{\bullet}}{g_{22}}\frac{\partial v^{\bullet}}{\partial \eta} + \frac{u^{\bullet} v^{\bullet}}{J}\frac{\partial g_{22}}{\partial \xi} - \frac{u^{\bullet 2}}{J}\frac{\partial g_{11}}{\partial \eta} =$$

$$-\frac{g}{g_{22}}\frac{\partial \zeta}{\partial \eta} - \frac{g v^{\bullet} \sqrt{u^{\bullet 2} + v^{\bullet 2}}}{C^2 h} + \frac{\varepsilon_t}{h}(\frac{1}{g_{22}}\frac{\partial A}{\partial \eta} + \frac{1}{g_{11}}\frac{\partial B}{\partial \xi}) - f u^{\bullet} \tag{6}$$

in which,

$$A = h(\frac{1}{g_{11}} \frac{\partial \overset{\bullet}{u}}{\partial \xi} + \frac{1}{g_{22}} \frac{\partial \overset{\bullet}{v}}{\partial \eta}), \quad B = h(\frac{1}{g_{11}} \frac{\partial \overset{\bullet}{v}}{\partial \xi} - \frac{1}{g_{22}} \frac{\partial \overset{\bullet}{u}}{\partial \eta}),$$

u^* (and v^*) is depth−averaged velocity along ζ (and η) direction in transformed plane,

$$u^{\bullet} = (ux_{\xi} + vy_{\xi})/g_{11}, \quad v^{\bullet} = (ux_{\eta} + vy_{\eta})/g_{22},$$

u (and v) is depth−averaged velocity along x (and y) direction in the physical plane, ζ and h are respectively water surface level and water depth, $h = \zeta - Z_b$, Z_b is river bottom elevation, ε_t is horizontal coefficient of eddy and molecular diffusivity, g and f are respectively the gravitational acceralation and the Coriolis coefficient, C is Chezy resistance coefficient, $C = \frac{1}{n} h^{1/6}$, J is the Jacobian relation of the transformation, $J = x_{\xi} y_{\eta} - x_{\eta} y_{\xi} = g_{11} g_{22}$ for the orthogonal coordinate transformation.

3.2 Numerical scheme

In orthogonal curvilinear coordinate system, numerical schemes well used at present can be adopted directly, on condition that relevant geometry feature is considered.

According to the feature of the convection−diffusion equation, a "staggered" grid system, in which the nodes for velocities and water surface level do not co-incide with each other, is set up in the computational domain. The following numerical scheme is based on the time−splitting method, with 3 steps: advection, diffusion and propagation along with source terms.

Advection step

$$\frac{\partial \overset{\bullet}{u}}{\partial t} + \frac{\overset{\bullet}{u}}{g_{11}} \frac{\partial \overset{\bullet}{u}}{\partial \xi} + \frac{\overset{\bullet}{v}}{g_{22}} \frac{\partial \overset{\bullet}{u}}{\partial \eta} = 0 \tag{7}$$

$$\frac{\partial \overset{\bullet}{v}}{\partial t} + \frac{\overset{\bullet}{u}}{g_{11}} \frac{\partial \overset{\bullet}{v}}{\partial \xi} + \frac{\overset{\bullet}{v}}{g_{22}} \frac{\partial \overset{\bullet}{v}}{\partial \eta} = 0 \tag{8}$$

This step is solved with an implicit characteristics method.

Diffusion step

$$\frac{\overset{\bullet}{\bar{u}} - \overset{\bullet}{\hat{u}}}{\Delta t} = \frac{\varepsilon_t}{h} (\frac{1}{g_{11}} \frac{\partial A}{\partial \xi} - \frac{1}{g_{22}} \frac{\partial B}{\partial \eta}) \tag{9}$$

486

$$\frac{\tilde{v}^{\bullet} - \hat{v}^{\bullet}}{\Delta t} = \frac{\varepsilon_t}{h}(\frac{1}{g_{22}}\frac{\partial A}{\partial \eta} + \frac{1}{g_{11}}\frac{\partial B}{\partial \zeta}) \tag{10}$$

where, Δt is the computational time step, $\hat{u}^{\bullet}$ and $\hat{v}^{\bullet}$ are the result of advection step, $\tilde{u}^{\bullet}$ and $\tilde{v}^{\bullet}$ will be the result of this diffusion step. This step is solved with an implicit spatial central difference method.

Propagation step

$$\frac{\zeta^{n+1} - \zeta^n}{\Delta t} + \frac{1}{J}[\frac{\partial(u^{\bullet} hg_{22})}{\partial \zeta} + \frac{\partial(v^{\bullet} hg_{11})}{\partial \eta}] = 0 \tag{11}$$

$$\frac{u^{\bullet n+1} - \tilde{u}^{\bullet}}{\Delta t} + \frac{u^{\bullet} v^{\bullet}}{J}\frac{\partial g_{11}}{\partial \eta} - \frac{v^{\bullet 2}}{J}\frac{\partial g_{22}}{\partial \zeta} =$$

$$-\frac{g}{g_{11}}\frac{\partial \zeta}{\partial \zeta} - \frac{gu^{\bullet}\sqrt{u^{\bullet 2} + v^{\bullet 2}}}{C^2 h} + fv^{\bullet} \tag{12}$$

$$\frac{v^{\bullet n+1} - \tilde{v}^{\bullet}}{\Delta t} + \frac{u^{\bullet} v^{\bullet}}{J}\frac{\partial g_{22}}{\partial \zeta} - \frac{u^{\bullet 2}}{J}\frac{\partial g_{11}}{\partial \eta} =$$

$$-\frac{g}{g_{22}}\frac{\partial \zeta}{\partial \eta} - \frac{gv^{\bullet}\sqrt{u^{\bullet 2} + v^{\bullet 2}}}{C^2 h} - fu^{\bullet} \tag{13}$$

where, ζ^{n+1}, $u^{\bullet n+1}$ and $v^{\bullet n+1}$ are respectively ζ, $u^{\bullet}$ and $v^{\bullet}$ at time $t^{n+1} = (n+1)\Delta t$, ζ^n is ζ at time $t^n = n\Delta t$. In this last step, a relation between $u^{\bullet n+1}$ and ζ^{n+1} and a relation between $v^{\bullet n+1}$ and ζ^{n+1} are first derived from the equations (12) and (13), then combining the relations with the continuity equation(11) and solving the equation we can obtain the water surface level at time t^{n+1}, i.e., ζ^{n+1}, last, the velocities at time t^{n+1} ($u^{\bullet n+1}$, $v^{\bullet n+1}$) can be solved.
In every time step(Δt), the above procedure is carried out and we can know the time varying velocity fields and water surface level.
The computation shows that the effect of momentum diffusion is unimportant for intense tidal current and it can be ignored.

4. Application

The above mathematical model is employed to simulate the tidal current in Hwuangpu River, China. The computational domain and the boundary–fitted orthogonal curvilinear meshes in physical plane are shown on fig.1(a). The computational meshes in transformed plane is displayed on fig.1(b).

The boundary conditions for the model are given. On the upstream and the downstream boundaries, water surface level is specified (of course, we can also specify the velocity instead of the water level). On the bank boundaries, unpenetrable condition is used, however, if any tributary joins with the river, the velocities at the joint will be specified according to discharge and geometry feature of the tributary.

The initial conditions for the model are approximately determined, in which water surface level at interior node is obtained by interpolating between the water surface level on the upstream and the downstream boundaries at time $t = 0$ and velocity at interior node are always taken as zero. During the computation, the impact of initial condition on the output will be eliminated quickly for the tidal current simulation.

Time step(Δt) equals 120s and the maximum Courant number is about 60 (i.e.,

$$(\Delta t \sqrt{gh} \sqrt{\frac{1}{\Delta x^2} + \frac{1}{\Delta y^2}})_{max} \approx 60)$$ and it is far greater than the number reached

by some other numerical schemes.

The above model is calibrated with field data during 5 tides and then verified with field data during 12 tides. The field measurements include water surface level at one site, velocities at five sites and discharges through two cross sections within the basin. Fig.2, 3, 4 and 5 are respectively displayed the computational results and field data of water surface level, velocity and discharge during last 9 tides.

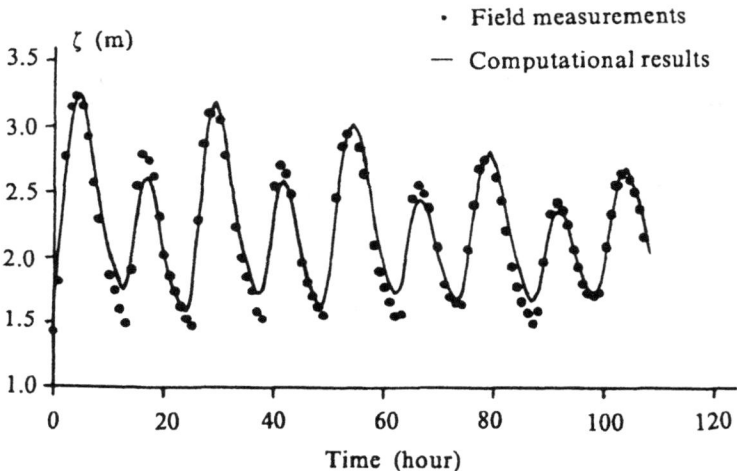

FIGURE 2 *Water surface level at HJ3*

488

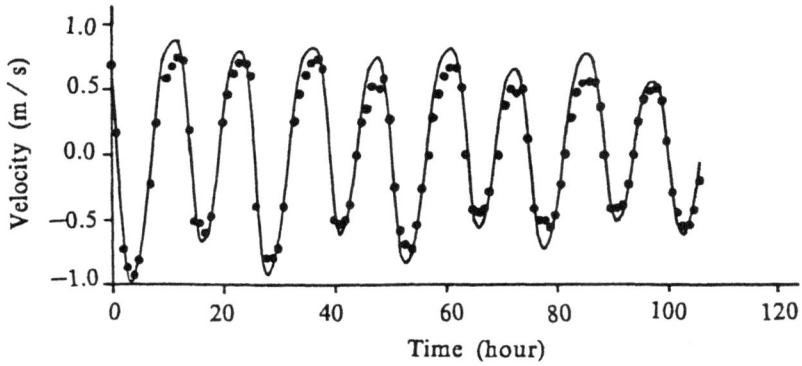

HJ2

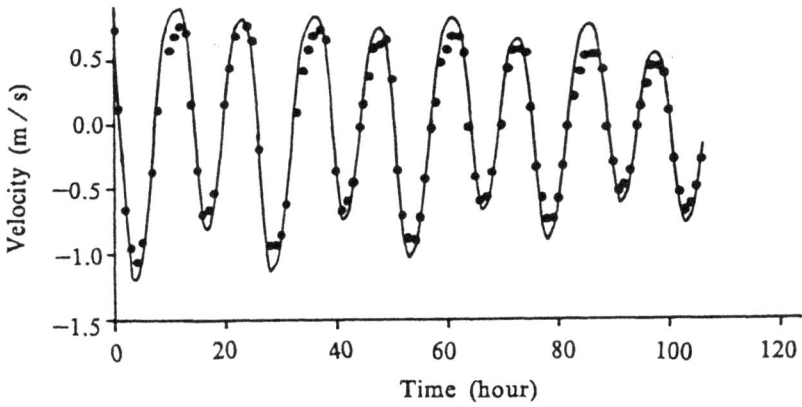

HJ3

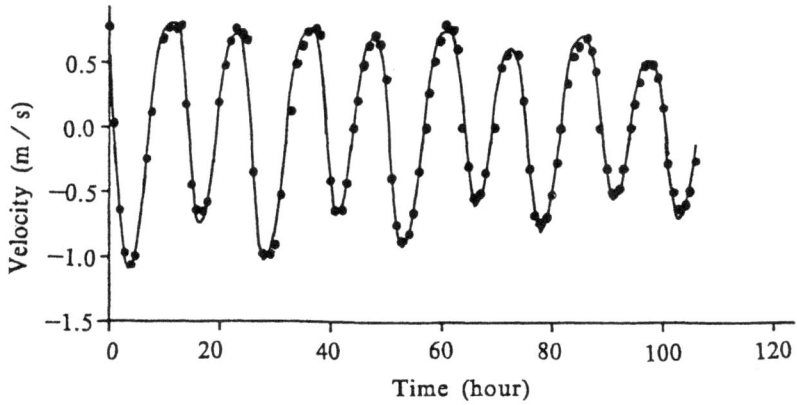

HJ4

FIGURE 3 *Velocity at HJ* (• Field measurements, —Computational results)

489

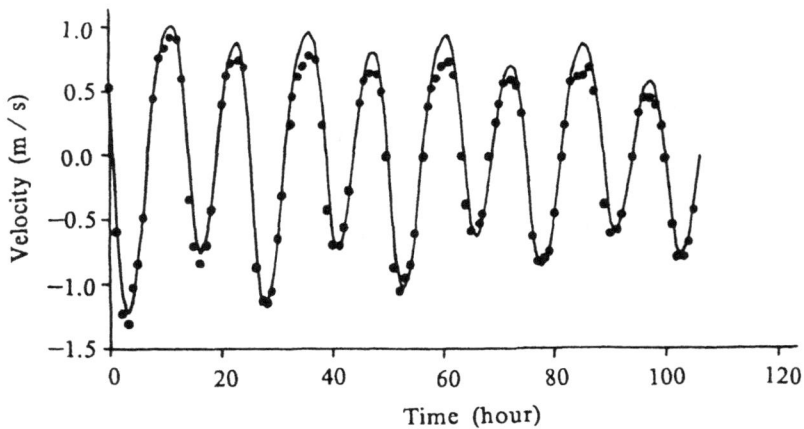

NR2

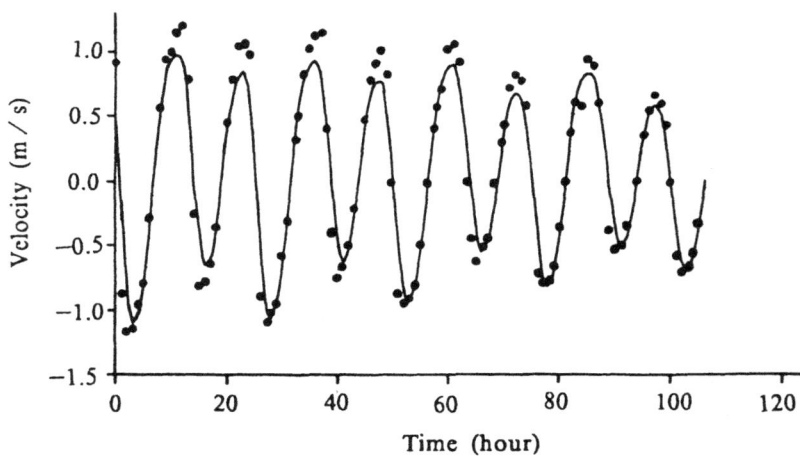

NR3

FIGURE 4 *Velocity at NR* (· Field measurements, — Computational results)

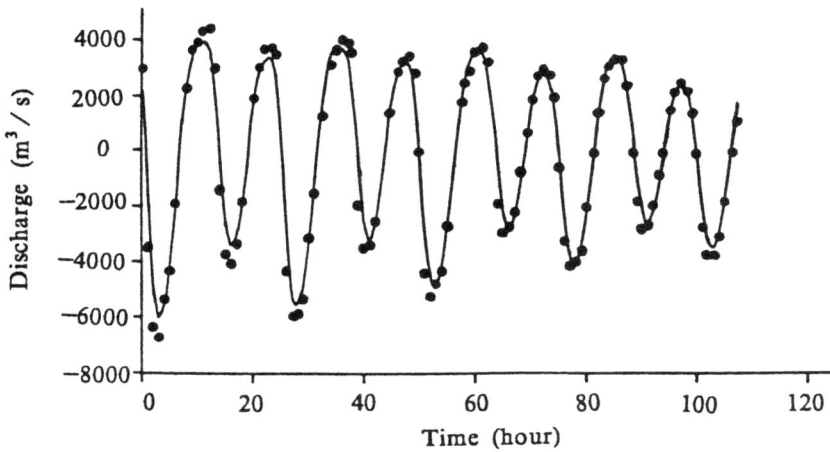

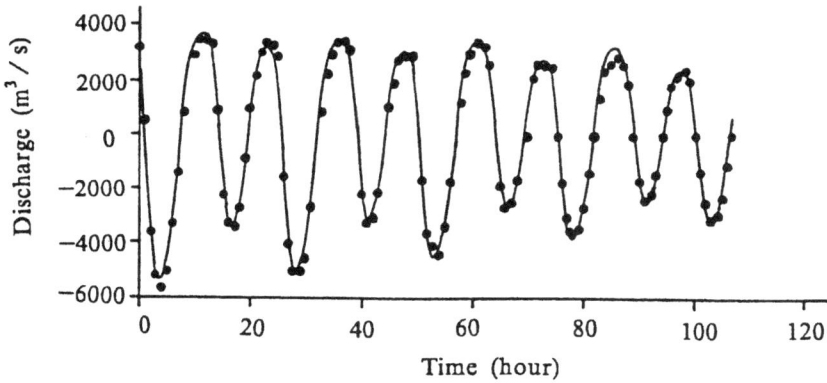

FIGURE 5 *Discharge* (· Field mesurements, — Computational results)

The above four figures show that the computational results by present model coincide fairly well with the field measurements on water surface level, velocity
and discharge.

5. Conclusion

The difficulties resulted from irregular geometry shape of the computational domain can be removed by the boundary—fitted coordinate transformation, especially the orthogonal coordinate transformation, and the finite difference method can be conveniently used again in numerical computation of nature flow. The problem with moving boundary can be turned into one with established boundary and the computational cost will be economized. Based upon the boundary—fitted orthogonal coordinate system and the grid "block" technique, a general mathematical model for depth—averaged unsteady current in tidal river can be built. The model is applied to a tidal river with intense tidal current in China. The computational results show that the present model can correctly simulate the depth—averaged unsteady current in tidal river.

REFERENCE

[1] Zhang,S.N., (1988), *Environmental Hydrodynamics*, Hohai University Press(in Chinese).
[2] Sheng,Y.P., (1986), Numerical Modelling of Coastal and Estuarine Processes Using Boundary—Fitted Grids, *Proceeding of Third International Symposium on River Sedimentation*, The University of Mississippi, March 31—April 4.
[3] Cheng,W.H., Wang,C.H., (1988), The Calculation of Flow Pattern in Rivers by the Orthogonal Curvilinear Coordinates and "Condensation" Technique, *Journal of Hydraulic Engineering*, No.6(in Chinese).
[4] Thompson,Joe F., et al, (1982), Boundary—Fitted Coordinate Systems for Numerical Solution of Partial Differential Equations——A Riview, *Journal of Computational Physics*, Vol.47.
[5] Thompson,Joe F., et al, (1985), *Numerical Grid Generation——Foundations and Applications*, Elsevier Science Publishing Co.,Inc., New York.
[6] Chen,C.J., Obasih,K.M., (1986), Numerical Generation of Nearly Orthogonal Boundary—Fitted Coordinated System, *Advancements in Aerodynamics, Fluid Mechanics and Hydraulics, Proceedings of Specialty Conference*, Minneapolis, Minnesoda, June 3—6.
[7] Yu Liren, et al, (1988), Generation of Orthogonal Body—Fitted Coordinated System, *Journal of Hohai University*, Vol.16, No.5(in Chinese).
[8] Patankar,S.V., (1980), *Numerical Heat Transfer and Fluid Flow*, Hemisphere Publishing Corporation and McGraw Hill Book Company.

41 Numerical model for estuary with compound cross-section

C. J. Lai and J. W. Yen

ABSTRACT

A depth averaged numerical model using depth controlled boundary fitted grids, k-ε model of turbulence, is formulated for simulating tidal flows in the Taitui Estuary. The tidal range at the estuary is around 4 meters and the cross-section of the Estuary are composite shapes. A proposed boundary judgement scheme in conjunction with the depth control grid are used for the tidal flow computation. The calculations are compared with a field measured data. The pattern computed are found to be in good agreement with field results in both ebb and flood tide. The depth averaged k-ε model of turbulence with standard coefficients is further proved can be used in the practical scale.

1. Introduction

Taitui Estuary is situated in the central part of Taiwan. Due to the refraction of tidal waves in Taiwan Strait, the tidal range at the nearby regions are around 4-5 meters. In the last decade several engineering works have been constructed in this region, and the Estuary has subjected noticeable morphological and environmental changes. Since more works are to be constructed in the north and south of the Estuary, their effects on the coastline and Estuary thus have to be fully understood. The impact of the engineering works on flow pattern is one of the major interests of engineers and the effective way to obtain those is to use numerical model computation.

Judging from the irregular boundary, composite cross-section and flow complexity of the Estuary, the steady state model "DIM-21S" reported by Lai, et.al(1991) was extended for the computation. DIM-21S is a depth averaged model using boundary fitted coordinates and model of turbulence for the grid system and eddy viscosity respectively. In extension of the model with unsteadiness, the variation of domain during different tidal stages caused inconvenient in the computation, and the normal boundary fitted grid could work only at high stages. This problem was eventually overcomed by introducing depth controlled grids and a boundary judgement scheme. The model thus was successful, very stable and produced good results for both ebbing and flooding tides.

Use this model, the tidal flow in Taitui Estuary were computed for different tidal cycles. The model, a general description on the Estuary, the field measurement of Lai, et.al (1990) and the computed results for one cycle are presented in this paper.

2. Description of Field Measurements

In 1989, Tainan Hydraulic Laboratory (THL) worked jointly with Danish Hydraulic Institute(DHI) on a thermal discharge project. THL performed field measurements at the Taitui Estuary at Feb. 14, Feb.24 and July 1-2 on salinity, Suspended sediment concentration Lagrangian velocity and bed materials grain size distributions. The cross-sections and depth of the channel were surveyed and typical examples are given in Figure 1. The cross-sections in the Figure show the Estuary is a single deep channel estuary with a broad tidal flat extend from the deep channel to dikes at both sides. The deep channel meanders within the dike and the bed topography are complex.

From the observation of THL, it was concluded that the tide within the estuary varied synchronously with the nearby Taichung Harbor which has four major astronomical constituents as:

TABLE 1 *Tide Constinents of Taitui Estuary*

	K1	O1	M2	S2.
amp.	26.47	21.13	172.17	59.74 cm
Phas.	197.13	100.76	120.08	9.58 degree.

The tide form ratio F defined by (K1+O1)/(M2+S2) was 0.2052 that the tide pattern here is semi-diurnal dominated. The bed material were in general sand and mud with median grain sizes range from 0.06 to 0.4 mm, and distributed over the deep and shallow water regions respectively. The Lagragnian velocity using drogue tracking showed the local velocity might reach 1.9 m/s in the deep channel at entrance of the Estuary. Typical examples for ebb and flood tides are given in Figure 2. Velocity and suspended profiles measurement suggested the estuary is partially mixed, more mixed at flooding stage. The normalized sediment concentration profiles did not vary much with tide, but the mean value for the normalizing increased with tide elevation.

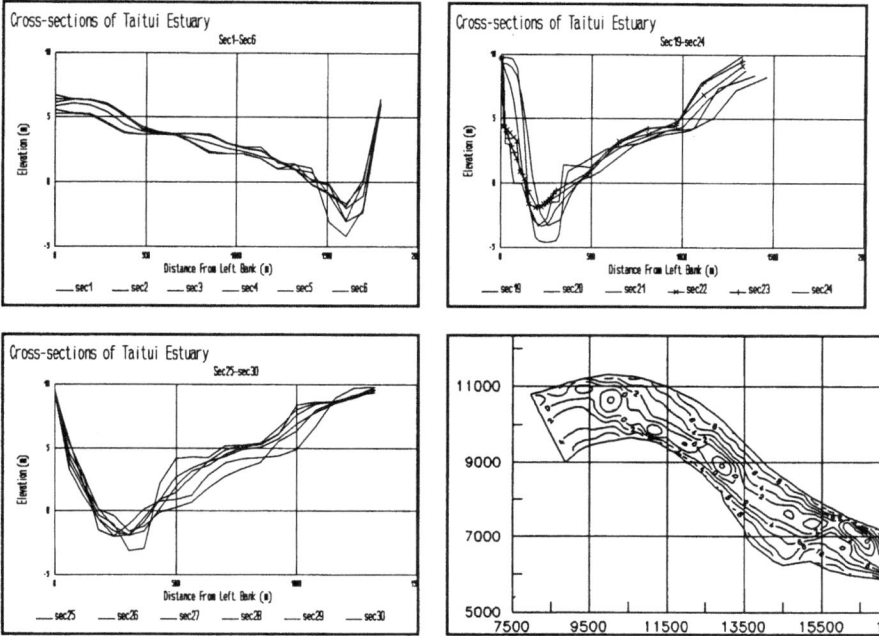

FIGURE 1 *Cross-sections and Contours of Taitui Estuary*

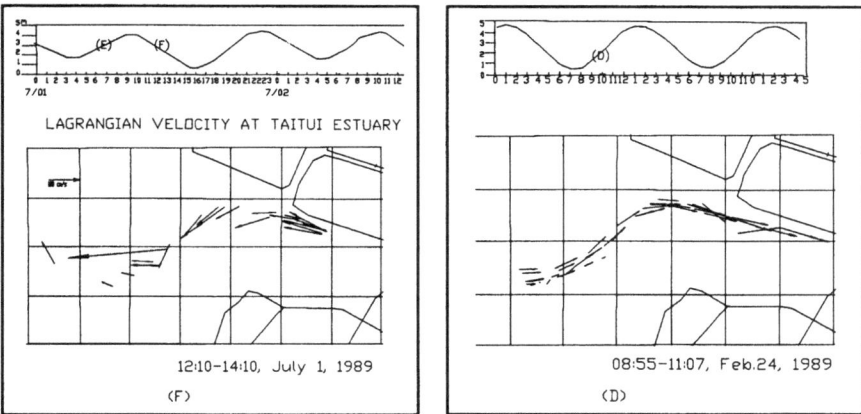

FIGURE 2 *Observed Langargian Velocities in June, 1989[4]*

3. The Computation Model

495

3.1 The governing equations

Since Taitui Estuary has a channel with a rather large aspect ratio, i.e. width to depth ratio, any flow property, Φ, may be sufficiently described by a depth averaged value ϕ defining by

$$\phi = \frac{1}{h} \int_{z_b}^{z_b+h} \Phi \, dz \tag{1}$$

The transport equation for ϕ in cartesian coordinates has the form as:

$$\frac{\partial}{\partial t}(\rho h \phi) + \frac{\partial}{\partial x^i}(\rho h v^i \phi) = \frac{\partial}{\partial x^i}(h \Gamma^\phi \frac{\partial \phi}{\partial x^i}) + S^\phi, \qquad i = 1, \ 2. \tag{2}$$

The equation is transformed into the boundary fitted coordinates and in two dimensions written as:

$$\frac{\partial}{\partial t}(\sqrt{g}\rho\phi) + \frac{\partial}{\partial \xi^i}(\sqrt{g}\rho h V^i \phi) = \frac{\partial}{\partial \xi^i}(\sqrt{g}g^{ii}\rho h \Gamma^\phi \frac{\partial \phi}{\partial \xi^i}) + S^\phi \tag{3}$$
$$i = 1,2 \quad ; \phi = U, \ V, \ k, \ \epsilon \quad ; \ \xi^1 = \xi \ ; \quad \xi^2 = \eta$$

$$S^\phi = \sqrt{g}R^\phi + \frac{\partial}{\partial \xi}(\sqrt{g}g^{12}h\Gamma^\phi \frac{\partial \phi}{\partial \eta}) + \frac{\partial}{\partial \eta}(\sqrt{g}g^{21}h\Gamma^\phi \frac{\partial \phi}{\partial \xi}) \tag{4}$$

R^ϕ are source terms and their values for different properties, such as u,v, k and ϵ are given in Table 2. ϕ value in 1 representing the continuity equation. The Jacobian, metric components for transformation and the contravariant velocities U and V are respectively given by:

$$\sqrt{g} = x_\xi y_\eta - x_\eta y_\xi \ ; \quad g^{11} = \frac{x_\eta^2 + y_\eta^2}{g} \ ; \quad g^{22} = \frac{x_\xi^2 + y_\xi^2}{g}$$
$$g^{21} = g^{12} = -\frac{x_\xi x_\eta + y_\xi y_\eta}{g} \ , \tag{5}$$
$$U = \frac{u y_\eta - v x_\eta}{\sqrt{g}} \ ; \quad V = \frac{v x_\xi + u y_\xi}{\sqrt{g}}$$

Based on Equation (3), the numerical discretization can be performed. Details of the diffusion and sources terms are also given in Table 2.

3.2 The Turbulence Model

The eddy viscosity approximation in depth averaged form proposed by McGuirk & Rodi(1978) is used to describe turbulent diffusion process. The eddy viscosity and turbulent stresses are respectively given by:

$$\mu_t = \rho C_\mu \frac{k^2}{e} \quad ; \quad \tau_{ij}^t = \mu_t (\frac{\partial u_i}{\partial x_j} + \frac{\partial u_j}{\partial x_i}) \quad ; \quad i, j = 1, 2 \quad (6)$$

Depth averaged k and ϵ value are found through the solution of k, ϵ transport equations. Standard values are used in the turbulence model and they are listed in Table.2

TABLE 2 *Diffusion, Source Terms and Turbulence Model Coeffcients in Transport Equations*

ϕ	Γ^ϕ	S^ϕ
1	0	0
u	μ_e	$\sqrt{g}R^u + \rho Gh(y_\xi h_\eta - y_\eta h_\xi) + C^u$
v	μ_e	$\sqrt{g}R^v + \rho Gh(x_\eta h_\xi - x_\xi h_\eta) + C^v$
k	μ_t/σ_k	$\sqrt{g}(hP_k - C_D \rho h\epsilon + \rho hP_{kv}) + C^k$
ϵ	μ_t/σ_ϵ	$\sqrt{g}(\epsilon/k)[C_1 hP_k - C_2 \rho h\epsilon + C_3 h(P_k^2/\rho\epsilon - P_k)] + \rho hP_{ev} + C^\epsilon$

C_μ	C_D	C_1	C_2	σ_k	σ_ϵ
0.09	1.00	1.44	1.92	1.00	1.30

$$R^u = \frac{\partial}{\partial x}(h\mu_e \frac{\partial u}{\partial x}) + \frac{\partial}{\partial y}(h\mu_e \frac{\partial v}{\partial x})$$

$$R^v = \frac{\partial}{\partial x}(h\mu_e \frac{\partial u}{\partial y}) + \frac{\partial}{\partial y}(h\mu_e \frac{\partial v}{\partial y})$$

$$P_k = \mu_t[2(\frac{\partial u}{\partial x})^2 + 2(\frac{\partial v}{\partial y})^2 + (\frac{\partial u}{\partial v} + \frac{\partial v}{\partial x})^2]$$

3.3 Numerical Discretization of the Governing Equations

Equation (3) is discretized into the computational domain using the finite control volume concept. The non-staggered arrangement to the variables. u, v, h, k, ϵ is adopted in the present model. The power-law scheme of Partankar (1979) is used in treating the convective and diffusive fluxes, and fully implicit scheme is used for the time domain. Use these procedure, the discretized equations take the form of:

$$A_p \phi_p^{n+1} = \Sigma A_{nb} \phi_{nb}^{n+1} + B_\phi^{n+1} \quad (7)$$

and are ready to be solved by iterative procedure.

3.5 Grid System and Methods of Generation

The elliptic generation system propose by Thomsen, et.al (1985) and the control function determination process of Middlecoeff & Thomas (1979) are adopted to generate the boundary fitted grids. For the depth control grid, deep channel and shallow zones were generated separately and the grids fits the deep and shallow channels reasonable well. Figure 3 shows grid system of the Estuary generated with and without depth control.

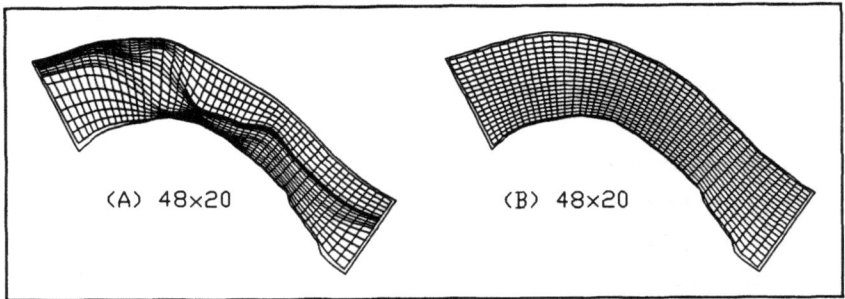

FIGURE 3 *Boundary Fitted Grid Systems for Computation:*
(A) Without Depth Control, (B) With Depth Control, Normal

3.6 Water Depth Correction and Boundary Judgement

During the iteration process of finding solution at every time step, the water depth is initially guessed and subsequently corrected until a converged value is achieved. The guessed water levels are corrected by assuming a hydrostatic pressure distribution in vertical and using "pressure linked" procedure of SIMPLE scheme. In avoiding the pressure oscillation due to the non-staggered grid system used in this model, Rhie & Chow's (1983) scheme is further adopted. At every iteration after the correction of water depth, the water levels on the left and right of the main channel are then used and compared to the bed elevation. Then the boundaries are set at the grids where their adjacent (left on right) grids have negetive water depth. The computation domain at the next iteration thus is fixed within positive water depth and the procedure works fine with the depth controlled grids. However, the normal grid tend to have larger boundary variations between two adjacent cross-section. Large gradient exists between fixed boundary and flow and the procedure does not work well at low stage and causes unstable in computation.

3.7 Boundary and Initial Conditions

498

On the wall boundary The wall boundaries vary at every iteration and time step and these solid boundaries may exist on east,west,north and south faces of one control volume. On these boundaries, the gradients of u,v,k,ε are generally high and require a so call modification procedure to the computation. The wall function suggested by Launder & Spalding (1974) is used in the first gird point. At the control volume surface, the depth is extrapolated from surrounded control grid points and no flux penetrating conditions are given at these solid boundaries.

On the upstream and downstream boundaries The downstream boundary condition is specified at the entrance of the Estuary at which water elevation are specified by the tidal curve generated from the K1,O1,M2 and S2 constituents in Table 1. At the upstream river side the v,k,ε and h values are extrapolated from the adjacent grid points. A constant river flow in 0.25 m/s is set to the longitudinal velocity u.

On bed boundary The universal resistance law is used. This law gives bed shear stress as

$$\tau_b^i = \rho C_f v^i q , \quad i=1,2 \tag{8}$$

where C_f is the friction coefficient depending on bed roughness, it is determined by

$$C_f = \frac{n^2 g}{h^{1/3}} \tag{9}$$

,where a value of 0.025 is used for the Manning coefficients in the equation.

Initial condition The steady state values within the computation domain based on the first time step. i.e., T=0.0 hr is used as the initial condition.

4   Model application and Result

4.1   Model computation

With model and the judgement scheme constructed, the flow in the Estuary was computed on a Hp 920 work station. The time step was 1800 sec with the criteria or convergence as 5×10^{-3} for the maximum residual. The numbers of iteration for different time steps were usually less than 500, but at lowest tide, the computation took more iteration to achieve convergence. The computation was proceeded to T=25 hrs, and the total computational time was 8 hrs.

4.2   Time history of water level in Estuary

Figure 4 shows the computed water level within the Estuary, in which the bed topography of the main channel is also plotted. The bed topograph shows high irregularities at various cross-section. The ebb stage occurs at T=1 hr. to T=5.5 hr and water level decreases from EL+3.9 to 0.65. The flood stage occurs between T=6.0 hrs to T=11.5 hrs and the water level rises from +0.68 to +4.03. The rise of the bed elevation in main channel at location

6 KM (cross-section 5)from the entrance of Estuary introduces a control sections to the flow. One can observe the water surface between T=3.0 and 7.5 hrs, the water surface upstream of the control point forms a M2 curve and tide does not affect the flow. This can also observed in Figure 4(c) in which the water level at sections 5 and 6 keep unchanged between T=3.5 hrs to 7.0 hrs, and between T=16 hrs to 20 hrs.

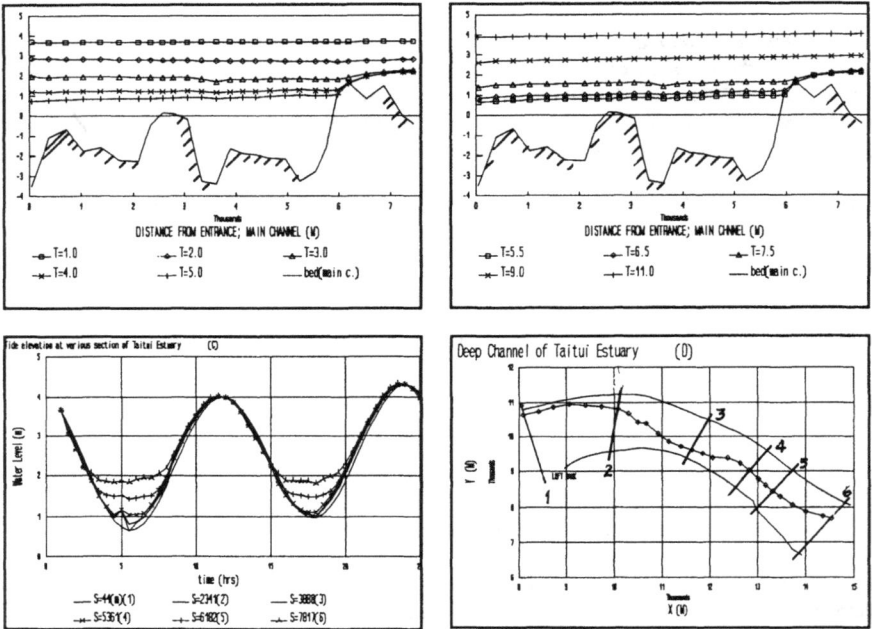

FIGURE 4   Computed Water Level at various time steps

4.3   Flow at Ebb Tide

The computed flows at ebb tide are shown in 1 hr interval in Figure 5. One interesting pattern can be noted at cross-section 3 at which flow pass through the sand bar and produces reverse flow with high velocities. The strongest current occurs at T=3.0 hrs. At a high stage, this pattern dispears.

4.4   At Flood tide

The flows in the Estuary are shown in 1 hour interval in Figure 6. The reverse and complex flow field at section 3 exists till 8 hrs. The tide flows from the sea into the Estuary through the main channel and at T=9.0 hrs, it begins to flow over the shallow region. The enlarged plots for t=9 and 10 hrs shown on Figure 7 indicate the velocity difference between deep and shallow channel produces shear layer, and vortex street at the edge of the deep channel.

500

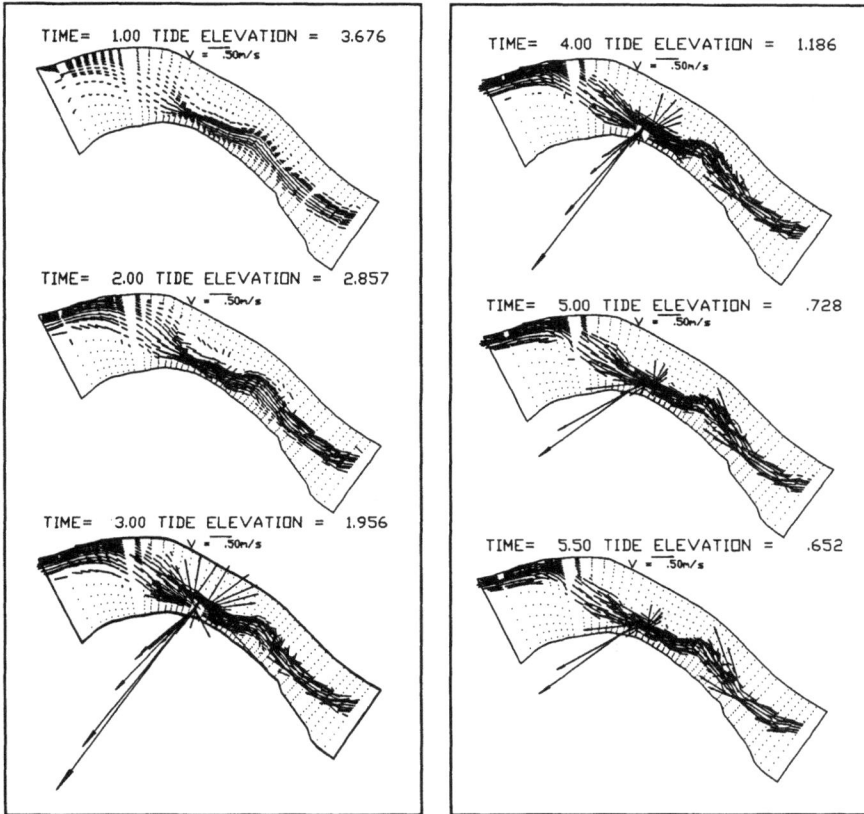

FIGURE 5 *Ebb Tide Current from T=1.0 Hrs to T=5.5 Hrs*

4.5 *Comparisons for Results of Normal and Depth control Grids*

Figure 8(A)(B) show the flow fields for T = 11.50 hrs computed using models with the normal and depth controlled grids respectively. The flow in the normal grids tends to have gentle gradient between deep and shallow zones. The circulation observed in the normal grid between section 3,4 is not so clear in the depth control grid since a stronger longitudinal velocity in the deep channel reduces the strength of circulation. On the contrary, the side inflow to the deep channel observed at a region between sections 2 and 3 in results of depth control model is stronger than that of the normal grid. Since there is no data to assess which is the correct result, intuitively the depth control grid seems to give pattern more close to the real case.

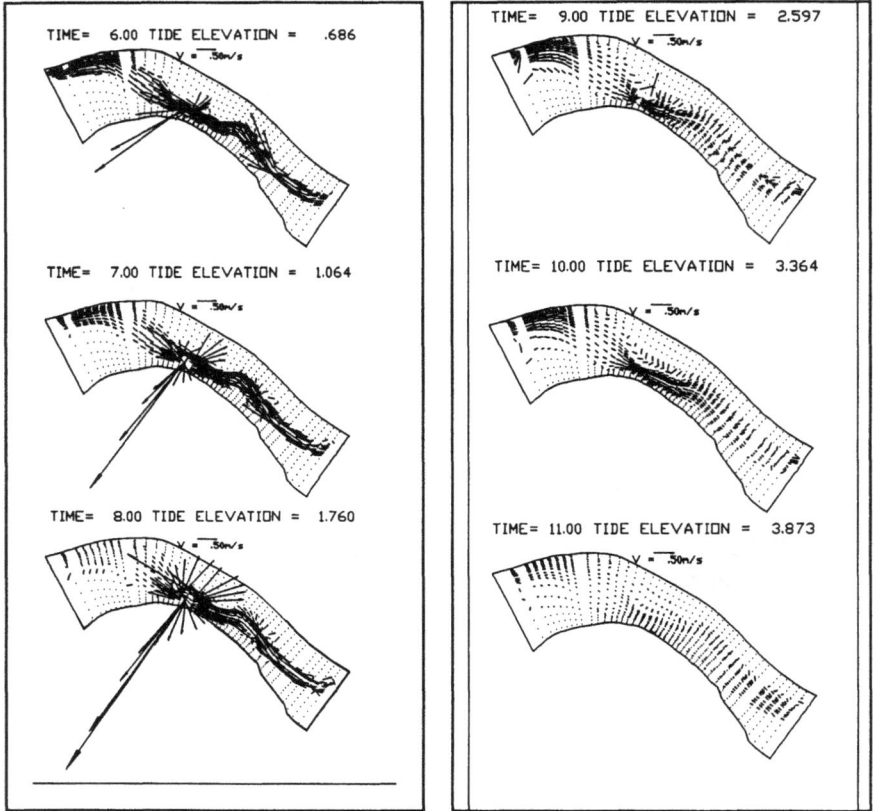

FIGURE 6 *Flood Tidal Current from T=6.0 Hrs to T=11.0 Hrs*

4.6 *The Residual (Long Time Averged) Current and Eddy Viscosity*

The residual current is defined as the velocity averaged over tidal cycles. Using the computed velocity field at every time step between T=0 to 13hrs, the residual currents for the Taitui Estuary are obtained and shown in Figure 9. Also on the left, the eddy viscosity distribution are presented. The eddy viscosity plot shows the high values of ν_t occur at the shallow zone, at which the velocity gradients are high. The residual currents appear to move mainly toward the seaside, and large values remain in the main channel. This is due to the constant 0.25 m/s river flow from the upstream. With this residual current, any pollutant can transport to the sea, so to keep a minimum fresh water inflow to the Estuary is necessary for the sake of pollution control.

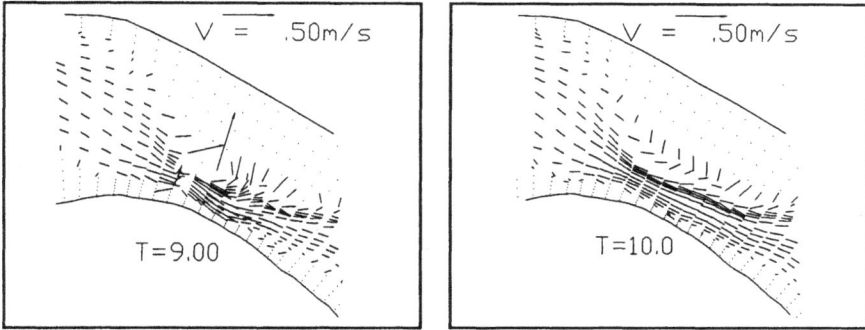

FIGURE 7   *Enlarged Vector Plots showing vortices*
            (A) T=9.0 Hrs (B) T=10.0 Hrs.

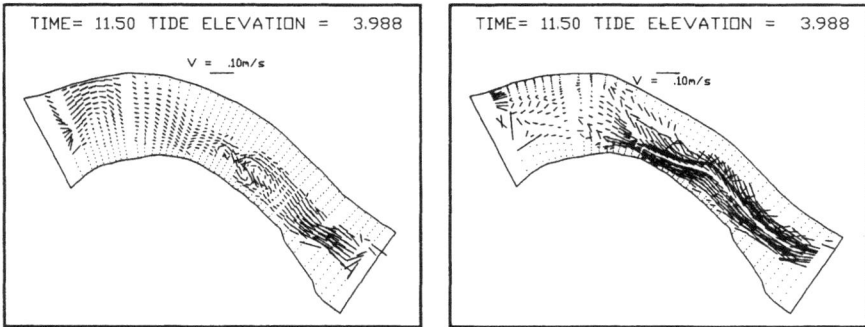

FIGURE 8   *Comparison of Flow Pattern for T=11.5 Hrs*
            (A) Normal Grids (48x20) (B) Depth Controlled (48x20)

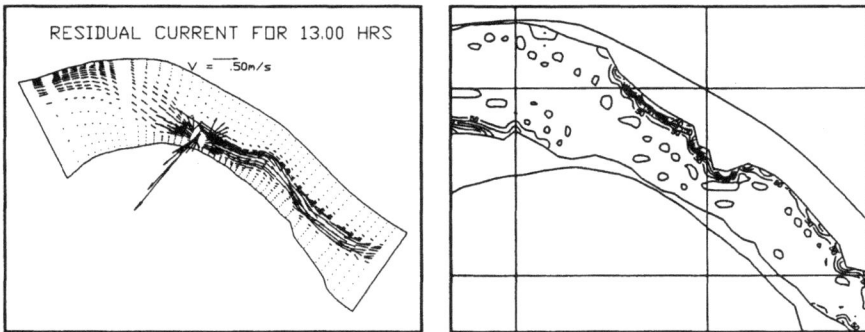

FIGURE 9   *Long Time averaged velocity and Eddy Viscosity for T=11.5 hrs.*

503

4.7 Discussion

Before the completion of present work, the k-ε model of turbulence have only been used by the authors to the flow in model scales. This paper further one step and applies the model to flow in scales of an Estuary with complex boundary. The velocities in the Estuary, although not fully verified over the whole estuary, the values and direction of the flow in entrance of the estuary are comparable with the field measurements. The use of depth averaged k-ε model is practical scale seems to be promising. The eddy viscosity values given in Figure 8 are computed from Equation 6. If ν_t is computed from the empirical equation proposed by Fisher et. al. (1979), the pattern is not the same due to the complexity of the bottom topography. Besides, different e* values produce different ν_t, so the adjustment of e* is required and more tasks are needed in the verification. In comparison, the only empirical value needed in the present model is the Manning coefficient which can be determined directly from the median size of the bed material based on experience. This seems to be more convinent in practice.

In our model, the depth control grids together with the boundary judgement scheme produced a much stable computational procedure. This is because the water depth in the present model is corrected at every iteration, and also the wall function is used at all grids adjacent to solid boundaries. Duringthe development of model we found the failure of computation usually came from a non-realistic gradient at the first grid, this spurious value affected all diffusion terms and calculation becomes unpredictable. The depth control grids greatly reduces the boundary irregularity in computational domain than that of the normal grids, and computation thus is much stable.

In this model, the density was taken vertically homogeneous so the effect of stratification did not considered. A multi-layered model which specifies density at every layer would expect to overcome this drawback of present model.

5. Concluding Remarks

Using the depth averaged model with depth controlled, boundary fitted grid, and k-εmodel of turbulence, the tidal flows in the Taitui Estuary, which has channel cross section in composite shapes, are computed. For a stable computation, the depth control grids and the boundary judgment scheme proposed in this paper shoued both be used. The comparison of the calculated and field results suggests the model is suitable for unsteady 2-D turbulent estuary flow calculation for both laboratory and practical scales. The computed results show there is a control section at low tide within the estuary. At low tide, this control completely stops the tidal flow further upstream. The shear layer usually exists between the deep channel and shallow zones occurs at mid-tide can also be calculated.

ACKNOLEDGEMENT

The data collection for the channel depth and the completion of the depth control grid were performed by Miss. S.F.Tasi. Part of the work in this paper was financed by the National Science Council under contract NSC80-E006-21, they are acknowledged.

REFERENCE

[1]   Djuric M., Kapor R. & Pavlovic R.N.,(1986), Depth-Average Model with Body

Fitted Coordinates for the Calculation of Strongly Curved Elliptic Flows , International Conference Hydro-soft, p.179-193.

[2]   Fischer,H.B.,List,E.J., Koh,R.C.Y., Imberger,J., Brooks,N.H.(1979), *Mixing in Inland and Coastal Waters*, Academic Press, New York p104-114.

[3]   Keller R.J. & Rodi W., (1988), Prediction of Flow Characteristics in Main Channel/Flood Plain Flows, Joul. Hydraulic Research, Vol.26, No.4, p.425-441.

[4]   Lai,C.J.,Leu,J.M.,Kao,R.C.,Hwung,W.C.   (1990),Observation   of   Salinity, Suspended Sediment and Lagrangian Velocity in Taitui Estuary, Proc. 12th Conf. Ocean Eng., Taiwan, p.810.

[5]   Lai,C.J.,Yen,C.W.,(1991), A Versatile Turbulent Flow Numerical Model for Hydraulic Engineering, Proc. Int. conf. Computer Appli. Water Res., July, Taipei.

[6]   Launder B.E. & Spalding D.B.,(1974),The Numerical Calculation of Turbulent Flows, Computer Methods in Applied Mechanics and Engineering, Vol.3, p.269-289.

[7]   McGuirk J.J.& Rodi W.,(1978), A Depth-averaged Mathematical Model for the Near Field of Side Discharges into Open-Channel Flow, Journal of Fluid Mechanics, Vol.86, Part.4,PP.761-781.

[8]   Middelecoff, J.F., Thomas, P.D. (1979), Direct control of the Grid Point Distribution in Meshes generated by Elliptic Equations , Proc. 4th. AIAA CFD Conf. Williamsburg., VA.,p175-179.

[9]   Nielsen P.,Skovgaard, O.,(1990), A Scheme for Automatic generation of Boundary-fitted Depth- and Depth-gradient-dependent Grids in arbitrary Two-dimensional Regions, Int. J. Num. Methods in Fluids., v.10,p741-752.

[10]   Patankar S.V.,(1980), *Numerical Heat Transfer and Flow Fluid*, Hemis Phere Publ.

[11]   Rastogi A.K.& Rodi W.,(1978), Predictions of Heat and Mass Transfer in Open Channels , Joul. Hyd. Div. Proc. ASCE., Vol.104, No.HY3, March, p.397-421.

[12]   Rhie, C.M.,Chow W.L.,(1983), Numerical Study of the Turbulent Flow past an Airfoil with Trailing Edge Separation , AIAA J.,V.21,No.11,p.1525-1532.

[13]   Thompson J.F., Warsi Z.U.A. & Mastin C.W.,(1985),*Numerical Grid Generation*

NOTATIONS:

C_1, C_2,
C_μ, C_ϵ: Coeiffs. for Turbulence model
C_f : Friction Coefficient
G : Gravitational Acceleration
g^{ii} : Metric tensor
$\sqrt{g}$: Jacobin
h : Water depth
·k : Turbulent kinetic energy
R^* : Source term of ϕ,physical plane
S^* : Source term of ϕ,computa. plane
U, V : Contravariant vels.,Computa.plane
u, v : x,y velocities in physical domain
U_* : Friction velocity
U/U_∞ : Non-dimensional velocity
z : Elevation related to datums

Γ : Diffusion coefficients
ϵ : Turbulent kinetic energy dissipation rate
ξ, η : General coordinates
μ_e : Effective viscosity
μ : Molecular viscosity
μ_t : Eddy viscosity
ρ : Fluid Density
τ_o : Overall shear stress
τ_w : Boundary shear stress
τ_{bx} : Bed shear stress in x dir.
τ_{by} : Bed shear stress in Y dir.

42 A finite difference method for the shallow water equations with conformal boundary-fitted mesh generation

S. N. Chandler-Wilde and B. Lin

ABSTRACT

An implicit finite difference scheme is developed for the depth-averaged shallow water equations using a conformal boundary-fitted mesh. A boundary-fitted mesh provides an accurate representation of the curved boundaries of natural flow regions, avoiding the approximate 'staircase' representation associated with models based on a regular, rectilinear finite difference grid. The use of numerical conformal mapping to generate the mesh leads to a particularly simple finite difference scheme, a modification of a scheme widely used for regular grids.

1 Introduction

Finite difference numerical models for solving the depth-integrated shallow water equations in estuaries and coastal waters have traditionally employed a regular, rectilinear finite difference grid [1,2]. The advantage of this approach is that finite difference equations on a uniform Cartesian grid are particularly simple. A disadvantage is that the treatment of the boundary conditions on curved boundaries of flow regions is necessarily approximate, the curved boundary replaced by a staircase of grid points. To provide an improved representation, the finite difference grid may be modified adjacent to the boundary as described in [3]. An alternative, more exact approach is to employ a curvilinear grid, with lines of the grid coinciding with the coastline, (though in practice small boundary features may not be modelled). One can view this method as the application of a transformation of the curved flow region onto a simpler domain with rectilinear sides on which a finite difference method can be applied using a regular grid. In transforming the region the equations governing the flow are also transformed, becoming more complex, the degree of complexity depending on the type of transformation. This is a disadvantage of the method.

While the use of boundary-fitted meshes is well known in computational fluid dynamics [4], the technique has not been extensively applied in computational hydraulics. Wijbenga [5,6] and Willemse et al. [7] at Delft Hydraulics Laboratory have developed a numerical scheme based on an orthogonal transformation which leads to an orthogonal curvilinear grid. More recently, Borthwjck and Barber [8] have employed a non-orthogonal transformation which leads to considerably more complicated transformed equations, but gives some extra control over the distribution of the grid points in the flow region.

In this paper the use of a *conformal transformation* (generated numerically) is discussed. This transformation, a special type of orthogonal transformation, changes the original flow equations the least, with the advantage of computational efficiency in the finite difference discretisation, and that standard Cartesian codes are easily modified to solve the transformed equations. However, it gives less control than a general orthogonal or non-orthogonal transformation over the grid point distribution in the flow region.

In Section 2 of the paper the transformed equations are derived and their finite difference discretisations presented: modifications of the methods in Falconer [1]. Section 3 discusses the generation of the conformal transformation. Section 4 presents preliminary results, comparing predictions of tidal flows in the Bristol Channel from the new model with those from a regular Cartesian grid code, and comparing both with field data.

2   The Original and Transformed Shallow Water Equations

For a constant density turbulent fluid flow on a rotating earth the depth-integrated Navier-Stokes equations, in the x and y horizontal directions are [9]

$$\frac{\partial q_x}{\partial t} + \beta\left[\frac{\partial(Uq_x)}{\partial x} + \frac{\partial(Uq_y)}{\partial y}\right] - fq_y + gH\frac{\partial\zeta}{\partial x} - (\rho_a/\rho)C^*W_xW + \frac{gU|U|}{C^2}$$

$$\qquad\qquad 1 \qquad\qquad\qquad 2 \qquad\qquad\qquad 3 \qquad 4 \qquad\qquad 5 \qquad\qquad\qquad 6$$

$$- \epsilon H\left[\frac{\partial^2 U}{\partial x^2} + \frac{\partial^2 U}{\partial y^2}\right] = 0 \qquad (1)$$

$$\qquad\qquad\qquad\qquad\qquad\qquad 7$$

$$\frac{\partial q_y}{\partial t} + \beta\left[\frac{\partial(Vq_x)}{\partial x} + \frac{\partial(Vq_y)}{\partial y}\right] + fq_x + gH\frac{\partial\zeta}{\partial y} - (\rho_a/\rho)C^*W_yW + \frac{gV|U|}{C^2}$$

$$- \epsilon H\left[\frac{\partial^2 V}{\partial x^2} + \frac{\partial^2 V}{\partial y^2}\right] = 0 \qquad (2)$$

508

where U and V are the x and y components of the depth mean velocity U, $|U| = (U^2+V^2)^{\frac{1}{2}}$, H is the total depth of flow, $q_x = HU$, $q_y = HV$, t is time, β is the correction factor for non-uniformity of the vertical velocity profile (= 1.016 for assumed seventh power law profile), f is the Coriolis parameter (= $2\omega\sin\varphi$, where ω is the angular velocity of the earth's rotation and φ is the geographical latitude - taken as 51° 20" for the Bristol Channel study in Section 4), ζ is the water elevation above Chart Datum, g is the gravitational acceleration, ρ and ρ_a are the water and air densities, C^* is the air-water interfacial resistance coefficient (2.6×10^{-3}), W_x and W_y are the x and y components of wind velocity, $W = (W_x^2+W_y^2)^{\frac{1}{2}}$, and ϵ is the depth mean eddy viscosity ($\epsilon = \sqrt{g}|U|H/C$ is assumed in the calculations in Section 4). C is the Chezy coefficient which, for the predictions in Section 4, is determined from the Colebrook-White equation [9]. In equation (1) the various terms are the depth integrated local acceleration (term 1), advective accelerations (2), Coriolis force (3), pressure gradient (4), wind shear force (5), bed shear resitance (6), and turbulence induced shear force (7).

The depth integrated continuity equation is

$$\frac{\partial \zeta}{\partial t} + \frac{\partial q_x}{\partial x} + \frac{\partial q_y}{\partial y} = 0. \tag{3}$$

Suppose now that a one-to-one transformation exists, between D and a region Ω of simple geometrical shape (the *computational region*) so that, to each point (x,y) in D there corresponds a unique point, with Cartesian coordinates (ξ,η), in Ω. As discussed in the introduction, the transformations in this paper are *conformal* (angle-preserving) transformations. Such transformations are most conveniently expressed in terms of complex variables. Let z = x+iy, w = ξ+iη, where i = $\sqrt{-1}$. Suppose that w = F(z), z = G(w), where F is a one-to-one mapping from D to Ω which is analytic in D, and continuous up to the boundary of the region, ∂D, and G := F^{-1} is its inverse mapping, analytic in Ω, and continuous up to its boundary $\partial\Omega$. For the case in which D and Ω are both simply connected (and with mild regularity conditions on the boundaries ∂D and $\partial\Omega$), the existence of such mappings F and G is guaranteed by the Riemann Mapping Theorem [10]. F will map the boundary of the physical region, ∂D, onto the boundary of the computational region, $\partial\Omega$.

The computational region is chosen so that its boundary consists of arcs on which either ξ = constant or η = constant. Then the lines ξ = constant and η = constant are mapped by G to form a boundary fitted curvilinear coordinate system on D. Since G is analytic this system is orthogonal. The mapping is also conformal: a small triangle in Ω is mapped onto a similar small triangle in D, the linear dimensions increased by the scale factor

$$h = |G'(w)| = ((\partial x/\partial \xi)^2+(\partial y/\partial \xi)^2)^{\frac{1}{2}} = ((\partial x/\partial \eta)^2+(\partial y/\partial \eta)^2)^{\frac{1}{2}}, \tag{4}$$

the equivalence of these equations for h following from the Cauchy-Riemann equations satisfied by G. The area of the triangle in D is a factor $J = h^2$ larger, J the Jacobian of the transformation.

Expressing, through the relationship w = F(z), equations (1)-(3) in terms of the new independent variables ξ and η, and introducing new dependent variables U_ξ, U_η and q_ξ, q_η, the ξ and η components of $\underline{U}$ and of $\underline{q} = \underline{U}H$, it can be shown that (1)-(3) are equivalent to

$$\underbrace{\frac{\partial q_\xi}{\partial t}}_{1} + \underbrace{\beta\left[\frac{1}{J}\left\{\frac{\partial}{\partial\xi}(U_\xi hq_\xi) + \frac{\partial}{\partial\eta}(U_\xi hq_\eta)\right\}\right.}_{2a} + \underbrace{\frac{U_\eta}{J}\left\{q_\xi\frac{\partial h}{\partial\eta} - q_\eta\frac{\partial h}{\partial\xi}\right\}\right]}_{2b} - \underbrace{fq_\eta}_{3} + \underbrace{\frac{gH}{h}\frac{\partial\zeta}{\partial\xi}}_{4}$$

$$ - \underbrace{(\rho_a/\rho)C^*W_\xi W}_{5} + \underbrace{\frac{gU_\xi|\underline{U}|}{C^2}}_{6} - \underbrace{\frac{\epsilon H}{h}\left\{\frac{\partial}{\partial\xi}\left[\frac{1}{J}\frac{\partial}{\partial\xi}(hU_\xi)\right] + \frac{\partial}{\partial\eta}\left[\frac{1}{J}\frac{\partial}{\partial\eta}(hU_\xi)\right]\right\}}_{7a}$$

$$ - \underbrace{\frac{2\epsilon H}{Jh}\left[\frac{\partial U_\eta}{\partial\xi}\frac{\partial h}{\partial\eta} - \frac{\partial U_\eta}{\partial\eta}\frac{\partial h}{\partial\xi}\right]}_{7b} = 0 \qquad (5)$$

$$\frac{\partial q_\eta}{\partial t} + \beta\left[\frac{1}{J}\left\{\frac{\partial}{\partial\xi}(U_\eta hq_\xi) + \frac{\partial}{\partial\eta}(U_\eta hq_\eta)\right\} + \frac{U_\xi}{J}\left\{q_\eta\frac{\partial h}{\partial\xi} - q_\xi\frac{\partial h}{\partial\eta}\right\}\right] + fq_\xi + \frac{gH}{h}\frac{\partial\zeta}{\partial\eta}$$

$$ - (\rho_a/\rho)C^*W_\eta W + \frac{gU_\eta|\underline{U}|}{C^2} - \frac{\epsilon H}{h}\left\{\frac{\partial}{\partial\xi}\left[\frac{1}{J}\frac{\partial}{\partial\xi}(hU_\eta)\right] + \frac{\partial}{\partial\eta}\left[\frac{1}{J}\frac{\partial}{\partial\eta}(hU_\eta)\right]\right\}$$

$$ - \frac{2\epsilon H}{Jh}\left[\frac{\partial U_\xi}{\partial\eta}\frac{\partial h}{\partial\xi} - \frac{\partial U_\xi}{\partial\xi}\frac{\partial h}{\partial\eta}\right] = 0 \qquad (6)$$

$$\frac{\partial\zeta}{\partial t} + \frac{1}{J}\left[\frac{\partial(hq_\xi)}{\partial\xi} + \frac{\partial(hq_\eta)}{\partial\eta}\right] = 0. \qquad (7)$$

In equation (5) the numbering of terms corresponds to equation (1). Comparing equations (1) and (5) it can be seen that terms 1, 2a, 3, 4, 5, 6, and 7a in equation (5) are, apart from the introduction of the scale factor h, identical to terms 1-7 in (1). The additional terms (2b) and (7b) are associated with spatial rate of change of the scale factor.

2.1 The Finite Difference Equations

A finite difference discretisation of equations (1)-(3) in an alternating direction implicit form, centred in space and time, using a space-staggered grid scheme is presented in Falconer [1]. This discretisation is easily modified to obtain finite difference versions of equations (5)-(7). For the first half time-step the continuity equation (7) is replaced by

$$\zeta_{j,k}^{n+\frac{1}{2}} - \zeta_{j,k}^n + \frac{\Delta t}{2J_{j,k}}\left[\frac{1}{\Delta\xi}\left\{(hq_\xi)\Big|_{j+\frac{1}{2},k}^{n+\frac{1}{2}} - (hq_\xi)\Big|_{j-\frac{1}{2},k}^{n+\frac{1}{2}}\right\}\right.$$

$$ + \frac{1}{\Delta\eta}\left\{(hq_\eta)\Big|_{j,k+\frac{1}{2}}^n - (hq_\eta)\Big|_{j,k-\frac{1}{2}}^n\right\}\right] = 0, \qquad (8)$$

where the subscript j,k denotes a value at the grid point with coordinates $\xi = j\Delta\xi$, $\eta = k\Delta\eta$, and the superscript n indicates time $t = n\Delta t$. The ξ-direction momentum equation (5) is replaced by

$$
\begin{aligned}
& q_\xi\Big|_{j+\frac{1}{2},k}^{n+\frac{1}{2}} - q_\xi\Big|_{j+\frac{1}{2},k}^{n-\frac{1}{2}} + \frac{\beta\Delta t}{J_{j+\frac{1}{2},k}}\Big\{\frac{1}{\Delta\xi}\Big[(hU_\xi{}'q_\xi{}')\Big|_{j+1,k}^{n} - (hU_\xi{}'q_\xi{}')\Big|_{j,k}^{n}\Big] \\
& + \frac{1}{\Delta\eta}\Big[(hq_\eta)\Big|_{j+\frac{1}{2},k+\frac{1}{2}}^{n} U_\xi{}'\Big|_{j+\frac{1}{2},k+p}^{n} - (hq_\eta)\Big|_{j+\frac{1}{2},k-\frac{1}{2}}^{n} U_\xi{}'\Big|_{j+\frac{1}{2},k-1+p}^{n}\Big] \\
& + U_\eta\Big|_{j+\frac{1}{2},k}^{n}\ (q_\xi{}'\frac{\partial h}{\partial\eta} - q_\eta\frac{\partial h}{\partial\xi})\Big|_{j+\frac{1}{2},k}^{n}\Big\} - \Delta t\,fq_\eta\Big|_{j+\frac{1}{2},k}^{n} \\
& + gH_{j+\frac{1}{2},k}^{n}\frac{\Delta t}{2h\Delta\xi}\Big[\zeta_{j+1,k}^{n+\frac{1}{2}} + \zeta_{j+1,k}^{n-\frac{1}{2}} - \zeta_{j,k}^{n+\frac{1}{2}} - \zeta_{j,k}^{n-\frac{1}{2}}\Big] - \Delta t\,(\rho_a/\rho)C^*W_\xi W \\
& + g\Delta t\frac{(U_\xi{}^2+U_\eta{}^2)^{\frac{1}{2}}}{2C^2H}\Big|_{j+\frac{1}{2},k}^{n}(q_\xi\Big|_{j+\frac{1}{2},k}^{n+\frac{1}{2}} + q_\xi\Big|_{j+\frac{1}{2},k}^{n-\frac{1}{2}}) \\
& - \Delta t(\frac{\epsilon H}{h})\Big|_{j+\frac{1}{2},k}^{n}\Big\{\frac{1}{\Delta\xi^2}\Big[\frac{1}{J_{j+1,k}}\Big\{(hU_\xi{}')\Big|_{j+3/2,k}^{n} - (hU_\xi{}')\Big|_{j+\frac{1}{2},k}^{n}\Big\} \\
& \qquad\qquad - \frac{1}{J_{j,k}}\Big\{(hU_\xi{}')\Big|_{j+\frac{1}{2},k}^{n} - (hU_\xi{}')\Big|_{j-\frac{1}{2},k}^{n}\Big\}\Big] \\
& \qquad + \frac{1}{\Delta\eta^2}\Big[\frac{1}{J_{j+\frac{1}{2},k+\frac{1}{2}}}\Big\{(hU_\xi{}')\Big|_{j+\frac{1}{2},k+1}^{n} - (hU_\xi{}')\Big|_{j+\frac{1}{2},k}^{n}\Big\} \\
& \qquad\qquad - \frac{1}{J_{j+\frac{1}{2},k-\frac{1}{2}}}\Big\{(hU_\xi{}')\Big|_{j+\frac{1}{2},k}^{n} - (hU_\xi{}')\Big|_{j+\frac{1}{2},k-1}^{n}\Big\}\Big] \\
& + \frac{2}{J_{j+\frac{1}{2},k}}\Big[\frac{\partial h}{\partial\eta}\Big|_{j+\frac{1}{2},k}^{n}\frac{1}{2\Delta\xi}\Big\{U_\eta\Big|_{j+1,k+\frac{1}{2}}^{n} + U_\eta\Big|_{j+1,k-\frac{1}{2}}^{n} - U_\eta\Big|_{j,k+\frac{1}{2}}^{n} - U_\eta\Big|_{j,k-\frac{1}{2}}^{n}\Big\} \\
& - \frac{\partial h}{\partial\xi}\Big|_{j+\frac{1}{2},k}^{n}\frac{1}{2\Delta\eta}\Big\{U_\eta\Big|_{j+1,k+\frac{1}{2}}^{n} + U_\eta\Big|_{j,k+\frac{1}{2}}^{n} - U_\eta\Big|_{j+1,k-\frac{1}{2}}^{n} - U_\eta\Big|_{j,k-\frac{1}{2}}^{n}\Big\}\Big]\Big\} = 0 \quad (9)
\end{aligned}
$$

In this equation, $p = 1$ if $U_\eta^* \leq 0$, $= 0$ if $U_\eta^* > 0$, where U_η^* is the average of the four U_η values at time level n at $j,k+\frac{1}{2}$, $j,k-\frac{1}{2}$, $j+1,k+\frac{1}{2}$, and $j+1,k-\frac{1}{2}$. $\partial h/\partial\eta$ and $\partial h/\partial\xi$ are approximated by central difference formulae. The equations (8) and (9) are used iteratively in a two-stage predictor-corrector mode. For the first iteration, (8) and (9) are solved with the terms written with a prime expressed explicitly at time step $n-\frac{1}{2}$. At the second iteration these terms are expressed in time-centred form, at level n, using the mean of the values calculated at the previous time-step and at the end of the first iteration.

For the second half time-step, from level $n+\frac{1}{2}$ to $n+1$, similar implicit finite difference discretisations of the continuity equation (5) and of the η-component momentum equation (7) are solved for q_η and ζ at time level $n+1$.

From computations with various test examples, the above outlined scheme appears to be unconditionally stable. However, defining the Courant number C_r by

$$
C_r = \frac{\Delta t}{\Delta_{max}}\sqrt{(g/H_{max})},
$$

where Δ_{max} is the maximum over all grid squares of $max(h\Delta\xi, h\Delta\eta)$, it appears that a Courant number of not greater than approximately 8 is necessary for accurate predictions. These observations are also reported for the regular grid scheme in [9], to which this scheme reduces if h ≡ 1.

3 Generation of the Coordinate Transformation

A method for generating the conformal mapping G in the particular case when Ω is a rectangle is now described. The further development of the method to the case when Ω is an arbitrary simply connected region whose boundary is composed of coordinate lines ξ = constant and η = constant presents no problem in principle.

Papamichael [11] surveys methods for conformally mapping an arbitrary simply connected domain onto a rectangle. The method described below appears to be new; a related method, using finite elements, is implemented in [12].

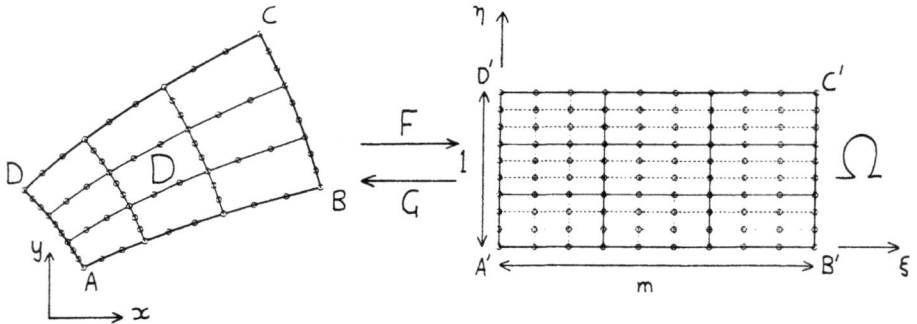

FIGURE 1   *The physical and computational domains.*

Figure 1 shows a typical physical domain D and the corresponding computational domain Ω, the correspondence between the two given by the mapping F and its inverse G. The height of the rectangle is essentially arbitrary. Thus fix the height of the rectangle as 1 and denote its width by m.

In Figure 1, A, B, C, D are four distinct points on the boundary ∂D, listed in counter-clockwise order, typically, though not necessarily, corners of ∂D. It can be shown that (see [10]), for any positions of the points A, B, C, D, there exists a unique analytic function F, and m > 0 such that F maps D onto Ω with A, B, C, and D mapped onto A', B', C' and D'.

To calculate this mapping note that, where $\xi = \xi(x,y)$ and $\eta = \eta(x,y)$ are defined by $\xi + i\eta = F(x+iy)$, η satisfies the following mixed boundary value problem for Laplace's equation:

$$\frac{\partial^2 \eta}{\partial x^2} + \frac{\partial^2 \eta}{\partial y^2} = 0 \quad \text{in D,} \tag{10a}$$

$$\eta = 0 \quad \text{on AB,} \tag{10b}$$

$$\eta = 1 \qquad \text{on CD}, \tag{10c}$$

$$\frac{\partial \eta}{\partial n} = 0 \qquad \text{on BC and AD}. \tag{10d}$$

Equations (10a) and (10d) follow from the Cauchy-Riemann equations satisfied by F. In particular, it follows from the Cauchy-Riemann equations that, on ∂D,

$$\frac{\partial \xi}{\partial s} = \frac{\partial \eta}{\partial n}, \quad \frac{\partial \xi}{\partial n} = -\frac{\partial \eta}{\partial s}, \tag{11}$$

where $\partial/\partial s$ and $\partial/\partial n$ denote the tangential and normal derivatives, s directed counter-clockwise around ∂D, and the normal n directed into D. Since $\xi = 0$ on AD, $\xi = m$ on BC, equation (10d) follows from (11).

For the generation of the boundary fitted coordinate system it will become apparent that it is crucial to compute F only on the boundary ∂D. For this purpose, using Green's theorem and the boundary conditions (10b-d), the above boundary value problem is reformulated as the following boundary integral equation. From [13], at each point $\underline{r} = (x,y)$ on the boundary which is not a corner point,

$$\tfrac{1}{2}\eta(\underline{r}) = \int_{\text{CD}} \frac{\partial \Phi(\underline{r},\underline{r}_s)}{\partial n(\underline{r}_s)} ds(\underline{r}_s) + \int_{\text{BC}\cup\text{AD}} \frac{\partial \Phi(\underline{r},\underline{r}_s)}{\partial n(\underline{r}_s)} \eta(\underline{r}_s) ds(\underline{r}_s) - \int_{\text{AB}\cup\text{CD}} \Phi(\underline{r},\underline{r}_s) \frac{\partial \eta}{\partial n}(\underline{r}_s) ds(\underline{r}_s) \tag{12}$$

where $\underline{r}_s = (x_s,y_s)$ is a point on the boundary and $ds(\underline{r}_s)$ is an element of arc-length at $\underline{r}_s$. $\Phi(\underline{r},\underline{r}_s)$ is a fundamental solution of Laplace's equation (10a), defined by $\Phi(\underline{r},\underline{r}_s) := -(1/2\pi) \ln|\underline{r}-\underline{r}_s|$, $\underline{r} \neq \underline{r}_s$. Equation (12) can be discretised by a simple boundary element method to find the unknowns in the equation, η on BC and AD and $\partial\eta/\partial n$ on AB and CD. In the results reported below a simple piecewise constant collocation method is used (see [13]).

Following the solution of (12), $\eta = \text{Im } F$ is known everywhere on the boundary ∂D, but $\xi = \text{Re } F$ only on AD and BC. However, ξ on the remaining part of the boundary is easily recovered from $\partial\eta/\partial n$: for example, it follows from (11) that, at an arbitrary point P on BC,

$$\xi = \int_B^P \frac{\partial \eta}{\partial n} ds, \tag{13}$$

the integral taken along that part of BC between B and P. In the computations below (13) is discretised using the composite midpoint rule.

The above describes the computation of the real and imaginary parts of the map F on the boundary ∂D. For the finite difference method decribed in Section 2.1, the value of $h = |G'(w)|$ is required at each point w of a

513

regular finite difference grid on the rectangle Ω. Also, to plot the results from these computations in the physical plane, the value of $G(w)$ at each point is required. To obtain these values an uniform finite difference grid is set up on the rectangle Ω, the grid on which the flow computations are to be performed a sub-grid of this new grid. From the above computations, $G = F^{-1}$ is known on a non-uniform grid of points on the boundary $\partial\Omega$. G can thus be determined, by linear interpolation, at every boundary point of the new grid, and then determined at every point of this grid, by solving a standard finite difference discretisation, using the 5-point formula, of Laplace's equation satisfied by the real and imaginary parts of g. This last computation can be performed very efficiently using the FFT [14]. Finally $G'(w)$ is calculated at each grid point using a 2nd order centred finite difference approximation.

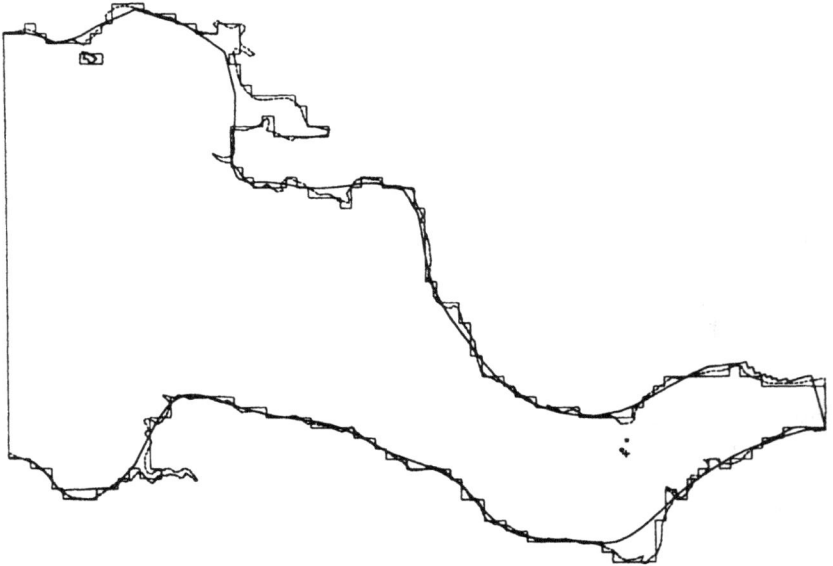

FIGURE 2   *The part of the Bristol Channel modelled (- - -), and the approximations to the true boundary used in the curvilinear and rectilinear numerical models.*

4   Computations of Tidal Flows in the Bristol Channel

To test the boundary fitted curvilinear code, it has been used to predict tidal flows in the Bristol Channel, the results from the new model tested by comparison with a rectilinear finite difference numerical model [1], and with field data.

Figure 2 shows the part of the Bristol Channel modelled together with the approximations to the true boundary adopted in the curvilinear and rectilinear models. For the curvilinear model a boundary fitted mesh (see Figure 3) is generated as described in Section 3.

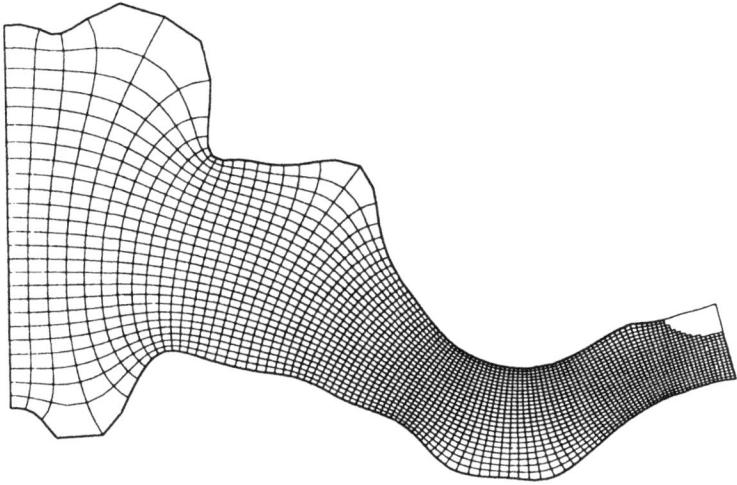

FIGURE 3 *The conformal boundary fitted mesh.*

Figure 4 shows predictions, from the rectilinear and curvilinear models, of flows at a typical point in the tidal cycle at Spring tide. Both predictions are driven by the same boundary data. No slip boundary conditions are imposed on the coastline. At the open boundaries the surface elevation is specified from tide tables, and by applying a Coriolis slope on the downstream boundary. Both models include flooding and drying effects using the scheme described in [15].

Comparing the two predictions it can be seen that the rectilinear model in general provides a more detailed and superior prediction at the downstream end of the estuary, in particular around the Gower peninsula. The curvilinear model represents the flow upstream in more detail, in particular the extensive flooding and drying in this area. Also, it avoids the generation of vorticity at artificial corners on the domain seen in the regular grid calculations. However, it is also significantly more computationally expensive to run, owing to the much more severe restiction placed, by the requirement $C_r \leq 8$, on the time step in the curvilinear model.

Figure 5 compares predictions of flow and surface elevation of the two models at the points shown in Figure 1. Comparisons are also made between the predicted depth-integrated velocities and the tidal stream data taken from the Admiralty Chart No. 1179. The data point is indicated by a cross in Figure 1: the comparison is made with the predicted values at the nearest grid points on the curvilinear (.) and rectilinear (*) grids.

It can be seen that agreement between the two models is good, and agreement is reasonable between the models and field data. The fit at other points examined is of similar quality, except that the agreement with field data deteriorates near the downstream open boundary. The representation of the downstream open boundary condition is currently being examined with a view to improving the fit at the downstream positions.

515

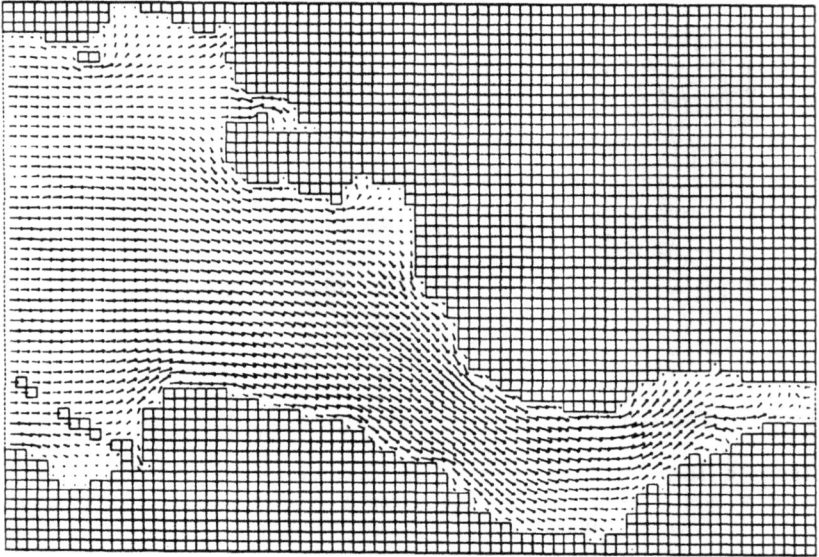

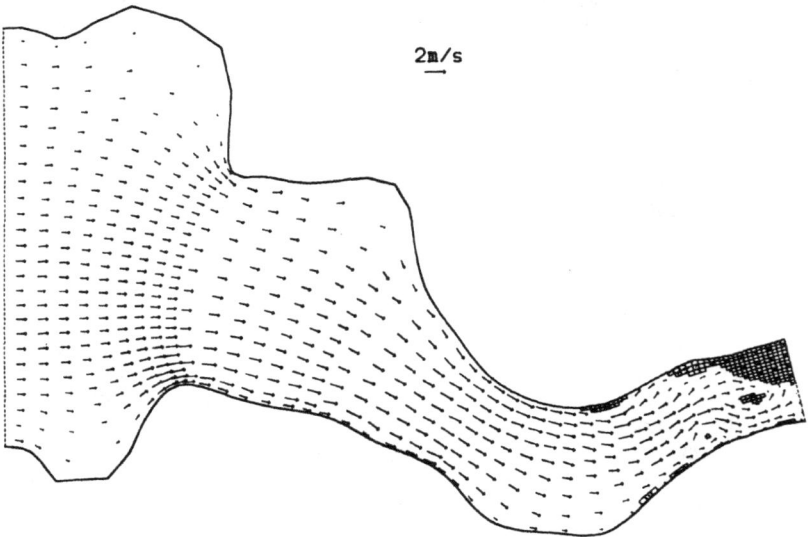

FIGURE 4   *Predictions of tidal flow in the Bristol Channel from the curvilinear and the regular mesh models.*

5   Conclusions

A finite difference method, utilising an orthogonal boundary fitted mesh, for representing accurately two-dimensional flows with curved boundaries, in rivers, estuaries, and coastal waters, has been described and its performance illustrated. The method employs an initial transformation from the flow region to a computational domain with rectilinear sides, this transformation being conformal so that the transformed flow equations are hardly changed. Thus standard codes, based on regular Cartesian grids, are easily modified to solve the transformed equations. A modification of the hydrodynamic part of the numerical model described by Falconer [1] has been reported.

Work is currently underway to include a depth-averaged advection-diffusion equation within the curvilinear model. The resultant numerical model should prove a valuable extension to existing methods for predicting flow and pollutant transport in regions with complex geometry.

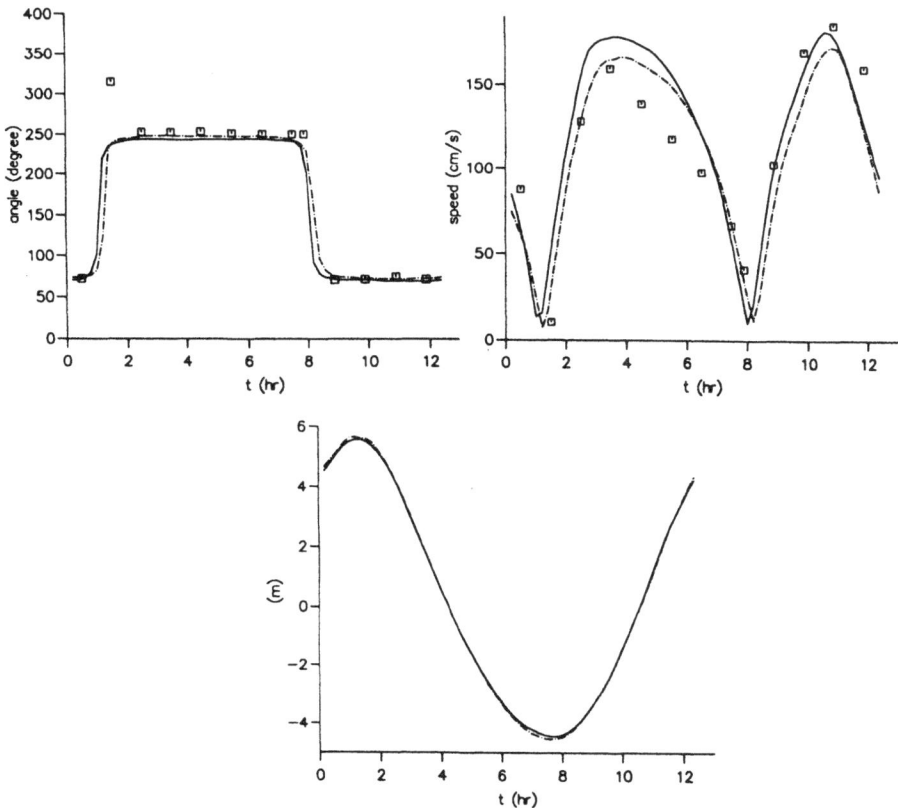

FIGURE 5   *Comparisons between field data (□), rectilinear (—·—·—) and curvilinear (——) model predictions, of speed and direction of flow at the position in Figure 1. Surface elevation predictions are also compared.*

517

REFERENCES

[1]   Falconer, R.A., (1986), A Two-Dimensional Mathematical Model Study of the Nitrate Levels in an Inland Natural Basin, *Proceedings of the International Conference on Water Quality Modelling in the Inland Natural Environment*, Bournemouth, England, June 1986, pp. 325-344.

[2]   Stelling, G.S. and Willemse, J.B.T.M., (1984), Remarks about a Computational Method for Shallow Water Equations that Works in Practice, *Colloquium Topics in Numerical Analysis*, J.G. Verwer (editor), C.W.I. Syllabus No. 5, Centrum voor Wiskunde en Informatica, Amsterdam, pp. 337-362.

[3]   Falconer, R.A. and Mardapitta-Hadjipandeli, L., (1987), Bathymetric and Shear Stress Effects on an Island's Wake, *Coastal Engineering*, Vol. 11, No. 1, pp. 57-86.

[4]   Thompson, J.F. and Warsi, Z.U.A., (1982), Boundary-Fitted Coordinate Systems for Numerical Solution of Partial Differential Equations - A Review, *Journal of Computational Physics*, Vol. 47, pp. 1-108.

[5]   Wijbenga, J.H.A., (1985), Determination of Flow Patterns in rivers with Curvilinear Coordinates, *Proceedings of the 21st Congress of the International Association for Hydraulics Research*, Melbourne, Australia, August 1985.

[6]   Wijbenga, J.H.A., (1985), Steady Depth Averaged Flow Calculations on Curvilinear Grids, *Proceedings of the Second International Conference on the Hydraulics of Floods and Flood Control*, Cambridge, England, September 1985, pp. 373-387.

[7]   Willemse, J.B.T.M., Stelling, G.S., and Verboom, G.K., (1986), Solving the Shallow Water Equations with an Orthogonal Coordinate Transformation, Delft Hydraulics Communication No. 356.

[8]   Borthwick, A.G.L. and Barber, R.W., to appear in *International Journal for Numerical Methods in Fluids*.

[9]   Falconer, R.A., (1991), Review of Modelling Flow and Pollutant Transport Processes in Hydraulic Basins, *Proceedings of the First International Conference on Water Pollution: Modelling, Measuring and Prediction*, Southampton, England, September 1991, pp.3-23.

[10]Henrici, P., (1974), *Applied and Computational Complex Analysis, Volume I*, John Wiley and Sons, New York.

[11]Papamichael, N., (1989), Numerical Conformal Mapping onto a Rectangle with Applications to the Solution of Laplacian Problems, *Journal of Computational and Applied Mathematics*, Vol. 28, pp. 63-83.

[12]Gaier, D., (1972), Ermittlung des konformen Moduls von Vierecken mit Differenzenmethoden, *Numerische Mathematik*, Vol 19, pp. 179-194.

[13]Jaswon, M.A. and Symm, G.T., (1977), *Integral Equation Methods in Potential Theory and Elasticity*, Academic Press, London.

[14]Dautry, R. and Lions, J.-L., (1988), *Mathematical Analysis and Numerical Methods for Science and Technology: Volume 2: Functional and Variational Methods*, Springer-Verlag, Berlin.

[15]Falconer, R.A. and Chen, Y.P., to appear in *Proceedings of the Institution of Civil Engineers*, Part 2, Research and Theory.

43 A characteristic method for nonlinear advective accelerations

L. Zhu and C. W. Li

Abstract

Under certain horizontal flow conditions in which the velocity variation is large, the accurate numerical representation of the nonlinear advective terms in the governing equations is essential in order to obtain a reliable solution. One of the successful numerical methods for horizontal flow is the split-operator schemes, in which the nonlinear terms for the advection of momenta are split from the remaining terms and the solution is obtained by a backward characteristics method, whereas the remaining terms are solved by standard numerical schemes. This work presents an efficient and accurate characteristics method for solving the nonlinear advection. The essence of the method is to locate the characteristic line based on a Lagrangian viewpoint, and determine the velocities by a quadratic interpolation over four nodes subjected to a minimax criterion. Test examples are used to illustrate the performance of the scheme.

Introduction

In certain horizontal flow situations the velocity variation can be large in some regions. Examples include side discharges into rivers, tidal flow in harbours with narrow entrances and flow behind islands. Numerical solution to these flow conditions may have problems. The use of central difference will cause oscillations, while the use of upwind difference will induce artificial diffusion. One of the successful approach to alleviate this difficulty is the split-operator schemes (e.g. [1]), in which the nonlinear terms for the advection of momenta are split from the remaining terms and the solution is obtained by a backward characteristics method, whereas the remaining terms are solved by standard schemes.

The error produced by the characteristics method can be divided into two parts. First is the error in the location of the characteristic lines. Second is the error in the estimation of the unknowns by interpolation. The first error does not appear in scalar transport problems in which the velocity field is known. In nonlinear advection of momentum, the determination of characteristics line is not so straightforward as the velocity is itself the quantity to be solved.

In this work, an efficient and accurate method in the location of the characteristic line is developed based on the Lagrangian viewpoint. For the estimation of the velocities at the end point of the characteristic line, the four-node minimax-characteristics method [2] is employed. Two test examples are used to illustrate the performance of the method.

Model Formulation and Solution Algorithm

The predominantly horizontal flow is described by the shallow water equations:

$$\frac{\partial \eta}{\partial t} + \frac{\partial q_x}{\partial x} + \frac{\partial q_y}{\partial y} = 0 \tag{1}$$

$$\frac{\partial q_x}{\partial t} + \frac{\partial}{\partial x}(uq_x) + \frac{\partial}{\partial y}(vq_x) = -g(H+\eta)\frac{\partial \eta}{\partial x} + D_x - \frac{\tau_{xb}}{\rho} \tag{2}$$

$$\frac{\partial q_y}{\partial t} + \frac{\partial}{\partial x}(uq_y) + \frac{\partial}{\partial y}(vq_y) = -g(H+\eta)\frac{\partial \eta}{\partial y} + D_y - \frac{\tau_{yb}}{\rho} \tag{3}$$

where ρ=fluid density; η=free surface elevation; H=still water depth; (u,v)=vertically averaged velocities in (x,y) directions respectively; $(q_x,q_y)=(u(H+\eta),v(H+\eta))$=volume fluxes per unit horizontal width; the bottom shear stresses (τ_{xb},τ_{yb}) are given by

$$\tau_{xb} = \rho g n^2 u \sqrt{u^2+v^2}/H^{1/3} \,, \quad \tau_{yb} = \rho g n^2 v \sqrt{u^2+v^2}/H^{1/3} \tag{4}$$

where n is the Manning's coefficient; the turbulent transport (D_x, D_y) are given by

$$D_x = \frac{\partial}{\partial x}(E_{xx}\frac{\partial q_x}{\partial x})+ \frac{\partial}{\partial y}(E_{xy}\frac{\partial q_x}{\partial y}) \,, \quad D_y = \frac{\partial}{\partial x}(E_{yx}\frac{\partial q_y}{\partial x})+ \frac{\partial}{\partial y}(E_{yy}\frac{\partial q_y}{\partial y}) \tag{5}$$

where E_{xx}, E_{xy}, E_{yx}, E_{yy} are the diffusion type coefficients. In the present model, these coefficients are assumed identical (=E) and are correlated to the bottom shear stress and the water depth [3]:

$$E=0.15 \sqrt{\frac{\tau_b}{\rho}} \, H \tag{6}$$

The split-operator approach is used in the solution. At each time step, the following three solution steps are carried out:

a) Advection

$$\frac{q_x^{n+1/3}-q_x^n}{\Delta t} + \frac{\partial}{\partial x}(uq_x)+\frac{\partial}{\partial y}(vq_x)=0 \tag{7}$$

$$\frac{q_y^{n+1/3}-q_y^n}{\Delta t} + \frac{\partial}{\partial x}(uq_y)+\frac{\partial}{\partial y}(vq_y)=0 \tag{8}$$

By using the continuity equation, the above equations can be transformed to

$$\frac{u^{n+1/3}-u^n}{\Delta t} + u\frac{\partial u}{\partial x} + v\frac{\partial u}{\partial y} = 0 \tag{9}$$

$$\frac{v^{n+1/3}-v^n}{\Delta t} + u\frac{\partial v}{\partial x} + v\frac{\partial v}{\partial y} = 0 \tag{10}$$

Details of the solution method is described in the next section.

b) Dispersion

$$\frac{q_x^{n+2/3} - q_x^{n+1/3}}{\Delta t} = \frac{\partial}{\partial x}(E_{xx}\frac{\partial q_x}{\partial x}) + \frac{\partial}{\partial y}(E_{xy}\frac{\partial q_x}{\partial y}) \tag{11}$$

$$\frac{q_y^{n+2/3} - q_y^{n+1/3}}{\Delta t} = \frac{\partial}{\partial x}(E_{xy}\frac{\partial q_y}{\partial x}) + \frac{\partial}{\partial y}(E_{yy}\frac{\partial q_y}{\partial y}) \tag{12}$$

As the diffusion terms are generally small in horizontal flow, the simple forward-time centred-space scheme [4] is used.

c) Propagation

$$\frac{\eta^{n+1} - \eta^n}{\Delta t} + \frac{\partial q_x}{\partial x} + \frac{\partial q_y}{\partial y} = 0 \tag{13}$$

$$\frac{q_x^{n+1} - q_x^{n+2/3}}{\Delta t} = -g(H+\eta)\frac{\partial \eta}{\partial x} - \frac{\tau_{xb}}{\rho} \tag{14}$$

$$\frac{q_y^{n+1} - q_y^{n+2/3}}{\Delta t} = -g(H+\eta)\frac{\partial \eta}{\partial y} - \frac{\tau_{yb}}{\rho} \tag{15}$$

A semi-implicit time marching scheme is used for Equations (13)-(15) except for the bottom friction and the nonlinear part of the pressure gradient terms, which are treated explicitly, while the continuity equation is time centred. The three equations are decoupled through two procedures. First, the unknowns q_x, q_y at time level n+1 are eliminated by differentiating Equations (14) & (15) with respect to x and y respectively and substituting the results into Equation (13). Second, essential boundary conditions are imposed. The procedure is similar to that used by Benqué et al. [1].

Nonlinear Advection

The nonlinear advection step is described in details in this section. For computational efficiency, the equations are further split into the following one-dimensional system:

$$\frac{\overline{u}^{n+1/3} - u^n}{\Delta t} + v^n\frac{\partial u}{\partial x} = 0 \tag{16}$$

$$\frac{\overline{v}^{n+1/3} - v^n}{\Delta t} + u^n \frac{\partial v}{\partial x} = 0 \tag{17}$$

$$\frac{u^{n+1/3} - \overline{u}^{n+1/3}}{\Delta t} + u \frac{\partial u}{\partial x} = 0 \tag{18}$$

$$\frac{v^{n+1/3} - \overline{v}^{n+1/3}}{\Delta t} + v \frac{\partial v}{\partial x} = 0 \tag{19}$$

From the Lagrangian point of view, these equations describe the pure advection of momenta (velocities). Hence the velocities should be invariant along the characteristic lines. Consequently, the location of the characteristic line can be obtained from

$$dx/dt = u_i^{n+1}; \qquad dy/dt = v_i^{n+1} \tag{20}$$

where u_i^{n+1}, v_i^{n+1} denote the velocities at the starting point of a characteristic line. This approach can be incorporated into equations (18),(19) to give an efficient and accurate scheme, but the resulting scheme will be complicated if this approach is incorporated into equations (16),(17). Instead, a simpler approach is adopted for equations (16),(17), the characteristic line is located by

$$dx/dt = u_*^n, \qquad dy/dt = v_*^n \tag{21}$$

where

$$u_*^n = u_i^n(1 - u_i^n \Delta t/\Delta x) + u_{i-1}^n u_i^n \Delta t/\Delta x \tag{22}$$

$$v_*^n = v_i^n(1 - v_i^n \Delta t/\Delta y) + v_{i-1}^n v_i^n \Delta t/\Delta y \tag{23}$$

for u_i^n, $v_i^n > 0$ and a linear variation of velocity in space. In the subsequent section, it will be shown that the commonly used method of location: $dx/dt = u_i^n$, $dy/dt = v_i^n$ is inferior than the present two approaches.

After the determination of the characteristic line. The values of u_i^{n+1} and v_i^{n+1} are obtained by interpolation of the nodal values at known time step n. However, the choice of interpolation function can have significant effect on the accuracy of the solution. From the consideration of both accuracy and efficiency, Li [2] proposed a quadratic interpolation over four nodal points subjected to a minimax constraint in one-dimensional case. The resulting schemes for equation (18) is as follows:

$$u_i^{n+1/3} = -0.25v(1-v)(\bar{u}_{i-2}^{n+1/3} + \bar{u}_{i+1}^{n+1/3}) + (1.25v-0.25v^2)\bar{u}_{i-1}^{n+1/3}$$

$$+ (1-0.75v-0.25v^2)\bar{u}_i^{n+1/3} \tag{24}$$

where $v=u_i^{n+1/3}\Delta t/\Delta x$ for $v\leq 1$. So we have a quadratic equation in $u_i^{n+1/3}$, which can be solved analytically. For equations (16), $v=v_*^{n+1/3}\Delta t/\Delta x$, similar expressions are obtained and $\bar{u}_i^{n+1/3}$ can be solved explicitly.

Test Examples

The first example is to study the effect of using different methods in locating the characteristic line. The following one-dimensional problem is considered:

$$\frac{\partial u}{\partial t} + u\frac{\partial u}{\partial x} = 0 \tag{25}$$

$x\in(0,1)$, $t\geq 0$

The initial condition is given by

$$u(x,0)=x/\alpha \tag{26}$$

where α is a constant. The analytical solution is given by

$$u = x/(t+\alpha) \tag{27}$$

As the interpolation function used in the minimax-characteristics method is quadratic, no interpolation error will occur in the solution which is linear in x. The error (if happen) is only due to the location of characteristic lines. Three methods of locating the characteristic lines are considered: $dx/dt=u_i^n$ (method I), $dx/dt=u_*^n$ (method II), and $dx/dt=u_i^{n+1}$ (method III). The ratio of the computed velocity to the exact velocity at x=0.5 is shown in Figure 1. Method III gives the exact solution and is the most accurate method. Method II produces a certain amount of error but is superior than Method I.

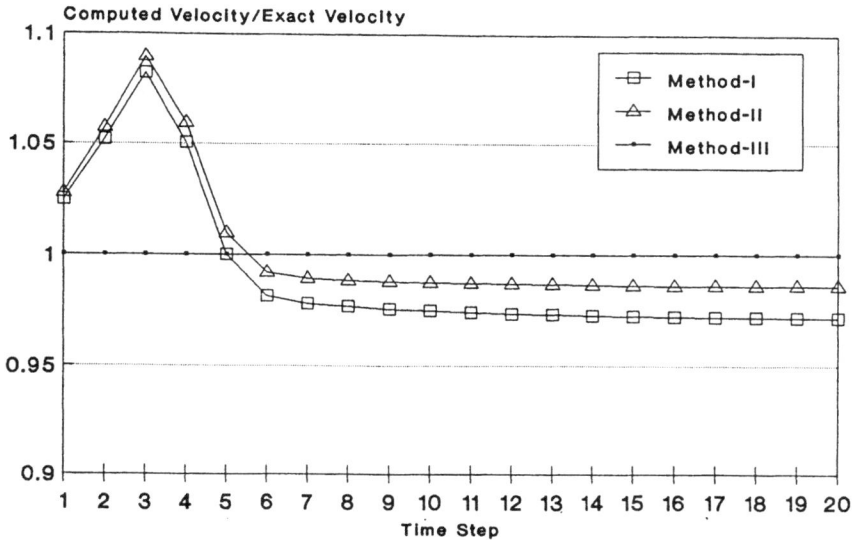

Figure 1 *Nonlinear advection of velocity.*

The second example is side discharges into an open channel flow. The present method and the upwind difference for the treatment of advective accelerations are studied. A typical computed flow field is shown in Fig. 2. The discharge jet is deflected by the channel cross-current, yet forcing the channel current to bend towards the far bank. The jet entrainment on the near-bank side is limited by the presence of the solid boundary, causing a recirculation region with low pressure (depressed surface elevation) to form behind the jet. The jet bends towards the near-bank and eventually attaches itself to the wall because of the low pressure. The size of the recirculation is mainly governed by the ratio of discharge momentum to channel flow momentum.

From the results shown in Figures 2 & 3, the upwind difference scheme produces a significant numerical diffusion effect (which is well known), while the present scheme performs better: the width of spreading of the jet is narrower, the peak velocity is higher, and the recirculation strength is stronger. It is also interested to note that the length and width of the recirculation eddy remain more or less the same for the two cases.

525

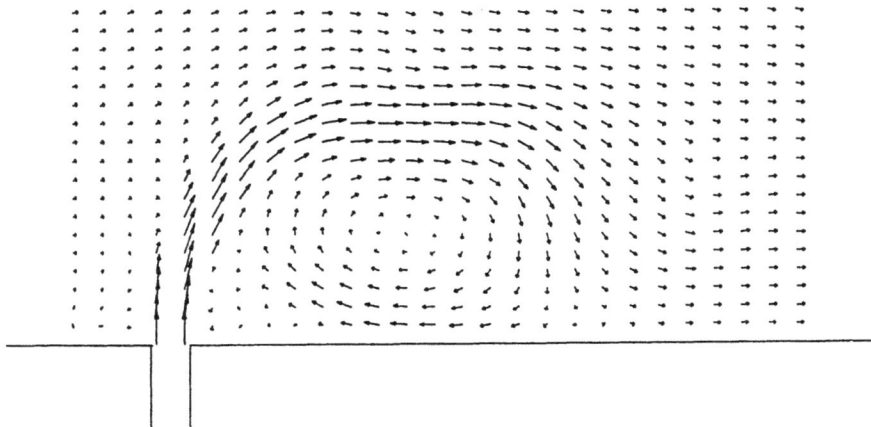

Figure 2   *Side discharge into open channel flow; minimax-characteristics scheme for advection*

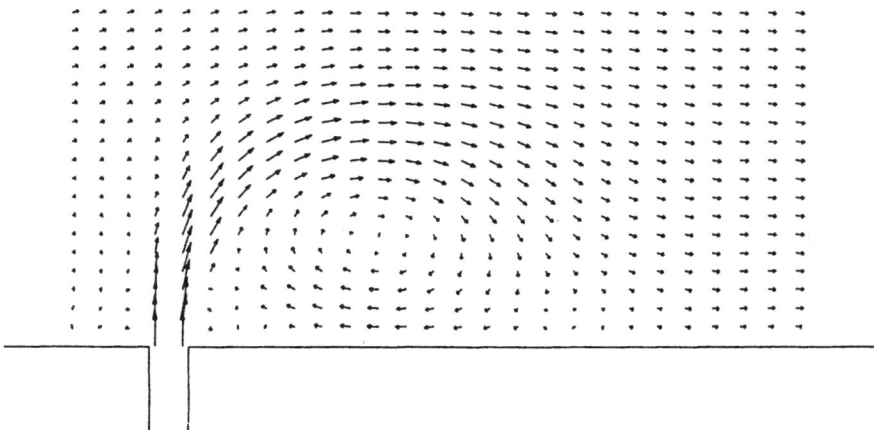

Figure 3   *Side discharge into open channel flow; upwind difference scheme for advection*

Conclusion

The accurate treatment of nonlinear advective terms in the shallow water equations is essential under certain flow situations in which the velocity variation is large. The use of minimax characteristics method with an accurate location of characteristic lines in treating the advective terms is efficient and accurate.

Acknowledgments

This work is supported by a grant from the Hong Kong Polytechnic.

References

[1] Benqué, J.P. et al., (1982), New Method for Tidal Current Computation, *Journal of Waterway, Port, Coastal and Ocean Division*, ASCE, Vol. 108, No. WW3, pp.396-417.

[2] Li, C.W., (1990), Advection Simulation by Minimax-Characteristics Method, *Journal of Hydraulic Engineering*, ASCE, Vol. 116, NO. 9, pp.1138-1144.

[3] Fischer, H.B. et al., (1979), *Mixing in Inland and Coastal Waters*. Academic Press.

[4] Roache, P.J., (1976), *Computational Fluid Dynamics*. Hermosa Publishers, Albuquerque, New Mexico.

44 Computation of incompressible fluid flows by an implicit fractional step scheme

J. Ye and G. Dou

ABSTRACT

The purpose of this paper is to introduce a new implicit fractional step method of two processes for the calculations of two − dimensional incompressible flows. It is found that the second step of propagation process in the present method has a similar feature to the velocities and pressure correction method procedure of the SIMPLE − like algorithm, and the SIMPLEC method is a special case of the proposed scheme. Three numerical examples of laminar and turbulent flows show that the rate of convergence can be significantly improved by the change of the optimal coefficient α.

1 Introduction

In numerical simulation of fluid flow problems, the SIMPLE (semi − implicit method for the pressure linked equations) algorithm of Patankar and Spalding[3] has become one of the most widely used implicit method. This particular procedure was developed to deal with the velocity − pressre coupling of fluid flows. Recently, several SIMPLE − like methods have been presented, e. g., the SIMPLER(SIMPLE revised) algorithm of Patankar[4] and SIMPLEC (SIMPLE consistent) algorithm of Van Doormaal and Raithby[5]. These variants are shown to have better convergence properties over the original SIMPLE method.

Another attractive technique in the fluid flow computation is the fractional step method which often applies to a time − discretized form of the transport equation. The principles of this method are quite simple: the equations are split up into simpler parts, allowing a special treatment to be applied to each of them; the whole transprot equations can be handled simultaneously. According to the investigation in Ref [1], the momentum equations and continuity equation are split into three processes, i. e. , convection, diffusion and propagation. The three steps are solved successively with appropriate schemes

In this paper, a new two−step implicit fractional step solution algorithm is presented for the calculation of the two−dimentional unsteady flow problems of incompressible fluid. Based on the fractional step technique, the proposed procedure includes the following two steps, the first step−the convection−diffusion process, the second step−the propagation process. A finite−volume method is employed in discreting the transport equations. In the first step, a general scheme is used for the solution of convection−diffusion equation as discussed in Ref.[3,4]. In the second step, an implicit−explicit weighting factor α is introduced for spatial derivations to accelerate the rate of convergence, and this step includes the continuity equation and the remainder of the momentum equations. It is found that the second step resembles the velocity−pressure correction procedure in the SIMPLE−like methods, and the SIMPLEC algorithm is a special case of the proposed method.

In the following sections, the SIMPLEC algorithm is first described briefly, then the proposed scheme is derived and discussed. Finally, the performance of the present method is studied by the calculations of the three test problems of laminar and turbulent flows.

2 Review of SIMPLEC Algorithm

2.1 Background

In this section, the basic ideas and notations in the well−known SIMPLE method of Patankar[4] has been used, and it is assumed that the reader is familiar with them.

The differential equation expressing the conservation of mass, momentum, concentration, etc. , of incompressible fluid can be written in the general form as

$$\frac{\partial \Phi}{\partial t} + \frac{\partial u_j \Phi}{\partial x_j} = \frac{\partial}{\partial x_j} (\Gamma \frac{\partial \Phi}{\partial x_j}) + S \qquad (1)$$

where Φ denotes the scalar variable or velocity component. The two terms on the left side of Eq. (1) are called the unsteady term and convection term, while the terms on the right side are known as the diffusion term and source term, respectively. Here, only two−dimensional situations are considered($j=1,2$). The source term is linearized as

$$S = S_c + S_p \Phi \qquad (S_p \leqslant 0) \qquad (2)$$

To discretize the flow domain, a staggered grid is used, each corresponded with a control

(a) control volume for p or Φ

(b) control volume for u_e

(c) control volume for v_n

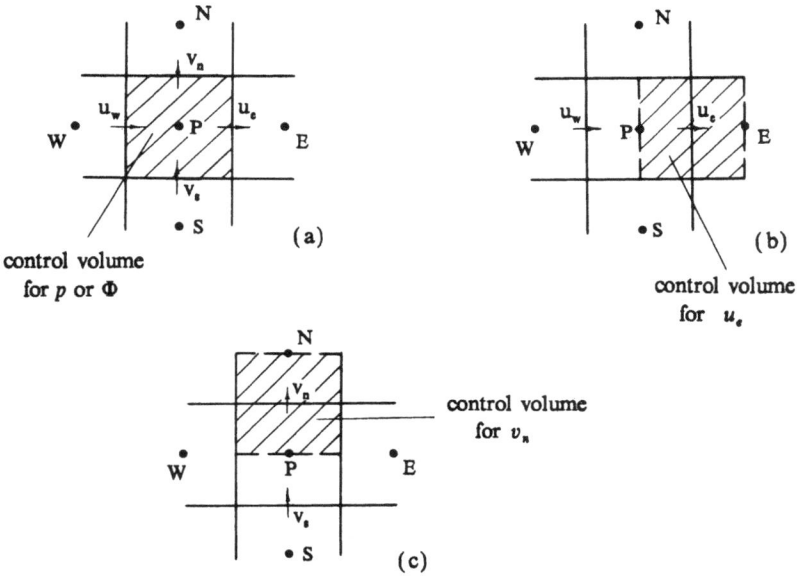

FIGURE 1   *Main and staggered control volumes*

volume. The scalar variables and pressure are stored at the grid points, while the velocities are stored at the staggered locations as shown in Fig. 1. A finite control volume method is used to integrate the general form of the convection – diffusion equation (1) over the suitable control volumes. The final discretization equation representing the integral conservation of Φ for the P control volume (Fig. 1a) becomes

$$a_p \Phi_p = \Sigma a_{nb} + b \tag{3}$$

where the subscript nb denotes the four neighbor grid points of P. The coefficient a_p and b are obtained by

$$a_p = \Sigma a_{nb} - S_p \Delta V + \Delta V/\Delta t \tag{4}$$

$$b = \Phi_p^o \Delta V/\Delta t + S_c \Delta V \tag{5}$$

the superscript o denotes the previous time step value, ΔV is the volume of the control volume, in two – dimensional situations of Cartesian coordinates, $\Delta V = \Delta x \Delta y$.

The steady form of Eq. (1) is also commonly used to obtain a steady—state solution. In this case, the equations are solved iteratively, and under—relaxation is introduced into the discretization equations to improve convergence. The distorted time step multiple E [5] is adopted here because of its direct physical interpretation, which results the following discretization equation

$$a_P(1 + \frac{1}{E})\Phi_P = \Sigma a_{nb}\Phi_{nb} + b \qquad (6)$$

where
$$a_P = \Sigma a_{nb} - S_p \Delta V \qquad (7)$$

$$b = \Phi_P^\circ a_P/E + S_c \Delta V \qquad (8)$$

The E factor is characteristically in the range of 1 to 10. Comparing the iterative approach with the time—marching approach, Eq.(6)~(8) will resemble the form of Eq.(3)~(5) by relation (9)

$$\Delta t = E\Delta t^* \qquad \text{where} \qquad \Delta t^* = \Delta V/(\Sigma a_{nb} - S_p\Delta V) \qquad (9)$$

It can be seen that the unsteady term in Eq. (1) and the under—relaxation practice have a similar effect on the finite—difference equation. It should be noted that Δt^* in Eq. (9) varies spatially, while in the time—marching formulation Δt is constant for all control volumes.

2.2 SIMPLEC algorithm

The SIMPLEC algorithm is described other than the original SIMPLE method since the latter produced poor results[4,5]. For incompressible fluid flows, the governing differential equations are

momentum equation

$$\frac{\partial u_i}{\partial t} + \frac{\partial u_j u_i}{\partial x_j} = -\frac{1}{\rho}\frac{\partial p}{\partial x_i} + \frac{\partial}{\partial x_j}(\varepsilon_{ij}\frac{\partial u_i}{\partial x_j}) + S_i \qquad (10)$$

continuity equation
$$\frac{\partial u_i}{\partial x_i} = 0 \qquad (11)$$

Using the staggered grid shown in Fig. 1b, the momentum equation in x direction for the control volume of e can be written as

$$a_e u_e = \Sigma a_{nb} u_{nb} + b_e + A_e(p_P - p_E) \qquad (12)$$

where A_e is the area of the face of the P control volume at e (in two—dimensional situations of Cartesian coordinates, $A_e = \Delta y$).

$$a_e = \Sigma a_{nb} - S_e\Delta V + \Delta V/\Delta t \qquad (13)$$

Similar equation can be derived for y direction. For a guessed pressure field p^*, the velocity u^* satisfy

$$a_e u_e^* = \Sigma a_{nb} u_{nb}^* + A_e(p_P^* - p_E^*) + b_e \qquad (14)$$

but will not generall satisfy the continuity equation (11). If Eq. (14) is subtracted from Eq. (12), the fully implicit velocity correction u' equation is obtained as

$$a_e u_e' = \Sigma a_{nb} u_{nb}' + A_e(p_P' - p_E') \qquad (15)$$

where
$$u' = u - u^*, \qquad p' = p - p^* \qquad (16\,a,\,b)$$

In the SIMPLEC algorithm, the term $u_e' \Sigma a_{nb}$ is subtracted from both sides of Eq. (15) and the term $\Sigma a_{nb}(u_{nb}' - u_e')$ is omitted, which leads to the velocity correction equation

$$u_e' = d_e(p_P' - p_E') , \qquad \text{where} \quad d_e = A_e/(a_e - \Sigma a_{nb}) \qquad (17)$$

The pressure correction equation is derived from the continuity equation (11) as

$$a_P p_P' = a_E p_E' + a_W p_W' + a_N p_N' + a_S p_S' + b \qquad (18)$$

where $a_E = A_e d_e$, $a_W = A_w d_w$, $a_N = A_n d_n$, $a_S = A_s d_s$, $a_P = \Sigma a_{nb}$ and $b = (u^* A)_w - (u^* A)_e$ $+ (v^* A)_s - (v^* A)_n$. Eq. (16), (17) and (18) are used to obtain the corrected p and velocity u, respectively.

3 Proposed Scheme

3.1 Principles and working equations

In this section, a new implicit fractional step solution algorithm is described for the computations of the unsteady Navier−Stokes equations, i. e. , Eq. (10) and (11). Similar to the split−operator approach according to the different transport processes[1], the proposed numerical procedure is divided into two steps

the first step : the convection−diffusion process $\Rightarrow u_i^{n+1/2}$

the second step: propagation process $\Rightarrow u_i^{n+1}, p^{n+1}$

where the superscript $n+1$ refers to time level $(n+1)\Delta t$, etc. , while the superscript $n+1/2$ is just a symbolic representing the intermediate variables betweenn the two step.

Denoting the relation

$$u_i' = u_i^{n+1} - u_i^{n+1/2} , \quad p' = p^{n+1} - p^n \qquad (19a,b)$$

The source term in the momentum equation (10) is expressing as

$$S_i = S_{ci} + S_{pi} u_i^{n+1} = S_{ci} + S_{pi} (u_i^{n+1/2} + u_i ')$$ (20)

In the propagation step, a coefficient of implicitization α is introduced for spatial derivations,

$$\frac{\partial p}{\partial x_i} = \alpha (\frac{\partial p}{\partial x_i})^{n+1} + (1-\alpha) (\frac{\partial p}{\partial x_i})^n = (\frac{\partial p}{\partial x_i})^n + \alpha \frac{\partial p'}{\partial x_i}$$ (21)

$$\frac{\partial u_L}{\partial x_i} = \alpha (\frac{\partial u_L}{\partial x_i})^{n+1} + (1-\alpha) (\frac{\partial u_L}{\partial x_i})^{n+1/2}$$ (22)

In practice, the known values at time $n\Delta t$ in the source term (Eq. (20)) and pressure gradient (Eq. (21)) of the momentum equation are solved in the first step, and the remainder of which combined with the continuity equation (11) are evaluated in the second step. Considering these, the working equations of the proposed scheme are as following

1st step : convection – diffusion process

$$\frac{u_i^{n+1/2} - u_i^n}{\Delta t} + \frac{\partial u_j^n u_i^{n+1/2}}{\partial x_j} = \frac{\partial}{\partial x_j} (\varepsilon_{ij} \frac{\partial u_i^{n+1/2}}{\partial x_j}) - (\frac{\partial p}{\partial x_i})^n + S_{ci} + S_{pi} u_i^{n+1/2}$$ (23)

2nd step : propagation process

$$\alpha (\frac{\partial u_L}{\partial x_i})^{n+1} + (1-\alpha) (\frac{\partial u_L}{\partial x_i})^{n+1/2} = (\frac{\partial u_L}{\partial x_i})^{n+1/2} + \alpha \frac{\partial u_L'}{\partial x_i} = 0$$ (24)

$$\frac{u_i^{n+1} - u_i^{n+1/2}}{\Delta t} = -\alpha \frac{\partial p'}{\partial x_i} + S_{pi} u_i'$$ (25)

3.2 Solution procedure and discussion

The method described in Section 2.1 is adopted to solved Eq. (23) in 1st step. The discretization equation in x direction with the control volume of u at e is

$$a_e u_e^{n+1/2} = \Sigma a_{nb} u_{nb}^{n+1/2} + b_e + A_e (p_P^n - p_E^n)$$ (26)

where

$$a_e = \Sigma a_{nb} - S_{pe} \Delta V + \Delta V / \Delta t$$ (27)

$$b = u_e^n \Delta V / \Delta t + S_{ce} \Delta V$$ (28)

534

In 2nd step, Eq. (25) is integrated in u_i control volume, e. g., in x direction, then

$$u_e' = d_e(p_P' - p_E') , \qquad \text{where} \qquad d_e = \alpha A_e / (\Delta V / \Delta t - S_{pe} \Delta V) \qquad (29)$$

Integrating Eq. (24) for the P control volume in Fig. 1a and Eq. (29) is substituted, which leads to

$$a_P p_P' = \Sigma a_{nb} p_{nb}' + b \qquad (30)$$

where $a_E = \alpha A_e d_e$, $a_W = \alpha A_w d_w$, $a_N = \alpha A_n d_n$, $a_S = \alpha A_s d_s$, and $b = (Au)_w^{n+1/2} - (Au)_e^{n+1/2}$ $+ (Av)_s^{n+1/2} - (Av)_n^{n+1/2}$.

The calculation procedure of the proposed scheme can be summerized as follows

1. the known values of u_i^n, p^n at time $n\Delta t$ are used to evaluate the coefficients of the momentum equation (23) and to get $u_i^{n+1/2}$
2. solve Eq. (30) for p'
3. using Eq. (19b) to obtain the pressure field p^{n+1}, using Eq. (19a) and (29) to obtain the velocity field u_i^{n+1}.
4. march to the next time step

It can be seen that the discretization equation of 1st step, i. e. Eq. (26), is just the same as Eq. (12); the finite-difference equations of 2nd step, i. e. Eq. (29) and (30), are the same as the velocity and pressure correction equations of the SIMPLEC algorithm, i. e., Eq. (17) and (18) if the coefficient of implicitization $\alpha = 1$. The resolution step of the present method is similar to that of the SIMPLE(or SIMPLEC) method. So it may conclude that the velocity-pressure correction procedure introduced in the SIMPLEC (or more generally SIMPLE-like) method is a kind of fractional step process, corresponding to the propagation step in the present method. The SIMPLEC algorithm can be considered as a special case of the proposed scheme (fully implicit situation). The application of boundary conditions discussed in Refs. [4][5], can be adopted directly in the present method. The conclusions above are the same to the steady state solution with an iterative approach if the E factor and Eq. (9) are used.

In the present computations, the power-law scheme of patankar[4] is chosen for approximating the combination of the convective-diffusive influence in the first step. The equation set is solved by an alternating direction implicit (ADI) scheme using the tridiagonal matrix algorithm (TDMA) on a grid line, as is commonly used. The convergence criterion adopted is that the summation of the absolute values of the mass residual in the entire calculation domain be less than a certain value γ_m.

	(a) laminar
	$$Rc = \frac{U_0 L}{\nu}$$
	$U_0 = 0.048$
	$L = 0.01$
	$\gamma_m = 2 \times 10^{-6}$
	grid: 37×37 nonuniform

(b)   laminar

$$Re = \frac{U_{in} H}{\nu}$$

$U_{in} = 0.015$

$H = 0.02$

$\gamma_m = 9 \times 10^{-6}$

grid: 36×16 nonuniform

(c)   turbulent

$$Re = \frac{U_{in} H}{\nu} = 30000$$

$U_{in} = 0.15$

$H = 0.2$

$\gamma_m = 9 \times 10^{-4}$

grid: 36×16 nonuniform

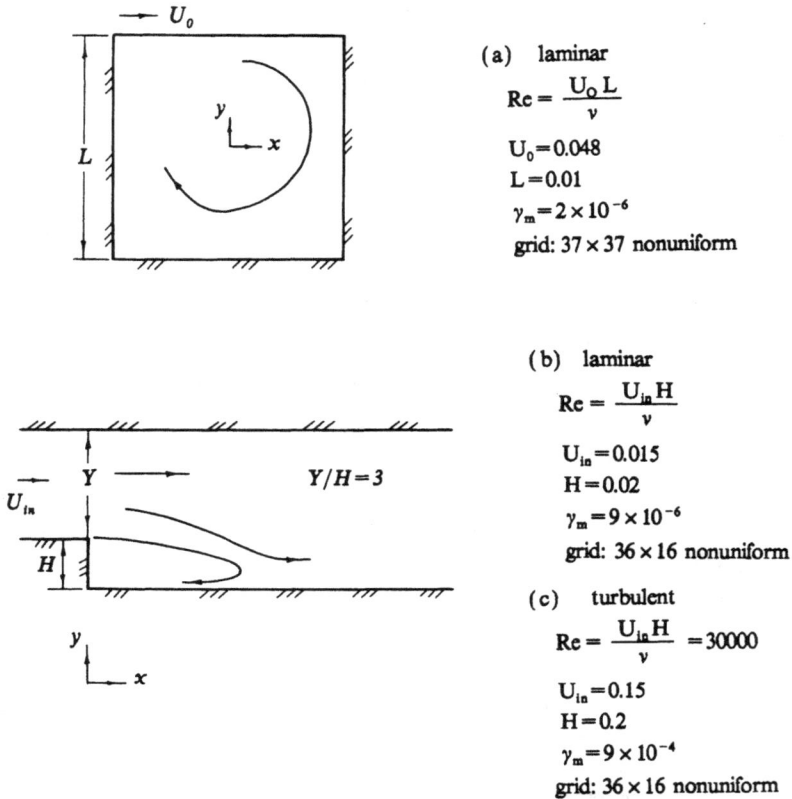

FIGURE 2   *The flow configurations of* (**a**) *laminar square cavity flow with a moving lid* (**b**) *laminar and* (**c**) *turbulent flows over a backward−facing step*

4 Application

The performance of the proposed method is investigated to the simulation with a different implicit coefficient α (usually $0.5 \leqslant \alpha \leqslant 1.0$), while the SIMPLEC method refers to the case of $\alpha = 1$. Three examples of laminar and turbulent flows are studied for this purpose. The configuration of the test problems with computational parameters is given in Fig. 2. Test 1 deals with the laminar flow in a square cavity with a moving lid (Fig. 2a). The other two tests are the laminar (Fig. 2b) and turbulent (Fig. 2c) flows over a backward−facing step. Results are obtained over a range of dimensionless time step $\Delta t U / L$ by the time−marching approach and driven to the same level of convergence. All calculations are carried out with the and 3, that with the optimal coefficients $\alpha = 0.75$ and 0.8 are 27% and 21% faster than the SIMPLEC method, respectively. The coefficient α less than its optimal value tends to result a unstable solution. So it seems that in the case of fully implicitization ($\alpha = 1$), the SIMPLEC method produces the most stable computation.

5 Conclusion

Based on the splitting up technique of different transport processes, a new implicit method of two steps is proposed for the computations of the unsteady flows of incompressible fluid. In the first step, the formulation for the combination of convection−diffusion problems described in Ref.[4] is used. In the second step, a coefficient of implicitization α is introduced for spatial derivations. It is concluded that the second step resembles the velocities and pressure correction procedure of the SIMPLE−like algorithms, and the SIMPLEC method is a special case of the present method (in the case of $\alpha = 1$). Although these two methods have a very similar formulation and resolution procedure, the principles of which are quite different. Three numerical examples of laminar and turbulent flows show that the computational costs can be significantly reduced by the change of the optimal coefficient α. The extension of the proposed method to compressible flows and the three−dimensional problems is straightforward.

References

1. Benque, J. P. , Cunge, J. A. , Feuillet, J. , Hauguel, A. and Holly, F. M. (1982), A New Method for Tidal Current Computation, ASCE *Journal of the Waterway*, *Port*, *Coastal and Ocean Division*, Vol. 108, WW3, pp. 396−417
2. Launder, B. E. and Spalding, D. B. (1974), The Numerical Computations of Turbulent Flows, *Com. Meth. in Applied Mech. and Eng.*, Vol. 3, pp. 269−289
3. Patankar, S. V. and Splading, D. B. (1972), A Calculation Procedure for Heat, Mass and Momentum Transfer in Three Dimensional Parabolic Flows, *Int. Journal Heat Mass Transfer*, Vol. 15, pp. 1787−1806
4. Patankar, S. V. (1981), A Calculation Procedure for Two Dimensional Elliptic Situations, *Numerical Heat Transfer*, Vol. 4, pp. 409−425
5. Van Doormaal, J. P. and Raithby, G. D. (1984), Enhancements of the SIMPLE Method for Predicting Incompressible Fluid Flows, *Numerical Heat Transfer*, Vol. 7, pp. 147−163
6. Ye, J. (1992), Improvement of Wall Function in Turbulence Modeling, *Proceedings of the 5th Asian Congress of Fluid Mechanics*, Daejon, Korea

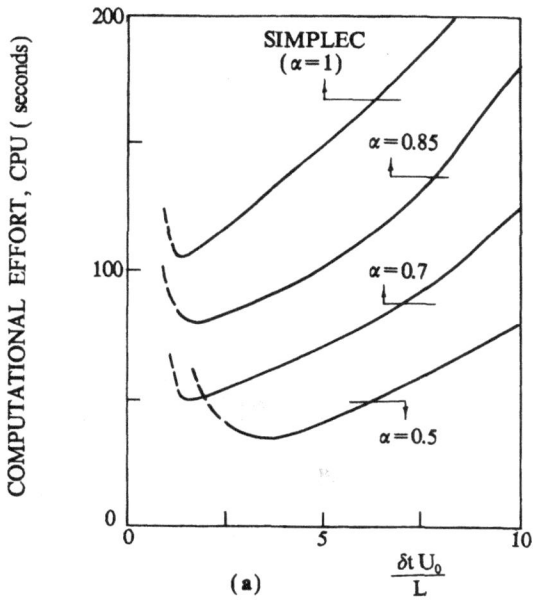

(a)

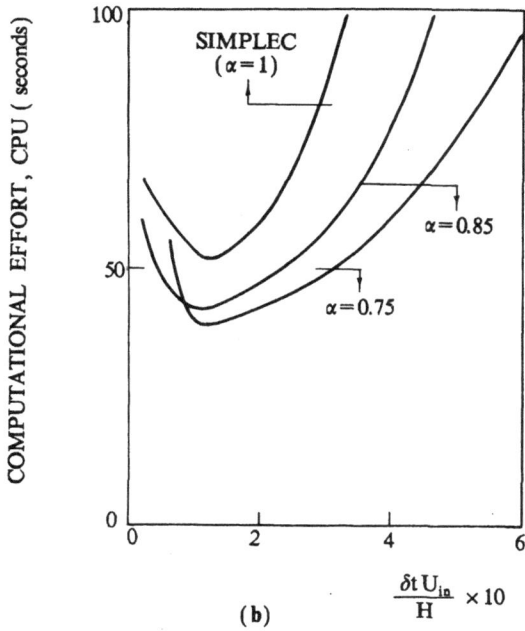

(b)

537

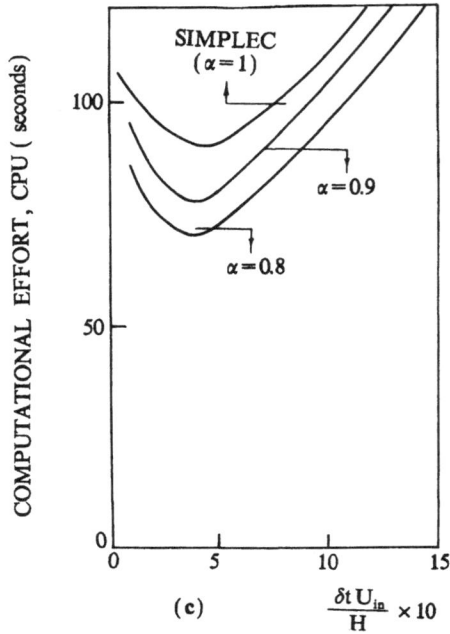

FIGURE 3   *Comparison of computational effort with the different coefficient α for*
(a) laminar square cavity flow , (b) laminar and (c) turbulent flows over a
backward – facing step

non – uniform grids and on an IBM 386/20 personal computer .

Test 3 is a problem of turbulent flow. The standard $k-\varepsilon$ model is used in which the differential equations for the turbulent kinetic energy k and its dissipation rate ε are solved to determine the turbulent diffusion. To account for the near – wall viscous effects, the refined wall function described in [6] is adopted in this work which was regarded as superior to the conventional log – law wall function.

Fig. 3 illustrates the computational effort required for different implicit coefficient α to satisfy the convergence criterion as a function of the time step for the three test problems. The trends of the computing effort for the three cases resembles each other . The SIMPLEC method (in the case of $\alpha=1$) produces the slowest execution time. The computational costs decrease continuously as the coefficient α becomes smaller. In test 1, the proposed method with the optimal coefficient $\alpha=0.5$ is about 65% faster than the SIMPLEC method, while in test 2

For Product Safety Concerns and Information please contact our EU
representative GPSR@taylorandfrancis.com
Taylor & Francis Verlag GmbH, Kaufingerstraße 24, 80331 München, Germany